RESPIRATORY DEFENSE MECHANISMS

LUNG BIOLOGY IN HEALTH AND DISEASE

Executive Editor: **Claude Lenfant**
Director, Division of Lung Diseases
National Institutes of Health
Bethesda, Maryland

RESPIRATORY DEFENSE MECHANISMS

(IN TWO PARTS)

PART I

Edited by

Joseph D. Brain

Harvard University School of Public Health
Boston, Massachusetts

Donald F. Proctor

The Johns Hopkins University
Schools of Hygiene and Public Health, and Medicine
Baltimore, Maryland

Lynne M. Reid

Harvard Medical School
Boston, Massachusetts

MARCEL DEKKER, INC.   New York and Basel

Library of Congress Cataloging in Publication Data

Main entry under title:

Respiratory defense mechanisms.

   (Lung biology in health and disease ; v. 5)
   Includes bibliographies.
   1. Respiratory organs. 2. Natural immunity.
3. Mucous membrane. 4. Cilia and ciliary motion.
I. Brain, Joseph D. II. Proctor, Donald F.
III. Reid, Lynne. IV. Series.
RC756.L83 vol. 5 [QP121]   616.2'4'008s
ISBN 0-8247-6381-5   [612'.2]   76-29567

MARCEL DEKKER, INC.
270 Madison Avenue, New York, New York 10016

Current printing (last digit):
10 9 8 7 6 5 4 3 2 1

PRINTED IN THE UNITED STATES OF AMERICA

CONTRIBUTORS

Ib Andersen, M.D., D.P.H.   Associate Professor of Hygiene, Institute of Hygiene, University of Aarhus, Aarhus, Denmark

Betsy G. Bang, AB   Research Associate in Pathobiology, The Johns Hopkins University School of Hygiene and Public Health, Baltimore, Maryland

Frederik B. Bang, M.D.   Professor of Pathobiology, The Johns Hopkins University School of Hygiene and Public Health, Baltimore, Maryland

Joseph D. Brain, Sc.D. in Hyg.   Associate Professor of Physiology, Harvard University School of Public Health, Boston, Massachusetts

Indrajit Das, Ph.D.†   Lecturer in Experimental Pathology, Cardiothoracic Institute, Brompton Hospital, London, England

Robert Frank, M.D.   Professor of Environmental Health, University of Washington School of Public Health and Community Medicine, Seattle, Washington

R. F. Hounam, B.Sc.   Aerosol Physicist, Inhalation Toxicology Group, Environmental and Medical Sciences Division, Atomic Energy Research Establishment, Harwell, Oxfordshire, England

Peter K. Jeffery, Ph.D.   Lecturer, Department of Physiology, Basic Medical Sciences Group, Chelsea College, University of London, England

E. E. Keal, M.D., FRCP.   Consultant Physician and Senior Lecturer in Experimental Pathology, Cardiothoracic Institute, Brompton Hospital, London, England

Maria Teresa Lopez-Vidriero, M.D.‡   Department of Experimental Pathology, Cardiothoracic Institute, Brompton Hospital, London, England

---

Present Affiliation
†FRIC Lecturer in Biochemistry, Medical Research Council Environmental Hazards Unit, St. Bartholomew's Hospital Medical College, London, England
‡Assistant Professor of Pathology, Harvard Medical School, Boston, Massachusetts

Gunnar Lundqvist, M.Sc.   Associate Professor of Hygiene, Institute of Hygiene, University of Aarhus, Aarhus, Denmark

Arthur Morgan, B.Sc.   Group Leader, Inhalation Toxicology Group, Environmental and Medical Sciences Division, Atomic Energy Research Establishment, Harwell, Oxfordshire, England

Michael S. Morgan, Sc.D.   Assistant Professor of Environmental Health, University of Washington School of Public Health and Community Medicine, Seattle, Washington

Donald F. Proctor, M.D.   Professor, Department of Environmental Health Sciences, The Johns Hopkins University School of Hygiene and Public Health; Professor, Department of Laryngology and Otology, The Johns Hopkins University School of Medicine, Baltimore, Maryland

Lynne M. Reid, M.D.†   Pathologist-in-Chief, Department of Pathology, The Children's Hospital Medical Center, Boston, Massachusetts

Michael A. Sleigh, D.Sc.   Professor of Biology, University of Southampton, Southampton, England

David L. Swift, Ph.D.   Associate Professor, of Environmental Health Sciences, Department of Environmental Medicine, The Johns Hopkins University, Baltimore, Maryland

------

Present Affiliation
†S. Burt Wolbach Professor of Pathology, Harvard Medical School, Boston, Massachusetts

FOREWORD

I would be hard put to say whether the expansion of research on the lung and on respiratory disease is a precursor or a result of increasing public awareness of the importance of pulmonary illness. Whatever the case, we have undoubtedly witnessed both phenomena within a relatively short period of time, and with them a rather sobering realization of the limitations in both scientific and general public knowledge of lung disease. New concerns about the impact of environmental and occupational hazards on lung function—impacts that may only manifest themselves in frank disease after decades of seemingly innocuous exposure—serve to remind us all that society pays a heavy price when knowledge lags behind action, and moreover that society looks to science for solutions.

This series of monographs is, therefore, most timely and important. The substantial increase in the number and variety of scientific reports on respiratory and pulmonary topics makes it all the more critical that work in this field be subjected to thorough and comprehensive review as a service to the scientist and the physician who find it virtually impossible to "keep up," let alone to assimilate and evaluate a rapidly growing body of knowledge in an area of human health that is of mounting importance.

Chronic and acute lung diseases are among the major causes of disability and death in all age groups. From a public health standpoint, prevention of these diseases is a goal that amply justifies an increased commitment of research resources. For it is clear that the most effective paths toward prevention will emerge out of disciplined study in many fields, from the physiology and biochemistry of the respiratory system to the pathology and therapy of respiratory disorders.

I feel sure that this series of publications will continue to make a substantial contribution to the science and practice of medicine and to the hopes we have for more effective concepts and methods of preventing a major segment of human disease.

Theodore Cooper, M.D.
Assistant Secretary for Health
Department of Health, Education,
and Welfare

PREFACE

The lungs have a unique proximity to the environment. The same thinness and
delicacy that qualify the air-blood barrier for the rapid exchange of oxygen and
carbon dioxide reduce its effectiveness as a barrier to inhaled microorganisms,
allergens, carcinogens, toxic particles, and noxious gases. Adults, depending on
their size and physical activity, breathe from 10,000 to 20,000 liters of air daily.
Contaminating particles and gases which enter the body with this air are potentially hazardous. The purpose of the following chapters (in two parts) is to explore the diverse physiological mechanisms which prevent the accumulation
and deleterious action of inhaled particles and gases. Thus, both the modification of the inspired air prior to its access to the lungs as well as the forces brought
to bear against noxious influences reaching the respiratory surfaces are discussed.

The prevalence of airborne disease provides ample evidence that this formidable array of respiratory defenses is not invulnerable. Perhaps a better understanding of the exact nature of these physiological mechanisms may lead to ways
of defending them against injury or even enhancing them and, thus, decreasing
human susceptibility to respiratory disease.

Unit One is devoted to the respiratory air conditioning system. In these
chapters, the overall objective is to describe the anatomical nature of the upper
air passages as it influences the flow of inspiratory air, and to consider how this
results in modification of the ambient air prior to access to the lungs. Unit Two
describes our knowledge of airway secretions and the cilia which move them.
The mucous membranes of the respiratory tract and the nature and action of
respiratory tract cilia are examined. This unit also covers the biochemical and
rheological properties as well as the pharmacological and physiological control
of airway secretions. Unit Three then integrates cilia and secretions and treats
the question of mucociliary transport. Basic mechanisms, environmental influences, and the impact of mucosal injury and repair are considered.

Unit Four describes other major factors involved in the removal of inhaled
deposited materials. Following a discussion of the overall clearance kinetics
usually seen, the importance of respiratory reflexes and cough is considered.
The unit concludes with a discussion of the pulmonary lymphatic system and the
nature and properties of the air-blood barrier. The final group of chapters, Unit

Five, emphasizes the role of phagocytic cells in defending the lung with special emphasis on the physiological and pathological roles of pulmonary macrophages.

After careful consideration of each subject area, the editors attempted to recruit authors who could summarize each topic and anticipate new directions of study. Each contributor has raised pressing questions against the background of a hard core of experimental facts. No obligation was felt to review all of the existing sprawling literature; rather, each author was encouraged to be selective. Inevitably, there is occasional overlap among some chapters; yet where this serves to preserve continuity in the development of each author's ideas, it has been considered important to maintain it.

Although the focus of these chapters is usually the adult human, we encouraged inclusion of comparative and developmental aspects of each topic. Similarly, although the series is not intended to be a handbook for clinicians, both pathological and clinical dimensions are included whenever appropriate. Overall, it is our hope that these chapters will be a chronicle of significant past contributions, a stimulus for their rapid and thoughtful application, and a guide for future work.

Joseph D. Brain
Donald F. Proctor
Lynne M. Reid

ACKNOWLEDGMENTS

Respiratory Defense Mechanisms is part of the Lung Biology in Health and
Disease series. Dr. Claude Lenfant, Director of the Division of Lung Diseases at
the National Heart,Lung, and Blood Institute, is the Executive Editor for the series.
We thank him for serving as a catalyst for our efforts and for his guidance. We
thank the many reviewers who critically examined the chapters and made count-
less valuable suggestions. We thank Ms. Judith Richards and Ms. Betty Hauser
for their essential role as editorial assistants. They orchestrated the review
process, made frequent suggestions regarding content and form, reviewed page
proofs, and generally provided an invaluable unifying influence. Finally, and
most important, the editors wish to thank the contributors for their heroic
efforts in writing and revising their chapters.

INTRODUCTION

The lung, an organ continuously exposed to the environment, has been recognized for almost five centuries to be susceptible to damage by substances carried in the air. As pointed out in Chapter 1, Leonardo da Vinci observed nearly 500 years ago that "inspired air could carry dust into and damage the lung." Thus, the integrity of the lung depends upon effective protective mechanisms against airborne hazards.

Many outstanding scientists have contributed to our knowledge of the Respiratory Defense Mechanisms. In 1953, Julius Comroe emphasized the "self-protective or defense mechanisms" of the lung in his Harvey Lecture on "The Function of the Lung." This, undoubtedly, was the beginning of a new wave of research that benefitted from advances in morphology, biophysics, and biochemistry.

This monograph provides an in-depth review and analysis of the defense mechanisms of the lung. It is a comprehensive report that includes a discussion of all known mechanisms. The authors represent many disciplines; hence, this most important function of the lung is considered from the viewpoint of the physiologist, the ultramicroscopist, the biochemist, the biophysicist, the microbiologist, the immunologist, and also the clinician. Although incomplete, this list points to a breadth of coverage that could only be accomplished through the extraordinary cooperation and enthusiasm of the editors who, although representing different disciplines, had a common goal.

As a personal note, I would like to acknowledge the dedicated and tireless efforts of Joseph D. Brain, Donald F. Proctor, and Lynne M. Reid, and to express my gratitude to them for editing such an original and excellent volume.

Claude Lenfant, M.D.
Bethesda, Maryland

CONTENTS

Cumulative Indexes Appear in Part II

CONTENTS OF PART II

RESPIRATORY DEFENSE MECHANISMS

AIR CONDITIONING
Modifications of Inspired Air

Respiratory defense mechanisms are directed against the pathogenic potential of airborne materials. They fall into three broad categories: those which affect access to the lungs, those which clear airway surfaces of foreign materials, and those which destroy or detoxify substances within lung tissues. Today we can also add to this list man-made devices which can minimize pathogenic influences in the ambient air. Unit I examines airborne threats to human health and the first group of respiratory defenses—those which modify inspired air during its passage to the lungs.

Our knowledge of the effects of indoor and outdoor climate on human health is limited; our opinions depend almost as much on folklore and "clinical impression" as they do on established fact. The continually changing character of ambient air makes the accumulation of data difficult. Little doubt exists, however, that climate affects both the incidence and the severity of disease. Yet, the mechanisms involved are imperfectly understood. Widespread use of artificial ventilation, heating, and cooling of indoor air offers some protection of the airways and increases the comfort of indoor life. But this indoor air may not always be conducive to human health. Population growth, rapid travel from one clime to another, and increasing urbanization and industrialization all further complicate the problem.

We know that the pathway and pattern of airflow through the nose or mouth affects the respiratory system's capacity to modify inspired air. The nose is the first line of defense against airborne threats. If inspired air arriving at the pharynx were close to body temperature, saturated with water vapor, and freed of all foreign gases and particles, airborne disease of the lungs would be a rarity. How effectively the nose can fulfill its defensive role depends on the nature, magnitude, and duration of the threat, the efficacy with which ambient air can be modified during its rapid nasal passage, and on the ability of the nose to withstand the effects of injurious inhaled particles and gases. Whenever the nose fails, the second line of defense, the tracheobronchial tree, must be brought into play.

The tracheobronchial tree can perform some of the air conditioning functions of the nose (as exemplified in tracheostomy), but at some cost. The tracheal mucosa may have impaired mucociliary clearance and there is an increased susceptibility to atelectasis, bronchitis, and pneumonia, especially in the young child.

A defense, not elaborated upon elsewhere in this monograph but connected with respiratory airflow, is adenoid and tonsillar tissue. The adenoids lie in the line of inspiratory airflow and in the path of mucociliary clearance from the nose. The cryptic surfaces of these tissues thus accumulate small samples of airborne materials. Such materials certainly include airborne viruses and bacteria as well as pollutants and allergens. In the lymphoid crypts these agents are intimately exposed to lymphoid tissue and a rich vasculature. Considerable evidence suggests that the tonsils and adenoids play an important early role in the development of immune reactions.

As in other fields, the jargon, familiar to a few, may be confusing to others. Although "air conditioning" implies cooling the indoor air, for lack of a broader term, we use it here to describe all modifications in the ambient air which can occur during respiration prior to gas exchange. The terms "nasopharynx" and "nasopharyngeal" are used by some to refer to the entire respiratory tract from nasal tip to larynx. We prefer to divide this region into the "anterior nares" (reaching back to ciliated nasal mucosa), the "main nasal passage" (from there to the squamous epithelium of the nasopharynx), the "nasopharynx" (from the back of the nose to the lower border of the soft palate), and the "pharynx" (extending to the pyriform sinuses and the larynx). The portion of the pharynx extending below the base of the tongue is often referred to as the "hypopharynx." It is common to refer to the major bronchi and trachea as the "upper respiratory tract." We suggest that this term be confined to the designation of the airways from larynx to nasal tip (or the lips in mouth breathing).

Donald F. Proctor

1

Historical Background

DONALD F. PROCTOR

The Johns Hopkins University
School of Hygiene and Public Health
Baltimore, Maryland

I. Introduction

The subject of respiratory defense mechanisms is so complex and involves so
many scientific disciplines that a search for the historical origins of our current
knowledge is an especially difficult task. An attempt to follow all the threads
which, when woven together, form the fabric of our present ideas makes a relatively disconnected series of tales. Since one of the objectives of this book is to
gather together bits and pieces of information into an integrated view of respiratory defenses, this chapter makes a good place to begin that effort.

Recognition that human health is related in some manner to climate and
weather and that there is a need for respiratory defenses began long before the
late seventeenth century, when a true understanding of respiratory function had
its beginnings. The fact that we sometimes find speculations on physiological
function and disease preceding factual knowledge based upon observation and

---

†Portions of this chapter have been published in a paper entitled "The
Nose, Ambient Air, and Airway Mucosa. A Pathway in Physiology," published
in *The Bulletin of the History of Medicine.* Material from that paper is used here
with consent of the Editor.

experiment is worthy of note. This is certainly true of our subject, and such speculations have often served to stimulate appropriate investigations at a later date. Prolonged delay between such stimulus and effect is usually related to inadequate communication between workers in different fields.

Even today, investigations of our ambient air, the artificial indoor climate, air pollution, airborne disease, and nasal and pulmonary physiology are carried out by scientists working in widely separated specialties. They attend different meetings and publish in their own journals, often writing in a jargon peculiar to each separate discipline. Only recently have international meetings on airborne infection, inhaled particles and vapors, respiratory tract secretions, etc. brought investigators together to discuss the points where their varied investigations may apply one to another for a better understanding of physiology and disease.

To achieve an integrated view of the historical background it will be easier to follow each of several topics separately, a process requiring jumping back and forth in time. We will begin with the subject which seemed to attract most ancient attention, our ambient air.

II. The Ambient Air

In the Hippocratic writings, effects of weather and climate were discussed, but it is difficult to link those considerations to any current concepts of either prevention or pathogenesis of respiratory disease. In contrast, nearly two millenia ago, Galen's observations are amazing forerunners of what has come to be clearly understood only in the past century. He wrote:

> the apertures of the nose, how marvelously they come next after the spongelike (ethmoid) bone placed before the cavities of the encephalon, and how the connection was cut through into the mouth at the palate in order that inspiration may not begin in a straight line with the rough artery (trachea) and that the air entering it may first be bent and convoluted, so to speak. For I think this should be doubly advantageous: the parts of the lung will never be chilled when often times the air surrounding us is very cold, and the particles of dust, ashes, or anything else of the sort very frequently mixed with the air will not penetrate as far as the artery (trachea). For the air can pass on its way in this winding path, but such particles are held back, encountering first the bodies (mucous membrane) which surround the bends and which are soft and moist, contain a sticky substance, and are capable for all these reasons of retaining what falls upon them. . . . those who wrestle in clouds of dust and those who travel a very dusty road . . . presently blow their noses and cough and spit out the

dust. If, however, the channels in the nose did not first pass straight up toward the head and then obliquely back toward the palate . . . , it is clear that nothing would keep all such materials from falling into the artery (trachea); for this is what happens if a person breathes through his mouth. (Ref. 25.)

Nearly 500 years ago Leonardo da Vinci [39] apparently observed some lung pathology in his dissections; he deduced that inspired air could carry dust into and damage the lungs. As shown in Fig. 1, he noted *polvere fa danno* (dust causes damage).

In 1733 John Arbuthnot recognized that studies of the ambient air might cast light on human disease, and published his "Essay Concerning the Effects of Air on Human Bodies" [2]. He recommended "the keeping of a journal of the Weather and reigning Diseases, as a thing which might be of singular use, especially to posterity." He also made the prophetic comments that "it seems preposterous that there should be so many inquiries about the Qualities of every drug which we take but seldom, and none into the Effects of a Substance that we take inwardly every moment", and "The air of Cities is not so friendly to the Lungs as that of the Country, for it is replete with Sulphorous Steams of Fuel . . . ." Nearly a century later James Clark, perhaps finally heeding the admonition of Arbuthnot, published "The Influence of Climate in the Prevention and Cure of Chronic Diseases" [15].

Clark, in addition to giving a contemporary description of consumption, provided extensive "tables on climates" in which he tabulated temperature, barometric pressure, rainfall and "weather" for many areas throughout Britain and Europe. Only two years later a book was published in which a "change of air" was extolled for its "moral, physical and medicinal influence" in addition to providing a "change of scene, foreign skies, and voluntary expatriation" [36].

Morrill Wyman, who received his M.D. degree at Harvard in 1837 and practiced medicine in Cambridge, divided his scientific endeavors between an interest in mechanical devices of use in medicine and the treatment of patients with respiratory disease. One contribution to medicine was his development of a suction pump and its application, with Dr. H. I. Bowditch (not to be confused with H. P. Bowditch shortly to be mentioned), to thoracentesis in acute pleural effusion [3]. He was perhaps the first to direct serious attention to the indoor climate [68], a subject still relatively neglected today. Wyman, also a student of hay fever, numbered among his patients suffering from that malady Daniel Webster and Henry Ward Beecher and, in 1876, published "Autumnal Catarrh (Hay Fever)" [69].

In 1878 Robert Briggs, an architect, followed the lead of Wyman with a series of lectures "on the relation of moisture in air to health and comfort" [13].

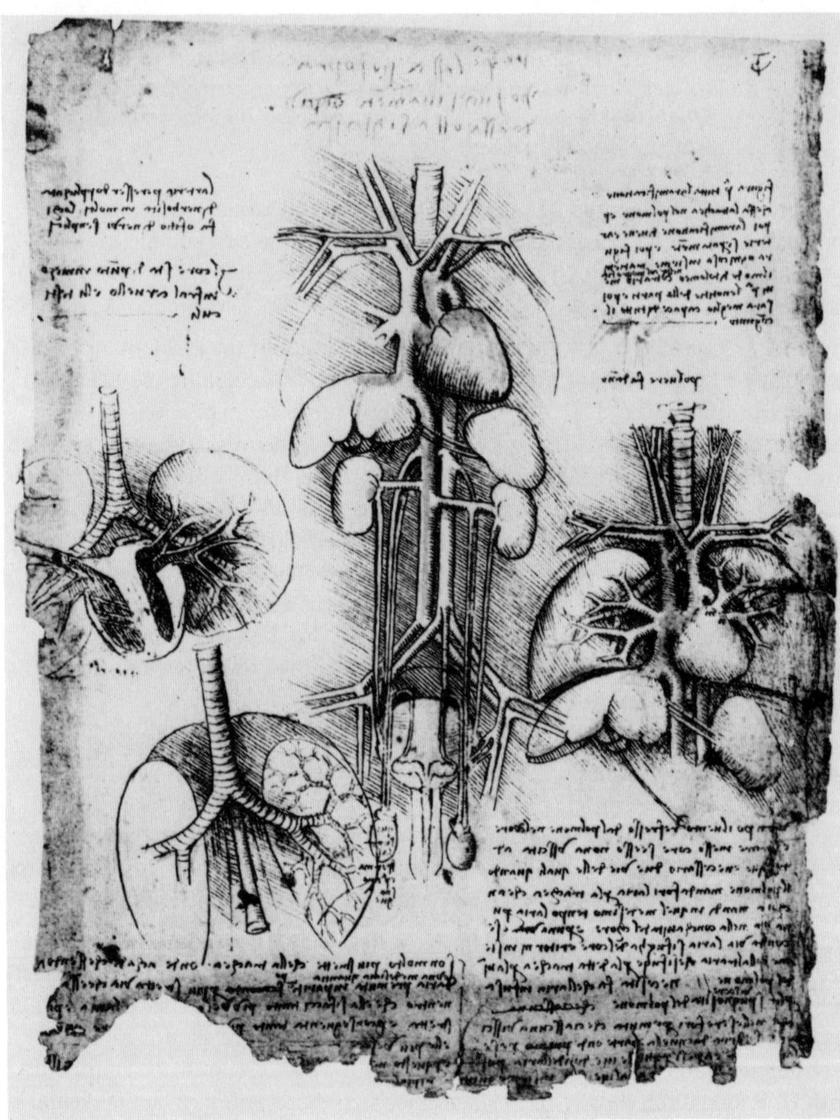

FIGURE 1 Just above the trachea in the drawing to the right appear the words "polvere fa danno" in mirror writing [39].

He made the statement that ". . . dry air, per se, of whatever temperature it may be found on the surface of the earth, is not unhealthy . . . ." In 1884 John Shaw Billings published "The Principles of Ventilation and Heating and their Practical Application" [8]. In his day he was accepted as the leading authority on ventila-

tion especially for hospital construction, assisting in the planning of both the
Johns Hopkins and the Peter Bent Brigham Hospitals. He, like Briggs, stated
that dry air in itself was innocuous, and he laid strong emphasis on good ventilation of hospital wards. Direct implication of the ambient air and the materials
carried in it in the causation of illness will be considered in Sect. IV. Clark [15]
opened his book by saying, "Although the power of different climates to produce
as well as to alleviate and cure diseases, is well established . . . perhaps, there is
nothing in science more unsatisfactory than the manner in which we are able to
explain this influence . . . ." Any such understanding was certainly impossible
at a time when the interrelationship of ambient air and the function of respiratory mucosal surfaces over which it passed was unknown. These mucous membranes are our next concern.

III. Mucous Membranes, Cilia, and Mucus

Mucous membranes were the subject of at least two interesting books by Bordeu
[9] and Bichat [6] in 1767 and 1802, respectively. Bichat (an army surgeon
who died at the age of 31 years) found time in his brief life to study and write
extensively on this subject. He was among the first to note the possible importance of the ambient air to nasal mucosa. Among the contributions of Francois
Magendie (founder of *Bulletin Physiologie,* the first journal devoted exclusively
to physiology) was his recognition that the nose warmed and moistened the inspired air, a matter to be the subject of extensive investigation ever since [47].
Recognition that nasal mucosa was in large part erectile tissue, similar morphologically to that found in the genital tract, was well known by 1855. This marked
a beginning of understanding of the ability of the nose to adapt as an air conditioning system to changing ambient conditions. Many of the problems, solutions
to which were essential to an understanding of man, his ambient air, and airborne disease became the subject of increasingly intense investigation in the latter
part of the nineteenth century. What was lacking was integration of the information available to form a unifying thesis on the basis of which new ideas and investigations could flourish. Essential to such an integration was recognition that
cilia could produce an organized motion in fluids contiguous to mucosal surfaces.
It is difficult to see how in spite of extensive studies of mucous membranes, the
availability of the microscope, and Leeuwenhoek's description of cilia, 150 years
were to pass between that description and recognition of their function.

In 1677 Leeuwenhoek sent the 18th of what were to eventually total 357
letters to the Royal Society in London. This letter, usually considered the first
description of cilia, is most commonly read in its abridged version [37]. The
original letter was sent to Henry Oldenburg who, in translating it for the *Trans-
actions,* condensed it to about half its original length. Those wishing to consult

a translation of the entire original letter (a rewarding experience) may turn to Dobell's book on Leeuwenhoek [20]. It is entitled, "Observations on Animalcula seen in Rain, Well, Sea and Snow-water; as also in Pepper-water." Leeuwenhoek did observe cilia in a variety of forms and described them repeatedly with remarkable clarity. But, it was not until 1830 that Sharpey introduced the concept of ciliary function with his delightful little paper "On a Peculiar Motion Excited in Fluids by the Surfaces of Certain Animals" [58]. Although this paper can only be considered a minor event in Sharpey's distinguished career, it was indeed a monumental advance in the understanding of the role of mucociliary function in respiratory physiology and disease.

William Sharpey was born in Arbroath, Scotland, April 1st, 1802. He attended the University of Edinburgh and completed his medical studies there, being admitted to the Edinburgh College of Surgeons in 1821. He continued his studies in London and Paris and evidently practiced medicine briefly in Arbroath; but, by the end of 1826 he became impatient to learn more pure science and spent most of the next three years traveling on the continent. In 1829 he returned to Edinburgh where he taught physiology under Dr. Allen Thomson, and it was evidently while there that he made his initial observations on ciliary activity at the age of 28.

In 1836, while he was still only 34 years of age, he was appointed to the Chair of Anatomy and Physiology at the University of London (now University College) (Fig. 2). This was the first recognition of physiology as a discipline in any English medical school. Sharpey continued in this capacity for 38 years during which time he was without doubt the major influence on physiology in Britain. In 1871 Edward Albert Schafer became the first Sharpey Scholar at University College. An indication of the affection and respect which Sharpey engendered in his students is the decision of Schafer to prefix Sharpey's name to his own. To this day he and his son, E. P. Sharpey-Schafer, are known under that name.

Sharpey opened his paper by saying, "In the course of some investigations on the development of the tadpole . . . I was accidentally led to observe, that the surface of that animal possessed the power of exciting currents in the water contiguous to it, in a constant and determinate direction." In cutting off one of the external gills he observed that ". . . globules of blood . . . were moved rapidly along its surface . . . in a constant and uniform manner" [57] (Fig. 2).

Sharpey did not immediately recognize the role of cilia in the fluid motion he observed, but he pursued this observation and published further investigations along with a translation of the report of Purkinje and Valentin of their similar findings at about the same time [55]. In this 1835 paper [59] in reference to his studies in rabbits he stated, "The ciliary motion was particularly strong in the nasal cavities, and, as noticed by Purkinje, continued longer in this than in other

FIGURE 2 Portrait of William Sharpey, provided through the kindness of Professor D. R. Wilkie, Jodrell Professor Physiology at the University of London. In the upper right hand corner is shown the drawing from Sharpey [58] of the direction of movement of the "globules of blood" along the gill.

situations." He noted anterior motion of particles in some parts of the nose and also observed motion toward the sinus orifice in the maxillary antrum.

One statement of Purkinje regarding cilia function is especially appropriate to our considerations. As translated by Sharpey [59], Purkinje said, "For, by its means, the secretions of those mucous membranes on which it occurs may be conveyed onwards, and many singular phenomena may perhaps be accounted for in this way. Thus, for instance, when the bronchial mucus accumulates during a long uninterrupted sleep, and is afterwards discharged, we do not bring it up from the interior of the lungs, but only from the larynx or top of the windpipe." Like Sharpey, Johann Evangeliste Purkinje was a Professor of Physiology (at

Breslau and Prague). He was the first to use a microtome, the discoverer of the Purkinje cells, the first to classify fingerprints, and was a friend of Goethe, who was apparently attracted to him after reading his graduation dissertation on subjective visual phenomena. As with Sharpey, his contribution to our field of interest was a relatively minor part of his career.

Joseph J. Lister reported interesting observations on cilia in 1834 [41]. Lister, (a London wine merchant and father of the great Lord Lister of antiseptic surgery fame) noted that the surfaces which he studied were, ". . . lined with closely set ciliae, which by their motion, cause the current of water . . . the effect upon the eye is that of delicately-toothed oval wheels revolving continually . . . but the ciliae themselves are very much closer than the apparent teeth, and the illusion seems to be caused by a fanning motion given to them in regular and quick succession, which will produce the apparent waves . . . ." Who today could write a better description of this phenomenon? These observations of Mr. Lister are referred to in Sharpey's 1835 paper. In 1841 Edward Forbes finished a brief but interesting report on the same subject by saying, "A minute inquiry into the nature of the involuntary vibratile cilia seen on mucous surfaces among the higher animals is most desirable" [24].

The work of Sharpey, Purkinje, and Valentin laid the ground for rapid advances, but there appears to have been a period of about 40 years in which mucociliary function was considered more a subject for curiosity than for intensive study. Claude Bernard did include observations on ciliary activity in one of his writings in 1866 [5]. Jeffries Wyman [69] (brother to the previously mentioned Morrill Wyman, and Professor of Anatomy at Harvard) was among those who devised interesting techniques for demonstrating ciliary activity to students.

In 1871 an army staff surgeon in India, Henry Veale, sent a brief but fascinating report to *The Lancet* [63]. Only 16 years after Garcia's initial report to the Royal Society on the use of indirect laryngoscopy, Veale was using the laryngeal mirror to observe mucociliary transport. Evidently descriptions of ciliary activity were unknown to him, but with the aid of what must have been a most cooperative patient and marvelous vision, he described in detail movement of discrete particles of mucus from the bronchi ascending the trachea in a circuitous manner. The details of his brief report leave no doubt that he actually observed this phenomenon. He noted that "the back of the larynx is . . . the part from which . . ." mucus was most easily cleared.

In 1876 Henry Pickering Bowditch (founder at Harvard of the first laboratory of physiology in the United States) published meticulous studies of the astounding capacity of cilia for work [12]. Using the frog's palate for his investigations, he devised a vehicle to be carried by cilia up an inclined plane and showed that 1 square centimeter of mucosa was able to do work at a rate equivalent to "6.805 grammillimeters per square cm per min."

An attempt at integrating and analyzing available information was made
as early as 1880 by Dr. Bryson Delevan [18]. Delevan gave an excellent description of nasal anatomy (although he missed the significance of the nasal entrance)
as well as analysis of nasal ciliary currents. Manley [48] recognized the potential
for defense and, speaking of mucous membranes in general, he wrote, "In the
healthy state . . . they provide formidable barriers against germ invasion or
against the introduction into the system of putrescent noxious elements. With
their serried, close columns of epithelial cells, they resist the ingress of, or, at all
events, neutralize or destroy in a large measure, lethal elements in the atmosphere
and in the food we consume. . . ."

In 1903 R. S. Lillie, working at Woods Hole, studied the relation of various
salts and ions to ciliary efficiency [40], and, in 1905, Maxwell at Harvard [49]
carefully repeated and confirmed the studies of Bowditch, investigated the question of ciliary fatigue, and also examined the effects of numerous salt solutions
on this function. In the same year Dixon and Inchley, working in pharmacology
in Cambridge, England, devised a most ingenious "cilioscribe" making it possible
to continuously record and quantitate the activity of cilia [19] (Fig. 3).

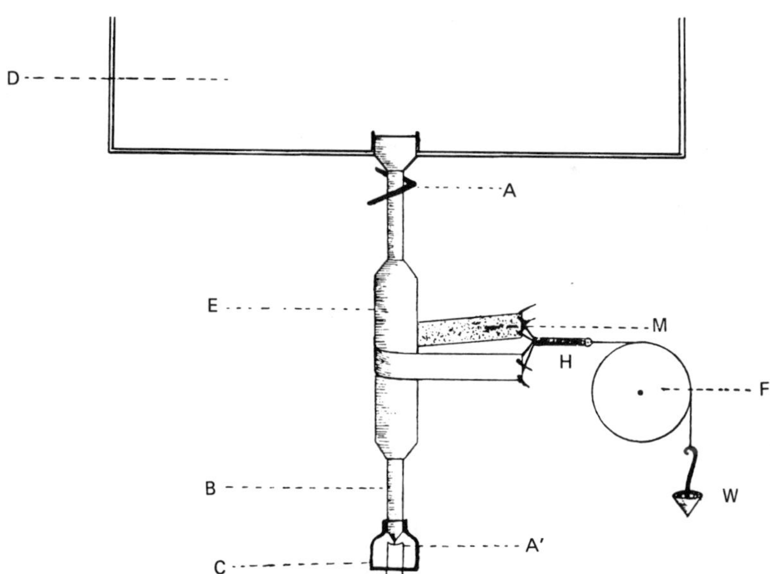

FIGURE 3 Device for quantifying mucociliary activity [19]. A, lateral bearing
supporting the spindle. A', agate cup supporting the spindle B. C, detachable
bell-cap. D, drum. E, ebonite cylinder on spindle B. F, pully wheel. H, hooks
holding the ciliated membrane M. W, weight.

Also, in 1905 George H. Parker (Professor of Zoology at Harvard) demonstrated a particularly interesting phenomenon [52]. Parker studied the action of the labial cilia of sea anemones, noted that they apparently were able to reverse the effective direction of their stroke, and suggested that they were able to discriminate between materials more or less nutritious. Using a minute piece of filter paper saturated with the test substance as bait, he showed that various organic substances (including meat, potassium, creatine but not creatinine, and glycogen) could all reverse the cilia beat to produce ingestion rather than expulsion. These findings have been both denied and supported; but yet to be done are studies designed to explore the possibility that disruption, incoordination, or reversal of ciliary beating may occur in respiratory disease in man.

Investigations of the mechanism of ciliary function continued to appear more or less regularly, however, the desire (apparent in the latter part of the 19th century) to apply this accumulating knowledge to the problems of human health had waned. In retrospect one wonders if scientists were looking to the growing specialty of otolaryngology to take the lead. No one in that field rose to accept the challenge until the advent of Arthur W. Proetz whose numerous papers, beginning as early as 1922, and whose book, *Applied Physiology of the Nose,* were to exert a wide influence on otolaryngologists [54].

One barrier hindering progress in an overall understanding of mucous membrane function was ignorance over the nature of mucus and its role in the clearance mechanism. From ancient times nasal mucus was assumed to be a discharge of excess fluid from the brain, especially from the pituitary gland, and it was not until 1655 that Schneider [57] (for whom the "Schneiderian membrane" term is applied to nasal mucosa) and later in 1670, Lower [42], showed that fluids could not travel from within the cranium to the nasal mucosa. Between them they put an end to recipes for "purging the brain." Schneider's elucidation of this admittedly controversial but important point was so long-winded that historian Jonathan Wright [66] would later say, "Never was the kernel of an important fact so wrapped up in the husks of verbosity." Even as late as 1751 Théophile de Bordeu commented on this earlier belief [9]. Perhaps the weighty character of these early deliberations on mucus accounts for the dearth of interest in the subject during the ensuing 250 years. Indeed, as late as 1888 Bosworth [10] denigrated the importance of mucus.

The major source of interest in mucus in the eighteenth century was the troublesome clinical problem of differentiating normal mucus from "purulent" secretions. In 1780 a prize-winning essay was published entitled "Experiments Establishing a Criterion Between Mucaginous and Purulent Matter" [16]. Its author was Charles Darwin, son of Erasmus Darwin, and uncle of Charles Darwin famed for his great work *The Origin of Species by Natural Selection.* Young Darwin had a short and tragic life. He cut himself performing an autopsy; that evening he suffered severe headache followed by fever, delirium, and petechiae, and died on the 15th of May, 1778, not quite 20 years of age, two years before his essay would be published. Darwin's experiments established chemical methods of some use in identifying mucus, but more penetrating investigations of this essential fluid were a long way off. At present we are only beginning to learn something of its biochemical makeup and rheological behavior. The factors which normally control its secretion in such a manner as to prevent flooding or drying of the airways and facilitate mucociliary transport are still largely unknown.

IV. Airborne Disease

The endpoint of the search for an improved understanding of respiratory defense mechanisms is the discovery of better methods to prevent or treat those respiratory diseases which result from injury caused by materials carried within the body by inspired air. The first reasonable concept of airborne disease came from Jacob Henle; in 1940, this culminated in his paper entitled "Von den Miasmen und Kontagien" [27]. At the age of 31 years this remarkable man (Professor of Anatomy at Zurich, and said to have been a skillful artist, poet, and musician) was able to precede Pasteur and the coming age of microbiology with a hypothetical description of the spread of contagious disease through living microorganisms. Some quotes from Henle reveal his understanding of the mechanisms of airborne infection:

> No fixed contagion . . . can act, when it is brought into contact with the unbroken skin. Its action is conditioned by its reception, either under the skin by inoculation, or on mucous membranes. . . . By and by the small cryptograms lose their water, dry up and change to a powder, which contains the germs and which at the slightest movement rises from the body on which it was produced and scatters in the air. . . . During epizootics one has even had the opportunity to measure the diameter of the infectious atmosphere which surrounds the sick individual . . . (under 12 feet). Miasmatic diseases . . . develop epidemically . . . where a large number of people are crowded together with a diminished circulation of air. . . . The first spots to suffer in almost all epidemic diseases are the conjuctiva, the nasal entrance, the pharynx, the oral cavity . . . [and] . . . the disease remains restricted to the nasal and pulmonary cavities most often. . . .

He clearly identified entrances to the air passages as targets of airborne pathogens, but failed to comprehend the natural defenses of those entrances.

The awakening interest in "infectious," "contagious," or "miasmatic" disease is also exemplified in Peter Panum's treatise on a measles outbreak in the

Faroe Islands [51]. In these early papers, written at a time when (in spite of Leeuwenhoek) microbiology was still essentially nonexistent, the exact meaning certain words had for them is somewhat obscure. Apparently "infectious disease" involved two possible means of transmission. "Miasmatic" transmission indicated pollution of the air itself, thus airborne. "Contagious" transmission indicated that a direct contact with the infected individual or with infected materials such as clothing was involved. Henle and Panum represent important contributions, but useful application of their ideas had to await developments in microbiology, aerobiology, epidemiology, immunology, and pulmonary, nasal, and mucosal physiology.

In the last quarter of the nineteenth century, Pasteur, Koch and others laid the groundwork of microbiology in identifying microorganisms as the causative factor in many diseases. Although Henle had already described how such microorganisms might be transmitted by air, application of these facts and theories to a full understanding of airborne disease and its prevention was then, and still is, incompletely fulfilled. With the growth of industry and the consumption of fossil fuels in the twentieth century, the need for study of problems related to nonviable airborne materials was added to the problem of airborne infection.

Analysis of what was in the air of cities, as well as in and about industrial plants, gave rise to groups devoting their attention to industrial hygiene and air pollution. Much of their efforts were bent quite properly toward identifying hazardous airborne materials, determining what concentrations of them were hazardous if inhaled, and seeking methods for reduction of air levels below that hazardous concentration. For a better understanding of those problems the practical question of what size particle could penetrate into the depths of the lungs was studied in theoretical models and later in man.

Lehmann addressed himself to the possibility that the susceptibility of miners to silicosis might be related to the efficacy with which the individual's nose filtered dust from the inspired air. He found a more efficient nasal filter among miners free of the disease [38]. Tourangeau and Drinker measured the filtering capacity of the nose for "finely divided" limestone dust and concluded that Lehmann must be wrong. Their findings suggested that the nose was a very ineffective filter [62]. We still do not have a satisfactory answer to this very practical problem, but Pattle, using much more refined techniques, did show that the capacity of the nose for removing particulates was indeed very high [53]. Pattle presented this paper at the First International Conference on Inhaled Particles and Vapours, a most fruitful meeting which has been held every five years since that time. The problem of airborne infection, in contrast to air pollution, has been attacked from many directions. One of these exactly parallels the studies of airborne nonviable particulates in raising the questions of what sort of particles carry infectious microorganisms in the air and where they are deposited in the respiratory tract. Wells introduced the concept of droplet nuclei in postulating that, when expired air is propelled from an infected subject at high velocity, some of the moist particles carried in it are rapidly dried to a small enough size to remain airborne for significant periods of time and be inspired by individuals in the appropriate neighborhood [65]. These droplet nuclei are probably of an aerodynamic size of 2-4 μm. This is the range of particle size about which we still have inadequate information regarding the likely site of deposition or factors which might influence the likelihood of their penetrating the nose into the lungs.

V. The Nose

Except for Galen's observations, mentioned earlier, little attention was paid to the role of the nose in respiratory defense until the nineteenth century. Since, in nasal breathing persons, it is obviously the first target of airborne hazards the nose must be the first line of defense. Among the earlier champions of this cause was a lawyer turned artist, explorer, and anthropologist. In 1832 George Catlin abandoned a law practice, studied painting at the Peale Museum in Philadelphia, and left for the upper Missouri River country to begin a life of study of North, and later South, American Indians. Catlin sought contact with Indians who remained in a primitive state, largely untouched as yet by the rapidly advancing white man's "civilization." He became enthralled by a sort of epidemiological study of these people. Admittedly handicapped by the lack of written records, he used a variety of imaginative methods to ascertain the prevalence of disease, especially the causes of infant and child mortality. He became convinced that, with the exception of accidental deaths, infant and child mortality was exceedingly low in comparison to that found in civilized communities of the time.

Catlin sought to explain their relatively healthy condition and in 1861 published a book in which he preached the doctrine of nasal breathing as the way to health and happiness, ending with the heartfelt exhortation, "And if I were to endeavor to bequeathe to posterity the most important Motto which human language can convey, it should be in three words—Shut - your - mouth" [14] (Fig. 4). Only in the past twenty years have serious scientific studies been directed at understanding the wisdom of those words.

The anatomy of the nasal passages in vivo is difficult to study, even today with our electrically lighted and radiographic devices. Although Leonardo da Vinci [39] and Vesalius [64] provided illustrations of the bony anatomy of the region, and Highmore [28] and duVerney [21], and still later Hilton [33], cast further light on the subject, those who gave any consideration to nasal

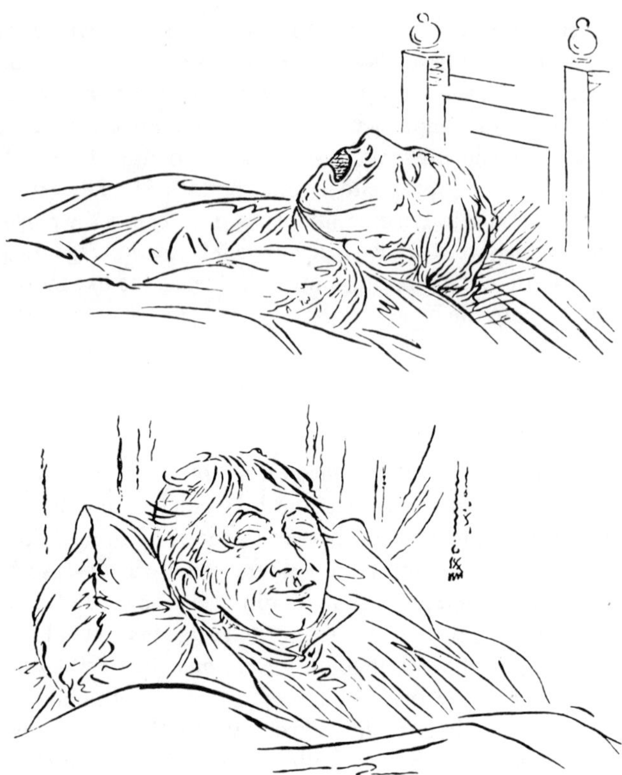

FIGURE 4   One page of Catlin's many drawings showing the horrors of mouth breathing and the joys of nasal breathing [14].

physiology seemed ignorant of the complexity of the nasal airway and the functional significance thereof until well into the nineteenth century. A partial exception was duVerney, although he saw the nose as the "organ of smelling." In a paper on nasal anatomy he observed that ". . . the cavities of the nose are filled with many cartilaginous laminae distinct from each other . . ." and that nature ". . . has furnished the said tunic of the nose with a great many small glands, which open into it, and so moisten it with a thick and slimy exudation, the better to entangle the dry odorous particles." Here was a recognition not only of the importance of the convoluted nasal airway, but of the role of its mucous membrane. Allen [1] and Delevan [18] provided excellent descriptive anatomy of the nasal chambers but their papers reveal no evidence of any interest in the physiological significance.

A connection between the nose and sexual function has long been a subject for conjecture. Sterne's Tale of Slawkenbergius in his Tristram Shandy is an amusing example of this observation, and Rostand's Cyrano de Bergerac is filled with allusions to the many marvelous ancillary functions attributed to the nose. John N. MacKenzie (Clinical Professor of Laryngology at the new Johns Hopkins Hospital) published a paper in 1898 on the Physiological and Pathological Relations between the Nose and Sexual Apparatus in Man [45]. It was not until the late 1940s that more serious attention and careful investigation was to be directed at the effects of emotional stimuli upon nasal function [34].

Except for Catlin's admonition to breathe through the nose, scant and periodic attention had been devoted to the nose as a defense organ until the last quarter of the nineteenth century when an increased interest in the physiology and disease of the nose was apparent. In 1875 Henry J. Bigelow, one of Boston's leading surgeons, linked the nature of nasal vasculature and environmental stimuli with common cold symptoms. Dr. Bigelow (professor of Surgery at Harvard and a witness to the first operation in which Morton administered ether for anesthesia), calling attention to the function of the nasal turbinates and their erectile tissue [7], pointed out that an understanding of the ". . . erectile tissue . . ." of the nose could ". . . elucidate, if it did not alleviate . . ." the symptoms of a cold. He also noted that ". . . a swallow of water, a pinch of snuff, a sudden start, mental or physical, . . . often clears the passage, to be again filled up." This was one of the early recognitions that the nose and its mucosa responded dynamically to a variety of environmental stimuli.

In 1888 Francke H. Bosworth, of the Bellevue Hospital in New York, drew together many observations into a surprisingly advanced view of nasal physiology [11]. He quoted Magendie's observation: "By successively passing the mouth, nasal cavities, pharynx, larynx, trachea and bronchi, the inspired air is charged with vapor which it carries away from the mucous membrane of the air-passages, and in this state always hot and humid, it arrives at the pulmonary lobules" [47]. Bosworth stated that the nose had three functions: olfaction, phonation, and respiration, and said that mucus is secreted ". . . only in such quantities as are sufficient to keep the membrane in a soft and pliable condition." He estimated that mucus was 93% water and 7% solids (in good agreement with findings today). He calculated the amount of water added to the inspired air each 24 hours ("5,000 grains" or 324 g) and attributed this to a transudation from the erectile tissues. He also quoted and agreed with Aschenbrandt's statement that ". . . all mechanical dust . . ." was ". . . deposited on the moist surfaces of the nasal membrane. . . ." These statements have only been documented in the past decade. Unfortunately, because the importance of this as a defense mechanism was not widely appreciated, he added that filtration was probably an adventitious function ". . . certainly it is a very unimportant one."

As far as I have been able to discover, Bosworth contributed little in the way of new investigation or even new ideas, however, he did fill that role which

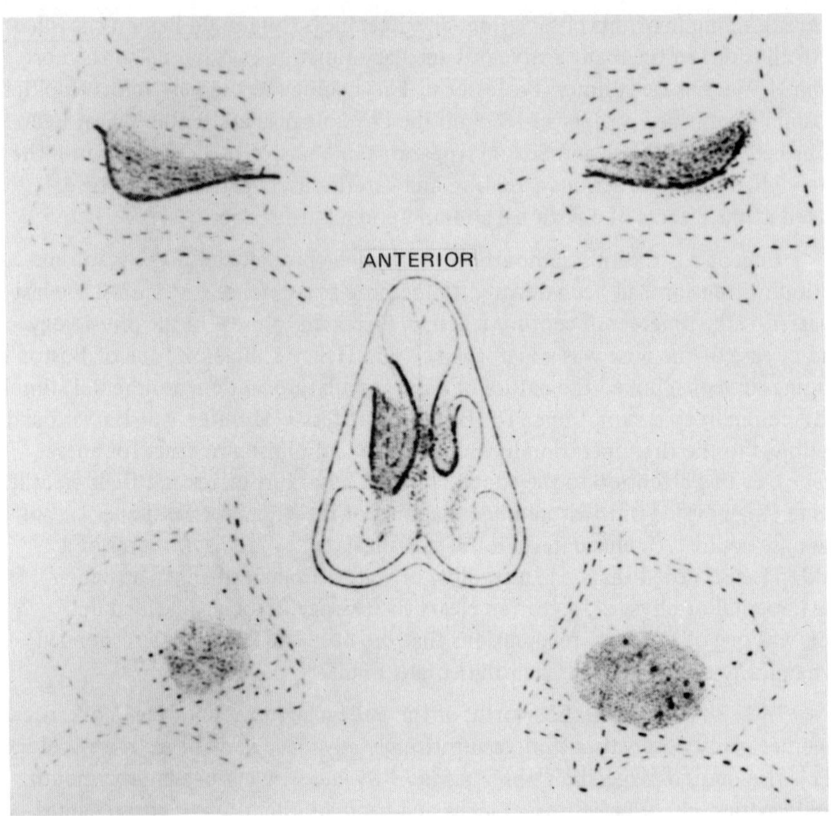

FIGURE 5 A page from Goodale's paper [26] showing his observations of nasal dust deposition: H. S., female, thirty-five, American. Tip of nose somewhat tilted; nostrils wider than normal, from slight development of constricting band. Turbinates essentially normal, except for anterior hypertrophy of both middle. Powder inhalation as by diagram. Practically a normal distribution.

science and especially medicine needs periodically: the student who reads widely and thinks broadly. He gathered together much of the then available information, organized it and presented it in a lucid and enjoyable format. One of his conclusions was, "This, then, is the great and prominent function of the nasal chambers, so to prepare the ingoing current of air that it shall exercise no injurious influence on the mucous membrane of the passage below."

Aschenbrandt [4] noted the role of the nose in heat elimination, and in 1888 Greville MacDonald used ingenious methods to study nasal air conditioning [44]. He wrote ". . . air passing through the nose is always saturated with water as well as raised to the temperature of the blood . . ." and added the observation that shrinking of the nasal mucosa reduced its efficiency in this regard. He also appears to have been the first to recognize the difference between nasal airflow patterns during inspiration and expiration. At this time Zuckerkandl's well illustrated monograph on normal and pathologic anatomy of the nose appeared [70].

Among the most interesting papers written at the end of the nineteenth century was one which won the Boylston Medical Prize of Harvard University for 28-year-old Joseph Lincoln Goodale, a laryngologist at the Massachusetts General Hospital [26]. In addition to describing ingenious experiments on the nose as an air conditioner he demonstrated nasal air currents at least as they were reflected by the deposition patterns of particulates. He clearly identifies these areas as the "anterior end of the middle turbinate" and the "cartilaginous (anterior) portion of the nasal septum", and noted that ". . . no deposit is discoverable posteriorly in the nose or pharyngeal vault" (Fig. 5). Goodale concluded that ". . . the normal nose, in an atmosphere not excessively dust-laden, is a very nearly, if not wholly, complete filter for the inspired air."

Then, as today, the practicing surgeon sometimes lagged behind his physiological colleagues in applying physiological knowledge to surgical therapy. At a time when the significance of the complex structure of the nasal passage was beginning to be recognized, J. N. MacKenzie delivered a paper on "some abuses in the intranasal surgery of today." In it he compares the ENT surgeon's ruthless attack on structures within the nose with the ". . . injunction of the Irishman at Donnybrook—If you see a head, hit it" [46].

VI. Integrated Ideas of Defense

By the twentieth century we appear to be on the doorstep of a better understanding of man's relations to his ambient air. Fundamental studies of climate and indoor heating and ventilation have been published. We not only have Henle's hypothetical analysis of infectious airborne disease but the new science of microbiology has provided it with a rational basis. The nature of respiratory mucus still eludes us, but primitive attempts at its analysis have already been made, and the observations of Sharpey and Purkinje on mucus transport have been followed by the development of other methods for observing mucociliary flow. Nasal anatomy and its relation to nasal air flow have at least been touched upon.

From this point there is increasing evidence of an attempt to integrate the findings in varied fields and apply them to practical problems of medicine and public health.

In 1916 Hill considered the possible relation between the indoor climate, nasal mucous membrane, and the "common cold" [32]. In discussing the ventilating system of the House of Commons he concluded that it resulted in "cold feet and stuffy heads . . . just the wrong conditions for legislators."

The physiological effects of ambient temperature received increasing attention and specific questions regarding mucosa and the microorganisms to which they were continually exposed were investigated. It became increasingly evident that an effective barrier did exist which in most people, most of the time, prevented pathogenic microorganisms on mucosal surfaces from penetrating and establishing disease. One clue to the nature of this barrier resulted from a cold in the nose which struck Alexander Fleming. When a drop of his nasal secretions fell upon a culture plate, he noted later a germ free zone about it; Fleming went on to investigate the material he named lysozyme [23].

The single term mucus as applied to the fluid lining the respiratory tract reflected the concept that it might be a homogeneous material. People began to speak of it as a "carpet" or "blanket." More detailed examination of the mucociliary system led to the suggestion that respiratory mucus was indeed an inhomogeneous and perhaps discontinuous material. Beneath was a relatively serous fluid in which the cilia beat, above a more viscid substance on the surface which was propelled along by the tips of the cilia [43]. The thinness of normal respiratory tract fluid ($\sim$10 μm) and the fact that it is dehydrated by most histological techniques have made resolution of these questions difficult. Only today are we beginning to understand the physical and biochemical nature of this essential material.

In recent years there has been a recognition that a major line of defense lay in the pulmonary macrophage. One of the coworkers with Pasteur began our understanding of this mechanism. Metchnikoff published a paper in 1881 on the subject of phagocytosis and proceeded to spend the next 35 years investigating this mechanism. In a lecture delivered in 1891 speaking of the pathology of pulmonary tuberculosis, he said, "The polynuclear cells englobe the tubercle bacilli readily but perish after a short time, and then with the microbes they contain, are eaten up by various mononuclear phagocytes which may be classed together under the term of macrophages" [50]. By 1915 the macrophage was more clearly identified [22] but its role in defense of the lungs remained somewhat obscure. Although most of our progress in this field is very recent indeed, by 1941 the importance of the pulmonary macrophage was clearly established [56].

In 1923 Tatum [61] discovered (first in experimental animals and later in man) the fact that asphyxia resulted in a widening of the nasal airway. Later studies have shown that oxygen demand influences nasal airflow resistance, thus enabling us to continue nasal breathing at the high ventilation rates of moderate exercise. Tatum also developed a primitive technique for rhinomanometry, a method of study of nasal airflow resistance to be extensively developed in succeeding years [60].

One line for further investigation comes from the finding of Davies that nasal morphology is related to the climate native to the region in which a race of man developed [17]. Might we learn something of the relation between form and function from such investigations? Another line of investigation is the question of whether we can use our current knowledge to protect humans from industrial and other atmospheric hazards. As has been mentioned, Lehmann demonstrated a relationship between the efficiency of the nasal filter and susceptibility to silicosis [38].

It would be unjust to omit from this chapter the monumental contributions of Anderson C. Hilding, whose work crowns the efforts of so many of his predecessors. His first report (on ciliary currents in the human nose) was published in 1931 [29]. A year later, and in 1963, he wrote papers on the subject of respiratory defenses which have served to stimulate the curiosity of students and investigators [30,31]. Hilding continues to introduce new and imaginative ideas into the field today (see Chaps. 11 and 12).

From 1930 to the present great strides have been made by increasing numbers of investigators. Much of the progress has been due to the application of modern technology [35]. In many instances the results have largely been confirmatory to conclusions drawn from cruder methods by earlier workers, but many new questions have arisen and remain to be answered. Now that the population explosion, rural-urban migration, rapid air travel, and increasing contamination of our atmosphere add daily to the risk of airborne disease, we can hope that this long story will reach fruition at an early enough date to save mankind from the increasing threat of airborne disease. We shall see in subsequent chapters how this historical background has been built upon to lead to our still incomplete but rapidly growing knowledge of respiratory defense mechanisms.

References

1. H. Allen, The anatomy of the nasal chambers, *Trans. Am. Laryngol. Assoc.*, **10**:76-84 (1889).
2. J. Arbuthnot, *An Essay Concerning the Effects of Air on Human Bodies*, London, J. Tonson, 1733.
3. E. C. Atwater, Morrill Wyman and the aspiration of acute pleural effusions (1850), *Bull. Hist. Med.*, **46**:235-256 (1972).
4. T. Aschenbrandt, *Die Bedeutung der Nase für die Atmung*, Wurzbutg, Stahel'schen Univers Buch und Kunstha, 1886.
5. C. Bernard, *Leçons sur les Propriétés des les Tissus Vivants*, Paris, Germer Bailliere, 1866, Leçons VI and VII, pp. 121-155.
6. M. F. X. Bichat, *Traité des membranes en général et diverse membranes en particulier*, Paris, chez Mme. Veuve Richard, 1802, pp. 54-55, LXIX.
7. H. J. Bigelow, Turbinated corpora cavernosa, *Boston Med. Surg. J.*, **92**:489-492 (1875).

8. J. S. Billings, *The Principles of Ventilation and Heating and Their Practical Application.* New York, The Sanitary Engineer, 1884.

9. T. de Bordeu, *Recherches Anatomiques sur la Position des Glandes et sur leur Action,* Paris, G. F. Quillau Pere, 1751.

10. T. de Bordeu, *Recherches sur le Tissu Muqueux, ou l'Organe Cellulaire et sur quelques Maladies de la Poitrine,* Paris, Pierre Franc, Didot le Jeune, 1767.

11. F. H. Bosworth, The physiology of the nose, *Med. News,* **53**:117-124, 1888.

12. H. P. Bowditch, Force of ciliary motion, *Boston Med. Surg. J.,* **95**:159-164 (1876).

13. R. Briggs, On the relation of moisture in air to health and comfort, *J. Franklin Inst.,* **75**:10-20, 82-93, 168-179, 251-255 (1878).

14. G. Catlin, *The Breath of Life; or Mal-Respiration, and Its Effects upon the Enjoyments and Life of Men,* New York, J. Wiley, 1861.

15. J. Clark, *The Influence of Climate in the Prevention and Cure of Chronic Diseases more particularly of the Chest and Digestive Organs,* London, Thomas & George Underwood, 1829.

16. C. Darwin, *Experiments Establishing a Criterion Between Mucaginous and Purulent Matter,* Lichfield, 1780.

17. A. Davies, A re-survey of the morphology of the nose in relation to climate, *J. R. Anthropol. Inst. Gr. Brit. Ire.,* **62**:337-359 (1932).

18. D. B. Delevan, On some points in the anatomy of the nasal fossae, *N. Y. Med. J.,* **32**:376-381 (1880).

19. W. E. Dixon and O. Inchley, The cilioscribe, an instrument for recording the activity of cilia, *J. Physiol.,* **32**:395-400 (1905).

20. C. Dobell, *Anthony van Leeuwenhoek and His "Little Animals,"* London, Harcourt Brace, 1932, pp. 112-166.

21. J. G. duVerney, Anatomical observations on the structure of the nose, *Phil. Trans.* (abridg.), **2**:432-433 (1678).

22. H. M. Evans, The macrophage of mammals, *Am. J. Physiol.,* **37**:243-258 (1915).

23. A. Fleming, On a remarkable bacteriolytic element found in the tissues and secretions, *Proc. R. Soc., London, S.B.,* **93**:306-317, 1921-1922.

24. E. Forbes, Note on the cause of ciliary motion, London and Edinburgh, *Month. J. Med. Sci.,* **1**:27-28 (1841).

25. Galen, *On the Usefulness of the Parts of the Body.* Translated by M. T. May, Ithaca, N.Y., Cornell Univ. Press, 1968, Vol. II, p. 525, Book XI, Section II, paragraph 4.

26. J. L. Goodale, An experimental study of the respiratory function of the nose, *Boston Med. Surg., J.,* **135**:457-460 (1896).

27. J. Henle, *Von den Miasmen und Kontagien [On Miasmata and Contagia],* 1840. Translated by G. Rosen, Baltimore, Md., Johns Hopkins Press, 1938.

28. N. Highmore, *Corporis Humani Disquisitio Anatomica,* The Hague, 1651.

29. A. C. Hilding, Ciliary activity and course of secretion currents of the nose, *Proc. Staff Meet. Mayo Clin.,* **6**:285-287 (1931).

30. A. C. Hilding, Four physiologic defenses of the upper part of the respiratory tract; ciliary action, exchange of mucin, regeneration and adaptability, *Ann. Intern. Med.,* **6**:227-234 (1932).
31. A. C. Hilding, Phagocytosis, mucous flow, and ciliary action, *Arch. Environ. Health,* **6**:61-73 (1963).
32. L. Hill, "Colds" and the influence of the atmosphere on the nasal mucous membrane, *Br. Med. J.,* **1**:541-543 (1916).
33. J. Hilton, *Notes on Some of the Developmental and Functional Relations of Certain Portions of the Cranium,* London, Churchill, 1855.
34. T. H. Holmes, H. Goodell, S. Wolf, and H. G. Wolff, *The Nose,* Springfield, Ill., Charles C Thomas, 1950.
35. S. Ingelstedt, Studies on conditioning of air in the respiratory tract, *Acta Otolaryngol. (Suppl.), (Stockh.),* **131**:1-80 (1956).
36. J. Johnson, *Change of Air,* New York, Samuel Wood & Sons, 1831.
37. A. v. Leeuwenhoek, Observations on Animalcula seen in Rain, Well, Sea, and Snow Water: as also in Pepper-water, *Phil Trans.,* **12**:821-831 (1677).
38. G. Lehmann, The dust filtering efficiency of the human nose and its significance in the causation of silicosis, *J. Industr. Hyg.,* **17**:37-40 (1935).
39. Leonardo da Vinci, *On the Human Body.* C. D. O'Malley and J. B. de C. M. Saunders, Henry Schuman, N.Y., 1952, p. 172.
40. R. S. Lillie, The relations of ions to ciliary movement, *Am. J. Physiol.,* **10**:419-443 (1903-1904).
41. J. J. Lister, Some observations on the structure and functions of tubular and cellular polypi and of ascidiae, *Phil. Trans.,* **2**:365-388 (1834).
42. R. Lower, *Dissertatio de origine catarrhi in qua ostenditus illum non provenire a cerebro (Tractato de Corde),* Londini, typ J. Redmayne, 1670, pp. 221-239.
43. A. M. Lucas and L. C. Douglas, Principles underlying ciliary activity in the respiratory tract, *Arch Otolaryngol.,* **20**:518-541 (1934).
44. G. MacDonald, On the mechanism of the nose as regards respiration, taste and smell, *Br. Med. J.,* **2**:1210 (1888).
45. J. N. MacKenzie, Physiological and pathological relations between the nose and sexual apparatus in man, *Bull. Johns Hopkins Hosp.,* **9**:10-17 (1898).
46. J. N. MacKenzie, Remarks on some abuses in the intranasal surgery of to-day, *N. Y. Med. J.,* Jan. 28, pp. 173-179 (1905).
47. F. Magendie, *Physiology,* 3rd ed. Translated by E. Millagan, Edinburgh, 1829, p. 387.
48. T. H. Manley, A few notes on the general and special features of the anatomy and histology of mucous membranes, *Med. Surg. Reporter,* **75**:735-737 (1896).
49. S. S. Maxwell, The effect of salt-solutions on ciliary activity, *Am. J. Physiol.,* **13**:154-170 (1905).
50. E. Metchnikoff, *Comparative Pathology of Inflammation,* Lecture X. Translated by F. A. and E. H. Starling. Dover, New York, 1968.
51. P. L. Panum, Observations Made during the Epidemic of Measles on the Faroe Islands in the Year 1846. Translated by A. S. Hatcher. In *Medical Classics,* Vol. 3, Baltimore, Md., Williams & Wilkins, 1939.

52.   G. H. Parker, The reversal of the effective stroke of the labial cilia of sea-anemones by organic substances, *Am. J. Physiol.*, **14**:1-6 (1905).

53.   R. E. Pattle, The retention of gases and particles in the human nose. In *Inhaled Particles and Vapours.* Edited by C. N. Davies. Oxford, Pergamon, 1961, pp. 302-309.

54.   A. W. Proetz, *Applied Physiology of the Nose,* St. Louis, Annals Publ., 1953.

55.   J. E. Purkinje and G. G. Valentin, Entdeckung continuerlicher durch Wimperhaare erzeugter Flimmer der Bewegungen, als eines allgemeine Phanomens in der Klassen der Amphiben, Vögel und Saugethiere. *Müller's Arch. Anat. Physiol.*, pp. 391-409 (1834).

56.   O. H. Robertson, Phagocytosis of foreign material in the lung, *Physiol. Rev.*, **21**:112-139 (1941).

57.   C. V. Schneider, *Liber Primus (Secundus, Tertius) de Catarrhis,* Wittebergae, T. Mevil & E. Schumacheri, 1660.

58.   W. Sharpey, On a peculiar motion excited in fluids by the surfaces of certain animals, *Edinb. Med. Surg. J.*, **34**:113-122 (1830).

59.   W. Sharpey, Account of the discovery by Purkinje and Valentin of ciliary motions in reptiles and warm-blooded animals; with remarks and additional experiments, *Edinb. New Phil. J.*, **19**:114-129 (1835).

60.   H. J. Sternstein and M. O. Schur, Quantitative study of nasal obstruction, *Arch. Otolaryngol.*, **23**:475-479 (1936).

61.   A. L. Tatum, The effect of deficient and excessive pulmonary ventilation on nasal volume, *Am. J. Physiol.*, **65**:229-233 (1923).

62.   F. J. Tourangeau and P. Drinker, The dust filtering efficiency of the human nose, *J. Industr. Hyg.*, **19**:53-57 (1937).

63.   H. Veale, On the movement of mucus in the trachea and larynx, *Lancet*, **2**:121-122 (1871).

64.   A. Vesalius, *De Humani Corporis Fabrica,* Venice, 1543.

65.   W. F. Wells and M. W. Wells, Airborne infection, JAMA, **107**:1698-1703, 1805-1809 (1936).

66.   J. Wright, *The Nose and Throat in Medical History,* St. Louis, Matthews, 1898.

67.   J. Wyman, Experiments with vibrating cilia, *Am. Naturalist,* **5**:611-616 (1871).

68.   M. Wyman, *A Practical Treatise on Ventilation,* Boston, James Munroe, 1846.

69.   M. Wyman, *Autumnal Catarrh (Hay Fever),* New York, Hurd & Houghton, 1876.

70.   E. Zuckerkandl, *Normale und Pathologische Anatomie der Nasenhöhle und ihren Pneumatischen Anhange,* Wien, Wilhelm Braumuller, 1882.

2

The Ambient Air

IB ANDERSEN

Institute of Hygiene
University of Aarhus
Aarhus, Denmark

I. Introduction

The atmosphere protects and sustains the earth and life upon it. It shields against cosmic radiation and meteorites, and transfers heat from equator to poles and water vapor from oceans to continents. It provides the oxygen essential to animal life and the carbon dioxide for photosynthetic processes necessary to vegetable life. At the same time, variations in the atmosphere may threaten life. Storms, floods, droughts, winds, heat, and cold have always posed a threat, but, in man's more recent history, the capacity of the atmosphere to carry noxious materials into the respiratory tract poses a special threat to human health. Any consideration of respiratory defenses must include an evaluation of what natural and artificial alterations in the ambient air are compatible with, optimal for, or hazardous to healthy life on earth.

The average adult male breathes about 15 kg of air each day. When we compare that to the daily intake of about 1.5 kg of food and 2 kg of water we see that, from a purely quantitative point of view, defenses against the air and airborne materials are more vital than those against materials we ingest. Since so large a quantity of air is respired by man, changes in its composition, physical properties, and even very small concentrations of pollutants may constitute a significant burden on the respiratory tract.

For all of these reasons, a definitive view of the earth's atmosphere and an analysis of those factors which influence its properties is an appropriate consideration for this volume. It should be borne in mind that the effect of the ambient air upon man will in part be influenced by the physical work load, which may increase pulmonary ventilation 20-fold and cause a change to mouth breathing instead of the usual nasal breathing.

II. Atmospheric Compartments

The air we breathe may be considered as a two-compartment system: outdoor and indoor air. For our purposes, the outdoor air will be considered as consisting of *ocean air* and *rural air,* both of which are very clean, and of *community air,* where the natural composition of the atmosphere is changed by high population densities, mechanization, and industrialization. The main subcompartments of indoor air are *home air* and *workroom air.* In view of the frequency with which we travel in motor-driven vehicles, we might add here *intravehicular air.* Indoor air is polluted by emissions from human subjects and from working processes, and high concentrations of pollutants may be reached due to the low rate of air change in comparison to that in the outdoor environment.

The importance of the different subcompartments varies widely. Generally, the importance of the indoor environment increases with increasing latitude and with increasing economic development, as the latter is followed by urbanization and industrialization. The majority of the population of Northern Europe and the United States spends 1 to 2 hours a day in the outdoor environment and 22 to 23 hours a day in the indoor environment, of which 8 hours are in the place of work. At lower latitudes, a larger part of the daily life takes place outdoors, both at work and at leisure. Scientists in the fields of environmental and occupational health have frequently studied the composition of community and workroom air in different situations. But this traditional approach must be supplemented with a more holistic one focusing on the human subject in a total environment, considered as a continuum of concentrations of substances and mixtures of substances. Much less is known about the natural atmosphere in the ocean and in the rural environment, and very limited information exists about the indoor home atmosphere.

This chapter will survey the important characteristics of each of the atmospheric compartments; then follows a systematic description of the atmospheric parameters: temperature, humidity, air velocity, gases and vapors, particles, radioactive substances, and ions.

TABLE 1 Some Gases in Clean Dry Air at Sea Level [107]

Constituent	Content by volume in ppm (cm^3/m^3)	
N_2	780,840 ± 40	
O_2	209,460 ± 20	
A	9,340 ± 10	
CO_2	330 ± 10	
Ne	18.18 ± 0.04	
He	5.24 ± 0.004	
CH_4	2	relatively constant
Kr	1.14 ± 0.01	
N_2O	0.5 ± 0.1	
Xe	0.087 ± 0.001	
H_2	0.5	relatively constant
O_3	0.01	variable
Rn	6.0×10^{-14}	variable

A. Ocean Air and Rural Air

The atmosphere in a rural, and especially a rural maritime, environment, undisturbed by human activity, is very clean. In addition to the mixture of gases considered pure air (Table 1), it contains a small amount of gases and particles which originate from natural processes, the so-called natural background pollution. Examples are: salt particles from atomized seawater; mineral dust and sulfur dioxide from volcanic eruption†; mineral dust from wind erosion; and organic particles from eroded vegetation or from living organisms such as bacteria, fungi, or pollen. Generally, the background pollution is lower over the oceans than in the countryside, and the lowest background is found over the oceans of the southern hemisphere. This is due to the concentration of human activity in the northern hemisphere and to the very slow air exchange between the two hemispheres.

---

†The magnitude of the volcanic eruptions at Krakatoa in 1883 and Mount Agung in 1963 was such that there were color changes of the sunsets all around the world.

B. Community Air

Community air, and particularly the atmosphere of large cities, is characterized
by man-made air pollution. This phenomenon is defined by the World Health
Organization as "situations in which the outdoor ambient atmosphere contains
materials in concentrations, which are harmful to man or to his environment"
[112]. Man-made air pollution has been recognized for centuries, but only since
the industrial revolution has it become a serious problem. Until World War II,
air pollution was usually considered to be sulfur dioxide and smoke, the so-called
"London type." This is characterized by incomplete combustion of sulfur-con-
taining coal in a vast number of open fires (e.g., for domestic heating) which produces sulfur dioxide, tarry droplets, and soot particles. During the 1950s meteorological conditions of inadequate horizontal and vertical mixing caused acute
air pollution episodes in London, causing thousands of deaths [69]. The British
Clean Air Act of 1956 attacked the problem of smoke and a change to more
efficient forms of domestic heating caused a spectacular reduction of pollution
by smoke (Fig. 1) [67].

The "Los Angeles type" of air pollution is mainly due to gasoline-powered
automobiles. It is found in warm climates where domestic heating contributes
only slightly to pollution. When meterological conditions are characterized by
low mixing and strong sunlight, the exhaust products undergo photochemical
reactions, and such oxidants as ozone, peroxyacetyl nitrate (PAN), and several
other substances are produced [2].

The atmosphere in most modern cities, depending on the latitude and the
degree of mechanization, lies somewhere between the extremes of "London"

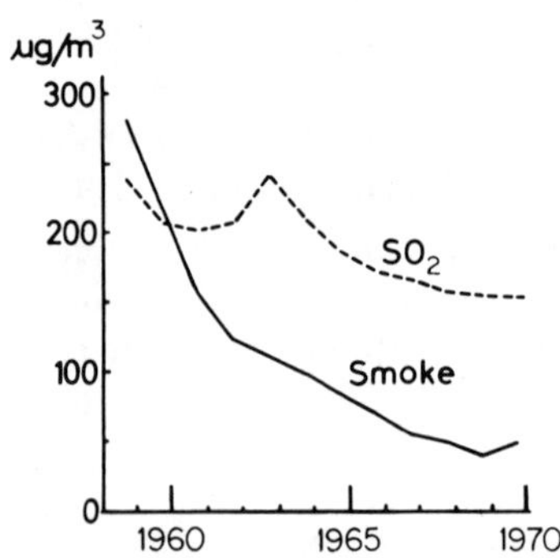

FIGURE 1   Annual mean concentration of smoke and sulfur dioxide at a group
of seven sites in Inner London from 1958 to 1970 [67]. (Courtesy of P. J.
Lawther, W. Heinemann Medical Books, Ltd., London.)

and "Los Angeles" type of pollution. Urban atmosphere is an ever-changing mixture of hundreds of different gases, vapors, and particles, and the great complexity
of their physical nature and chemical composition is the reason for the still im-
perfect description of the processes taking place in urban air. The few substances
monitored in our surveillance systems are only a small fraction of the pollutants
present. The common assumption that the measured levels of SO_2, CO, and particulates yield a measure of the overall pollution may be misleading.

C. Indoor Air

To protect against adverse climate and weather, early man used clothing, fire, and
shelter. Through several millennia, buildings have provided a favorable milieu
for productivity (workrooms) and for leisure (homes) by reducing fluctuations
in temperature and controlling wind velocity. But this has been achieved at the
expense of a reduced rate of change of air inside buildings, and consequently
pollutants produced by humans themselves, human activity, and by building
materials may accumulate. Our capacity to modify indoor air has become more
sophisticated, and new problems have arisen as to optimum temperature and
humidity.

1. Penetration of Pollutants from the Outside

Research on air pollution has to this day been focused on the outdoor air
with relatively little attention paid to modifications of the pollutants during
the passage from outdoors to indoors. The first studies in this area were performed in 1954 in the United States [44] and in 1958 in the U.S.S.R. [63],
but a recent, very complete review cites only 77 references [12]. Most of
these studies focused on the two main indicators of atmospheric pollution:
suspended particulate matter and sulfur dioxide, both of which are reduced
indoors. The reduction of sulfur dioxide is due to absorption by room surfaces and fixtures [90,91,110,111]. The reduction of suspended particulate
matter presumably is due to differences in outdoor and indoor sedimentation,
diffusion, and coagulation processes. Unreactive gases such as carbon monoxide penetrate building structures freely, without reduction in concentration [118]. The fact that the indoor environment provides a significant protection against some common outdoor pollutants has passed virtually un-
noticed, although some urban areas do broadcast warnings for persons with
respiratory ailments to stay indoors during pollution episodes.

III. Emissions into the Atmosphere

A. Outdoor Emission

Outdoor emission consists of the natural emission of gases and of particles from vegetation, volcanoes, forest fires, etc. and of man-made pollution. Although no effects on human health have been attributed to natural emissions, serious effects of man-made pollution are recognized. These problems will be dealt with in Sect. IV.

Vast geochemical cycles link the present atmosphere (total mass about 51 $\times 10^{17}$ kg [54]) to the other two strata—the lithosphere and the hydrosphere. Except during severe pollution episodes, the inputs and outputs of key constituents in these cycles have been in balance. The composition of the natural atmosphere has been stable and substantially the same at all parts of the earth [87]. This stability and the fact that several common pollutants (SO_2, NO_2, CO, and particles) are produced from natural sources, emphasizes that the atmosphere is able to scavenge the common pollutants emitted from urban sources. Some of the removal processes are atmospheric migration, biological removal by terrestrial and marine microorganisms and vegetation, and oceanic absorption and adsorption on surfaces.

B. Indoor Emission

Indoor emission consists of a background emission from room surfaces and furniture, and emissions from humans and activities taking place in the room.

Background emission is due to degassing from building materials, fixtures, furniture, textiles, paints, lacquers, etc. This emission is quite variable and is dependent on the age of the materials and conditions in the room. The problem of background emission was first identified in the confined spaces of atomic submarines and space capsules, where air is recirculated.

Emissions may also affect human health and comfort in normal dwellings [7,36,41]; for example, formaldehyde from particle board (made from wood chips and formaldehyde-urea glue) has been found in concentrations exceeding the threshold limit values (TLV) for workrooms [7].

Emission from humans consists of heat, water vapor, CO_2, particles (viable and nonviable), and odors. The magnitude of these emissions depends on physical activity, clothing, personal hygiene, and smoking patterns. With the exception of particles carrying pathogens which may spread disease, human emissions usually have little effect on health.

As a result of the emanation of body heat, the indoor temperature will tend to rise, and the humidity will increase due to moisture evaporated from the skin and exhaled from the lungs. The metabolic processes will cause a decrease in oxygen and an increase in CO_2 levels in the air proportional to the number of people and their level of physical activity. Other gases and vapors produced by humans are CO, acetone, ethanol, methanol, and amines (methylamine, dimethylamine, and trimethylamine) [41,65]. Individual body odor is related to diet, occupation, garments, activities, and personal hygiene, as well as enzymes and lipid breakdown. The apocrine glands are responsible for most body odors. When apocrine sweat first appears on the skin surface, it is normally odorless, but within a few hours its decomposition is initiated by the normal resident skin bacteria or transient organisms, and the proteins, carbohydrates, and lipids are converted to fatty acids (caproic, caprylic, butyric, etc.). Odorous substances are also present in intestinal gases (skatol, indol, etc.) and in saliva [79,105]. The most detailed studies of contaminants produced by man has been performed within a space cabin simulator [28].

Emission from indoor processes varies widely between homes and workrooms. In the domestic environment, the contaminants are mainly derived from heating and cooking devices, from cleaning materials, and from storage of foods and garbage. Carbon monoxide is a product of incomplete combustion, and in rooms heated by stoves burning gas or solid fuels, it may accumulate in dangerous quantities. Water vapor and odor are produced by heating and cooking devices, or generated during cleaning and drying processes. The storage of food and garbage mainly gives rise to odor. Cigarette smoking is discussed in Sect. IV.D and E.

In industry, a broad spectrum of gases, vapors, odors, and dust are generated (see Refs. 55, 56, and 77).

C. Ventilation—Exchange between Indoor and Outdoor Air

The air change in a building (or in a room) is caused by natural and/or mechanical ventilation [11,93]. Natural ventilation is produced by thermal forces and wind forces. These bring about the passage of air from the outside or from surrounding rooms through openings, leaks, and cracks, especially around the windows and doors. Cool and relatively dense air displaces warmer and lighter air, so that convection currents are set up. In a cold climate, where the indoor/outdoor air temperature differences may be 20-30°C or more, the contribution of this "stack effect" (particularly in multistory buildings) can be very significant. In mechanical ventilation systems, fans are the driving forces, utilizing either positive or negative pressure. In air conditioning systems, the incoming air is filtered, cooled or heated, and humidified or dehumidified.

Ventilating air dilutes and removes pollutants in room air. The number of air changes per hour, the size of the room, the rate of production of the pollutant in question and the concentration in the incoming air determine the concentration of pollutants in the air. The most accurate method for ventilation measurements is to monitor the dilution phase of a single dose of a component not usually found in the air. A convenient method utilizes a radioactive gas (^{85}Kr), but other gases, such as helium and hydrogen, are also commonly used.

IV. Air Quality

Now that we have described different processes determining the composition of outdoor and indoor air, we will deal with air quality in terms of temperature, humidity, velocity, and the concentration of gases and vapors, particles, radioactive substances, and ions.

A. Air Temperature

Air temperature is an important parameter because of its effect on the rate of chemical, physical, and biological processes; on upper airways where inspired air is cooled or heated to body temperature; on heat balance of the body; and on the physical and mental performance of humans.

Air absorbs little solar energy, so the outdoor temperature is mainly determined by the temperature of the surface of the earth or nearby large bodies of water. The annual and diurnal patterns of air temperature thus depend on the variations in surface temperatures. Since the effect of insolation is largely determined by latitude, there is a general decrease in temperatures from the equator toward the poles (Fig. 2). The distribution of land and water masses causes great variations in this general pattern. Land surfaces cool and warm more rapidly than great bodies of water, with the result that temperature changes in air masses are greater over land than over water. The range of air temperatures worldwide is between $-79°$C (measured at the South Pole) and $58°$C (measured in Libya).

Mean monthly values indicate the annual variation (Fig. 2 shows one location in each of the four climate zones). Detailed information about the temperature patterns for different locations are given in world maps of climatology (see Refs. 82, 98, and 106). Altitude also alters the air temperature. When an air mass rises it expands, and, therefore, it cools. The rate of temperature decrease is about $1°$C/100 m increase in altitude.

The air temperature in cities is 1-$2°$C higher than the temperature of the air in the surrounding countryside, producing the "heat island" phenomenon.

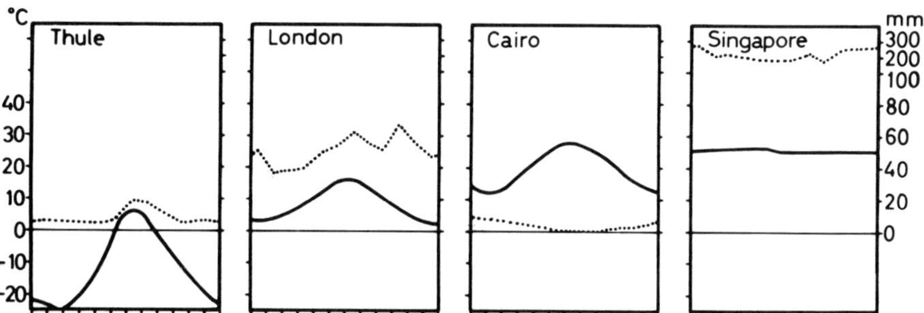

FIGURE 2   Variations in monthly average temperature (———) and precipitation (·······) from January to December on four locations. The polar zone: Thule, Greenland 77°N, 69°W; the temperate zone: London, Great Britain 52°N, 0°E; the hot-dry zone: Cairo, Egypt 30°N, 31°E; and the hot-wet zone: Singapore, Malaysia 2°N, 104°E. The yearly average temperatures are −11, 10, 21, and 27°C and the yearly average precipitation 64, 611, 26, and 2415 mm, respectively. All locations are only a few meters above sea level [106]. (Courtesy of VEB Fischer Verlag, Jena.)

Since buildings and streets are made of rocklike materials, which conduct heat faster than soil, more heat will be stored at the end of the day than in an equal volume of soil. The architecture of the city, with street valleys, walls, and roofs, functions as a mass of reflectors; almost the entire surface of the city thus accepts and stores heat. These large heated surfaces efficiently transfer heat to the air. Cities generate large amounts of heat, particularly during the winter when the heating systems are in operation. In summer, heat production is limited to that derived from automobiles, power stations, factories, etc. Rain in the city is quickly drained away, and snow is cleared quickly and efficiently by mechanical processes. In the countryside, precipitation will remain in the surface layers and, over a period of time, evaporate—a heat-consuming process which cools the surface. A comprehensive treatise on urban climates has been published by the World Meterological Organization [114].

Temperature inside buildings is influenced by the latitude, site, orientation, and design of the building, as well as the materials used. Solar radiation influences the indoor climate directly as heat penetrates windows, and indirectly by heating walls and roof. Incident radiation on windows is of three types: transmitted, absorbed, and reflected. Glass transmits most of the radiation in the range 0.4 to 2.5 μm, approximately the range of the solar spectrum. But glass is opaque to the radiation of longer wavelengths. Thus, glass permits solar radiation to penetrate into the building where it is absorbed by the internal surfaces and objects elevating their temperature. Heated surfaces and objects emanate radiation with wavelength of 10 μm which cannot be transmitted outward through glass, a process known as the "greenhouse effect."

The indirect thermal effect of solar radiation is determined by the properties of the materials on which it impinges. Some of the incident radiation is reflected, but the remainder is absorbed in the material, elevating its external surface temperature. Absorption of solar radiation is very dependent on color. A new whitewashed surface absorbs 10-15%, whereas a grey or a black surface absorbs 60-70% or 90%, respectively. Bricks and concrete absorb 70 to 75%. The magnitude of the heating depends on the thermal resistance and capacity of the building envelope. Control of the solar radiation may be achieved by the architectural design of the building, and especially the orientation in relation to the sun, the external color, and the shading of windows and other glazed areas [40,41,93].

In cold regions, it is important to have a high insulation value, both in walls and in windows to secure an interior surface temperature not much lower than the interior air temperature. This is necessary if discomfort due to radiation loss and condensation of water vapor is to be avoided. The latter brings about growth of molds, fungi, etc., on moist surfaces, and these microorganisms may cause allergic diseases. Minimization of heat loss through walls and windows is not only money-saving, but reduces community air pollution due to domestic heating. Insulation is also important in subtropical and tropical areas, especially in the roof, to reduce the heat load from the sun [62]. If artificial cooling is used, insulation reduces air pollution. Since World War II, there has been a trend in modern architecture for buildings all over the world to have a similar design without consideration of the local climate. To ensure comfort, energy-consuming air conditioners have been introduced to an extent that the cost of maintaining buildings has increased alarmingly. A more rational approach would be to design buildings in accordance with local climate. The concept of different housing forms in different climatic regions, adapted in a way that human physiological requirements are satisfied, will minimize the energy consumed [76].

Thermal comfort can be achieved over a wide temperature range since the heat balance of the human body is determined by environmental factors (air and radiation temperatures, humidity, and air movement) and personal factors (physical work level and clothing) [33]. Official temperature standards for dwellings are roughly the same in various countries, from 18 to 24°C [41]. The temperature required in workrooms such as offices, schools, etc., is very similar to that required in dwellings; but where human subjects perform physical work, the temperature should be lower.

B. Humidity

Water vapor affects the heat balance of the body and many chemical, physical, and biological processes, especially in the upper airways where inspired air becomes saturated. Humidity is best expressed in terms of absolute humidity, either as water vapor pressure (gas pressure in mmHg) or as the weight of water vapor—the mass of water vapor per unit weight (g water/kg dry air) or per unit volume (g water/m^3). These definitions of absolute humidity are equivalent for any given temperature and barometric pressure. The relative humidity (RH) expresses the ratio of the actual quantity of water vapor to the saturation quantity.

The capacity of air for water vapor increases progressively with temperature. Heating of an air mass with a constant absolute humidity, therefore, causes a decrease in the relative humidity. Conversely, cooling of this air mass will increase relative humidity and if the cooling continues, formation of fog, rain, or snow, or condensation on a surface will occur. Expressing relative humidity in percent without giving the air temperature thus is meaningless. For calculations a Mollier diagram is extremely useful (Fig. 3).

Water vapor enters the air by evaporation primarily from the surfaces of the oceans, but also from smaller bodies of water, vegetation, and soil. The percentage of water vapor by volume relative to all gaseous constituents of the air may range from 0.000002% to nearly 4 or 5%, the lowest amounts being either at high altitudes or over the antarctic plateau. Maximum amounts of water vapor have been reported in subtropical or equatorial regions on extremely sultry days at localities near shallow bodies of water. In middle latitudes, the average annual water vapor content usually lies in the range of 1-1.5% by volume [48]. At any particular location, water vapor content varies seasonally, usually higher in summer than in winter. Diurnal variations are small, but great variations may occur with changes in the meteorological conditions.

Widely diverging opinions on the water vapor content in city air are reported in the literature. But Chandler [23] concludes that cities have a lower absolute and relative humidity than surrounding rural areas. In cities, the number of foggy and rainy days and the amount of precipitation is 5-8% higher than in the surrounding countryside due to the high number of condensation nuclei in the former [24,64]. Precipitation in cities is quickly drained away through sewer systems. Further, the emission of water vapor by combustion processes has a seasonal variation. The higher temperature of the cities leads to lower relative humidities, in summer from 10 to 18% lower and in winter about 6% lower than in the countryside. Differences in relative humidity are maximum at night and in summer when heat islands are strongest.

Optimally, humidity at room temperature should not exceed 60% or be lower than 30% RH. A high relative humidity, particularly at elevated temperatures, has an adverse effect on the thermal equilibrium of humans as the evaporation of sweat is reduced [33]. The lower limit has been set to decrease evaporation from mucous membranes of the respiratory system [41]. Recent investigations

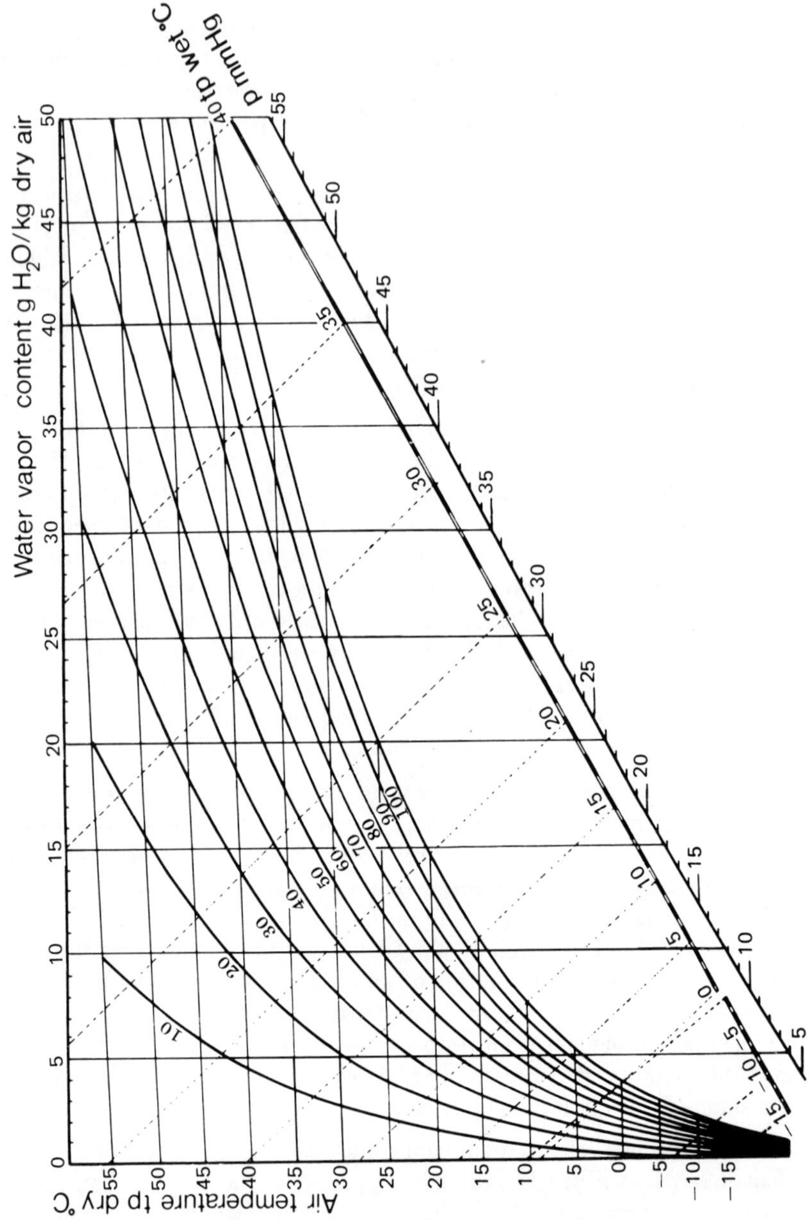

Water vapor content g H$_2$O/kg dry air
tp wet °C
p mmHg
Air temperature tp dry °C

FIGURE 3   Mollier diagram (i-x) for air at 760 mmHg. The horizontal lines from the y axis represent the air temperature in °C on a dry thermometer (tp dry) and the vertical lines from the x axis the water vapor content in g/kg dry air (at 20°C 1 m^3 air weighs 1.2 kg). The lower oblique scale gives the water vapor pressure in mmHg (p mmHg). Above this another oblique scale gives the temperature in °C on a wet thermometer (tp wet). The dotted lines ascending obliquely from the latter scale connect points with identical wet temperatures. Finally, the curved lines (marked 10 to 100) connect points with the same relative humidity (RH). 100% RH = saturation point.

The Mollier diagram may be used for: (a) estimation of RH from tp dry and tp wet (both values may be obtained from readings on a psychrometer); (b) estimation of the dew-point temperature (the temperature where condensation occurs); and (c) estimation of the need for humidification or dehumidification.

indicate that relative humidities as low as 9% at 23°C are safe for healthy subjects in clean air [6,8].

Humans produce about 40 g water vapor/hr, and processes of cooking, washing, and drying bring about a production of several kilograms water per day in the home. The main problem, therefore, is to remove water vapor from air in dwellings, and this is especially necessary in more temperate regions where condensation on cold surfaces (outer walls and windows) may create a favorable microclimate for mites, fungi, molds, and other potentially allergy-producing microorganisms. In temperate and polar regions, heating systems which draw (and subsequently heat) outdoor air produce a low relative humidity indoors. For this reason, humidifiers have gained popularity, but these devices have several drawbacks: Some are noisy, some produce heat, and some emit a microbial-loaded aerosol with gram-negative rods [1]. The latter is particularly dangerous in hospitals where humidifiers may serve as a reservoir for these microorganisms [14].

C. Air Velocity

The velocity of air influences the size of particles suspended in the air, affects the dilution of pollutants and affects the heat balance of the body. In the polar regions, wind speed is of vital importance as a few minutes outdoors may be lethal if cold weather is combined with high wind speeds (causing a huge heat loss by convection). Information about dangerous combinations of wind and dry bulb temperatures are found in a variety of charts, e.g., the wind-chill index [35].

During periods of relatively calm weather, the city heat island induces air circulation; country air drifts in at low levels toward the center of the city. Buildings often block the advance at street level, although there is a flow above the rooftops. The circulation is, therefore, not particularly effective in improving air quality [73]. Indoors, in occupied rooms without ventilation systems, air velocity is usually about 10-15 cm/min.†

D. Gases and Vapors

Today's atmosphere has been influenced by a continuous degassing from the interior of the earth throughout geological time. The bulk of the water vapor condensed and now forms the oceans. The main portion of carbon dioxide has been

---

†Higher air velocities may be felt as draft, and lower velocities may cause a sensation of oppression [41]. Combinations of comfortable air velocities and environmental and personal factors are listed by Fanger [33].

removed from the atmosphere by dissolving in water and precipitating in geological deposits. Nitrogen was released in much smaller quantities, but because of its chemical inertness it has accumulated in the atmosphere. Oxygen is a secondary product occurring mainly from photosynthetic consumption of carbon dioxide. Through geological time, this process has caused an increase in the atmospheric oxygen content even though great quantities have been consumed by oxidation of surface geological materials. The gases of pure, dry air are listed in Table 1. The largest portion, more than three-fourths of pure dry air by volume, is nitrogen, and about one-fourth of the air is oxygen. Together these make up about 99% of the air. Of the remaining 1%, argon is more than nine-tenths. Together with 0.033% carbon dioxide, these four elements make up 99.99% of dry air. The remaining gases, among these the other noble gases, are present only in very small quantities.

1. Clean Air

Nitrogen can be considered for the most part as an inactive gas, although a fraction is fixed every year. Most of this fraction is returned to the atmosphere within a few years through decay of plant material.

Oxygen is chemically very active and is consumed by oxidation of geological surface materials, by oxidation of metals, by fuel in burning and by metabolism in animals and man. These processes so far have been counterbalanced by the photosynthetic oxygen production. But this balance may be affected by pollution, land reclamation, etc.

Carbon dioxide, although present in only a small quantity in the atmosphere, is an extremely important gas because of its heat-absorbing properties. It is produced by oxidation of recently grown organic materials and by burning of fossil fuels; the consumption of CO_2 is mainly due to photosynthetic processes.

It is known that the CO_2 concentration in outdoor air is gradually rising. Data collected in the nineteenth century show that the level in the atmosphere at that time was close to 530 mg/m³ (290 ppm), and it now stands at 585 mg/m³ (320 ppm). This increase has been confirmed in the past decade by two series of measurements made at locations far away from local sources of pollution, the South Pole and the summit of Mauna Loa in Hawaii. The upward trend at both locations amounts to 1.28 mg/m³ (0.7 ppm) per year over an 11-year period. This increase is superimposed on a marked annual fluctuation. Carbon dioxide absorbs little of the sun's radiation, but the surface of the earth reradiates much of the energy in the invisible infrared region of the spectrum, where carbon dioxide absorbs a significant part of the energy. An increased carbon dioxide content, therefore, prevents the escape of heat into space, and trapped radiation will warm up the atmosphere ("greenhouse effect"). Although the present increase in atmospheric CO_2 since 1860 is no immediate cause for alarm, there is certainly need for future study. An increase in global temperature of 1-2°C might occur in the next century if fossil fuels continue to be the main energy source [64,71,85].

Indoors, humans cause an increase in the CO_2 concentration. A man engaged in light work exhales about 0.023 m³ CO_2/hr. In 1881, Pettenkofer suggested that CO_2 should be used as an index of the human contamination of air, and that the concentration limit should be 0.1% [41]. The CO_2 measurement and the 0.1% limit are still used as an easy way of estimating the air quality and the level of human occupancy in a room.

2. Pollutants

Pollutants added to the atmosphere by human activities come mainly from the combustion of fuel (especially coal and oil), generation of energy, heating and cooling, power, and transport. A complete combustion of pure hydrocarbon fuels produces only water vapor and carbon dioxide. Due to the shortcomings of the present techniques for domestic heating, generation of power, and propulsion of motor vehicles, significant amounts of incompletely oxidized by-products are also produced, i.e., aldehydes, ketones, and carbon monoxide. Further, most fuels contain impurities such as sulfur components, which are added to the air in a more or less oxidized form. Besides these common pollutants, specific industrial pollutants are added to the air. In this chapter, only sulfur dioxide, carbon monoxide, photochemical oxidants, and nitrogen oxides will be discussed [112]. For other pollutants see Table 2 and major handbooks about air pollution [92].

Sulfur dioxide causes mucous membrane irritation and reflex bronchoconstriction with increased airway resistance. Sulfur dioxide is produced by combustion of sulfur compounds present as impurities in many coals and heavy oils. Some of the sulfur in these fuels may also be released as H_2S, or may be further oxidized during the combustion process to sulfur trioxide; about 5% of the sulfur is emitted in the latter form. Sulfur trioxide will, at normal humidities, be converted into sulfate, and sulfur is thus transformed from gas phase to particulate phase. By comparing emission of sulfur within different regions with the deposited amounts, it has been estimated that the sulfur remains in the air for an average of 2-4 days, and that it is usually transported more than 1000 km before being deposited on the ground. Sulfur dioxide in the atmosphere is oxidized photochemically or by catalytic oxidation (where metal salts act as catalyzers). Sulfur dioxide in the atmosphere accounts for the low pH value (down to 3.0) of rain. This may acidify lakes and rivers, causing damage to metals, painted sur-

TABLE 2 Some Pollutants Found in the Air of the City of London [67]
(Arranged in descending order of highest concentrations observed since 1954)

Pollutant	Highest hourly concentration (μg/m^3 air)
Carbon dioxide	2,520,000
Carbon monoxide (street sample)	68,700
Particulate matter (smoke)	9,700
Sulfur dioxide	5,650
Nitric oxide	1,350
Sulfate (water-soluble)	666
Sulfuric acid	680
Nitrogen dioxide	470
Chloride (water-soluble)	410
Calcium	82
Hydrogen sulfide	70
Ammonia (gas)	56
Sodium	33
Iron	25
Zinc	24
Lead	22
Nitrate	5
Manganese	5
Chromium	2
Vanadium	2
Copper	2
Tin	2
3,4-Benzpyrene	2
Nickel	1
Cadmium	1
Titanium	1
Molybdenum, cobalt, bismuth, antimony, beryllium, barium	< 1

faces, buildings, and vegetation. Natural emissions of sulfur, in the form of H_2S, are about 30% greater than estimated industrial emissions of SO_2 and H_2S. The background concentration of SO_2 measured in very remote areas is about 0.6 $\mu g/m^3$ (0.2 ppb) [87].

In the United States, sulfur dioxide levels recorded in six large cities during 1962 through 1967 show mean annual concentrations ranging from 29 to 510 $\mu g/m^3$ (0.01-0.18 ppm); during 1% of the study period, averages exceeded 260 and 1950 $\mu g/m^3$ (0.09 and 0.68 ppm) [100]. Generally speaking, in polluted areas the concentration of SO_2 in the atmosphere varies between 60 $\mu g/m^3$ (20 ppb) and 6000 $\mu g/m^3$ (2 ppm) [21]. The potentiation of toxic responses to SO_2 by particulates has been observed under conditions promoting the conversion of SO_2 to sulfuric acid. Aerosols of soluble salts of ferrous iron, manganese, and vanadium enhance the conversion of SO_2 to sulfuric acid. Increased incidence of bronchitis may be associated with community exposures to SO_2, where annual levels rise to 95 $\mu g/m^3$ (32 ppb) and 15 $\mu g/m^3$ suspended sulfates [104]. A short-term rise in the levels of suspended sulfate may aggravate symptoms of cardiopulmonary and asthmatic disease [104]. The World Health Organization [112] has set SO_2 limits as an annual mean of 60 $\mu g/m^3$ (20 ppb) with 98% of the observations below 200 $\mu g/m^3$ (67 ppb). The permissible 2% observations over this limit may not fall on consecutive days.

Carbon monoxide causes the formation of carboxyhemoglobin, thus reducing the oxygen transport capacity of blood. This is important in tissues with impaired and high oxygen uptake, i.e., a cardiac muscle with coronary disease. There is no respiratory defense mechanism against CO inhalation. The main source of CO is automobiles, but other internal combustion engine sources, (airplanes, trains, ships) are also significant. The CO concentrations in heavy traffic in city streets are almost three times the CO levels in the central urban areas, and five times the CO levels in residential areas. CO easily penetrates from outside to inside, and in buildings constructed over highways, concentrations exceeding 100 mg/m^3 (87 ppm) have been measured [101,118]. Indoor CO also may derive from the use of heating and cooking devices and other domestic processes. Natural sources, such as forest and prairie fires, contribute insignificantly. The mean residence time in the atmosphere is between 1 month and 5 years; in the atmosphere, CO is oxidized to CO_2.

Endogenous formation of CO is about 0.42 ml/hr in normal, nonsmoking adults [26,101], but smokers will, even when not smoking, degas much larger amounts of carbon monoxide from the CO-Hb. "Passive smoking" is the absorption of CO by nonsmokers. Several investigations [47,84] of this problem conclude that the average concentration of CO in rooms where smokers are present normally will not exceed the ambient air quality standard [56 mg/m^3 (50 ppm)]. This is valid even if about five times as much CO is generated in the cigarette sidestream smoke as in the mainstream smoke. The smokers themselves, however, will be exposed to high concentrations of CO, and, even if the smoke is diluted by air during inhalation, the average concentration in smoke inhaled into the lungs is about 460 mg/m^3 (400 ppm). This may result in the loss of up to 10% of the capacity of the blood of smokers to transport oxygen [83]. The long-term goal for CO in urban air recommended by the World Health Organization [112] is an 8-hr average of 10 mg/m^3 (9 ppm) and 1-hr maximum of 40 mg/m^3 (35 ppm).

Photochemical oxidants arise from a complex series of atmospheric reactions of hydrocarbons and nitrogen oxides in the sunlight, producing ozone, peroxyacetyl nitrates (PAN), aldehydes, and other complex chemical compounds. The different compounds may be measured separately, but are often expressed as total oxidants. Ozone is formed naturally in the atmosphere by electric discharge and in the stratosphere by solar radiation. Maximum ozone levels of 20 to 100 μg/m^3 (0.01 to 0.05 ppm) have been recorded in nonurban areas. In the United States [102], an analysis of oxidant concentration data for 4 years at 12 stations showed a daily maximum of 1-hr average concentration equal to or exceeding 290 μg/m^3 (0.15 ppm) up to 41% of the time. Maximum 1-hr average concentrations ranged from 250 to 1140 μg/m^3 (0.13 to 0.58 ppm) with short time peaks as high as 1310 μg/m^3 (0.67 ppm). Systematic indoor measurements of these pollutants have not been performed, but it appears that the indoor levels depend on the rate of ventilation. Ozone decays rapidly; PAN is more persistent [94].

The photochemical oxidants are able to injure vegetation and damage polymer materials, especially rubber and cellulose. The adverse health effects of ozone appear to be primarily due to irritation. Increased resistance to airflow in the respiratory tract, impairment of the diffusing capacity of the lungs, and a decrease in effective function of alveolar macrophages have been reported [102].

The long-term goal of the World Health Organization [112] for urban concentrations of these photochemical oxidants (as measured by the natural buffered KI method and expressed as ozone) is an 8-hr average of 60 μg/m^3 (0.03 ppm) and a 1-hr maximum of 120 μg/m^3 (0.06 ppm).

Nitrogen oxides, on a global basis, are generated by natural sources in an amount exceeding that from man-made sources. Natural scavenging processes keep background levels in nonurban areas low, on the order of 8 μg/m^3 (4 ppb) of NO_2 and 2 μg/m^3 (2 ppb) of NO. In urban areas, however, the levels are frequently higher since fuel combustion is the major source of nitrogen oxide air pollution. In the United States [103], peak values of NO above 1.2 mg/m^3 (1 ppm) are widespread, but NO_2 concentrations have rarely been measured at this level. The latter in urban areas rarely exceeds 0.94 mg/m^3 (0.05 ppm).

The adverse health effects of nitrogen oxides include increased airway resistance and structural changes of the lung proteins. Also, a general relationship between levels of nitrogen oxides and incidence of respiratory disease has been demonstrated [103]. Nitrogen oxides may cause acute necrotic leaf injury on vegetation. Also, a fading of several textile dyes, a deterioration of cotton and nylon fibers, and corrosion of metals have occurred.

No systematic indoor measurements have been made of NO or NO_2. For nitrogen oxides in the urban atmosphere, there is no long-term goal set by the World Health Organization [112].

E. Particles

1. General Considerations

Atmospheric air is an aerosol, i.e., an aerocolloidal system in which small solid and liquid particles are dispersed among the molecules. The particles are of various dimensions with a diameter ranging from 6×10^{-4} to 2×10^{1} μm, excepting raindrops (Table 3). Two general processes are involved in the formation of aerosols: dispersion and condensation. In the former, relatively coarse matter is split up into fine particles by abrasion, grinding, etc., whereas in the latter, clusters of molecules coalesce.

The fate of particles in the air is dependent on the size of the particles and the air velocity. In still air, particles with a diameter larger than the mean free path of the molecules of air (6.53×10^{-2} μm at 20°C and 760 mmHg) will rapidly attain a constant settling velocity. The sedimentation of spherical particles of this size is described by Stokes' law, which relates size and density of spherical particles to terminal velocity:

$$U_t = \frac{\delta g d^2}{18\gamma}$$

where

  U_t = terminal velocity of falling particle

  g = gravitational constant

  d = particle diameter

  δ = particle density

  γ = air viscosity

TABLE 3 Approximate Size Ranges of Airborne Particles[a]

Substances	Diameter (μm)	
	Minimum	Maximum
Raindrops	500	5000
Pollen	10	100
Plant spores	10	35
Cement dust	3	100
Foundry dust	1	1000
Insecticide dust	0.5	10
Bacteria	0.3	35
Salt nuclei (from seawater)	0.03	0.5
Tobacco smoke	0.01	1
Normal impurities in quiet air	0.01	1
Combustion nuclei	0.01	0.1
Viruses and proteins	0.003	0.5
Gas molecules	0.0001	0.0006

[a]Modified from Ref. 61.

When the particle diameter approaches or is smaller than the mean free distance between the air molecules, resistance to fall decreases, and therefore, a correction has to be applied to the equation. The terminal settling velocity of spherical particles in still air (760 mmHg, 20°C) for particles with diameters of 100, 10, 1, and 0.1 μm are 24.7, 0.3, 0.003, and 0.00009 cm/sec, respectively. These values apply to spherical particles of any density when diameter is expressed in terms of equivalent unit density spheres. For irregularly shaped dust particles such as asbestos, quartz, clay, coal, and the like, the measured diameter is larger than the equivalent aerodynamic diameter. The aerodynamic size is the diameter of the unit density sphere having the same settling velocity as the particle in question of whatever shape and density [49].

Particles with a radius less than 0.1 μm are often referred to as Aitken nuclei or condensation nuclei, as these particles form the nucleus of water drops during condensation processes in the air. Contact between particles may cause adhesion due to mass attraction, and this creates larger particles (coagulation). Other processes, important for the stability of particles in the air, are diffusion, impaction, electrostatic fields, and thermal gradient. Reviews of the physics of aerosols are given in several publications [20,42,86,92].

The atmosphere contains nonviable and viable particles of various dimension. In this discussion only particles less than 2×10^1 μm will be described, as the larger particles settle very rapidly, and, therefore, are not inhaled. This dust fall constitutes a nuisance, but is not a health hazard.

Suspended particulate matter in air is a major pollutant with pronounced effects on our health and well-being, especially when particles 0.5 μm or less in diameter are deposited in the respiratory system [99]. Nonviable particles in clean, country air mainly rise from volcanoes in eruption, from grass and forest fires and minerals from wind erosion of land surfaces, salt crystals from breaking waves, etc. The viable fraction includes bacteria, viruses, algae, fungi, and pollen. The present rate of addition of suspended particulate matter into the world atmosphere by human activities is estimated to be 2×10^7 metric tons per year [87]. Over cities, where the particle production is high, the particles can form a layer which screens out some sunlight. This effect is due to a back-scattering to space of the incoming solar radiation, and causes an increase in the earth's albedo, which may result in a decrease of the global temperature.

At present there are only rather crude estimates of the natural annual dust load of the atmosphere from soil erosion and sand storms, volcanic eruptions, etc. There is an indication that the concentration of suspended particulate matter is slowly rising. But assessment of this rise must consider two factors: (a) total particulate loading of the atmosphere by human activity is only beginning to approach the long-term average stratosphere dust-loading by volcanic activity; and (b) the volcanic load has been increasing much more rapidly during the past quarter-century than the human-derived particle loading [87]. However, locally, in industrial and population centers, there may be important effects of human activity.

For nonurban areas in the United States, the annual geometric mean concentration of suspended particulate matter is typically between 10 and 60 μg/m^3. In urban areas, the annual geometric mean ranges from 60 to $\sim$ 200 μg/m^3.

In the United States and Great Britain yearly average mass mean diameter varies from 0.40 to 0.83 μm [68]. The maximum 24-hr average concentration is about three times the annual mean, but values of seven times the annual mean do occur. There is a daily cycle with two maxima, one at 8 to 9 a.m., and the other about 6 p.m. Also there is a weekly cycle, where weekend concentrations are about 15% lower than weekday levels [92]. In the temperate and polar regions of the world there also is an annual change with maximum in the heating season [21]. The long-term goal of the World Health Organization is an annual mean of 40 μg/m^3 with 98% of observations below 120 μg/m^3.

The Aitken nuclei are found in a concentration of less than 1×10^3 nuclei/cm^3 on the west coast of Ireland, but 5×10^4 nuclei/cm^3 in the Dublin air [30,

34]. In Schenectady, New York $\sim 1 \times 10^4$ nuclei/cm^3 on the windside of the town was found as compared with 2.5×10^4 nuclei/cm^3 downwind of the town center [78]. Skala [88] states that a typical concentration of nuclei over sea areas is $\sim 5 \times 10^2$ nuclei/cm^3, as against $\sim 5 \times 10^5$ nuclei/cm^3 in urban areas. These values correspond to findings by Israël [57,58], who adds that the values to be found in rural air may range from 5×10^3 to 1×10^4 nuclei/cm^3. Volatile emanations from vegetation may undergo a gas-to-particle conversion under the action of sunlight. This formation of condensation nuclei by polymerization of vapor of terpenes and other compounds is especially intense over forests and causes decreased visibility [30].

Few systematic measurements of the indoor dust concentrations have been performed in homes, but it appears that indoor aerosol has a more narrow size range, reflecting on indoor removal of large particles. In simultaneous outdoor and indoor measurements [68], 22-62% of the outdoor aerosol was less than or equal to 0.5 μm diam compared to 48 to 75% of the indoor aerosol. Indoor particles contain more organic material than outdoor particles, which indicates that significant sources are found within the home. Cigarette smoke adds significantly to the concentration of particulate suspended matter, thus a "passive smoking" problem similar to that described for carbon monoxide exists. One cigarette generates about 25 mg dry particulate matter in the mainstream smoke and about 60 mg in the sidestream smoke. The mean median diameter has been reported from 0.16 to 0.5 μm and the range from 0.01 to 1 μm. But the size will depend among other factors on ambient humidity [49,67]. Concentrations of particulate matter in room air range from 6 mg/m^3 at one cigarette and eight air changes per hour to 17.4 mg/m^3 at three cigarettes and one air change per hour if absorption on surfaces, coagulation, and settling are disregarded. In the former situation, four minutes would be required for the nonsmoker to obtain an exposure equivalent to that permitted during 24 hours in outdoor air (25μg/m^3) [17].

In industry, the concentration of inert particles has to be lower than 10 mg/m^3. Inert particles have little effect on the respiratory system. For other chemical substances the level is lower. Threshold limit values are listed in *Documentation of the Threshold Limit Values for Substances in Workroom Air* [3] and in the yearly published abbreviated list *Threshold Limit Values for Chemical Substances and Physical Agents in the Workroom Environment* [4].

The suspended particulate matter in the air may be divided into nonviable and viable particles.

2. Nonviable Particles

The chemical composition of suspended particulate matter has been analyzed extensively in nonurban and urban locations [92,112]. Some of the important toxic substances found are lead, mercury, cadmium, beryllium, manganese, arsenic, asbestos, and organochlorine pesticides (Table 2).

3.   Viable Particles

Viable particles in the air are of three types: microorganisms, pollens, and insects. Microorganisms include bacteria, viruses, algae, molds, yeasts, fungi, rusts, and spores. They originate from man, animals, birds, plants, and soil.

Airborne bacteria are ubiquitous and have been found up to altitudes of 30 km [18]. In marine air, the greatest number of different genera of both bacteria and fungi are obtained at and near the coast. Fungi of the genera *Alternaria, Hormodendrum, Penicillium,* and *Aspergillus* and bacteria of the genera *Micrococcus, Bacillus,* and *Corynebacterium* [38]. In midocean air, there are few bacteria, 1/2 m^3. Air of terrestrial origin normally has a higher content, 4 to 11 bacteria per cubic meter have been found on the summit of Mt. Blanc [92]. The total bacteria content in country air varies according to the season, reaching the highest concentration during the summer and autumn. Low concentrations approaching zero are found during the winter, when snow covers the ground [9]. Diurnal variation appears to be largely dependent on wind and humidity, as well as on the local disseminating activity. Local sources may be sewage treatment plants, garbage destruction plants, or slaughterhouses.

The number of airborne bacteria is considerably higher in the city than in the surrounding countryside. The first systematic study of bacteria in cities was performed by Miquel [70], and his work remains one of the most extensive in this area. Another classic study is that of Colebrook and Cawston [27], who studied air at different locations in a city for a year. At roof level the average count was 59/m^3, whereas the count doubled or trebled at street level. The predominant types grown were (in order of decreasing frequency): *Staphylococcus albus, Sarcinae,* gram-positive spore-bearing bacilli, such as *Bacillus cereus* and *B. subtilis.* A few colonies of *E. coli* and *Staphylococcus aureus* were found. Anaerobic cultivation yielded about 28% of the colonies grown on the comparable aerobic plates; *Clostridium perfringens* was much the most common here.

Outdoor air appears to have a bactericidal activity due to substances known collectively as "the open air factor." The properties of this factor are similar to those exhibited in the laboratory by ozone-olefin products [32]. Indoors, very high bacteria concentrations may be found in some localities (Table 4), but as a whole, only a few studies have been made outside hospitals, and most of them have focused on one or a few species. Generally speaking the indoor concentration of airborne bacteria is determined by the physical activity: talking, coughing, and sneezing in the room and the room ventilation. Viable and nonviable particles are released constantly from human surfaces and clothing. During

TABLE 4 Bacterial Concentrations in Indoor Air in Various Localities [9]

Sampling place	Bacteria (per m^3 air)	Coliforms (per m^3 air)
Sewage treatment plant	700,000	850
Garbage destruction plant	13,000	480
Chicken slaughterhouse	30,000	
Printing office	50,000[a]	
Sawmill	14,000	
Laboratory	200	
Animals room	900	

[a]The high numbers in the printing office were caused by a heavily contaminated air humidifier of fan type.

sneezing, coughing, and phonation, droplets of saliva from 1 to 1000 μm in diameter are projected into the air. Virtually no particles are released during quiet breathing; more particles are produced by talking and coughing; and sneezing produces by far the largest number. Whether or not a particle contains a living microbe depends on the concentration of microbes in respiratory surface fluids. Only a small proportion of the droplets less than 5 μm in diameter contain microbes, but viruses (especially those of the common cold or flu) may easily be concentrated in these small droplets. The airborne range of the largest droplets ejected is about 2 m, whereas droplets smaller than 100 μm in diameter quickly evaporate to give a nucleus of solids, including the bacteria and viruses. These particles (droplet nuclei), which are less than 10 μm in diameter and mostly in the 2-4 μm range, remain suspended in air for long periods, and can therefore be carried over long distances [74,108,109].

From the surface of the skin there is a continuous desquamation of small skin scales, which are released due to the friction between the skin surface and clothing, or, in naked subjects, simply by movements of the body [46]. The skin scales are small and are capable of penetrating the weaving of most fabrics. The natural convective boundary layer flow around the body has air velocities capable of transporting particles up to 50 μm diam upward into the rising plume of warm air above the head. The particles are then widely disseminated by air currents in the room [25]. A large quantity of bacteria are dispersed in this way. In some individuals, a liberation of 10^6 bacteria-carrying particles during undressing and dressing in a 2-min period has been reported [75]. Bacteria from human sources often contaminate a secondary environmental reservoir such as bedclothes and carpets [15], and dust from these reservoirs will be made airborne by the movement of people, by sweeping and bedmaking, and by drafts produced by open windows. Due to the large size of the dust particles, however, they settle in a few minutes [80]. An extensive survey of the normal microbial flora of man is edited by Skinner and Carr [89].

There is no general agreement as to the best index of the bacteriological purity of air. Therefore, only some important bacteria will be mentioned: M. tuberculosis, streptococci, staphylococci, and anthrax [16,41,80,95]. Several comprehensive surveys on airborne infection exist [31,43,51,80,109]. Transmission problems in hospitals are treated extensively by Hers and Winkler [51].

4. Viruses

Some viruses (e.g., foot and mouth disease) can be dispersed by the wind over distances up to 100 km and still infect [116], but generally the concentration of viruses pathogenic for man never reaches a level sufficient for outdoor transmission [51]. For measles and chickenpox there is reasonable evidence for spread on droplet nuclei, but indoors, determination of viruses in the air around human patients has been difficult, even with large-volume samples [10]. Probably direct droplet infection during talking, coughing, and sneezing is more common than actual droplet nuclei airborne transmission. Recent investigations indicate that the rhinoviruses, which cause a large percentage of common colds in adults, are able to survive several hours on skin, fabrics, and environmental surfaces, and that transfer by hands to an intermediate surface or directly to the fingers of another person is common [50].

Room air disinfection may be accomplished by ultraviolet radiation or chemical means. Ultraviolet radiation cannot be used for direct radiation of the air in occupied rooms because of the risk of conjunctivitis, but may be employed if only the upper part of the room is irradiated. As the method relies on natural air turbulence to carry airborne microorganisms into the irradiated zone, a high convective air exchange in the room is very important; an effect equivalent to 83 equivalent air changes per hour in the part of the room occupied by human subjects may then be obtained [81]. The use of chemical disinfectants has not yet provided encouraging results. Triethylene glycol, hexylresorcinol, and α-hydroxy-α-methylbutyric acid are the substances most thoroughly tested [109].

5. Pollens

Pollens may act as allergens in susceptible individuals. Due to their small size, they remain airborne for long periods and are carried by the wind far from the point of origin. Generally, tree pollen is shed in early spring before the foliage is developed, whereas grass pollen is found in high concentrations during the summer [72]. However, each regional pattern differs and is determined by the local vegetation. The air in the center of large cities is relatively free of airborne pollen and is dominated by local sources (e.g., trees and grass in parks). Indoor pollen counts reflect the general pattern of the outdoor spora, but in much lower concentrations.

6. Insects

In outdoor air, the concentration of insect debris is high during the late summer, with a daily maximum just before dusk [45]. Indoors, the 300 μm-long *Dermatophagoides* mite is most prevalent, which seems to be the main allergic component in house dust and may cause "house dust asthma." The *Dermatophagoides* mite feeds on keratin, mainly in the form of skin scales, therefore it is abundant in the dust from bed linen. In mattress dust, a mean density of 296 mites/100 μg dust has been reported [13]. Mite feces and debris are found in the air; indeed, the mite itself may become airborne. During bedmaking, 1 mite/5 m^3 of air has been measured [29]. The mite thrives at 25°C and in high humidities, but not at low humidities. It is present in household dust in most parts of the world [37].

F.   Radioactive Substances in the Air

Environmental radiation is divided into two types, natural background and man-made radiation, and both are divided into a number of components. Man-made radiation comes mainly from the use of diagnostic x-rays. About 20 mrem/year is average for the United States and Western Europe; radioactive fallout, electronic devices, etc. add about 15 mrem/year. Thus the total man-made radiation amounts to 35 mrem/year [113]. Radiation of natural origin is the largest source of human exposure to ionizing radiation. In the United States, it contributes a dose equivalent to 80-200 mrem/year. Sources of natural origin include cosmic radiation, radiation from naturally occurring radionuclides in the earth, or in materials in man's immediate environment, and radiation from radionuclides within the body. Although the cosmic ray intensity varies slightly with latitude, the terrestrial γ-radiation exposure is strongly influenced by geology. Over large freshwater lakes there is virtually no terrestrial γ-radiation, but very high values (up to 350 mrem/year) have been found in rocky areas [117].

The natural radiation dose to the respiratory tract is largely due to inhalation and deposition of the gas radon (222 Rn) with half-life of 3.8 days and its short-lived (minutes) daughter products of solid state (^{218}Po, ^{214}Pb, ^{214}Bi, also named RaA, RaB, RaC, respectively). Existing data are inadequate for making accurate calculations of lung dose from natural radioactivity due to radon and radon daughters [39], but it is known that the radiation dosage delivered to the bronchial epithelium of the lung results primarily from the daughters and not from radon itself.

High occupational exposure to radioactive gases occurs in uranium mines, and control requires huge amounts of fresh air for dilution-ventilation, or recirculation through electrostatic precipitators [115]. In dwellings, the decay of natural radioactive substances in building materials produces radioactive gases, and measurements have been performed on a rather extensive scale [53,96,117]. The building materials, as well as the rate of ventilation, significantly influence the concentration of radon and radon daughters. In timber houses, the concentration is lower than in masonry houses. From 4 to 15 times higher concentrations are found in unventilated basement rooms compared to the first floors of the same house. People living in poorly ventilated homes, especially in basement apartments, are thus experiencing an increased exposure greater than that expected from many man-made sources [96]. The variations throughout the day are considerable, the lowest values occurring at about 8 p.m. and the highest in the early hours of the morning. The concentration varies inversely with the atmospheric pressure, which influences the rate of degassing from the walls. The average RaA activity is 4 picocuries/liter [59]. This is higher than the internationally accepted maximum level of activity of 222 Rn in radioactive equilibrium with its daughter products for continuous exposure of the general public. The differences in dose rate for persons living in buildings of different construction materials are great, and a screening of materials to eliminate those with excessive radiation may reduce the population dose equivalent to the projected increase from development of nuclear power.

G. Ions

Atmospheric air contains carriers of electric charge, which may be divided into two main classes according to mobility: (a) light gas ions (mobility > 0.6 cm^2 V^{-1} sec^{-1}) and (b) heavy gas ions (mobility < 0.6 cm^2 V^{-1} sec^{-1}). The latter classification is determined by the velocity of ions in an electric field in still air. From time to time the question arises as to whether light gas ions are of any importance for biological systems, but so far no well-substantiated and reproducible effects have been reported. The postulated favorable effect of negative ionized air on the mucociliary function in the conductive zone of the airways has recently been rejected [5]. The physical aspects of air ions are dealt with in several reviews [22,57,58].

V. Air Quality Standards

The establishment of air quality standards is an extremely complex matter, as the probability of acquiring diseases due to air pollutants depends on the susceptibility of the subjects and on the environmental load.

The susceptibility of the subjects depends on inherited factors (e.g., airway size, mucociliary clearance rates, α_1-antitrypsin level); on acquired factors (e.g., related to age and disease); and on the habits of the subjects (e.g., nasal or mouth respiration, ventilation rates, and smoking). Due to this wide variation in individual susceptibility, a small percentage of persons may be affected at any concentration level. Generally, the most susceptible subjects are infants, old people, and persons with severe cardiac and pulmonary diseases. Air quality standards for the community must include the requirements of these very susceptible persons, which are not considered in the air quality standards for the workroom environment. In both cases, however, the decision is based on statistical criteria.

The environmental load consists of a mixture of substances which is different for each of the three compartments of air to which man is exposed: the home environment, the working place, and the outdoor air. At present there is a rather extensive system of standards and control programs for the work-place air and for urban air, whereas no standards exist for indoor air in homes or offices. The effects of tobacco smoking should also be included in the standards. It is estimated that the relative contribution to chronic bronchitis of air pollution alone ranges from one-third to one-seventh of that of cigarette smoking. The quantitative contribution of occupational exposure is somewhat larger than that of air pollution [104].

For workrooms, the first threshold limit values were published in the Soviet Union in 1922 listing three substances; a corresponding list was published in the United States in 1937. Today most countries have air quality standards for workrooms listing up to 700 substances. The standards differ considerably from country to country, but are grouped around two different concepts of health criteria. In the United States and Western Europe no serious threat to health is considered to exist, so long as the level of exposure does not introduce a demonstrable disturbance of a kind predictive of ill health [4,97,113]. In contrast, the USSR and Eastern Europe consider a potential threat to health to exist as soon as the organism undergoes any detectable change from its normal state [60,113]. It is therefore not surprising that the permissible levels in the latter countries are the lowest (Table 5) [113]. It should be remembered, however, that a discussion of the magnitude of air quality criteria has only academic interest if the standards are not enforced by control measurements.

TABLE 5 Maximum Allowable Concentration of Some Harmful Substances in Various Countries, in mg/m^3 [113]

Country	Lead	Carbon dioxide	Carbon monoxide	Toluene
USSR (1970–1971)	0.01	—	20	50
USA (1971–1972)	0.15	9000	55	375
Czechoslovakia (1970–1971)	0.05	9000	30	200
Poland (1970)	0.05	—	30	100
Federal Republic of Germany (1971–1972)	0.2	9000	55	750
United Kingdom (1972)	0.15	9000	55	375
Switzerland (1971)	—	9000	55	380

For urban air pollution, the air quality standards have to be lower than the threshold limit values for workrooms. The time spent outside the working place is longer than the time spent in the working place, and the population as a whole (due to high-risk groups such as children, old, and sick subjects) has a higher susceptibility for inhaled substances than the working population. Several countries (e.g., United States, West Germany, and Soviet Union) have norms for a limited number of components, and the World Health Organization recommends long-term goals for four pollutants: sulfur oxides, suspended particulates, carbon monoxide, and photochemical oxidants [112].

Air quality standards for the air inside dwellings do not exist in any country. The need for standards in this area increases rapidly as new synthetic materials are widely used without studies of their implication for health. In many cases these materials give rise to a gas-off of volatile components [7,41].

Information is also urgently needed about the interaction of different atmospheric pollutants in the outdoor and indoor environments. The combined presence of two or more substances in the air inhaled can significantly change the degree of the toxic effect that is characteristic for each component of the mixture. The effect of the combined action may be summation, potentiation, or antagonism. A further complication is that the direction of the combined actions of the toxic substances may be different at different quantitative levels [19].

VI. Summary and Conclusions

The atmosphere may be considered as a two-compartment system, outdoor air and indoor air, each consisting of several subcompartments. Two important subcompartments in outdoor air are rural air and community air; in indoor air, workroom air and home air.

The natural atmosphere in the human scale of time has had a stable composition in balance with the lithosphere and the hydrosphere. A background pollution of gases and particles originating from natural sources exists, but in general it can be considered innocuous to human health. Community air is characterized by man-made pollution with hundreds of gases and particles from domestic and industrial combustion and from automobiles. Community air pollution is a local threat to human health and well-being. Legislative pressure and technical improvements, however, have been able to either halt increase in pollutant concentrations, or even reverse the trend. Our current understanding of the physical and chemical interactions between the different substances in community air is poor, as is our knowledge of the health effects of combinations of gases and/or particles. This especially applies to the effects of low concentrations. The present air quality standards for outdoor air cite only few pollutants; low-level and long-term exposure studies, as well as epidemiological investigations, are needed to improve and expand these standards.

Inside buildings the rate of change of air is low in comparison to that in the outdoor environment, and a buildup of pollutants produced by human subjects, human activities, and building materials takes place. A broad spectrum of gases and particles are found in workroom air and generally in higher concentrations than in community air. Air quality standards for workroom air cite about 700 substances. The standards are based on industrial experience and on experimental human and animal exposure studies. Again, very little is known about the effects of combinations of pollutants. Very limited information exists about the indoor home atmosphere, and no air quality standards exist for that environment. The long periods of time spent in the home environment and the high susceptibility of some persons renders further studies necessary.

No disease group is so definitely related to the environment as is respiratory disease. Without doubt, genetic and other factors are important, but the environment is primary because it can be manipulated. Recent studies indicate that the role of cigarette smoking in the development of chronic respiratory disease far exceeds the contribution from occupational exposure and from community air pollution. Also it appears that synergistic effects are caused by a combination of these different environmental loads. Therefore, the fight against respiratory disease has to rest upon a holistic view of man-environment interactions and on measures continuously adjusted to the ever-changing relative importance of the different subcompartments of the atmosphere.

References

1. T. Airoldi and W. Litsky, Factors contributing to the microbial contamination of cold-water humidifiers, *Am. J. Med. Technol.*, **38**:491-495 (1972).
2. A. P. Altshuller and J. J. Bufalini, Photochemical aspects of air pollution: A review, *Environ. Sci. Technol.*, **5**:39-64 (1971).
3. American Conference of Governmental Industrial Hygienists. *Documentation of the Threshold Limit Values for Substances in Workroom Air*, Cincinnati, 1971.
4. American Conference of Governmental Industrial Hygienists. *Threshold Limit Values for Chemical Substances and Physical Agents in the Workroom Environment*, Cincinnati, 1975.
5. I. Andersen, *Mucociliary Function in Trachea Exposed to Ionized and Non-ionized Air*, Aarhus, Denmark, Akademisk Boghandel, 1971.
6. I. Andersen, G. R. Lundqvist, and D. F. Proctor, Human nasal function under four controlled humidities, *Am. Rev. Resp. Dis.*, **106**:438-449 (1972).
7. I. Andersen, G. R. Lundqvist, and L. Mølhave, Indoor air pollution due to chipboard used as a construction material, *Atmospheric Environment*, **9**:1121-1127 (1975).
8. I. Andersen, P. L. Jensen, G. R. Lundqvist, and D. F. Proctor, Human response to 78-hour exposure to dry air, *Arch. Environ. Health*, **29**:319-324 (1974).

9. R. Andersson, B. Bergström, and B. Bucht, Outdoor sampling of airborne bacteria. In *Airborne Transmission and Airborne Infection.* Edited by J. F. Hers and K. C. Winkler. Utrecht, Oosthock Publ., 1973.

10. M. S. Artenstein and W. S. Miller, Air sampling for respiratory disease agents in army recruits, *Bacteriol. Rev.,* **30**:571-572 (1966).

11. T. Bedford, *Basic Principles of Ventilation and Heating,* London, H. K. Lewis, 1964.

12. F. B. Benson, J. J. Henderson, and D. E. Caldwell, *Indoor-Outdoor Pollution Relationships—A Literature Review,* U.S. Environmental Protection Agency, Publ. no. AP-112, 1972.

13. M. E. Blythe, J. D. Williams, and J. Morrison Smith, Distribution of pyroglyphid mites in Birmingham, *Clin. Allergy,* **4**:25-33 (1974).

14. G. J. Bonde, Water problems in anaesthesiology, *Acta anaesthesiol. Scand.,* **1**:88-92 (1966).

15. G. J. Bonde, Bacterial flora of synthetic carpets in hospitals, *Health Lab. Sci.,* **10**:308-318 (1972).

16. P. S. Brachmann, S. A. Plotkin, F. H. Bundford, and M. A. Atchinson, An epidemic of inhalation antrax, *Am. J. Hyg.,* **72**:6-23 (1966).

17. D. P. Bridge and M. Corn, Contribution to the assessment of exposure of non-smokers to air pollution from cigarette and cigar smoke in occupied spaces, *Environ. Res.,* **5**:192-290 (1972).

18. C. W. Bruch. In *Airborne Microbes.* Edited by P. H. Gregory and J. L. Menteith. Cambridge, University Press, 1967.

19. K. A. Bustueva and I. V. Sanotsky, On the interaction of atmospheric pollutants. CEC-EPA-WHO Symposium on Environmental Health, paper 105, Paris, 1974.

20. R. D. Cadle, *Particle Size,* New York, Reinhold, 1965.

21. P. Camner (ed.), Air quality criteria and guides for Sweden in regard to sulphur dioxide and suspended particulates, *Nord. Hyg. T. suppl. 5. Stockholm,* (1973).

22. J. A. Chalmers, *Atmospheric Electricity,* London, Pergamon, 1957.

23. T. J. Chandler, Absolute and relative humidities in towns, *Bull. Am. Meterol. Soc.,* **48**:394-399 (1967).

24. S. A. Changnon, Recent studies of urban effects on precipitation in The United States. In *Urban Climates,* Geneva, World Meteorological Organization, 1970, Technical Note no. 108.

25. R. P. Clark and R. N. Cox, Dispersal of bacteria from the human body surface. In *Airborne Transmission and Airborne Infection.* Edited by J. F. Hers and K. C. Winkler. Utrecht, Oosthock Publ., 1973.

26. R. F. Coburn, W. S. Blakemore, and R. E. Forster, Endogenous carbon monoxide production in man, *J. Clin. Invest.,* **42**:1172-1178 (1963).

27. L. Colebrook and W. C. Cawston, Microbic content of the air on the roof of a city hospital, at street level, and in the wards. In *Studies in Air Hygiene,* London, MRC Special Report Series no. 262, HMSO, 1948.

28. J. P. Conkle, W. E. Mabson, J. D. Adams, H. J. Zeft, and B. E. Welch, *A Detailed Study of Contaminants Produced by Man in a Space Cabin*

Simulator at 760 mm Hg, SAN-TR-67-16, Texas, USAF School of Aerospace Medicine Brooks Air Force Base, 1967.

29. A. M. Cunnington and P. H. Gregory, Mites in bedroom air, *Nature,* **217**: 1271-1272 (1968).

30. C. N. Davies, Particles in the atmosphere—Natural and man-made, *Atmos. Environ.,* **8**:1069-1079 (1974).

31. R. L. Dimmick and A. B. Akers (eds.), *An Introduction to Experimental Aerobiology,* New York, Wiley, 1969.

32. H. A. Druett, In *Airborne Transmission and Airborne Infection.* Edited by J. F. Hers and K. C. Winkler. Utrecht, Oosthock Publ., 1973.

33. P. O. Fanger, *Thermal Comfort,* New York, McGraw-Hill, 1972.

34. V. P. V. Flanagan and T. C. O'Connor, Ionization equilibrium in aerosols, *Geofis. Pura Appl.,* **50**:148-154 (1961).

35. J. Folk, *Introduction to Environmental Physiology,* Philadelphia, Pa., Lea and Febiger, 1966.

36. R. S. Foote, Mercury vapor concentrations inside buildings, *Science,* **177**: 513-514 (1972).

37. A. W. Frankland, Asthma and mites, *Postgrad. Med. J.,* **47**:178-180 (1971).

38. J. D. Fulton, Microorganisms of the upper atmosphere, *Appl. Microbiol.,* **14**:241-244 (1966).

39. A. C. George, Indoor and outdoor measurements of natural radon and radon decay products in New York City air. Natural Radiation Environment Symposium, Houston, Texas. Aug. 7-11, 1972.

40. B. Givoni, *Man, Climate and Architecture,* Amsterdam, Elsevier, 1969.

41. M. S. Goromosow, *The Physiological Basis of Health Standards for Dwellings,* Geneva, World Health Organization, Public Health paper no. 33, 1968.

42. H. L. Green and W. R. Lane, Particulate clouds: Dusts, smokes and rusts, Spon, London (2nd ed.), 1964.

43. P. H. Gregory and J. L. Monteith (eds.), *Airborne Microbes,* London, Cambridge University Press, 1967.

44. C. W. Gruber and E. L. Alpaugh, The automatic filterpaper sampler in air pollution measurement program, *Air Repair,* **4**:143-146 (1954).

45. E. D. Hamilton, Studies on the air spora, *Acta Allergol. (Kbh.),* **13**:143-173 (1959).

46. R. Hare and C. G. A. Thomas, The transmission of *Staphylococcus aureus, Br. Med. J.,* **2**:840-844 (1956).

47. H. P. Harke, Zum Problem des Passivrauchers, *Int. Arch. Arbeitsmed.,* **29**: 312-322 (1972).

48. L. P. Harrison, Fundamental concepts and definitions relating to humidity. In *Humidity and Moisture,* Vol. 3. Edited by A. Vexler. New York, Reinhold, 1965.

49. T. F. Hatch and P. Gross, *Pulmonary Deposition and Retention of Inhaled Aerosols,* New York, Academic Press, 1964.

50. J. O. Hendley, R. P. Wenzel, and J. M. Gwaltney, Transmission of rhinovirus colds by self-inoculation, *N. Engl. J. Med.,* **288**:1361-1364 (1973).

51. J. F. Ph Hers and K. C. Winkler (eds.), *Airborne Transmission and Airborne Infection*, Utrecht, Oosthock Publ., 1973.

52. F. N. Hodgson and J. V. Pustinger, Gas-off studies of cabin materials. Proceedings of the 2nd Annual Conference on Atmospheric Contamination in Confined Spaces. Aerospace Medical Research Laboratories, Wright-Patterson Air Force Base, Ohio, AMLR-TR-66-120, 1966.

53. B. Hultqvist, Studies on naturally occurring ionizing radiations, *Kungl. Svenska Vetensk. Akad. Handl. (Stockh.)*, ser. 4, 6(3) (1956).

54. W. J. Humphreys, *Physics of the Air*, New York, Dover, 1964.

55. D. Hunter, *The Diseases of Occupations*, London, English Universities Press, 1969.

56. International Labour Office, *Encyclopaedia of Occupational Health and Safety*, Geneva, 1971.

57. H. Israël, *Atmosphärische Elektrizität*, teil 1, Leipzig, Geest und Portig, 1957.

58. H. Israël, *Luftelektrizität und Radioaktivität*, West Berlin, Springer-Verlag, 1957.

59. N. Jonassen and E. I. Hayes, The measurement of low concentrations of the short-lived Radon-222 daughters in the air by alpha spectroscopy, *Health Phys.*, 26:105-110 (1974).

60. D. Kahle, *Arbeitshygienische Normen und MAK-werte*, Berlin, Verlag Tribüne, 1969.

61. M. Katz, *Measurements of Air Pollutants*, Geneva, World Health Organization, 1969.

62. O. Koenigsberger and R. Lynn, Roofs in the warm humid tropics, London, Architectural Association, 1965.

63. C. P. Kruglikova and V. K. Efimova, Atmospheric sulfur gas as a source of air pollution in residential dwellings, *Gigiena i Sanit.*, 23:75-78 (1958).

64. H. E. Landsberg, Controlled Climate (Outdoor and Indoor). In *Medical Climatology*. Edited by S. Licht. New Haven, Connecticut, 1964.

65. B. T. Larsson, Gaschromatography of organic volatiles in human breath and saliva, *Acta Clin. Scand.*, 19:159-164 (1965).

66. P. J. Lawther, Air pollution and its effects on man, *Proc. Roy. Inst. Gt. Br.*, 44:74-94 (1970).

67. P. J. Lawther, Air pollution and tobacco smoke. In *Clinical Aspects of Inhaled Particles*. Edited by D. C. F. Muir. London, Heinemann, 1972.

68. R. E. Lee, Jr., The size of suspended particulate matter in air, *Science*, 178:567-575 (1972).

69. Ministry of Health, Mortality and Morbidity during the London Fog of December 1952, Reports on Public Health and Related subjects no. 95, London, England, 1954.

70. M. P. Miquel, *Les Organismes vivants de l'Atmosphere*, Paris, Gautier-Villars, 1883.

71. I. M. Mitchell, Jr., A preliminary evaluation of atmospheric pollution as a cause of the global temperature fluctuation of the past century. In *Global Effects of Environmental Pollution*. Edited by S. F. Singer. Dordrecht, Holland, Reidel, 1970.

72. D. C. F. Muir, *Clinical Aspects of Inhaled Particles,* London, Heinemann, 1972.

73. R. E. Munn, *Descriptive Micrometeorology,* New York, Academic Press, 1966.

74. W. C. Noble, O. M. Lidwell, and D. Kingston, The size distribution of airborne particles carrying micro-organisms, *J. Hyg. (Camb.),* **61**:385-391 (1963).

75. W. C. Noble and R. R. Davies, Studies on the dispersal of staphylococci, *J. Clin. Pathol.,* **18**:16-19 (1965).

76. V. Olgyay, *Design with Climate,* Princeton, N.J., Princeton University Press, 1963.

77. F. A. Patty (ed.), *Industrial Hygiene and Toxicology,* Vols. 1-3, New York, Wiley-Interscience, 1962.

78. T. A. Rich, L. W. Pollak, and A. L. Metnieks, On the time required for aerosols to reach electrical equilibrium, *Geofis. Pura Appl.,* **51**:217-224 (1962).

79. V. J. Richter and J. Tonzetich, The application of instrumental technique for the evaluation of odoriferous volatiles from saliva and breath, *Arch. Oral Biol.,* **9**:47-53 (1964).

80. R. L. Riley and F. O'Grady, *Airborne Infection,* New York, Macmillan, 1961.

81. R. L. Riley, S. Permutt, and J. E. Kaufman, Room air disinfection by ultraviolet irradiation of upper air, *Arch. Environ. Health,* **23**:35-39 (1971).

82. E. Rodenwaldt and H. J. Jusatz, *World Maps of Climatology,* Berlin, Springer-Verlag, 1965.

83. Royal College of Physicians of London, *Smoking and Health Now,* London, Pitman, 1971.

84. M. A. H. Russel, Absorption by non-smokers of carbon monoxide from room air polluted by tobacco smoke, *Lancet,* **1**:576-579 (1973).

85. J. S. Sawyer, Man-made carbon dioxide and the "Greenhouse" effect, *Nature,* **239**:23-26 (1972).

86. L. Silverman, C. E. Billings, and M. W. First, *Particle Size Analysis in Industrial Hygiene,* New York, Academic Press, 1971.

87. S. F. Singer (ed.), *Global Effects of Environmental Pollution,* Dordrecht, Holland, Reidel, 1970.

88. G. F. Skala, A new instrument for the continuous measurement of condensation nuclei, *Anal. Chem.,* **35**:702-706 (1963).

89. F. A. Skinner and J. G. Carr (eds.), *The Normal Microbial Flora of Man,* London, Academic Press, 1974.

90. D. J. Spedding, The fate of sulphur-35/sulphur dioxide released in laboratory, *Atmos. Environ.,* **3**:341-346 (1969).

91. D. J. Spedding, Sulphur dioxide uptake by limestone, *Atmos. Environ.,* **3**:683 (1969).

92. A. C. Stern (ed.), *Air Pollution,* Vol. 1, 2nd ed., New York, Academic Press, 1968.

93. J. F. van Straaten, *Thermal Performance of Buildings,* Amsterdam, Elsevier, 1967.

94. C. R. Thompson, E. G. Hensel, and G. Kats, Outdoor-indoor levels of six air pollutants, *J. Air Pollut. Control Assoc.,* 23:881-886 (1973).

95. J. C. Torrey and M. Lake, Air streptococci and colds, *JAMA,* 117:1425 (1941).

96. A. Tóth, Determining the respiratory dosage from RaA, RaB and RaC inhaled by the population in Hungary, *Health Phys.,* 23:281-289 (1972).

97. R. A. Trevethick, *Environmental and Industrial Health Hazards,* London, Heinemann, 1973.

98. U.S. Dept. of Commerce, *Climatic Atlas of the United States,* Washington, D.C., 1968.

99. U.S. Dept. of Health, Education and Welfare, *Air Quality Criteria for Particulate Matter,* Washington, D.C., Public Health Service, 1969.

100. U.S. Dept. of Health, Education and Welfare, *Air Quality Criteria for Sulfur Oxides,* Publ. AP-50, Washington, D.C., Public Health Service, 1970.

101. U.S. Dept. of Health, Education and Welfare, *Air Quality Criteria for Carbon Monoxide,* Publ. AP-62, Washington, D.C., Public Health Service, 1970.

102. U. S. Dept. of Health, Education and Welfare, *Air Quality Criteria for Photochemical Oxidants,* Washington, D.C., Public Health Service, 1970.

103. U.S. Dept. of Health, Education and Welfare, *Air Quality Criteria for Nitrogen Oxides,* Washington, D.C., Environmental Protection Agency, 1971.

104. U.S. Environmental Protection Agency, *Health Consequences of Sulfur Oxides: Report from CHESS,* Research Triangle Park, N.C., National Environmental Research Center, 1970-1971.

105. W. Viessman, Ventilation control of odor. In *Recent Advances in Odor.* Edited by R. L. Kuchner. *Ann. N.Y. Acad. Sci.,* 116:357-746 (1964).

106. H. Walter and H. Lieth, *Klimadiagramm, Weltatlas,* Jena, VEB, Gustav Fischer, 1966.

107. R. C. Weast (ed.), *Handbook of Chemistry and Physics,* 51st ed., Cleveland, Ohio, The Chemical Rubber Co., 1971.

108. R. E. O. Williams, O. M. Lidwell, and A. Hirch, The bacterial flora of the air of occupied rooms, *J. Hyg. Camb.,* 54:512-523 (1956).

109. R. E. O. Williams, Inhaled microbes. In *Clinical Aspects of Inhaled Particles.* Edited by D. C. F. Muir. London, Heinemann, 1972.

110. M. J. G. Wilson, Indoor air pollution, *Proc. Roy. Soc. (Lond.),* **A 300**:215-221 (1968).

111. H. C. Wohlers, H. Newstein, and D. Daunis, Carbon monoxide and sulphur dioxide adsorption on—and desorption from glass, plastic and metal tubing, *J. Air Pollut. Control Assoc.,* 17:753-756 (1967).

112. World Health Organization, *Air Quality Criteria and Guides for Urban Air Pollutants,* Geneva, WHO, Tech. Rept. Ser. 506, 1972.

113.   World Health Organization, *Health Hazards of the Human Environment,*
       Geneva, WHO, 1972.
114.   World Meteorological Organization, *Urban Climates,* Geneva, Tech. Note
       no. 108, 1970.
115.   M. E. Wrenn, M. Eisenbud, C. Costa-Ribeiro, A. J. Hazle, and R. D. Siek,
       Reduction of radon daughter concentrations in mines by rapid mixing
       without make up air, *Health Phys.,* **17**:405-414 (1969).
116.   P. B. Wright, Effects of wind and precipitation on the spread of foot-and-
       mouth disease, *Weather,* **24**:204-213 (1969).
117.   D. B. Yeates, A. S. Golden, and D. W. Moeller, Natural radiation in the
       urban environment, *Nucl. Safety,* **13**:275-286 (1972).
118.   J. E. Yocom, W. L. Clink, and W. A. Cote, Indoor/outdoor air quality
       relationships. 63rd Annual Meeting of the Air Pollution Control Associa-
       tion, St. Louis, Missouri, 1970.

3

Access of Air to the Respiratory Tract

DAVID L. SWIFT and DONALD F. PROCTOR

The Johns Hopkins University
School of Hygiene and Public Health
Baltimore, Maryland

I. Introduction

In the preceding chapter the physical characteristics and constituents of the ambient air have been discussed. Modification of this air begins the moment it enters the body. The efficacy with which this is accomplished depends upon the dimensions of the conducting airway, the air-to-surface contact, and the nature of the epithelium on that surface. The airway which provides access of ambient air to the lungs is of significance in relation to those characteristics determining its capacity to modify the inspired air, and in respect to its ability to withstand injurious airborne influences [38,39]. During nasal breathing adjustment of inspired air temperature begins at the nostril and, even when breathing very warm (39°C) or very cool (7°C) air, the temperature has reached 28 to 37°C in the nasopharynx. Further adjustment occurs during air passage from that point on into the bronchi, but even in third-division bronchi it is not quite at body temperature. Final adjustment to the temperature of the alveolar epithelium probably occurs in the most peripheral bronchi. Addition of water vapor to the inspired air parallels the temperature change (see Chap. 4).

Under ordinary circumstances inspired air moves through the nose, nasopharynx, pharynx, and larynx. When nasal obstruction occurs, or when ventilatory demands are such as to call for airflows greater than those compatible with nasal breathing, the alternative airway is the mouth, oropharynx, and larynx. But, when the mouth is open, part of the airflow may still take place through the nose and the size and shape of the oropharynx is susceptible to wide variation. In disease it may be necessary to bypass all of these upper airways and provide an artificial access through a naso- or orotracheal tube or tracheostomy.

This chapter is directed at an analysis of airflow through the nose and other airways of access to the intrapulmonary airways as it relates to function, some of the factors which determine the choice of nasal or mouth breathing, some of the effects of altered nasal airflow, and the employment of an alternative airway of access. The tracheobronchial tree is capable of adjusting air temperature and humidity, but not with the efficiency of the nose and at some cost in terms of injury to epithelium. The smallest peripheral airways do provide narrow airstreams exposed to a wide surface area, but they have relatively few mucus-secreting cells and there is some evidence that mucociliary clearance of their surfaces is relatively slow.

In the main nasal passage we have the combined effect of nasal secretions (supplemented by those from the paranasal sinuses), a narrow airstream, relatively low linear velocity, and a highly vascular mucosa and submucosa capable of rapid adjustment to changing ambient air demands. Special conditions prevail at the nasal entrance, to be discussed below. An architect asked to design a ventilating system to protect a room full of delicate instruments would undoubtedly plan one which removed and disposed of particulates and foreign gases in the air before its entry, and assured constant temperature and humidity in the face of any conceivable ambient conditions. We shall see that the normal upper airway fulfills these specifications remarkably well.

Diagrammatic views of the respiratory tract as a whole are provided in Figs. 1 and 2. Although most of the anatomical and physiological features of the respiratory tract related to defense are discussed in subsequent chapters in detail, it may be useful here to seek some general orientation as to the airways and their surfaces as a whole. Whereas respired air traverses the entire airway, clinicians and respiratory physiologists often focus their attention on one part or another, sometimes resulting in a rather restricted or distorted view of the remainder.

In Fig. 1 a representation of the entire airway is accompanied by diagrams delineating the special characteristics of the lining epithelium at various levels; and in Fig. 2 an indication is given of cross-sectional dimensions and resultant linear velocities in the airstream, of obvious importance to exchanges between respired air and airway surfaces.

From the nasopharynx to the tracheal bifurcation there is a single airway, while in the nose the airway is not only double but convoluted (Fig. 3). Beyond the trachea the bronchial tree divides repeatedly. It is of passing interest that every part of the airway except the upper cervical trachea is susceptible to narrowing or even total obstruction by physiological influences. Because of the complete ring of the cricoid cartilage (Fig. 4) and the nature of the cartilaginous attachments of the smooth muscle in the trachea the extrathoracic trachea is always open, and even severe smooth muscle constriction only slightly narrows its lumen. Sufficient vascular congestion can block one or both sides of the nose, the nasopharynx can be closed by constriction of the palatal muscles, the glottis is closed by intrinsic laryngeal muscles, major bronchi and the lower trachea may be collapsed by sufficiently high intrapleural pressure as in cough, and smaller bronchi can be closed by smooth muscle constriction.

It is tempting to ask how this capacity for narrowing airways may be related to defense. Obviously the ability to close the glottis and nasopharynx during swallowing is essential to prevent aspiration of ingested materials into the nose and lungs. Nasopharyngeal (velopharyngeal) closure is also necessary for the production of certain speech sounds. Relative vascular congestion and decongestion in the nose is probably a necessary factor in controlling air conditioning. Collapse of the major airways during cough assures the high velocity of airflow necessary for expelling material from the lungs, and the local closure of some of the smaller bronchi could protect an area of the lung from an irritant gas or shunt tidal air away from a region of poor blood flow. Relative changes in bronchial smooth muscle tone may affect the depth of the lining fluid, perhaps influencing the efficacy of cough or mucociliary clearance.†

Thus, although it would appear a matter of the first importance to keep airways open, it is easy to see how partial transitory obstructions may serve a useful purpose. It is perhaps only an unfortunate corollary that in disease these useful mechanisms may be excessive, resulting in what can only be harmful, i.e., prolonged, severe, or total obstruction in either the nose or the lungs.

II. Anatomy

Although any part of the airways above the respiratory bronchioli could be called "upper" airways, we believe confusion can be avoided by confining the term "upper airways" to the larynx and above, and using the term 'tracheo-

---

†This possible function of bronchial smooth muscle was suggested to us by Dr. Jere Mead, Harvard School of Public Health.

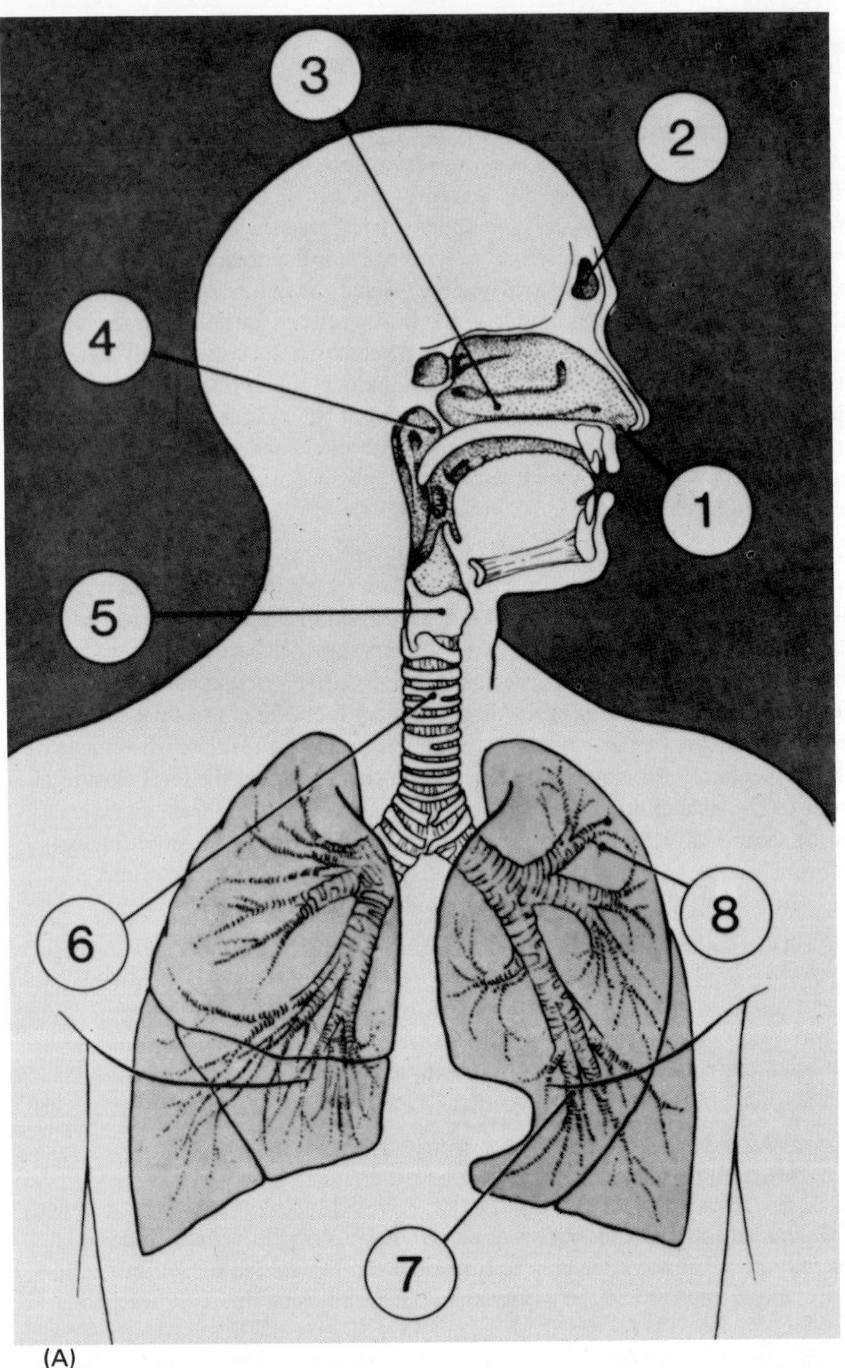

FIGURE 1 Diagram of the entire respiratory tract. The numbers indicate areas of distinctive or transitional epithelium (e.g., 4, the nasopharynx).

66

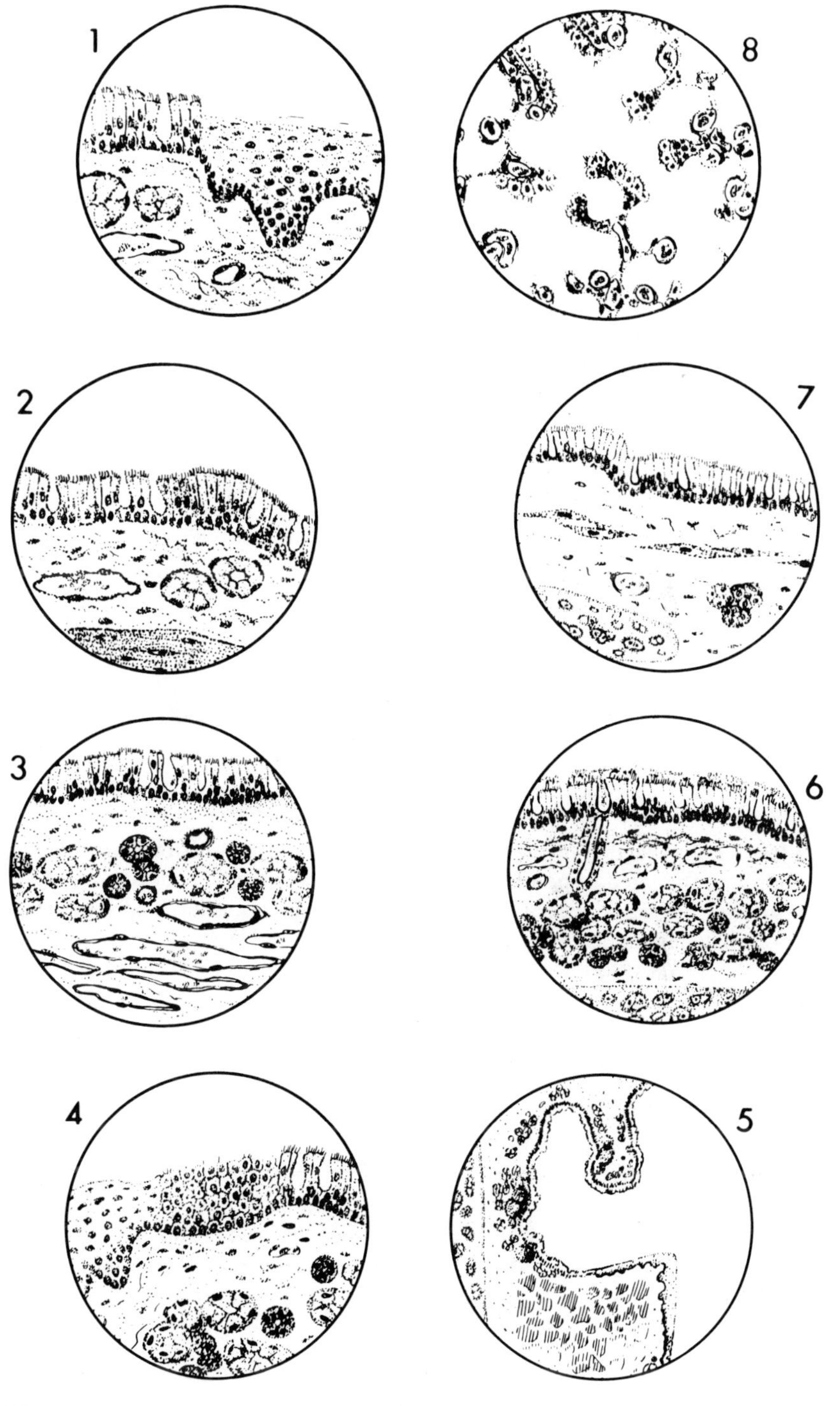

1
8
2
7
3
6
4
5
(B)

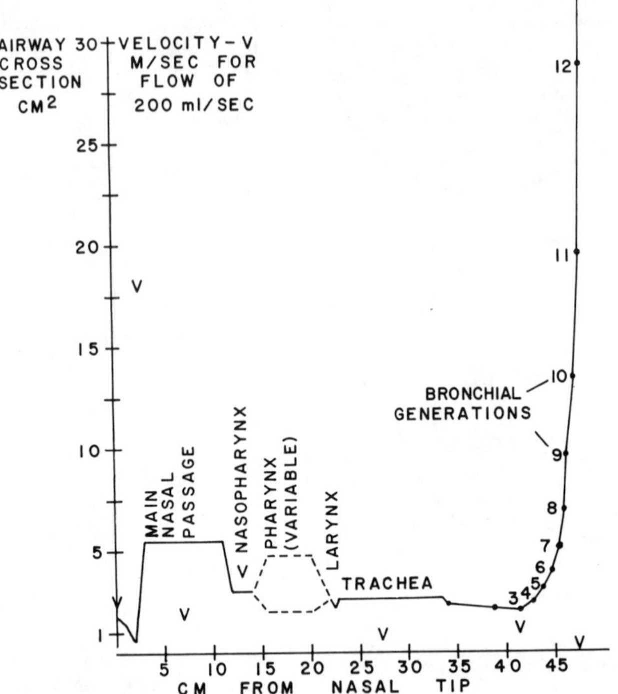

FIGURE 2   Diagram of the changing total airway cross-sectional area (vertical coordinate) related to points along the airway (horizontal coordinate). The resulting changes in linear velocity of the airstream also shown on the vertical coordinate, are indicated above by the letter V.

bronchial tree" (or some subdivision thereof) for the conducting airways below the larynx. A useful terminology for referring to the subdivisions of the upper airways is suggested in Fig. 5.

Most adults are preferentially nasal breathers, but nasal obstruction in some results in chronic or periodic mouth breathing. In the young infant nasal breathing, except during crying, is compulsory [37]. Nasal obstruction in early infancy can result in death. This compulsory nasal breathing in infancy is probably purely functional and not a result of apposition of epiglottis to palate as is the case in many animals. Certainly some infants born with congenital choanal atresia have died before its recognition, while simple insertion of an oropharyngeal airway will save their lives.

Seen from the point of view of inspired air, after entering the nostrils it passes upward, backward, and downward. The airway consists of two narrow, collapsible constrictions (Fig. 6) opening into the main convoluted nasal passages,

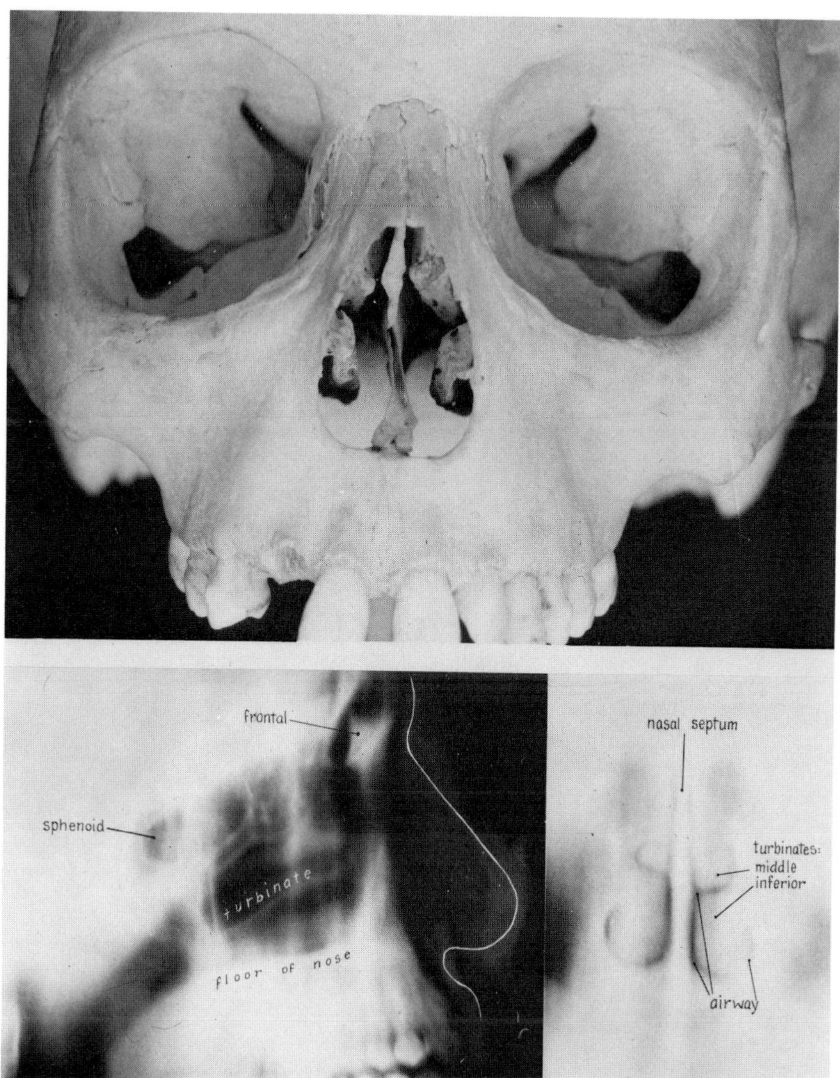

FIGURE 3 Frontal view of the skull (top) showing the bony nasal aperture, the folds of the turbinates and the nasal septum. When clothed with other tissues, the contour of the airway in the main nasal passage becomes as shown in the tomogram to the bottom right. (This figure and Fig. 4 are used with the courtesy of the Publisher of the *Handbook of Physiology Respiration I* [38].)

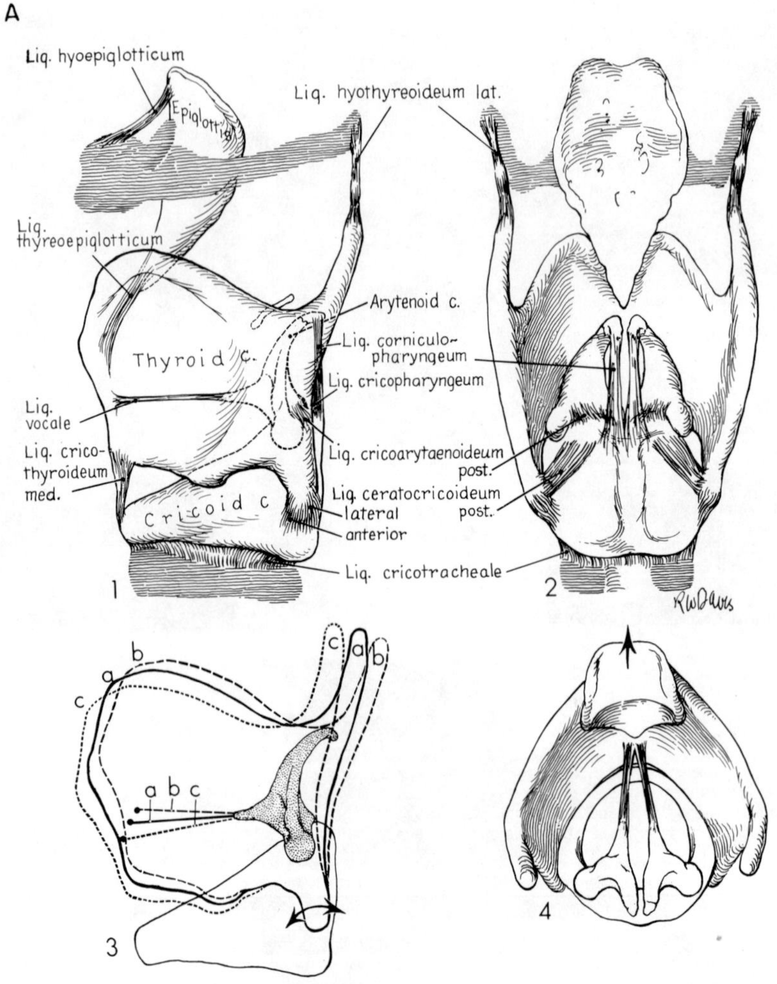

FIGURE 4   (A) cartilages and ligaments, and (B) intrinsic muscles of the larynx. Note the complete ring of the cricoid cartilage.

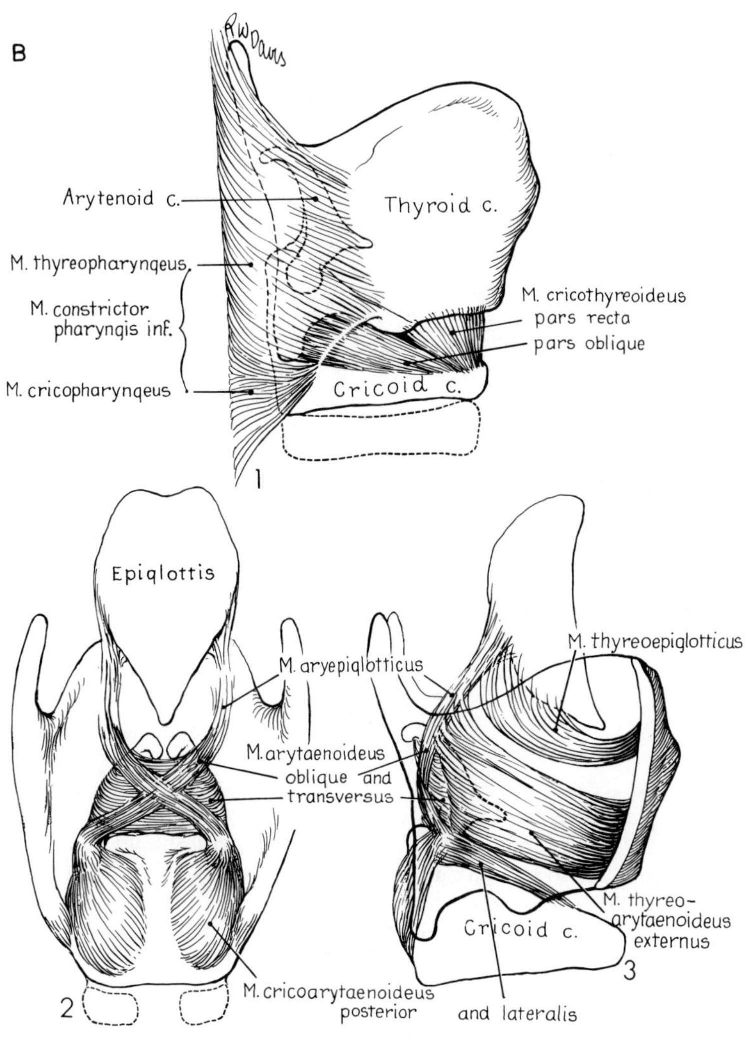
B
RWDavis
Arytenoid c.
Thyroid c.
M. thyreopharynqeus
M. constrictor pharyngis inf.
M. cricothyreoideus
pars recta
pars oblique
M. cricopharynqeus
Cricoid c.
1
Epiqlottis
M. aryepiqlotticus
M. thyreoepiqlotticus
M. arytaenoideus oblique and transversus
M. thyreo-arytaenoideus externus
Cricoid c.
M. cricoarytaenoideus posterior
and lateralis
2
3

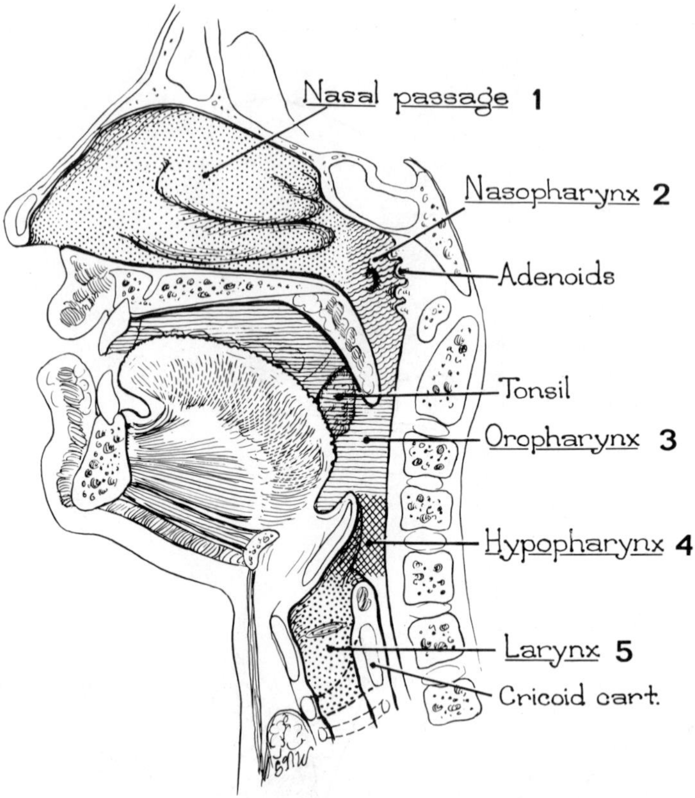

FIGURE 5 Suggested terminology for the upper airways. Number 3 should read "pharynx." The term oropharynx or oropharyngeal should be confined to the airway from lips to pharynx during mouth breathing. [Used with the courtesy of the publisher, *Bacteriological Reviews,* from D. F. Proctor, Airborne disease and the upper respiratory tract, *Bacteriol. Rev.,* **30**:498-513 (1966).]

converging and bending into the nasopharynx, and, finally, funneling into the glottis and trachea [3]. Alternatively, under certain circumstances, the more or less open mouth partially bypasses the nose and nasopharynx. During mouth breathing some air continues to pass through the nose, unless it is totally obstructed. Nearly surrounding the main passage and communicating therewith through small openings are the air spaces of the paranasal sinuses. The mastoid air spaces are in intermittent communication with the nasopharynx through the Eustachian tubes. Although the necessity for the latter is generally appreciated, the role of the paranasal sinuses has frequently been misunderstood, minimized, or ignored. These air spaces in the bones of the face provide protection of the

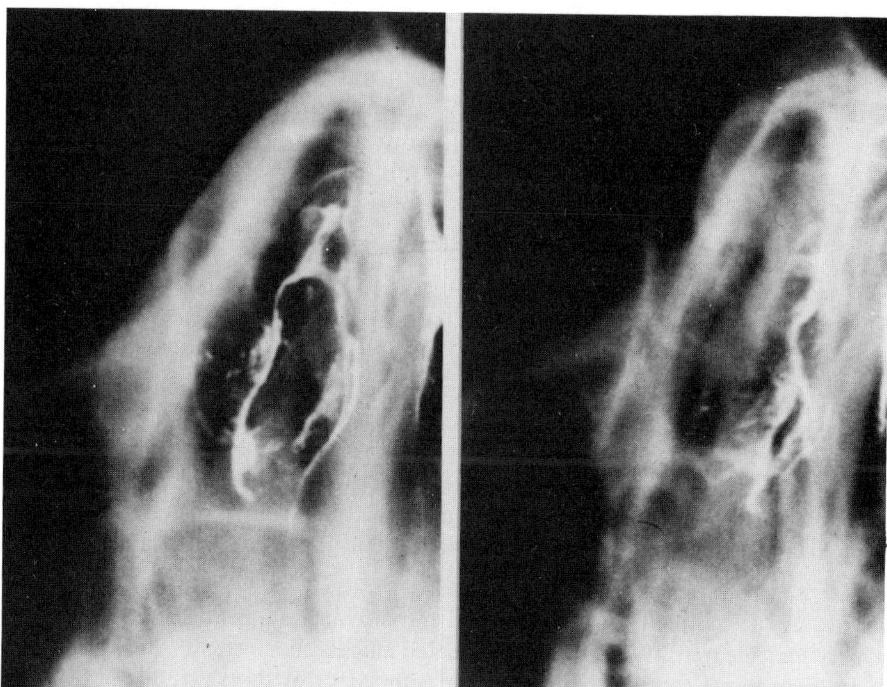

FIGURE 6 The nasal opening shown from below. Radiographs with tantalum dust coating the airway, during quiet breathing (left) and during forced inspiration (right). [Used with the courtesy of the publisher, *Archives of Internal Medicine,* from D. F. Proctor et al., Clearance of inhaled particles from the human nose, *Arch. Intern. Med.,* **131**:132-139 (1973).]

brain from blows on the front of the head. From our point of view they may act as insulators for the nasal cavity and supplement nasal secretions. The sinuses are lined with ciliated epithelium and the fluid covering its surface is continually being propelled into the nose all along the principal line of inspiratory airflow. Thus, if the nasal secretions themselves are modified by ambient air influences, this supplemental source of fluid is always available for their replacement.

Seen from the side, the nasal airway starts in an upward direction, bends to run horizontally backward until, at the nasopharynx, it turns downward and runs vertically in a more or less straight line into the trachea (Fig. 7). When the mouth is open, depending upon the position of the tongue and palate, the passage runs curvilinear horizontally to bend rather sharply at the pharynx downward to the trachea. When the head is thrust forward, the chin tilted upward, and the tongue depressed, the oropharyngeal airway is wide and bends much less sharply to reach the vertical axis of the trachea [39].

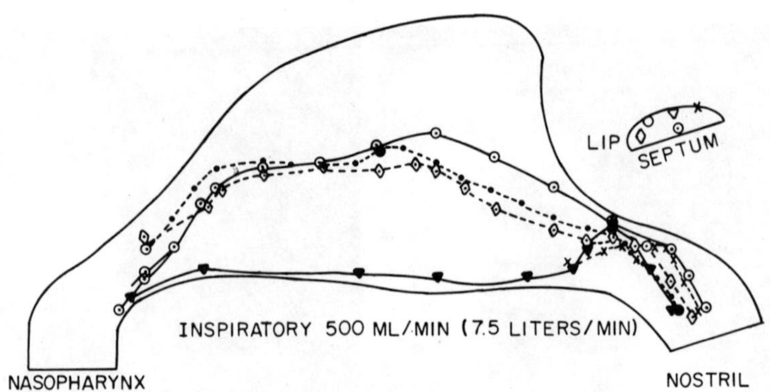

FIGURE 7   Principle lines of inspiratory airflow as derived from model studies. The mainlines flow between the middle meatus and the septum.

The dimensions and shape of these airways determine the resistance which they offer to airflow [11,17,31,40,51] and their physiological effectiveness in providing relatively easy access of air to and from the lungs and the modification of inspired air during its passage.

Squamous epithelium at the nasal entrance makes a gradual and somewhat variable transition to columnar ciliated mucus-secreting epithelium just posterior to the anterior tips of the nasal turbinates. With the exception of the olfactory areas, the main nasal passage, paranasal sinuses, Eustachian tubes, and middle ears are lined with this ciliated epithelium. At the nasopharynx there is a gradual transition back to squamous epithelium which, except for the nasopharyngeal and pharyngeal lymphoid tissue, lines the passage from there to the upper trachea. The nasopharyngeal lymphoid tissue (adenoids) is probably important to the development of immune mechanisms in defense against airborne pathogens. More will be said of this subsequently. Although the ciliated mucosa is everywhere superficially similar there are differences between that in the main nasal passage, that lining the sinuses, middle ears, and Eustachian tubes, and that in varying portions of the tracheobronchial tree. This matter is discussed in Chap. 7.

The erectile nature of the submucosal vasculature in the main nasal passage is of prime importance [13,30,32]. Its capacity for quick and profound change can alter the width of the main nasal passage or obstruct one or both sides of the nose. Changes in the width of the airstream produced through modifications in this vasculature, and perhaps changes in surface temperature and nasal secretion are factors which contribute to the efficiency with which the nose conditions inspired air over wide ranges of ambient temperature and humidity.

The important anatomical characteristics of the airway in analysis of nasal breathing thus are:

1. A narrow constriction just beyond the nasal entrance

2. A bend at or just beyond this constriction

3. A large cross-sectional area of narrow width along the main nasal passage

4. Communications with the paranasal sinuses along the main nasal passage

5. A further bend and slight constriction at the nasopharynx

6. The presence of adenoid tissue lining the posterior naso-pharyngeal wall

7. A relatively wide, but variable, cross section from the naso-pharynx to trachea with some constriction at the vocal cords.

III. Airflow

During quiet nasal breathing the inspired air enters the nostrils at an angle of about 60° to the horizontal floor of the nose. About 1.5 to 2.5 cm beyond the entrance the airstream passes through its narrowest point and, at the same time, bends to travel horizontally backward along the floor of the nose. Most of the moving airstream seems to be adjacent to the middle meatus and no part of it is much more than 1 mm from the surface, while in the trachea and major bronchi the center of the airstream is 5 to 10 mm from the airway surfaces.

With each inspiration some air from the sinuses joins the nasal stream and with each expiration the reverse occurs [15]. At the nasopharynx there is a bend as the two nasal airstreams join and move vertically downward [8]. Between these two bends the length of the main passage is approximately 5-8 cm.

The shape and narrowness of the nasal airway make it impossible to introduce even small measuring instruments into the airstream without interfering with airflow. Therefore we have resorted to the use of transparent models constructed from casts made at autopsy. It is of interest that, although casts and models have varied from one another, as do human noses, basic shapes seem to vary little since the main characteristics of airflow are very similar from model to model. From these models of one entire side of the nasal airway and from some studies possible in vivo, we have concluded the following.

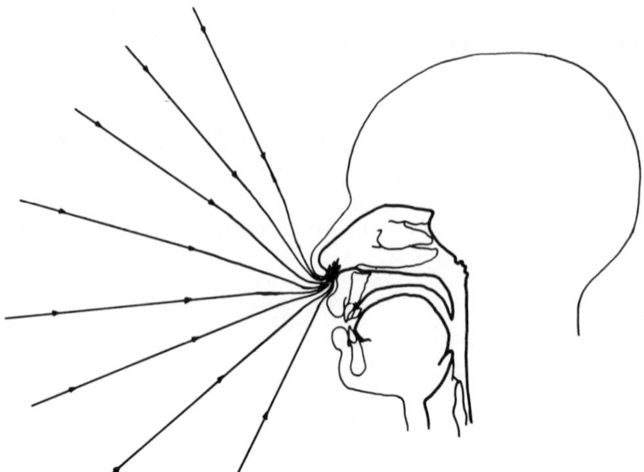

FIGURE 8 Diagrammatic presentation of airflow external to the nose drawn to approximate those of a point source on a plane.

The nature of airflow external to the nose during inspiration is shown by the streamline diagram in Fig. 8. At distances greater than several centimeters from the nostril the air velocities must approximate those of a point source on a plane, i.e., if the total nasal flow rate is Q_T then the velocity at any distance, r, from the nasal entrance is

$$v = \frac{Q_T}{2\pi r^2}$$

Thus the air velocity becomes significantly large only very close to the nasal entrance. As can be seen in Fig. 8, the simple point source is no longer appropriate in the near vicinity of the nasal entrance because the nostril faces downward rather than horizontally, requiring a significant streamline directional change. In addition there are two nasal entrances.

The two nostrils, as discussed above, lead into parallel and similar nasal passages. Since in the model studies we used one of these two passages, it is convenient to define the flow rate through one passage as the "half-passage flow rate" and denote it as Q_H.

At the nostril itself the average air velocity $\bar{V}_N$ for a half-passage flow rate Q_H can be computed simply as

$$\bar{V}_N = \frac{Q_H}{A_N}$$

where A_N is the cross-sectional area of one nostril. For resting breathing, a reasonable value for Q_H is 12.5 liters/min. There is, of course, considerable variation in A_N among adults, but a reasonable average value is 0.9 cm^2. From this the velocity is found to be 2.3 m/sec. In the upright position the nostrils act as "inverted samplers" and are incapable of sampling particles whose gravitational falling velocity is >2.3 m/sec. A sphere of diameter 575 μm (1 μm = 10^{-4} cm), density 1 g/cm^3, has this falling speed in air at 25°C; thus particles of this size and larger cannot be nasally inspired under these conditions (for further discussion of particles in the nasal passage, see Chap. 5).

We have carried out experimental studies of the streamline patterns in air in an entire half-nasal passage model. The model was constructed of clear polyester resin to permit viewing and made to exact scale (Fig. 9). Waterflow was used in some studies instead of air to facilitate dye injection streamline marking. The results from the waterflow experiments can be directly applied to airflow as long as the waterflow rate is chosen to meet the condition of "hydrodynamic similarity," in this case the equality of the Reynolds number for the two flows. The Reynolds number is defined as

$$R_e = \frac{lv\rho}{\eta}$$

where

    l = a system dimension (e.g., a pipe diameter), cm

    v = fluid velocity, cm/sec

    η = fluid viscosity, g/cm-sec

    ρ = fluid density, g/cm^3

Dye injection at various locations outside the nostril not only allowed visualization of the direction of flow (the streamlines) but indicated if and where the flow underwent transition from laminar to turbulent conditions.

According to the model studies, the flow at the nostril is laminar at a half nasal flow rate of 12.5 liters/min and is undoubtedly so even at higher flows because of the inherent stability of convergent flow.

The boundaries of the nasal passage continue to converge for the next 1.5-2 cm until the end of the vestibular region is reached. At the same time the passage bends toward the horizontal so the streamlines change from vertical at the nasal entrance to horizontal beyond this constriction. This anterior constriction, called the ostium internum, has the smallest cross-sectional area of the nasal passage, about 0.32 cm^2 per half-passage. The computed average velocity

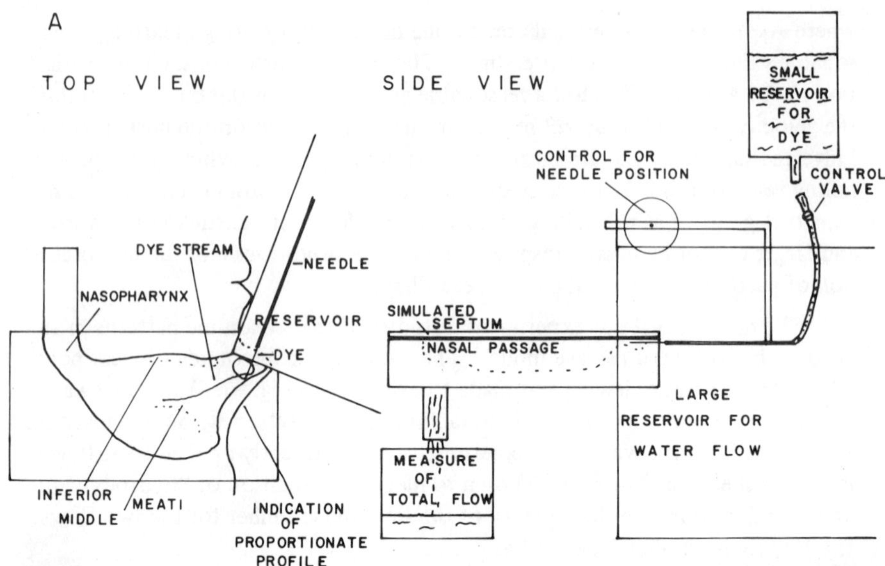

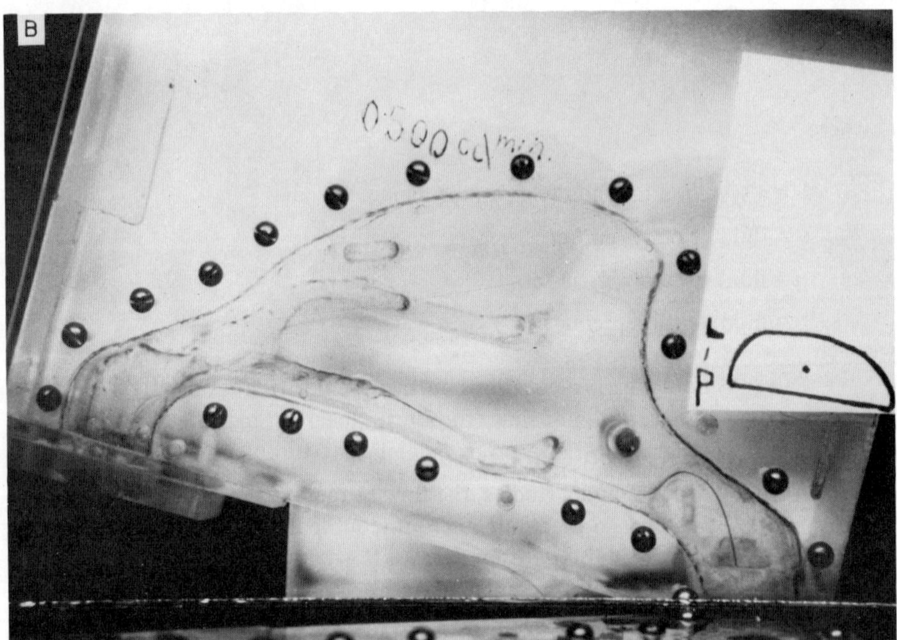

FIGURE 9 (A) Diagram of the apparatus used in some of the model studies, seen from above on the left and from the side on the right. [Used with the permission of the Editor, *Archives of Environmentsl Health,* from D. F. Proctor et al., **18**:671-680 (1969), copyright 1969, American Medical Association.] (B) Photograph of one of the lines of flow.

through the ostium internum at a half-passage flow rate of 12.5 liters/min is 6.5 m/sec. This is approximately six times the largest velocity in the tracheobronchial tree, which occurs at the segmental bronchus.

There is evidence from the model studies using airflow that the actual velocity at the ostium internum is significantly greater than 6.5 m/sec. A miniature Pitot device which converts air velocity into static pressure difference indicated that the maximum velocity was 18.6 m/sec. A static pressure measurement in a human subject breathing quietly gave -2.0 cm H_2O corresponding to a velocity only slightly lower than this surprisingly high figure. It is not unreasonable that the peak velocity should be greater than the average, since a boundary layer of viscous retarded fluid builds up as the air passes from the nasal entrance to the ostium internum so that the major part of the flow is confined to a small central core.

The inspired air passes through the ostium internum and into the main nasal chamber containing the turbinate structures. There is a very abrupt increase in cross-sectional area at this point. The significant decrease in air velocity and the viscous retardation of the air by the large surface area of the chamber give rise to turbulent flow, typical for such divergent channel flows [8].

Waterflow model studies demonstrate this turbulent flow in the region of the turbinated structures. With a waterflow rate corresponding to a half nasal airflow rate of 7.5 liters/min the streamlines do not break up immediately downstream of the ostium internum. It is possible to trace the streamlines into the main chamber to determine the pathway of the majority of the air.

The major proportion of air passes between the middle meatus and the nasal septum above the inferior turbinate. A small amount of air entering along the septal wall of the nostril, undergoes a significant change of direction and follows the floor of the main chamber (Fig. 9). The model studies have also indicated that a small fraction of the mainstream breaks off upward and forms a standing eddy at the top of the chamber, where the olfactory area is located (Fig. 10). This is probably sufficient to provide for olfaction on inspiration but not so great a quantity or velocity as to injure the olfactory mucosa. This eddying over the olfactory area is undoubtedly increased with sniffing.

The velocity along the middle meatus has been measured at a flow of 12.5 liters/min in the model and found to be 2-3 m/sec. At the rear of the main nasal chamber the septum wall ends and the flows from the parallel nasal passages join and funnel into the nasopharynx, again changing direction, this time from horizontal to vertical. The nasopharyngeal passage decreases in cross-sectional area so that the velocity increases to 4 m/sec. Turbulence introduced downstream of the ostium internum at the main nasal passage entrance persists throughout the remainder of the passage, but the nasopharynx geometry does not introduce fresh flow instability. In fact, the convergence has the effect of reducing the level of turbulent motion.

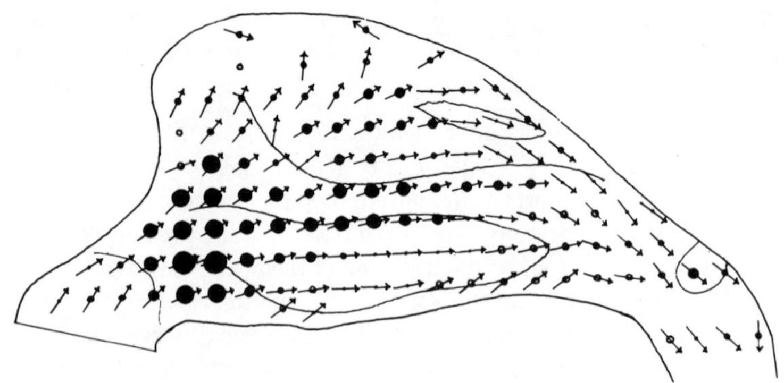

FIGURE 10 Linear velocity of the inspiratory nasal airflow as derived from model studies. The nostril is to the left. The size of each dot indicates velocity and the arrow direction. Note the indication of an eddy above.

Static pressure measurements at various locations in the nose of normal volunteer subjects indicate that there is essentially no pressure change with distance beyond the first 1.5 cm of the main nasal chamber, i.e., just beyond the ostium internum. Thus the entire transnasal pressure drop is produced in the first 1.5 cm of the passage. This may appear surprising in view of the 8-10 cm length of the entire nasal passage, but it is quite consistent with the concept of orifice flow produced by the ostium internum. Leading up to this constriction the pressure drop is accelerative, i.e., essentially all due to the velocity increase and thus recoverable in principle. But, the turbulent flow just anterior to the constriction is "dissipative," and thus essentially none of the pressure drop is recovered as the mean velocity decreases.

The magnitude of the pressure drop for a given flow varies considerably, not only from person to person but from time to time for the same person and at any time between the two passages. Although we shall discuss the "nasal cycle" below, suffice it to say here that for normal subjects the flow distribution may be as great as 20-80% between the two passages without any subjective awareness of a "blocked" nasal passage.

The curvilinear nature of the experimental flow vs. pressure drop relationship for the nasal passage is entirely consistent with the concept of "orifice flow" turbulence discussed above. At very low flow rates, say 3 liters/min, the relationship is nearly linear, and the flow is undoubtedly laminar throughout. Over most of the region of interest, covering resting and mild exercise breathing, the transnasal pressure drop is proportional to the square of the total nasal flow rate, which is typical of turbulent flow.

The inspiratory flow rate does not increase indefinitely with decreasing nasopharyngeal pressure but reaches a plateau or limiting flow rate. This is a result of the collapse of the anterior passage walls (probably just at the anterior end of the inferior turbinate) when the negative static pressure at the ostium internum produced by high flow exceeds the supporting forces in the wall which tend to keep the passage open. This maximal flow phenomenon usually occurs in the heavy exercise range of flow rates and undoubtedly causes a change from nasal to mouth breathing.

During quiet breathing, the resistance of the nasal passage represents about one-half of the entire respiratory tract resistance. During mouth breathing, which we shall consider later, the oral passage accounts for about one-fifth of the total respiratory resistance. Thus, nasal breathing requires about 50% more effort than mouth breathing under normal conditions. The fact that under normal conditions we choose nasal breathing despite its increased resistance to breathing may appear paradoxical. Nevertheless, it is fortunate that our "instinct" is to do that which is best from the point of view of air conditioning and particle removal.

One might ask whether the preference for nasal breathing is closely related to the obligatory nasal breathing characteristic of infants to the age of 6 months (or more). These are questions which should be answered by future investigations.

In contrast to inspiratory flow, during expiration the air seems to move in roughly equal parts all along the passage between the septum and the turbinates. Thus, the olfactory area has little contact with inspired air but is washed by the expiratory stream. But, since we are able readily to detect the odor of water-soluble gases which are removed in the main nasal passage, sufficient air must reach the olfactory area in the eddy mentioned above.

Thus during quiet breathing (0.15-0.25 liter/sec) we see that the linear velocity of the inspiratory air rises to 6-18 m/sec at the anterior constriction, falls to about 2 m/sec in the main passage, and rises again to about 3 m/sec at the nasopharynx. These figures should be compared to a linear velocity of about 1 m/sec (the highest velocity elsewhere in the respiratory tract during quiet breathing) which is found at the level of the third generation of bronchi. Beyond this, as the bronchi funnel out into the multitude of fine peripheral tubes, linear velocity falls to very low levels (Figs. 2 and 10).

The normal nose has a built-in flow-limiting device. During rapid inspiration if airflow reaches about 1-1.25 liter/sec (more than triple the flow in quiet breathing) the pressure drop just beyond the anterior narrow constriction, as mentioned above, is sufficient to produce collapse of the airway, and additional effort produces no further increase in airflow (a fact clearly evident to anyone indulging in a sniff). Because of these characteristics there is a high impaction rate of particles in the anterior nares [14,18,22,36], ample opportunity for exchange of gases, vapors, and temperature during the slow main nasal passage, and a further opportunity for particle deposition in the nasopharynx.

Although there is slight constriction of the airway at the vocal cords, ordinarily airflow through this region is essentially undisturbed [46]. It is important to note that the larynx plays a vital role in the defense of the lungs in its tight closure during swallowing and before cough. When this function is impaired, aspiration of respiratory tract fluids and ingested materials occurs.

Closure of the glottis is of importance at the onset of coughing. Glottic closure accompanied by extreme expiratory effort produces a high pleural pressure. When this is followed by sudden glottic opening, both a burst of rapid expiratory airflow and collapse of the major airways result (see Chap. 15).

Much has been made in the rhinological literature of the so-called "nasal cycle" [25,47]. There seems to be little doubt that relative vascular congestion and decongestion occurs in some individuals with a degree of rhythm, sometimes resulting in what appears to be a systematic raising and lowering of resistance to airflow first on one side of the nose and then on the other. Observation of this phenomenon has led some to believe that the resulting cycling of the main airstream from side to side has important physiological implications.

In considering the nose as a defense organ, a built-in mechanism to assure that neither side of the nose is continuously burdened with the task of conditioning and cleaning the air is an attractive idea. But, it is difficult for us to visualize such a physiological mechanism and especially difficult to accept the many other functions attributed by some to this "nasal cycle."

Teleologically, one would like to believe that the changes which undoubtedly occur serve something other than a purely local function. A particularly attractive idea is that unilateral nasal obstruction on the down side during sleep occasions turning in bed and could assure optimum pulmonary ventilation and circulation going from lung to lung. What seems more likely is that this cycling results either from simple changes in hydrostatic pressure [41,44] or from some environmental air effect. Since the veins of the head and neck are largely valveless, the extensive venous network of the nose is especially prone to congestion related to change in position [5]. That some ambient air influence is involved in this cycling phenomenon is suggested by Keuning's finding [25] that no cycle occurs in laryngectomees. In any event, it is unlikely that any such cycling is a vital part of nasal function since it is not found in all persons; and, when it occurs, the number of hours involved in a cycle is highly variable.

Thus, the nose serves as an admirable air conditioner and trap for inspired foreign materials. Since it is lined with ciliated mucus-secreting epithelium, materials deposited upon its surfaces can be cleared in large part to the nasopharynx whence they are dispatched to the stomach at the next swallow. This clearance mechanism is discussed in detail in Units II and III of this volume.

IV. Choice of Nasal or Mouth Breathing

In circumstances where nasal resistance to airflow is sufficiently increased, during talking, singing, or yawning, and when ventilation demands are high, we resort to mouth breathing. The former circumstance may exist as a result of anatomical abnormality [19], disease, or certain environmental influences producing vascular congestion. The last is principally encountered as a result of voluntary exercise or work loads [6]. We also employ mouth breathing for inspiration between phrases when talking or singing [39]. Usually, during mouth breathing, the tongue lies close to the palate and the resulting airway is nearly as efficient in air cleaning as is the nose. This is especially true of the mouth breathing employed in conversation or light work loads. During heavy exercise, as a result of severe work loads, or during singing, inspiration occurs through a wide open oropharynx. The cigarette smoker probably also employs this wide airway beyond the lips, but we really know very little of the airway employed in inhaling cigarette smoke. In those circumstances, and where an artificial airway such as tracheostomy is employed, ambient air reaches the trachea relatively unaltered. Evidence from laryngectomees shows that the tracheobronchial mucosa does not escape unscathed [20].

In singing, because of the need for quick deep breaths between phrases, inspiration occurs through a wide oropharyngeal airway. Evidence that this bypass of the nasal defenses may be hazardous comes from reports of the spread of tuberculosis from one individual to others singing in the same chorus [4,21,29, 50]. A similar problem may exist in those resorting to mouth breathing while doing heavy work as in mining [28,43]. Fortunately there is some evidence that nasal airflow resistance is reduced in exercise [35,42,45], but we need further information to document the circumstances under which one might expect the individual to abandon his nasal airway. Marathon runners are said to stick to nasal breathing throughout the race.

We know too little about the capacity of an individual to maintain nasal breathing when increased pulmonary ventilation is needed. In one of our studies, directed at the effects of prolonged exposure to dry air, eight normal subjects were subjected to exercise to the limit of nasal breathing and their nasal airflow resistance measured before and afterward. In seven of these eight subjects nasal airflow resistance decreased during the 20-minute exercise on a bicycle ergometer. The exercise was sufficient to occasion an average minute ventilation over 70 liters/min, and nasal breathing was maintained throughout. In one subject nasal airflow resistance increased slightly, while in the other seven there was a marked decrease of nasal airway resistance. It is of interest that this change, even under dry ambient air conditions, did not impair nasal mucociliary function [1].

Baumann [6] also found that nasal airflow resistance decreased as the work load increased. He concluded that "oxygen demands of the organism rather than local conditions provide an adequate stimulus for widening the nose during physical effort." At what point this widening of the nasal airway, permitting continuance of nasal breathing, reduces the effectiveness of the nasal air conditioner and cleaner to a point no better than the oropharyngeal airway has not been determined.

What happens when we combine such influences as cold air, exercise, emotional stress, irritant ambient gases, etc.? What factors, other than pathological, influence this choice between an effective nasal airway, a somewhat less effective narrow oropharyngeal airway, and the ineffective wide oropharynx? It could indeed be useful to be able to predict in an individual worker at what work load he could no longer count on the protection offered by his nose.

The nasal turbinates contain a rich vasculature of an erectile character similar to that in the genitalia, and similarly responsive to emotional and other stimuli. Although useful in air conditioning, when vascular congestion becomes sufficient to limit nasal airflow to less than about 0.4 liter/sec (in the adult) it may lead to at least partial resort to mouth breathing. Simple increased nasal airflow resistance per se is not the sole deciding factor in choosing mouth breathing. Takagi [48] found, in comparing voluntary hyperventilation with the involuntary hyperventilation associated with CO_2 breathing, that the CO_2 produced a reduced nasal airflow resistance and permitted a higher maximum nasal airflow. But the subject abandoned nasal for mouth breathing at a lower peak nasal pressure drop than was the case with voluntary effort. This choice seems to be based upon a complex mixture of physical, reflex, and psychological influences.

There is a wide range in nasal airflow resistance between normal individuals and in an individual from time to time. The nose may account for anything between a quarter and over half of total airway resistance and nasal airflow resistance may vary by more than a factor of 2 without producing symptoms. A key point in determining the ability to continue nose breathing in the face of increasing ventilatory demand is the peak inspiratory flow which can be achieved.

The anterior nasal passage acts as a collapsible tube (Fig. 6) in such a manner that, in most individuals, increasing the pressure drop beyond a certain point produces no further increase in volume flow during inspiration. The peak flow is normally determined by the nature of this collapsible segment and to a lesser degree by nasal vascular congestion. Cold air (about 5°C) (Fig. 11) increases nasal congestion and decreases peak flows [48]. Ambient SO_2, even at 1 ppm, has a similar effect (see Chap 12). There are probably many other imperfectly understood environmental influences which either decrease peak nasal flows, or possibly result in shrunken nasal mucosa and an overly wide airway decreasing the efficacy of the nasal air conditioning system [7,49]. On assuming a recumbent

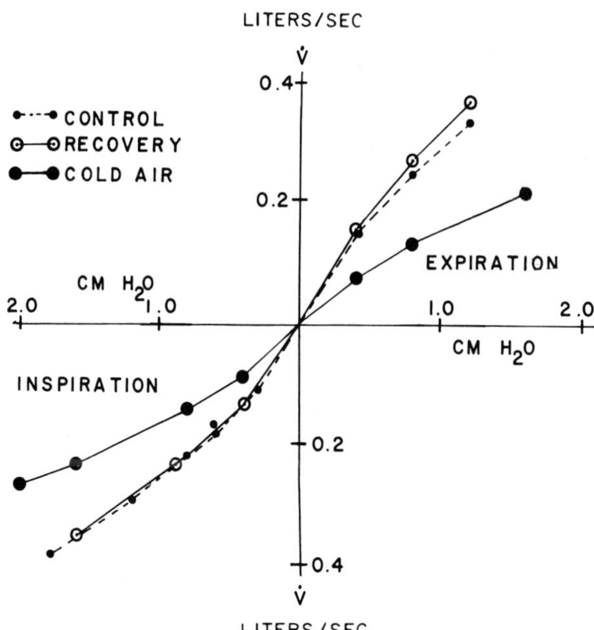

FIGURE 11   Increased resistance to nasal airflow on exposure to cold air (5°C), and return to normal 30 min afterward. (Used with the courtesy of the publisher, *Annals of Otology* [48].)

posture the slight increase in hydrostatic pressure within nasal vessels results in vasodilatation and an increase in nasal resistance.

V. Rhinomanometry

Measurement of nasal airflow resistance can be accomplished in several different ways [2,10,12,16,23,26]. The objective is simply to relate the nasal pressure drop to nasal airflow. The most commonly employed methods are:

1.   Measurement of mouth pressure as a function of nasal airflow through the use of a pneumotachometer for airflow and a tube held in the mouth for pressure. Some individuals have difficulty in keeping the mouth in communication with the pharynx and an alternative method may be required.

2.   Blowing of air through the nose and out of the mouth at known rates while measuring the pressure required. This offers the dis-

   advantage of resorting to a form of nasal airflow not wholly
   representative of normal breathing.

3. Measurement of pressure at one nostril as a function of airflow through the other nostril. Its disadvantage is the possible distortion of the breathing nostril and the fact that
  directing the entire airstream through one side of the nose
  is not physiological.

4. Body plethysmography. Nolte, in an excellent study, has demonstrated that either number 1 or 4 is a reliable technique,
  while numbers 2 and 3 are subject to serious error [33].

A particularly informative technique for rhinomanometry was developed by
Bridger [9], a modification of method 1 above. In this technique the subject
is instructed to breath out to near-residual volume and then inspire with steadily
increasing effort. Airflow is measured with a nasal pneumotachometer and related to mouth pressure. The sides of the nose may be measured separately by
occluding one nostril with a small piece of tape. The resulting record of pressure
related to flow shows a curvilinear increase in flow as the pressure drop across
the nose is increased, finally reaching a plateau where no further increase in flow
results from increasing the driving pressure. In the normal nose almost all of the
nasal resistance to airflow is produced in the anterior 2-3 cm. Collapse of this
region when the pressure drop between ambient air and the interior of the nostril
occasions flow limitation is seen in the plateau.

  If resistance in this anterior constriction is increased the pressure drop
across the nostril is greater and the maximal flow will be lower and occur at a
lower total pressure drop across the nose (Fig. 12 above). When the reverse
occurs, decreased resistance in the anterior segment, the maximal flow will be increased and flow limitation occur at a higher total pressure drop (Fig. 12 above).

  When resistance in the main nasal passage is increased, maximal flow will
remain the same, but a higher total pressure will be required to produce it (Fig.
12 below). When the reverse occurs, decreased main nasal passage resistance,
again maximal flow is unchanged but it is reached at a lower total pressure
drop (Fig. 12 below).

  This method immediately tells us what peak nasal flow an individual
can achieve nasally and what pressure is required for its production. By an
artificial induction of resistance at the nostril, or its reduction through dila-
tion of the nostril, or artificial induction of vasodilation or constriction in
the main passage we can determine the relative contribution to resistance of
the anterior nasal constriction and the main nasal passage (Fig. 12).

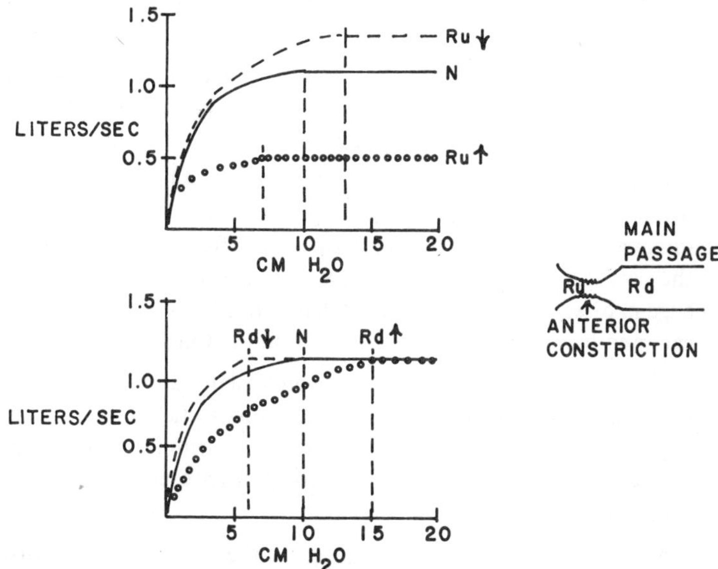

FIGURE 12   Inspiratory nasal airflow (vertical coordinate) related to pressure drop across the nose (horizontal coordinate) during inspiration with steadily increasing force. Note that under all conditions above a certain pressure no further increment in flow develops with increasing pressure drop. Also note effects of changing resistance at the nostril, Ru (above) and changing resistance in the main passage, Rd (below). (Data taken from Bridger [9].)

VI.   Oral Airflow

Discussion of the nature of airflow in the oral passage must be based almost entirely on the dimensional characteristics of the passage, since there appear to be no adequate velocity measurements in the oral passage of a subject (or in a suitable model) reported in the literature. The dimensional characteristics of the oral passage are subject to significant variability depending on the position of the lips, jaw, and tongue and palate. We shall first consider quiet oral breathing as during conversation, then oral breathing at exercise where the nasal "flow plateau" is exceeded.

The external flow in oral breathing is similar to that of nasal breathing, except that the velocity at any point is less in mouth breathing. This is because the mouth is a "slit" source rather than a "point" source. When the head is erect, the oral entrance is a "horizontal slit sampler" rather than the "inverted sampler" of the nostril; thus the exclusion of rapidly settling particles from the respiratory tract does not take place in oral breathing.

In oral breathing at rest or in exercise, the passage can be conveniently divided into three regions: (a) the entrance consisting of lips, front teeth, and leading edge of the tongue; (b) the middle region, an arching "channel" bounded by the tongue and the hard palate; and (c) the mixing region or oropharynx where the passage joins the nasopharynx and the flow becomes vertical.

At rest, assuming a peak inspiratory flow rate of 0.4 liter/sec, the average velocity at the entrance is estimated to be approximately 1.6 m/sec. At this velocity and in an airway having the approximate configuration of a horizontal channel, the flow is laminar. As the air passes into the middle channel it undergoes a change in direction of such a degree that impactive removal of some large particles can (and have been observed to) occur on the tongue.

The central passage between the tongue and palate can be quite narrow during quiet breathing, (1-2 mm) but the overall cross-sectional area is similar to that in nasal breathing and the flow continues to be laminar. The abrupt changes in cross section which occur in the nasal passage do not occur in quiet oral breathing.

As the air passes from the central passage to the oropharynx, the airway changes from a rectangular channel to a circular duct, and the direction of flow becomes vertical. The posterior wall of the oropharynx is a second site of particle deposition by impaction, confirmed by observation.

During exercise, the jaw is usually dropped and the tongue is flattened to increase the airway cross section in the forward two sections. It is not likely even in this condition that the flow becomes turbulent, since the average velocity does not significantly increase. The criterion for flow type (laminar, transition, or turbulent), the Reynolds number, is estimated for both quiet and exercise oral breathing to be ~500, which for pipe or channel flow in regular cross sections is well in the laminar region.

VII. Laryngeal Airflow

Air entering either nose or mouth passes through the larynx to reach the tracheo-bronchial tree. Model studies of larynx flow in our own laboratory and studies with normal subjects [24] indicate that the larynx behaves fluid-mechanically like an orifice in which vocal cords act as the constriction [45].

Upstream of the cords (in the pharynx) the flow is convergent and laminar. The cords are the constriction and the immediate increase in downstream area leads to turbulent flow, similar to the turbulence beyond the nasal valve. The intensity and persistence of this turbulence is dependent upon the ventilatory flow rate through the larynx.

The pressure drop through the larynx, ΔP_L, is found experimentally to be related to the flow rate, Q_T, by the equation

$$\Delta P_L = k(Q_T)^a$$

where the pressure across the larynx varies from 1.2 to 2.2 cm H_2O for different gases [24]. It is convenient to use the concept of an orifice discharge coefficient C_d, defined as

$$C_d = \frac{Q_T}{A_L} \left(\frac{\rho}{2\,\Delta P_L}\right)^{1/2}$$

where

ρ = gas density

A_L = cross-sectional area at vocal cords

because the effect of gas density, velocity, and viscosity can be shown for the larynx by a single plot of C_d vs. Reynolds number.

The significance of the turbulence downstream of the vocal cords is that it establishes the entrance conditions for the tracheal and bronchial flows for at least two generations downstream. Thus a model study of the bronchial flow carried out with a smooth pipe leading to the trachea is an incorrect entrance condition and will give erroneous velocity information at the bifurcations and in the branching airways.

VIII. Implications for Clinical Medicine and the Public Health

From a practical point of view, to what conclusions do these considerations lead us?

1. Nasal breathing offers superior protection against airborne hazards because of the nature of airflow and the air-to-surface contact.

2. Wide open oropharyngeal breathing (or the use of artificial airways) offer maximum hazard of airborne infection to the lungs, again because of the nature of airflow and the relative lack of air-to-surface contact above the bronchioles and alveoli.

3. Ability to maintain nasal breathing makes the individual less prone to environmental injury, but certain factors may im-

> pair the efficacy of the nasal defense and others limit the
> ability to maintain nasal breathing.
>
> 4. The need to resort to mouth breathing (or artificial airways)
>   should raise a red flag of warning that cleaning or conditioning
>   of air prior to respiratory access, or the provision of some arti-
>   ficial intervention between the air and the respiratory tract
>   may be needed.

Some of the resulting implications for clinical medicine and the public health
are more fully discussed later. For the physician faced with a patient suffering from nasal obstruction it should be clear that relief of such obstruction is
desirable. If surgery is required, its performance should not eliminate those
physical characteristics of the nose which enable it to carry out air filtration and
conditioning. There is some indication that nasal airflow resistance may influence
the resistance of the lower airways [27,34].

For those involved in industrial or occupational medicine, thought should
be given to screening employees working in hazardous areas for normal nasal
function. Adequate respirators are particularly important for those with impaired function who cannot be transferred to areas free of airborne hazard. It
would be of special importance in those exposed to potential airborne carcinogens or agents responsible for the pneumoconioses to test the efficacy of the
nasal filter and to determine the nature of mucociliary clearance. Such tests
might be done before employment and periodically as long as hazardous exposure
continues. Where the employment involves heavy work, it would be good to
know whether the subject can maintain nasal breathing at the required higher
level of pulmonary ventilation.

For physicians caring for patients with tracheostomies or endotracheal intubation simple humidification of the inspired air is not enough. Care should be
taken that the added water aerosol is not contaminated with pathogenic microorganisms. More is said of this in Chap. 4.

The mucosa of the trachea and major bronchi is relatively ill-equipped for
withstanding the effects of direct access of unmodified inspired air. Patients
with long-standing tracheostomies develop squamous metaplasia in the lower
trachea and main bronchi, lose their ability to clear secretions to the tracheal
stoma through mucociliary action, and are dependent upon a chronic cough for
such clearance [20].

Otolaryngologists contemplating nasal surgery should not be content with
measurement of nasal airflow resistance (even though done with an accurate
method) and demonstrating a reduction in resistance postoperatively. Especially
those operations which involve alterations in the nasal entrance may seriously

impair the nasal filter. It should also be borne in mind that there are wide variations in resistance to nasal airflow among individuals with no nasal complaint. The more important characteristic is the level at which nasal flow limitation occurs. A relatively "high" resistance is not sufficient reason for operation in a patient whose complaint may be on a neurotic basis.

References

1. I. Andersen, G. R. Lundqvist, P. L. Jensen, and D. F. Proctor, Human response to 78 hours exposure to dry air, *Arch. Environ. Health,* **29**:319-324 (1974).
2. G. Aschcan, B. Drettner, and H. E. Ronge, A new technique for measuring nasal resistance to breathing, *Ann. Acad. Reg. Sci. Upsal.,* **2**:111-126 (1958).
3. W. Bachmann, Die Topographie des anatomischen Ostium internum der Nase im Hinblick auf seine funktionelle Bedeutung, *Z. Laryngol. Rhinol. Otol.,* **48**:263-270 (1969).
4. J. H. Bates, W. E. Potts, and M. Lewis, Epidemiology of primary tuberculosis in an industrial school, *N. Engl. J. Med.,* **272**:714-717 (1965).
5. O. V. Batson, The venous networks of the nasal mucosa, *Ann. Otol.,* **63**: 571-580 (1954).
6. A. Baumann and H. Masing, Über der Einflutz körperlicher Arbeit auf den Nasenwiderstand, *Z. Laryngol. Rhinol. Otol.,* **49**:264-270 (1970).
7. A. J. Bentley and R. T. Jackson, Changes in the patency of the upper nasal passage induced by histamines and antihistamines, *Laryngoscope,* **80**:1859-1870 (1970).
8. G. Birnmeyer, Der Verlauf des Inspirationsstromes Kehlkopf, *Arch. Klin. Exp. Ohren. Nasen Kehlkopfheilkd.,* **174**:369-374 (1959).
9. G. P. Bridger and D. F. Proctor, Maximum nasal inspiratory flow and nasal resistance, *Ann. Otol.,* **79**:481-488 (1970).
10. E. A. Brown, The measurement of the resistance of the nasal passage to the movement of air, *Rev. Allergy,* **21**:567-581, 852-857 (1967).
11. L. J. Cass, Measurement of total respiratory and nasal airflow resistance, *JAMA,* **199**:396-398 (1967).
12. B. M. Cohen, The interrelationship of the respiratory functions of the nasal and lower airways, *Bull. Physiopathol. Respir. (Nancy),* **7**:895-911 (1971).
13. J. Dankmeijer, Applied anatomy, vascularization and innervation of the nose, *Int. Rhinol.,* 237-43 (1964).
14. W. L. Dennis, The effect of breathing rate on the deopsition of particles in the human respiratory system. *Inhaled Particles III,* Vol. 1. Edited by W. H. Walton, Surrey, Unwin, 1971, pp. 99-103.
15. B. Drettner, Pressure recordings in the maxillary sinus, *Int. Rhinol.,* **3**:13-18 (1965).

16. B. G. Ferris, Jr., J. Mead, and L. H. Opie, Partitioning of respiratory flow resistance in man, *J. Appl. Physiol.*, **19**:653-658 (1968).

17. R. Fischer, Das Strömungsprofil der Respirationsluft in der nase bei physiologischen Atmung, *Arch. Klin. Exp. Ohren Nasen Kehlkopfheilkd.*, **188**: 404-408 (1967).

18. F. A. Fry and A. Black, Regional deposition and clearance of particles in the human nose, *Aerosol Sci.*, **4**:113-124 (1973).

19. L. Gray, Deviated nasal septum. III. Its influence on the physiology and disease of the nose and ear, *J. Laryngol.*, **81**:953-985 (1967).

20. T. E. Griffith and S. A. Friedberg, Histologic changes in the trachea following laryngectomy, *Ann. Otol.*, **73**:882-892 (1964).

21. R. Horton, R. D. Champlin, E. F. H. Rogers, and R. F. Korns, Epidemic of tuberculosis in a high school, *JAMA*, **149**:331-334 (1952).

22. R. F. Hounam, A. Black, and M. Walsh, The deposition of aerosol particles in the nasopharyngeal region of the human respiratory tract. *Inhaled Particles III*, Vol. 1, Edited by W. H. Walton. Surrey, Unwin, 1971, pp. 71-79.

23. S. Ingelstedt, B. Jonson, and H. Rundcrantz, A clinical method for determination of nasal airway resistance, *Acta Otolaryngol. (Stockh.)*, **68**:189-200 (1969).

24. R. J. Jaeger and H. Motthys, The pattern of flow in the upper human airways, *Respir. Physiol.*, **6**:113-127 (1968).

25. J. Keuning, On the nasal cycle, *Int. Rhinol.*, **6**:99-136 (1968).

26. A. E. Kortekangas, Significance of anterior and posterior technique in rhinomanometry, *Acta Otolaryngol. (Stockh.)*, **73**:218-221 (1972).

27. G. Lacourt and G. Polgar, Interaction between nasal and pulmonary resistance, *J. Appl. Physiol.*, **30**:870-873 (1971).

28. G. Lehmann, The dust filtering efficiency of the human nose and its significance in the causation of silicosis, *J. Industr. Hyg.*, **17**:37-40 (1935).

29. R. G. Loudon and R. M. Roberts, Singing and the dissemination of tuberculosis, *Am. Rev. Respir. Dis.*, **98**:297-300 (1968).

30. K. G. Malcolmson, The vasomotor activity of the nasal mucous membrane, *J. Laryngol.*, **73**:73-98 (1959).

31. H. Masing, Patho-physiology of the nasal airflow, *Int. Rhinol.*, **5**:63-67 (1967).

32. A. F. Millonig, H. E. Harris, and W. J. Gardner, Effect of autonomic denervation on nasal mucosa, *Arch. Otolaryngol.*, **52**:359-368 (1950).

33. D. Nolte and I. Lüder-Lühr, Comparing measurements of nasal resistance by body plethysmography and by rhinomanometry, *Respiration,* **30**:31-38 (1973).

34. J. H. Ogura, Physiologic relationships of the upper and lower airways, *Ann. Otol.*, **79**:495-498 (1970).

35. G. A. Patrick and G. R. Sharp, Oronasal distribution sof inspiratory flow during various activities, *J. Physiol., (Lond.)*, **206**:228-238 (1970).

36. Pattle, R. E., The retention of gases and particles in the human nose. In *Inhaled Particles and Vapours.* Edited by C. N. Davis. London, Pergamon, 1961, pp. 302-309.

37. G. Polgar, Opposing forces to breathing in newborn infants, *Biol. Neonate,* **11**:1-22 (1967).

38. D. F. Proctor, Physiology of the upper airway. In *Handbook of Physiology: Respiration I.* Edited by W. Fenn and H. Rahn. Washington, D.C., American Physiological Society, 1964, Chap. 8.

39. D. F. Proctor and D. L. Swift, The nose—A defense against the atmospheric environment. In *Inhaled Particles III.* Vol. 1. Edited by W. H. Walton. Surrey, Unwin, 1971, pp. 55-69.

40. A. W. Proetz, Air currents in the upper respiratory tract and their clinical significance, *Ann. Otol.,* **60**:439-467 (1951).

41. S. Rao and A. Potdar, Nasal airflow with body in various positions, *J. Appl. Physiol.,* **28**:162-165 (1970).

42. H. B. Richerson and P. M. Seebohm, Nasal airway response to exercise, *J. Allergy,* **41**:269-284 (1968).

43. V. I. Rodin, The importance of nasal septal deformation in miners in the development of respiratory diseases due to dust inhalation, *Vestn. Otorinolaringol.,* **22**:36-40 (1960).

44. H. Rundcrantz, Postural variation of nasal patency, *Acta Otolaryngol. (Stockh.),* **68**:435-443 (1969).

45. F. Saibene, P. Mognoni, and G. Sent'Ambrogio, Lavoro respiratiorio e ventilazione al passagio dalla respirazione nasale a quella orale, *Boll. Ital. Biol. Sper.,* **41**:1550-1552 (1965).

46. H. Schiratzki, The oral and laryngeal components of the upper airway resistance during mouth breathing, *Acta Otolaryngol. (Stockh.),* **60**:71-82 (1965).

47. H. Stoksted, The physiologic cycle of the nose under normal and pathologic conditions, *Acta Otolaryngol. (Stockh.),* **42**:175-179 (1952).

48. Y. Takagi, D. F. Proctor, S. Salman, and S. A. Evering, Effects of cold air and carbon dioxide on nasal airflow resistance, *Ann. Otol.,* **78**:40-48 (1969).

49. M. Taylor, An experimental study of the influence of the endocrine system on the nasal respiratory mucosa, *J. Laryngol.,* **75**:972-977, 1048-1055 (1961).

50. C. W. Twinam and A. S. Pope, Pulmonary tuberculosis resulting from extra-familial contacts, *Am. J. Public Health,* **32**:1215-1218 (1942).

51. M. Uddströmer, Nasal respiration: critical survey of some of the current physiological and clinical aspects on respiratory mechanism with description of new method of diagnosis, *Acta Otolaryngol. [Suppl.] (Stockh.),* **42** (1940).

4

Temperature and Water Vapor Adjustment

DONALD F. PROCTOR and DAVID L. SWIFT

The Johns Hopkins University
School of Hygiene and Public Health
Baltimore, Maryland

I. Introduction

Whether or not an air-breathing warm-blooded animal could live if the site of
respiratory gas exchange were the same as the site of primary exposure to ambient
air it may be impossible to say. An atmosphere at 37°C and relative humidity
close to 100% is quite uncomfortable for man. Although such ambient conditions would obviate the need for temperature and water vapor adjustment of inspired air, they would eliminate the possibility for dissipation of body heat either
through sweat evaporation or convective or radiative heat loss through body surfaces. Even without air pollution or airborne infection maintenance of water
balance and core temperature compatible with life could be a costly process if the
entire pulmonary circulation were exposed to unaltered ambient air with only
the thin alveolar membrane intervening. The beginning of air conditioning in
the nose and its completion as the inspired air passes through the conducting
airways assures not only that air reaching the alveoli is close to the temperature
of the blood but is also nearly saturated with water [18-20, 32,33,59]. We
know also that, during expiration much of the body heat and water added to inspired air is recovered, and that both of these processes are usually in large part
carried out in the nose [13,48-50,52]. In this chapter we will discuss the mech-

anism which enables the nose to act as an efficient air conditioner, to rapidly and accurately adjust to changing ambient conditions, and we will consider some of the problems when the nasal system fails or must be bypassed.

II. The Air Conditioning Mechanism

The key factors in the capacity of the nose to serve as an effective air conditioner are the shape of the airway, the character of airflow, the nature of the mucosa and its vasculature, and the possibility for change related to ambient demands.

A. The Airway and Airflow

Some aspects of the anatomy of the nasal airway have been described in Chap. 3. We need only emphasize here how its structure fits this function. The relatively large cross-sectional area of the main nasal passage, combined with its narrow width (resulting from the division of the airway into two by the nasal septum and further division of each side of the nose by the folds of the turbinates) assures maximum contact between the air and mucosal surfaces. Because of the relatively large cross section the linear velocity of the airstream is low; but, in spite of this, there is probably some turbulence even in quiet breathing. The low linear velocity assures longer contact with mucosa, the narrowness of the airway brings all parts of the air close to the surface, and turbulence produces mixing [49]. When higher ventilation rates are required, as in exercise, the linear velocity must increase, thus shortening the time for nasal passage. But, at the same time, turbulence increases. We have recently shown that increasing nasal airflow, at least in the anesthetized dog, increased the efficacy of absorption of soluble vapors by the nasal mucosa [1]. Presumably the efficacy of heat and water exchange will be similarly enhanced.

The capacity of the nose to humidify inspired air is echoed in measurements of the removal of water-soluble gases during nasal passage [2,4,12,14,15, 22,24,54,64]. This function is related not only to solubility in water but also to the vapor pressure of the dissolved gas. More will be said of this in Chap. 6.

One indication of the relative capacity of the nasal air conditioner is seen in the effect of breathing dry air upon mucociliary clearance which is dependent upon a moist surface. Dry ambient air (9% relative humidity at 23°C for 78 hr) had no effect upon nasal clearance in man [3,5]. In contrast, at least in the anesthetized dog, even 50% RH at 32°C impaired tracheal clearance [30].

Changes in the airway from swelling or shrinking of mucosa, anatomical abnormality, or disease will influence airflow characteristics. A marked increase in the width of the airstream (as with profound vasoconstriction) probably reduces the efficiency of air conditioning. Sufficient narrowing of the airway to force resort to mouth breathing will obviously nullify the nasal conditioner (see Chap. 3).

B. Mucosa and Vasculature

Nasal mucosa, like that in the tracheobronchial tree, is ciliated and mucus-secreting. But in the main nasal passage the mucosa and submucosa combine to form a relatively thick membrane, while in the paranasal sinuses the lining mucosa is extremely delicate and thin. These differences reflect the fact that, while a surface fluid is requisite for ciliary function, in the main nasal airway a capacity for wide variations both in the surface fluid and in blood flow are necessary for adaptation to ambient change.

A number of studies have been directed at the nasal vasculature and its neural control [23,26,27,29,35]. Anyone accustomed to examining the nasal passage can testify to the wide range of variability in mucosal color, widening or narrowing of the airway, and moistening or drying of the surface. It is probable that the most important changes associated with adaptation to changing ambient conditions are:

1. Change in shape and size of the airway through relative engorgement of the erectile vessels

2. Change in blood flow through capillary beds near the surface

3. Possibly changes in interstitial fluid, and certainly changes in the amount and water content of fluids reaching the surface.

Some extremes must be considered pathological (for instance, the pale grey swelling of allergic rhinitis and the bright red swelling of acute inflammation), but differences exist between normal individuals and changes occur in the individual from time to time. These changes seem to be under autonomic control and chiefly involve the septal and turbinate mucosal vessels [21,35]. The secretory activity of mucosa is probably also, at least in part, under autonomic control [28].

It is of some interest to speculate on mechanisms by which the nose accomplishes self-adjustment appropriate to changing ambient conditions. Presumably in some manner nasal blood flow, airway dimensions, and quantity and quality of nasal secretions are all subject to very fine adjustment, but the actual reflex arcs are unknown. If there are sensors of inspired air temperature and humidity linked to nasal function they have yet to be identified. Theoretically the skin of the face or nostrils should be equipped to sense both temperature and moisture and relay the message to the nasal interior.

One of us, in the practice of otolaryngology, has frequently noted that the skin about the nostrils is peculiarly sensitive. It is also generally known that firm pressure on the upper lip just below the nostrils can suppress the sneeze reflex. Perhaps the sensors of ambient air are located in that region. In any event the mechanism is capable of remarkably effective air conditioning even when presented with sudden changes between hot and cold or wet and dry air. We are all familiar with the rapid alteration of the quantity and quality of nasal secretions as seen on exposure to an irritant such as onion fumes. In that case it serves to dilute and wash away the offending irritant. Less extreme changes are called upon when very dry air must be humidified.

Swindle [56] has done a particularly interesting study of nasal vasculature using the heads of reindeer and kangaroo for his experiments. Interesting in itself is his choice of these two species as representatives of opposite extremes in nasal vascularity and erectility. We can speculate that the extreme vascularity and erectility of the reindeer nasal mucosa has developed as adaptation to the rigors of the climate to which he is native.

Swindle's principal findings in the reindeer were as follows. He described the arteries as composed of two interconnected and intricately interlaced networks, one thin-walled and very "inflatable" (which he called "vein-like") not giving rise to the usual arteriocapillary anastomoses found elsewhere in the body, and the other more or less like typical arteries elsewhere. On inflating the "vein-like" arteries the underlying veins were compressed trapping the liquid within them. To quote from his paper,

> Upon inflating the vein-like arteries and the venous traps to cause erection of the mucous lining, liquid which is already in the perivascular tissue is driven into the nasal passages as a viscous discharge, . . . if . . . pressure is varied in a rhythmical manner, the viscous discharge is followed by a watery one, as water is filtered through the walls of some blood vessels and then urged into the nasal passages.

If such a chain of events occurs in vivo and in man, it may in part explain the adaptability of the nose to changing ambient demands in regard to both temperature and water vapor.

Whether or not the relationship between inspiratory airflow and blood flow provides a countercurrent is not certain but seems likely. This could be of importance to the efficacy of heat exchange; but, if this relationship were optimum for inspiratory flow, the reverse might be the case for expiration. It is conceivable the anterior airflow-blood flow relationship might be optimal for inspiration and the reverse true of the posterior nasal region.

C. The Paranasal Sinuses

The paranasal sinuses nearly surround the nasal passages and communicate with the airway through small openings all along the main path of the inspiratory airstream (see Fig. 1 in Chap. 12). Although these mucosa-lined air spaces may serve as insulators helping to maintain a relatively constant temperature within the nose, it is probably of greater importance that their secretions are constantly passing into the nose and serve as useful reservoirs when added moisture is needed along the path of inspired air. Other suggestions for the function of the sinuses include the fact that their air spaces serve to lighten the head.

In view of the extreme demands of some climates one might expect an evolutionary adaptation in nasal anatomy to have occurred in natives of arctic and tropic climates commensurate with their prevalent weather. Anthropologists have uncovered information that such adaptation has indeed occurred [25,39, 63]. The dimensions of the bony nasal aperture have been found to differ significantly between such races as those native to arctic and equatorial regions [25, 63]. The size of the paranasal sinuses are also said to differ in a similar manner [39]. How these differences result in functional changes affecting the air conditioning system is not clear.

III. The Efficiency of Nasal Air Conditioning

Early work on nasal air conditioning is reviewed in Chap. 1. Until more recent times methodology was not available for appropriate and highly accurate measurements, but the fact of nasal function in this regard was recognized long ago. Even today any analysis of the system providing for heat and water exchange is complex and experimental proof substantiating such an analysis difficult. A theoretical approach to the problem follows.

A. Physical Principles

The physical principles of heat and water vapor transport which we shall now discuss have been mathematically described and experimentally investigated by physicists and engineers (particularly chemical engineers) for many years. For simple flow patterns and regularly shaped conduits, such as laminar flow in long straight circular cylinders, exact solutions for the transport equations exist. But for many practical situations, such as the heat and moisture transport in the upper respiratory tract, resort must be made to empirical approximations and simplifications of the conduit shapes.

1. Transport in the Gas Phase

For both heat and moisture transport it is important to understand that the transport occurs both transverse and parallel to the direction of air motion. In inspiratory flow of air whose temperature is $<37°C$, heat and water vapor flow from the walls of the airways to the bulk of the incoming air in response to a negative gradient of temperature and water vapor pressure and are swept distally by the airflow.

The two processes which must be considered for heat transfer are molecular conduction and convection. For each process there is an equation of heat flux (amount of heat transferred per unit cross-sectional area per unit time) defined in terms of a fluid property and the spatial distribution of temperature (T). At any location in the fluid the equation for conduction in the x direction is:

$$F_x = -k \frac{dT}{dx}$$

F_x = flux of heat, $\dfrac{cal}{cm^2\ sec}$

k = thermal conductivity of the fluid, $\dfrac{cal}{°C\ cm\ sec}$

$\dfrac{dT}{dx}$ = temperature gradient (spatial derivative) in the x direction, $°C/cm$

The equation for convective flux at any location in the direction x is:

$$F_x = \rho S v_x T$$

where

ρ = air density g/cm^3

S = air specific heat, $\dfrac{cal}{g\ °C}$

v_x = air velocity in x direction, cm/sec

An overall equation of conservation of energy can be constructed from the conductive and convective heat fluxes in all directions, and the solution to this equation, using appropriate wall (or boundary) conditions of temperature gives the complete spatial distribution of temperature. Practically speaking, we need to know the average temperature of the air at the outlet of the heat exchange region (say the nasal passage) for a given inlet temperature and wall temperature distribution, but this can ordinarily be obtained only by solving for the entire temperature distribution.

The equations for water transport are very similar in form to those for heat transport; the two processes are molecular diffusion and convective mass transport. The equation for mass flux, J, by diffusion is:

$$J_x = -D \frac{dC}{dx}$$

where

J_x = mass flux in x direction, $\dfrac{g(H_2O)}{cm^2 \ sec}$

D = diffusion coefficient for water vapor in air, $\dfrac{cm^2}{sec}$

C = water vapor concentration, $\dfrac{g}{cm^3 \ air}$

$\dfrac{dC}{dx}$ = concentration gradient in x direction

Similarly, the equation for mass flux by convection in this x direction is:

$$F_x = V_x C$$

In a similar manner to the heat transfer case, the fluxes can be combined into an overall mass conservation equation which can, in theory, be solved to yield the spatial distribution of concentration. The concentration of water vapor in air, expressed in the units of g water vapor/m^3 air instead of g/cm^3 air, is called the absolute humidity. At any temperature the ratio of the actual absolute humidity to the absolute humidity of saturated air is the relative humidity, so that the concentration is uniquely related to the relative humidity.

In general, the conservation equations for heat and water vapor cannot be isolated from each other and solved; they are, mathematically speaking, "coupled." There are several reasons for this: (a) When water vapor enters the air, it does so by evaporation from the mucosal surface and this process has an attendant heat loss. (b) The boundary condition for water vapor concentration (at the mucosal surface) is assumed to be saturation, and this is a function of the surface temperature. (c) The specific heat of air is a function of the humidity (vapor concentration) which is itself a function of air temperature. The specific heat change going from any ambient temperature 37°C and any humidity to 37°C saturated is $\leqslant 8\%$, so it is entirely appropriate to use an average value of S and assume it constant. Likewise for ambient temperatures $\geqslant 20°$C the density change going to 37°C is $\leqslant 6\%$, for which an average value is also appropriate. If the further assumption is made that adequate heat can be supplied to the mucosal surface so that it always remains at 37°C, then the equations can be uncoupled. The validity of this assumption will be considered when we discuss heat and moisture transport on the liquid side of the boundary walls.

If the surface temperature is 37°C, the entering air cannot be heated above 37°C nor can it be brought to contain more water vapor than saturation at 37°C. At any location along the respiratory tract we may define factors describing the degree of equilibrium attainment for heat and moisture. For heat transfer this factor is:

$$E_H = \frac{\overline{T} - T_{in}}{37°C - T_{in}}$$

where

$\overline{T}$ = average temperature at a location

T_{in} = ambient (inlet) temperature

This factor is a convenient measure of equilibrium attainment going from zero at the entrance to one, if and when equilibrium at 37°C is attained. A similar factor for water vapor equilibrium attainment is:

$$Em = \frac{\overline{C} - C_{in}}{C_{37°} - C_{in}}$$

where

$\overline{C}$ = average water vapor concentration at some location

C_{in} = ambient inlet water vapor concentration

$C_{37°}$ = equilibrium concentration at 37°C

For the assumption of constant wall temperature (37°C), along narrow rectangular duct geometry, and known flow, an equation for the equilibrium attainment factor, E_H, for any distance X in the flow direction, can be obtained by non-dimensionalizing the uncoupled heat transfer equation, yeilding dimensionless parameter Φ:

$$\Phi = \frac{kX}{V_x \, \rho Sh^2}$$

where

k   = thermal conductivity of air

V_x = average air velocity

ρ = air density

S   = air specific heat

h   = the duct width

The solution of this equation can be represented symbolically as $E_H = F(\Phi)$ showing that changes in these properties influence E_H only through the parameter Φ.

We have applied appropriate values for air and assumed laminar flow in the main nasal chamber, assuming the walls are at a constant temperature and that the chamber is a rectangular duct. We obtain from this calculation $E_h = 0.68$, at the nasopharynx, i.e., the temperature has gone 70% toward thermal equilibrium. This is a conservative estimate in view of the fact that the flow in the main nasal passage is actually turbulent; this is known to markedly increase heat transfer.

The similarity of the transport equations for both heat and vapor has been discussed above. For water transport there is a similar dimensionless parameter, Ψ, which is of the form:

$$\Psi = \frac{DX}{V_x h^2}$$

and for the water vapor-air system $\Psi = \Phi$; this fact leads directly to the convenient conclusion that $E_H = E_m$. Thus we need only investigate one process in detail to draw appropriate conclusions for both.

2. Heat Transfer on the Tissue Side of the Walls

In the simplified analysis above, the assumption was made that the walls of the upper airways were at core temperature, 37°C. But, we know from experiment that this is not so. A temperature gradient exists between the mucosal surface and the tissue depths supplied by blood. This implies that, in order to understand the heat exchange, we must consider the overall transport from blood to air.

Considering the nasal passage, we have air entering the passage at a mass flow rate W_A g/sec and temperature T_{Ai}. If the air leaves the passage at T_{Ao} the total heat gained in transit per second is:

$$H_A = W_A S_A (T_{Ao} - T_{Ai}) = W_H G$$

where

H_A = heat transferred to air, cal/sec

W_A = air flow rate, g/sec

S_A = specific heat of air, cal/g $^\circ$C

W_H = evaporation rate of water, g/sec

G = latent heat of vaporization for water, cal/g

and this must be equal to the heat lost by the blood which passes along the walls:

$$H_A = H_B = W_B S_B (T_{Bi} - T_{Bo})$$

Where

W_B = blood flow rate, g/sec

S_B = specific heat of blood cal/g $^\circ$C

T_{Bi} = "inlet" blood temperature

T_{Bo} = "outlet" blood temperature

Although we do not have any experimental way to measure W_B, we can see that for any particular heat demand, H_A, the product of W_B and $(T_{Bi} - T_{Bo})$ will be constant. If we wish $(T_{Bi} - T_{Bo})$ to be small, then W_B must be large.

The concept of blood moving along the walls of the nasal passage is not unreasonable anatomically and is convenient in analysis. It is a well-established principle in engineering that for a given exchange area, more efficient heat exchange takes place when the fluids are moving countercurrently. There is no direct evidence that the movement of blood adjacent to the nasal mucosa is strictly countercurrent with respect to inspiratory air flow, and if it is so, it is then cocurrent with respect to the heat transfer at expiration.

If we assume that the nasal passage is a countercurrent heat exchanger at inspiration, the overall driving force for transport at any point along the passage is the difference between the blood temperature and the bulk air temperature. We may express the transport at any point by an Ohm's law equation:

$$F_H = \frac{(T_B - T_A)}{R_{BA}}$$

where

F_H = heat flux across air-mucosa boundary, cal/sec cm^2

T_B = blood temperature

T_A = air temperature

R_{BA} = overall transport resistance

The convenience of this concept is that the overall resistance can be divided into two components on the air and tissue sides of the boundary, and these component resistances can be obtained from empirical equations if the flow properties are known.

The observation that the surface temperature of the nasal mucosa is significantly below 37°C (see below) means that the component resistance of the tissue side of the boundary is significant and is subject to changes with blood flow rate. This conclusion that the heat transfer rate is significantly alterable with blood flow is entirely consistent with the observed rapid adaptability of the conditioning mechanism with changing ambient conditions.

B. Gradients in Air Temperature and Water Vapor Pressure

It is clear that, whenever ambient air temperature differs from body temperature and water vapor saturation is less than 100%, a gradient in these two physical parameters must extend between the nostrils and the alveoli. "Body" temperature is a variable figure not only from time to time, from individual to individual, in health and disease, but also in various portions of the body. From our point of view the significant temperature may be considered to be that of the alveolar air.

The exact temperature and water vapor content of alveolar air are difficult to determine and probably vary somewhat between individuals and in the individual from time to time. Alveolar air temperature must be very close to that of the blood (itself a variable). Christie [16] has said that the "vapor pressure of water in alveolar air is lower by some 2 mm than the generally accepted value of 47 mmHg." He bases this statement on the fact that the alveolar fluid lining is not pure water. Were it pure water, the vapor pressure at 37°C would be 47.067 mmHg, while at 36°C it would fall to 44.563 and 38°C rise to 49.692. In a patient with a fever of 40°C the vapor pressure would go to 55.324 mmHg. The significance in terms of actual grams H_2O may be appreciated by reference to Fig. 1.

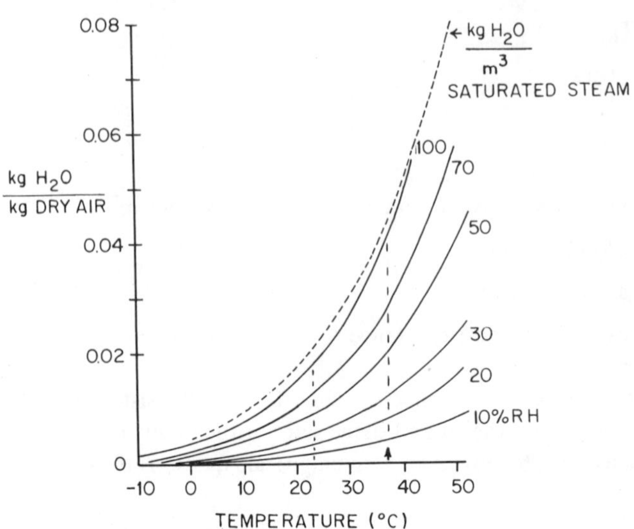

FIGURE 1 Water content of air at various temperatures and various relative humidities shown in relation to weight of dry air (except in dashed line which relates to cubic meters in saturated steam). Vertical lines indicate usual room air temperature 23°C and body temperature 37°C.

If we think in terms of 100% relative humidity (RH) in the alveoli (recognizing that this is in slight error) the amount of water which must be added to inspired air is a function of its temperature and relative humidity on entrance to the body and the core body temperature, for instance 36.8 g/m^3 (BTPS) in going from 23°C at 50% RH to 37°C at 100% RH. Since the entire respiratory tract is lined with a fluid surface which is largely water, it is obvious that the addition of water to inspired air may occur in any or all portions of the airway, but the amount of water which can be added will be limited by the air temperature, in turn a function of surface temperature in the airway and distance from airway surface to center of airstream. We must bear in mind that only in the nose and peripheral bronchi is the distance from airstream center to mucosa small (1 mm or less). In the trachea and major bronchi this distance in the adult human is 3-10 mm.

1. Inspiratory Gradients

There are no adequate data for mapping either mucosal surface or air temperature along the entire airway, but let us examine what data are available. The amount of water which can be contained in the air will be limited by its temperature and thus can be inferred from temperature information.

In a study conducted with Dr. Ib Andersen and associates in Aarhus, Denmark, 15 normal subjects were subjected to changes in ambient air temperatures ranging from 7 to 39°C. Measurements were made of the temperature of respired air just within the nostrils and in the nasopharynx. The plan of this study is described in Chap. 12 (Fig. 7) and will not be further discussed here.

As is shown in Figs. 2-4, when ambient air was at a comfortable 23°C almost no change was noted at the nostrils but a rise to 32.4 ± 0.8°C was noted in the nasopharynx during inspiration. Among the 15 subjects there was considerable variation in nasopharyngeal inspiratory temperature, the range being between 29 and 37°C. When ambient air was kept at 23°C, but inspired air raised or lowered to 31 or 15°C we see that the temperature at the nostrils fell well short of that at ambient control consitions (23°C) (that is, cooler than warm inspired and warmer than cool inspired); but, by the time air reached the nasopharynx it closely approached the control temperature (31 ± 2.1°C for the cool and 32.3 ± 1.3°C for the warm).

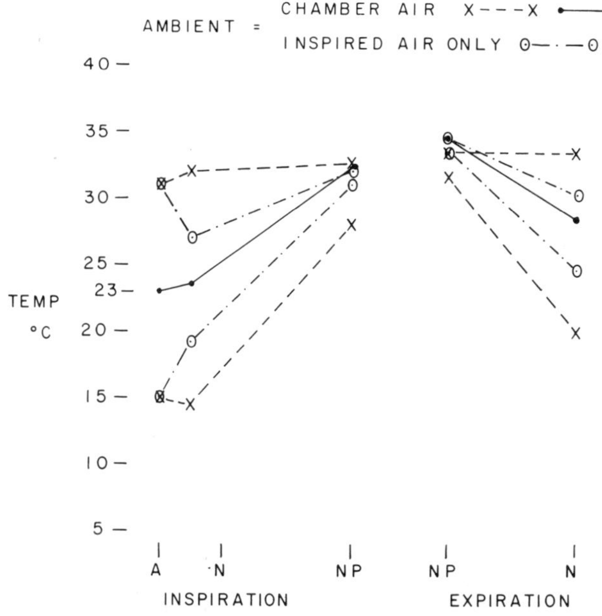

FIGURE 2 Data from the unpublished study of Andersen et al. Air temperatures, during inspiration and expiration, in the ambient (or inspired) air A, within the nostril N, and in the nasopharynx NP. The dots indicate control studies, the circles studies when inspired air temperature alone was altered, and the X when chamber air was altered. Figures are averages of measurements on 15 normal young men. The nasal adjustment of air temperature is obviously different depending upon total ambient air change compared to change in inspired air only.

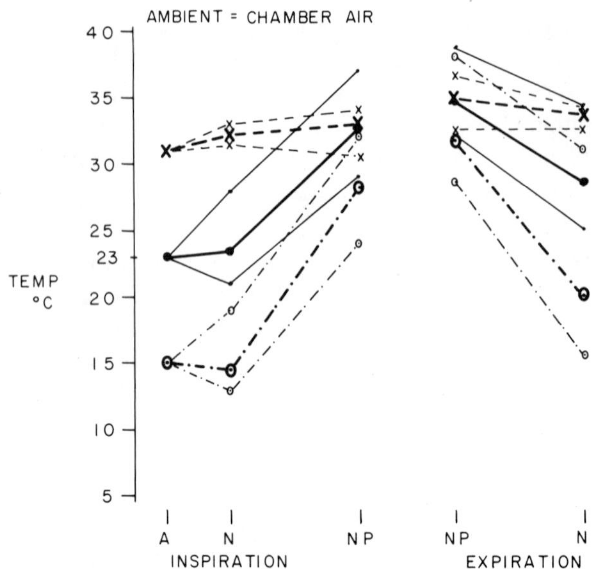

FIGURE 3 From the study referred to in Fig. 2. These data refer to air temperatures when total chamber air was changed. The large symbols are the averages
from 15 subjects and the small symbols show the range of temperatures within
the group. Both in the nostrils and in the nasopharynx, and in inspiration and
expiration there is a wide range of temperature adjustment. Larger differences
are seen with cool ambient air than with warm.

When ambient air itself was altered to 31 or 15°C the approach of nasopharyngeal air temperature to that found in control conditions was less complete (28.1 ± 2.7°C for the cold and 32.8 ± 1°C for the warm). Thus, there is a
suggestion that warming or cooling the whole body in comparison to warming or
cooling inspired air alone decreases somewhat the efficacy of the nasal air conditioning system (in this study the subjects were not nude but were clothed in a
standard suit worn at all temperature conditions).

It is also of interest that, on exposure of the subjects to more extreme
temperatures (inspired air temperature only), nasopharyngeal air temperature
closely approximated that found in control conditions (32.7 ± 1.3°C for the 7°C
inspired air and 33.1 ± 1.4°C for the 39°C inspired air). In the case of the inspiration of air at 39°C the temperature first fell at the nostrils to 29.4°C before
rising again toward the nasopharyngeal level of 33.1°C.

The effects of rapid breathing with exercise are shown in Table 1. A definite reduction in the efficacy of nasal temperature adjustment is seen, although
even at the extremes of 7 and 39°C inspiratory nasopharyngeal temperatures
were 28.4 ± 1.3°C for the cold and 32.8 ± 1.2°C for the warm.

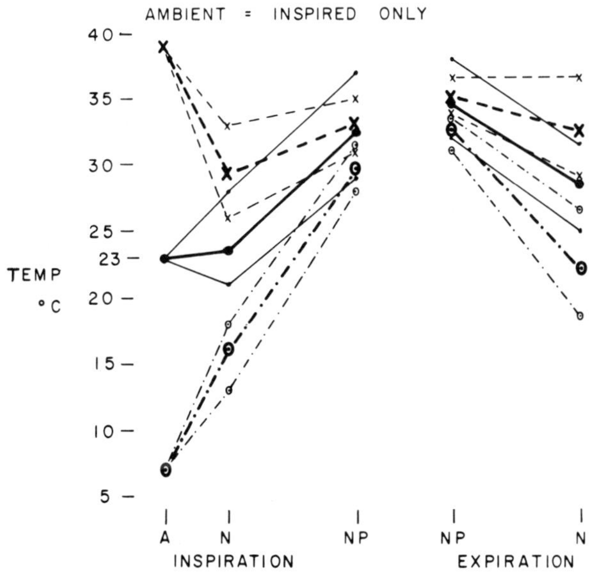

FIGURE 4 From the study referred to in Figs. 2 and 3. These data refer to air temperatures when chamber temperature remained at control conditions and inspired air temperature alone was changed (cooled or warmed air conducted to a hood covering the head). As in Fig. 3 the average temperatures are shown by the large symbols and the small symbols show the range among 15 subjects. In contrast to Fig. 3, here the wider range is with warm air. Even with 7°C the lowest temperature found in the nasopharynx (NP) was 28°C. When total chamber air was lowered to only 15°C the average in the nasopharynx was 28°C and the lowest 24°C (see Fig. 3).

When we consider changing demands (such as going from an atmosphere at 40°C with 90% RH on a hot humid day to an air-cooled room with a temperature of 27°C and 20% RH or vice versa) it is readily apparent that large changes must take place in the supply of water to the airway surfaces if either drying or flooding is to be prevented. Such ambient changes occur in the time needed to pass through a doorway. In this respect, the variability of the nasal airway and its blood supply seem better equipped to make appropriate adjustments than other portions of the respiratory tract. Similar considerations apply to the problem of bringing inspired air to the temperature of the body. Here also the process could be accomplished in any part of the airway (least effectively in the trachea and main bronchi), but adaptation to ambient change seems most feasible in the nose.

The capacity of the nasal airway to accomplish most of this air conditioning has been clearly demonstrated by a number of investigators, especially Cole [18-20], Ingelstedt [33], and Seeley [53]. Ingelstedt has shown that even with

  D. F. Proctor and D. L. Swift

TABLE 1 Air Temperatures in the Nose in Relation to Ambient[a]

		Temperatures (°C)		
Condition		Ambient	Within nostrils	Nasopharynx
Control I		23	23.5 (21-28)	32.4 (29-37)
Control F_I		23	23 (21-28)	31.5 (28-35)
Control E		23	28.5 (25-34)	34.5 (32-38.5)
Control F_E		23	30.4 (25-34.5)	35 (33-38)
A	I	15	19.2 (17-21)	31 (26.5-33.5)
A	F_I	15	18.9 (16.5-21)	30.1 (26-32.5)
A	E	15	24.7 (19-32)	33.5 (30.5-37)
A	F_E	15	25.3 (21-32)	33.9 (31-37)
A	I	7	16.1 (13-18)	29.7 (28-31.5)
A	F_I	7	15.6 (12-17.5)	28.4 (26-30)
A	E	7	22.1 (18.5-26.5)	32.6 (31-33.5)
A	F_E	7	22.5 (18.5-28)	33.2 (31.5-34)
B	I	19	17.2 (14.5-20)	30.9 (28-35)
B	F_I	19	17.3 (14-20)	30.1 (27-35)
B	E	19	23.8 (19-31)	34.1 (31-38.5)
B	F_E	19	25.1 (21.5-31.5)	34.6 (31-39)
B	I	15	14.5 (13-19)	28.1 (24-32.5)
B	F_I	15	14.7 (12-19)	26.9 (21-32)
B	E	15	20 (15.5-31)	31.7 (28.5-38)
B	F_E	15	22.2 (15-32)	32 (28.5-38)
C	I	27	28.3 (27-30)	32 (30-33.5)
C	F_I	27	28.2 (27-30)	31.1 (29-33)
C	E	27	31.8 (29.5-33.5)	34.1 (33-35)
C	F_E	27	32.5 (30.5-34)	34.5 (33-35.5)
C	I	31	32.1 (31.5-33)	32.8 (30.5-34)
C	F_I	31	32 (31-32.5)	32.2 (30-34)
C	E	31	33.5 (32.5-34)	34.7 (32.5-36.5)
C	F_E	31	33.6 (33-34.5)	35.1 (33-37)
D	I	31	32 (31-35)	32.3 (31-35)
D	F_I	31	27.6 (25-31)	31.7 (31-33)

TABLE 1 (continued)

| | | \multicolumn{3}{c}{Temperatures (°C)} | | |
Condition		Ambient	Within nostrils	Nasopharynx
D	E	31	34.3 (33.5–36.5)	34.5 (33–36.5)
D	F_E	31	31.3 (27–33.5)	34.9 (33–36.5)
D	I	39	29.4 (26–33)	33.1 (31–35)
D	F_I	39	30.2 (27–33)	32.8 (31–34)
D	E	39	32.4 (29–36.5)	35.1 (33.5–36.5)
D	F_E	39	33.1 (30–37)	35.5 (34–37)

[a]Conditions B and C signify experiments in which entire chamber air was changed. In conditions A and D chamber air remained at 23°C and only inspired air was changed. I signifies inspiration and E expiration. F_I or F_E signifies forced nasal breathing as with exercise. The temperatures are the means for a group of 15 subjects with the ranges given in parentheses. See Fig. 7, Chap. 12 for overall design of this study.

oronasal breathing of cold air most of the adjustment of temperature and humidity has occurred before the air enters the trachea. This adjustment was more effective with nasal breathing (subglottic inspired air temperature 32.1°C compared with 24.3 to 23°C with oronasal breathing). Seeley, with himself as a subject and measuring at three points along the nasal airway, found a rise from an ambient of -7 to +17°C by the time air reached the turbinates and to +25°C halfway through the nasal passage. Similarly, an ambient temperature of +54°C was reduced to +39°C at the turbinates and +38°C halfway through the nose (Fig. 5). Adjustments in relative humidity followed the same general pattern.

Hair et al [32] studied laryngectomees (whose noses may have lost some of their adaptability through prolonged lack of exposure to ambient air) and found that water content of inspired air went from 0.028 to 0.035 g/liter during nasal passage. Webb [60] like Seeley measured air temperature at three positions along the nasal airway. Resting subjects in the cold raised air temperature from an ambient ranging between -20 and -31°C to -3°C at 1 cm along the passage, +15°C at 5 cm, and +25.4 ± 3.4°C at 9 cm (close to the nasopharynx). In subjects breathing ambient air at +24°C the temperature at 9 cm was +31.7 ± 2.1°C. Measurements such as our own and those of Seeley and Webb may under- or overestimate changes produced in the nose owing to disturbance of airflow conditions produced by the presence of the measuring device.

2. Expiratory Gradients

Somewhat different considerations come into play in the recovery of water and heat from expired air. This process must take place near the exit from the airway

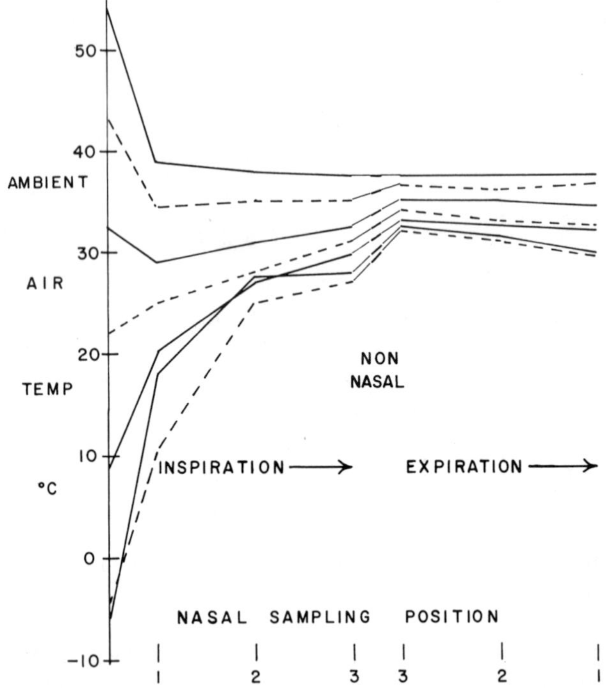

FIGURE 5　Chart redrawn from Seeley [53] showing changes in temperature as air at indicated ambient conditions moves in and out of the nose. The numbers 1, 2, 3 indicate the three positions at which he sampled during inspiration to the left and expiration to the right.

as conditions approach ambient. The area of transition from body to ambient conditions as seen by expired air will vary with the degree to which ambient air differs from body temperature, with the nature of the airway near the point of exit, and with the surface temperatures along the airway. Thus, for instance, with cold ambient air and mouth or tracheostomy breathing one might expect relatively cooler surfaces near the airway exit than would be the case with nasal breathing. Although we do not have sufficient data on airway surface temperatures under varying conditions, expired air temperatures may be a good reflection of those of the surface over which air is passing (see expiratory NP and N temperatures in Figs. 2, 3, and 4). The cooler the surface temperature near the exit, the larger will be the fall in expired air temperature and the greater the loss of water (regain for the body) to airway surfaces.

According to the estimates of Cole [18-20] in normal man in a temperate climate, the net result of all this is a loss of 300-400 ml H_2O and 250-350 kcal/24 hr in the conditioning of respired air. These figures will be subject to large change in extreme ambient conditions or when core body temperature is altered as with fever. Cole has made the interesting suggestion that the mucosa of the upper respiratory tract may have a thermoregulatory function similar to that of the skin. How this could be accomplished without disturbing the air conditioning system (except in panting) is not clear.

Liese et al. [40,41] concluded from mass spectrometer measurements of water content of end tidal expired air that the nose contributes only 3 of 30 g H_2O added to m^3 of inspired air (10%). This conclusion, drawn from a comparison of nasal and mouth breathing, seems to us to be erroneous and only to indicate that portions of the respiratory tract other than the nose are capable in short experiments of adding 90% of the H_2O ordinarily coming from nasal surfaces. Such an experiment tells us nothing of what the nose ordinarily contributes relative to the remainder of the respiratory tract during nasal breathing. Other studies, as mentioned above, clearly show that normally most of this water is added to the inspired air during its brief passage through the nose.

The temperature of air during expiration is probably a reflection of surface temperature gradients along the airways. If this is so, we may assume that both air and surface temperature in the alveoli is around 37°C. Marsh (personal communication) has found evidence that temperature adjustment is nearly complete at the level of third generation bronchi (study in five patients during bronchoscopy). He found surface temperatures ranging from 32.6 to 36.1°C at the carina to 34.8 to 36.8°C in a subsegmental bronchus. These temperatures rose rapidly toward core temperature when the patient's ventilation was interrupted. From the data in Ingelstedt's study [33] it appears that air temperature in the upper trachea is not grossly different from that in the nasopharynx. Thus we have a gradient of some 3-5°C within the bronchial tree.

From our data, and from that of Webb, we may conclude that in extreme ambient conditions surface temperatures in the nasopharynx may be from 8°C below to slightly above those in the alveoli (Figs. 2-4). The major gradient, as reflected by expired air temperature, is along the nasal passage itself. Here, breathing air at 23°C, the gradient is some 5°C, but, when breathing very cold or very warm air, the gradient may be as much as 15°C or more. From this one might judge that the surface temperature near the nostril is most susceptible to change with ambient conditions. It is interesting to note, from our data, that in warm ambient air (39°C) the air at the nasal entrance is at a lower temperature than in the nasopharynx (26-33°C on inspiration and 29-36.5°C on expiration).

From all of these considerations, it appears that, during nasal breathing and in the face of the various ambient air conditions which have been studied, the greater part of air conditioning takes place within the nose. But, it is equally clear that a small additional change must take place further along the airways even under such ordinary air conditions as 23°C and 50% RH. It is also of interest that among normal subjects there is a surprising variability in the degree to which the nose accomplishes this function.

IV. Influences Which May Affect the System

A. Environmental Factors

We might expect the efficacy of the air conditioning system to be influenced by extreme ambient demands or changes, by alterations in the nasal airway, or by the use of an alternate airway of access. Ambient air temperature extremes seem to have little effect upon the lower, pulmonary airways [8,37,46,47], but they might influence the overall system through a reflex effect on or from the nasal mucosa. A number of studies have been directed at these problems [27,55,60]. Cold air does indeed cause a nasal vascular congestion resulting in increased resistance to airflow [51,57]. This could force resort to mouth breathing (especially during exercise) thus bypassing the superior nasal air conditioner. As one might expect, humid air seems to have no effect on the airways or the efficacy of air conditioning. The widespread belief that dry air or wet air or cold air may be detrimental to respiratory health has motivated many studies, but not led to substantiation of that belief. As already mentioned, we have been unable to uncover any effect of even prolonged exposure to dry air in normal subjects (see also Chap. 12).

It seems clear that the average normal nose is capable of adjusting the temperature and humidity of inspired air even in extreme ambient conditions. Large variations from such average function might be related to individual differences in susceptibility to airborne disease. One indication of the magnitude of these variations is found in Webb's study, mentioned above [60] and in our own data from the Aarhus study (Figs. 3 and 4). At an ambient temperature of -20 to -31°C, Webb found that inspired air temperature in the nasopharynx averaged 25.4°C, but the range among 10 subjects was from 18.7 to 32.2°C, over 25% ± variation (Fig. 6). This study also showed the differences in cooling of nasal surfaces as indicated by the temperature of expired air both in the nasopharynx and even as little as 1 cm from the nasal entrance. Here again, with the ambient of -20 to -31°C among 10 subjects, expired air temperature ranged between 29.4 and 34.9°C at the nasopharynx and 18.8 to 32.7°C, 1 cm from the nostril (Fig. 6). Even at the moderate ambient temperatures of 23-28°C variations among 8 subjects were large, ±10 to 12%. Obviously the water vapor content of both inspired and expired air must also vary with the temperature.

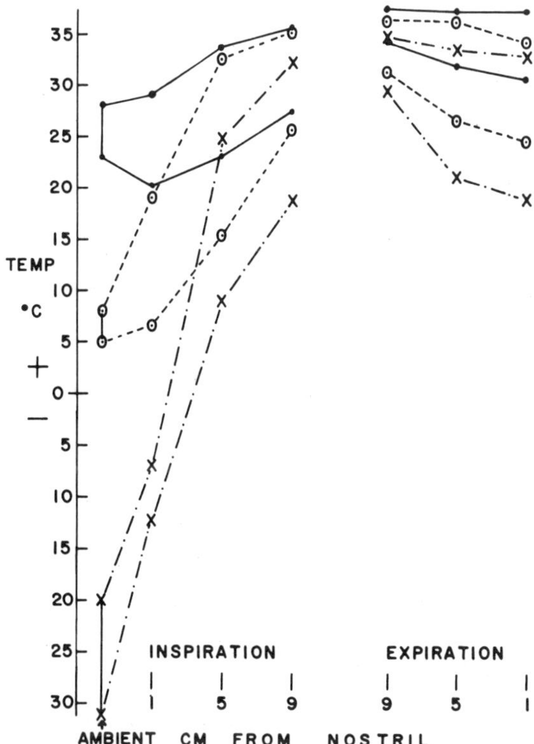

FIGURE 6 Redrawn from Webb's data [60] to emphasize the range of variation in respired air temperature adjustment among normal subjects. Eight subjects were used for the higher ambient temperature, 10 for the cold, and 6 for the medium.

Thus it would appear that in normal subjects, the burden of completing the air conditioning function in the lower respiratory tract will be significantly greater in some subjects than in others. Such added burden will to some degree fall to the smaller peripheral airways, since the width of the airstream from nasopharynx through the larger bronchi precludes optimum exchange of temperature and water vapor. To our knowledge, no attempt has been made to find a possible correlation between relatively efficient or inefficient nasal air conditioning and later development of such diseases as chronic bronchitis, pneumonia, etc.

Other factors which may directly or indirectly affect the nasal air conditioner include cooling or heating the body surfaces. Gusic [31] has found edema and vasodilatation in the nasal mucosa as well as an appearance of dryness among workers in an electric light factory exposed to ambient temperatures between 30 and 51.5°C (RH 37 to 55%). Bacteria counts in the upper respiratory tract were said to have risen.

Exposure to irritant materials [4] and a variety of pharmacological agents taken systemically or applied topically may affect the system, but little satisfactory work has been done in this area. It is generally accepted that sufficient widening of the nasal airway, as with profound vasoconstriction, will reduce the effectiveness of the system. Dehydration, as with water deprivation for 48 hours, has been shown in the experimental animal to bring nasal mucus clearance to a halt [9]. Such an effect would result in drying of the surface and quickly stop the usefulness of the nasal air conditioner.

Although it is quite possible that failure to adapt to changing conditions may account for many respiratory symptoms and disease, in most individuals most of the time the adjustment is such as to prevent any apparent untoward effects. Many patients attribute symptoms to ambient conditions such as the room air conditioner, drafts, cold air, wet weather, unseasonable weather, etc. Attempts to document the relation between such ambient conditions and disease have commonly met with failure. As evidence of this, while many physicians attribute respiratory problems to a damp climate and recommend a dry climate for their alleviation, others insist that indoor air must be humidified for respiratory health. Indeed, many physicians sit on both sides of this fence. Although appropriate artificial indoor air conditioning undoubtedly can add to human comfort [11,17] evidence as to its beneficial or detrimental effect upon human health is lacking.

Some symptoms related to ambient conditions are not reflections of malfunction. The dripping of water from the nasal tip in very cold weather is not a result of excess secretion but merely of condensation of water from the expired air as it is suddenly cooled near the nostril.

B. Other Influences

If we understood the exact mechanisms involved in the control of the air conditioning system, we might also understand the factors which may interfere with its function and even the methods by which malfunction could be corrected. It is clear that emotional stimuli and certain hormonal malfunctions can interfere with nasal mucosal function. Some women find that nasal congestion (on examination found to be due to a very pale, swollen mucosa) is literally the first sign of pregnancy, not a symptom which develops after they find they are pregnant. The problem disappears immediately after delivery.

In a study carried out some years ago, Wolff et al. [61] found that all of the mucous membranes of the body accessible to observation underwent changes in response to a variety of emotional stimuli. Of special interest were two facts. In general the mucosa of the nose changed as much as or more profoundly than any other mucosa with any given stimulus. Similar emotional events led to different mucosal responses in different individuals. Thus the same stimulus could lead to mucosal blanching and drying in some and congestion and moistening in others. A more complete understanding of all the physiological, environmental, and disease factors influencing this vital system would surely lead to more logical management of many respiratory problems.

V. Non-nasal Air Conditioning

A. Mouth Breathing

Since, when we resort to mouth breathing, the system is less effective, it is important to consider circumstances which produce nasal congestion sufficient to reduce significantly the capacity for nasal airflow. Rhinitis, anatomical malformation, cold air, recumbent posture, and SO_2 even at 1 ppm significantly increase nasal airflow resistance [4,57]. Physical exercise decreases nasal airflow resistance. Undoubtedly a multitude of other factors also influence the nasal airway and many of these interrelate (allergy, emotional stress, pharmacological agents, etc.).

Sufficient alteration in the nasal airway on the basis of anatomical malformation or as a result of surgical intervention can result in either chronic mouth breathing or an ineffective nasal air conditioner. Severely deflected nasal septum, nasal polyps, enlarged adenoids, certain congenital nasal or nasopharyngeal deformities all can lead to mouth breathing. Many nasal septal deflections produce no difficulty. Those which interfere with the normal path of inspiratory airflow, between the middle meatus and the septum, are of greatest significance [44].

On the other side of the coin, atrophy of nasal mucosa or surgical removal of the middle turbinate excessively widen the airway and reduce the effectiveness of the nasal air conditioner. The widened inspiratory airstream no longer has full contact with mucosal surfaces and the reduced surface is prone to drying.

In each of these circumstances the efficacy of the nasal air conditioner is either reduced or bypassed. The alternative use of mouth breathing still assures the passage of inspired air over a fairly large moist surface. There is no really satisfactory study of chronic mouth breathers to determine what hazard this offers to the individual's health. But clinical experience leaves little doubt that access of air directly into the trachea, without the interposition of either nasal or oropharyngeal airway, is harmful.

An analysis of oropharyngeal air conditioning is even more difficult than that given above for the nose because of the highly variable size and shape of that airway [34,35]. It seems likely that a narrow airway produced by approximation of tongue and palate is nearly as effective as the nose but probably less adjustable. During rapid breathing through a wide oropharyngeal airway air will almost certainly reach the trachea relatively unmodified.

B. Tracheal Breathing

In medical practice we are frequently faced with a patient in whom breathing must take place through a tracheostomy or naso- or orotracheal tube. In each instance ambient air passes directly into the trachea and it is generally recognized that artificial air conditioning must be supplied for inspired air. Patients in these circumstances are prone to atelectasis (owing to drying of bronchial secretions and subsequent airway closure) and pulmonary infections (owing to the lack of the nasal filter and injury to bronchial epithelium). In some instances the infection is introduced through the artificial conditioning system. Such adverse effects are more serious in children with their smaller airways. Many patients live with tracheostomies for long periods of time, or, in the case of the laryngectomee, for life. But in such patients the lower respiratory mucosa does not escape unscathed. Most laryngectomees have chronic bronchitis and their tracheobronchial mucosa undergoes squamous metaplasia and the clearance system is impaired. Some adaptation occurs, since the incidence of complications and the difficulty of clearing secretions subsides somewhat after the first several days. But one indication of the handicap of tracheal breathing is seen in the fact that long-term tracheostomy in infants and young children is associated with a 10-40% mortality from respiratory complications.

C. Artificial Air Conditioning

In the management of the acute problem (during the early hours or days after tracheotomy or intubation), and indeed in some patients breathing through the nose but thought to suffer from inadequate inspired air humidification, artificial humidification is provided. Some years ago this was attained with steam led into a tent. With that system the air was not only saturated with water vapor, but its temperature approached that of the body due to the heat of the steam. Thus little or no further conditioning within the body was necessary. Because of the fear that hyperthermia could result, some years ago a nebulizer producing a cold fog of water aerosol particles was substituted for the steam tent. This seemed an eminently suitable change, allowing the patient to be in a pleasantly cool environment which, as anyone could see, was saturated with the fog. It was only after a surprisingly long time that the disadvantages of this system were first suspected and then documented. These disadvantages are related to three points: the water is in the form of an aerosol most of which is deposited in the first part of the airway providing no benefit for those airways beyond, the cooler air is

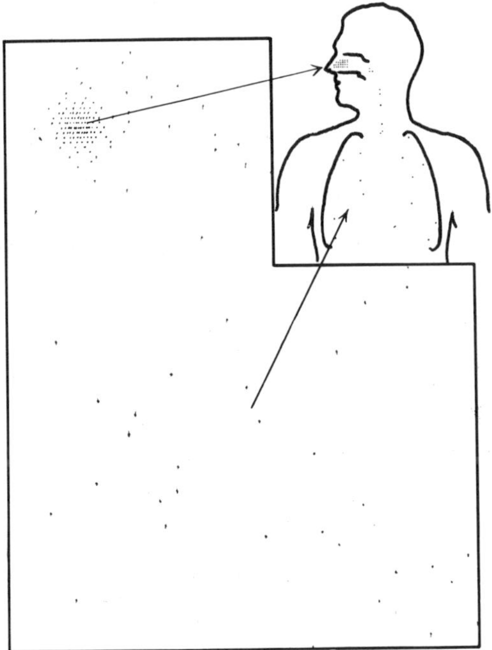

FIGURE 7   Scan following nasal breathing of labeled aerosolized water from jet nebulizer. Insignificant quantities pass beyond the nasal tip. (Used with the courtesy of the Editor of *Pediatrics,* [62].)

incapable of containing anywhere near the amount of water vapor required for full saturation at body temperature (see Fig. 1), and the reservoirs of water from which the aerosol is produced make excellent culture media for many pathogenic microorganisms [45].

Although only a minute amount of the aerosol need be deposited in the lungs to make the establishment of infection possible, to be of therapeutic benefit when a serious deficit in normal air humidification is present very large quantities of aerosol deposited along airway surfaces would be necessary. With nasal breathing insignificant amounts of water penetrate beyond the nasal tip [6,10, 62], and even with tracheostomy or tube breathing there is as yet insufficient evidence as to the penetration of the aerosol to deeper airways (Fig. 7). There is evidence in the experimental animals of increased lower airway resistance to airflow from excess deposition of water aerosols [42].

Added to the inefficiency of the nebulizer system of air humidification is the problem of respiratory infections traceable to inhalation therapy equipment. Numerous studies have now traced nosocomial infectious outbreaks to this source [7,43].

FIGURE 8   Toremalm's heat and moisture exchanger. This device is made in various sizes suitable for different ages and plugs directly into tracheostomy or orotracheal tube. It offers a resistance to airflow comparable to that of the nose.

To meet these problems Toremalm has developed what amounts to an artificial nose (Fig. 8) which, when properly used, seems the ideal substitute for the normal air conditioning system in patients with tracheostomies or tracheal tubes [38,58]. This device precipitates water from the expired air which can be picked up by the ensuing inspiration. At the same time the proximal portion of the device is warmed by the expired air and in turn warms the inspired air.

Whether or not a system of water nebulization can be developed which both guards against carrying infection into the lungs and assures adequate humidification of inspired air remains to be demonstrated. Meanwhile the Toremalm device, or some modification thereof, appears to us preferable to the currently employed water nebulizers.

VI. Summary

In summary, the nose in normal man is an excellent air conditioning system, but as yet, we know too little of factors which may impair its efficiency, environmental factors which may overcome it, or those influences which produce nasal congestion and lead to mouth breathing. It is possible for the oropharynx or the tracheobronchial tree to take over this function, but less effectively and at some cost to the organism.

References

1. E. F. Aharonson, H. Menkes, G. Gurtner, D. L. Swift, and D. F. Proctor, The effect of respiratory airflow rate on the removal of soluble vapors by the nose, *J. Appl. Physiol.*, **37**:654-657 (1974).

2. M. Ainsworth and R. J. Shephard, The intrabronchial distribution of soluble vapours at selected rates of gas flow. In *Inhaled Particles and Vapours*. Edited by C. N. Davies, Oxford, Pergamon, 1961, pp. 233-248.

3. I. Andersen, G. Lundqvist, and D. F. Proctor, Human nasal mucosal function under four controlled humidities, *Am. Rev. Respir. Dis.*, **106**:438-449 (1972).

4. I. Andersen, G. Lundqvist, P. L. Jensen, and D. F. Proctor, Human exposure to controlled levels of sulfur dioxide, *Arch. Environ. Health*, **28**:31-39 (1974).

5. I. Andersen, G. R. Lundqvist, P. L. Jensen, and D. F. Proctor, Human response to 78 hours exposure to dry air, *Arch. Environ. Health*, **29**:319-324 (1974).

6. T. Asmundsson, R. F. Johnson, J. H. Kilburn, and J. K. Goodrich, Efficiency of nebulizers for depositing saline in human lung, *Am. Rev. Respir. Dis.*, **108**:506-512 (1973).

7. M. E. Avery, M. Galina, and R. Nachman, Mist therapy, *Pediatrics*, **39**: 160-165 (1967).

8. A. M. Baetjer, Effect of ambient temperature and vapor pressure on cilia-mucus clearance rate, *J. Appl. Physiol.*, **23**:498-504 (1967).

9. B. G. Bang and F. B. Bang, Effect of water deprivation on mucous flow, *Proc. Soc. Exp. Biol. Med.*, **106**:46-49 (1961).

10. S. K. Bau, N. Aspin, D. E. Wood, and H. Levison, Measurement of fluid deposition in humans following mist tent therapy, *Pediatrics*, **48**:605-612 (1971).

11. L. L. Boyer, Jr., The meaning of universal comfort in buildings, *J. Physiol. (Paris)*, **63**:210-212 (1971).

12. J. D. Brain, The uptake of inhaled gases by the nose, *Ann. Otol.*, **79**:529-539 (1970).

13. G. E. Burch, Study of water and heat loss from the respiratory tract of man, *Arch. Intern. Med.*, **76**:308-314 (1945).

14. G. R. Cameron, J. H. Gaddum, and R. H. D. Short, The absorption of war gases by the nose, *J. Pathol.*, **58**:449-445 (1946).

15. L. J. Casarett, Distribution and excretion of Polonium-210, *Radiat. Res. [Suppl.]*, **5**:148-165 (1964).

16. R. V. Christie and A. V. Loomis, The pressure of aqueous vapour in the alveolar air, *J. Physiol. (Lond.)*, **77**:35-48 (1932).

17. S. Clemmesen, B. Ibsen, B. Jensen, M. Krogsgaard, and G. Werner, Air-conditioning and human comfort, *Dan. Med. Bull.*, **10**:217-224 (1963).

18. P. Cole, Some aspects of temperature, moisture and heat relationships in the upper respiratory tract, *J. Laryngol. Otol.*, **67**:449-456 (1953).

19.   P. Cole, Further observations on the conditioning of respiratory air, *J. Laryngol. Otol.*, **67**:669-681 (1953).

20.   P. Cole, Respiratory mucosal vascular responses, air conditioning and thermoregulation, *J. Laryngol. Otol.*, **68**:613-622 (1954).

21.   P. Cole, Nasal turbinate function, *Can. J. Otolaryngol.*, **2**:259-262 (1973).

22.   G. Cumming, J. Crank, K. Horsfield, and I. Parker, Gaseous diffusion in the airways of the human lung, *Respir. Physiol.*, **1**:58-74 (1966).

23.   A. Dahlström, The adrenergic innervation of the nasal mucosa of certain mammals, *Acta Otolaryngol. (Stockh.)*, **59**:65-72 (1965).

24.   T. Dalhamn and J. Sjohölm, Studies on SO2, NO2 and NH3: Effect on ciliary activity in rabbit trachea of single in vitro exposure and resorption in rabbit nasal cavity, *Acta Physiol. Scand.*, **58**:287-291 (1963).

25.   A. Davies, A resurvey of the morphology of the nose, *J. R. Anthropol. Inst. Gt. Brit. Ire.*, **62**:337-359 (1932).

26.   J. D. K. Dawes and M. M. L. Prichard, Studies of the vascular arrangement of the nose, *J. Anat.*, **87**:311-322 (1953).

27.   B. Drettner, Vascular reactions of the human nasal mucosa on exposure to cold, *Acta Otolaryngol. [Suppl.] (Stodkh.)*, **166**:1-109 (1961).

28.   R. Eccles and H. Wilson, The parasympathetic secretory nerves of the nosf of the cat, *J. Physiol. (Lond.)*, **230**:213-223 (1973).

29.   K. Flisberg and S. Ingelstedt, Nasal vascular responses to foot cooling in normal and allergic patients, *Acta Otolaryngol. (Stockh.)*, **55**:457-466 (1962).

30.   A. R. Forbes, Humidification and mucous flow in the intubated trachea, *Br. J. Anaesth.*, **45**:118 (1973).

31.   B. Gusic, Z. Krajina, Z. Poljak, V. Konić-Carnelutti, and I. Babić, The influence of heat on the upper respiratory tract, *Acta Otolaryngol. (Stockh.)*, **67**:150-157 (1969).

32.   G. E. Hair, N. D. Fischer, and M. J. Preslar, Humidification of air by the nasal mucosa, *Laryngoscope*, **79**:375-381 (1969).

33.   S. Ingelstedt, Studies on the conditioning of air in the respiratory tract, *Acta Otolaryngol. [Suppl.] (Stockh.)*, **131**:1-180 (1956).

34.   S. Ingelstedt and N. G. Toremalm, Aerodynamics within the larynx and trachea, *Acta Otolaryngol. [Suppl.] (Stockh.)*, **158**:81-92 (1960).

35.   S. Ingelstedt and N. G. Toremalm, Airflow patterns and heat transfer within the respiratory tract, *Acta Physiol. Scand.*, **51**:1-14 (1961).

36.   T. Ishii, The cholinergic innervation of the human nasal mucosa, *Pract. Otorhinolaryngol. (Basel)*, **32**:153-158 (1970).

37.   R. Jones, A. M. Baetjer, and L. Reid, Effect of extremes of temperature and humidity on the goblet cell count in the rat airway epithelium, *Br. J. Ind. Med.*, **28**:369-373 (1971).

38.   H. Koch, C. Allander, S. Ingelstedt, and N. G. Toremalm, A method for humidifying inspired air in post tracheotomy care, *Ann. Otol.*, **67**:991-1005 (1958).

39.   T. Koertvelyessy, Relationship between the frontal sinus and climatic conditions: A skeletal approach to cold adaptation, *Am. J. Phys. Anthropol.*, **37**:161-172 (1972).

40. W. Liese and R. Joshi, Effect of atropine on humidification of respired gases, *Naunyn Schmiedebergs Arch. Pharmacol.*, **274**:415-417 (1972).

41. W. Liese, R. Joshi, and G. Cumming, Humidification of respired gas by nasal mucosa, *Ann. Otol.*, **82**:330-332 (1973).

42. G. N. Melville, Water content level in inspired air on specific airway resistance in rats, *Respiration (Basel)*, **29**:127-134 (1972).

43. J. J. Mertz, L. Scharer, and J. H. McClement, A hospital outbreak of Klebsiella pneumonia from inhalation therapy with contaminated aerosol solutions, *Am. Rev. Respir. Dis.*, **95**:454-460 (1967).

44. R. Moe, The effect on the respiratory functions of the nose of the lumen-dilating nasal operations, *Acta Otolaryngol. [Suppl.] (Stockh.)*, **45** (1941).

45. H. L. Moffett and T. Williams, Bacteria recovered from distilled water and inhalation therapy equipment, *Am. J. Dis. Child.*, **114**:7-25 (1967).

46. A. R. Moritz, F. C. Henriques, and R. McLean, Effects of inhaled heat on air passages and lungs, *Am. J. Pathol.*, **26**:311-331 (1945).

47. A. R. Moritz and J. R. Weisiger, Effects of cold air on air passages and lungs, *Arch. Intern. Med.*, **75**:233-240 (1945).

48. V. E. Negus, *The Comparative Anatomy and Physiology of the Nose and Paranasal Sinuses*, Edinburgh, Livingstone, 1958.

49. B. Olariu and M. Teodorescu, L'écoulement biphasé au niveau des fosses nasales, *Acta Otorhinolaryngol.*, **27**:60-68 (1973).

50. A. W. Proetz, *Essays on the Applied Physiology of the Nose*, St. Louis, Annals, 1953.

51. S. Salman, D. F. Proctor, D. L. Swift, and S. A. Evering, Nasal resistance: Description of a method and effect of temperature and humidity change, *Ann. Otol.*, **80**:736-744 (1971).

52. J. H. Scott, Heat regulating function of the nasal mucous membrane, *J. Laryngol.*, **68**:308-317 (1954).

53. L. E. Seeley, Study of changes in the temperature and water vapor content of respired air in the nasal cavity, *ASHVE*, June, pp. 377-388 (1940).

54. F. E. Speizer and N. R. Frank, The uptake and release of SO_2 by the human nose, *Arch. Environ. Health*, **12**:725-728 (1966).

55. I. G. Spiesman, Vasomotor responses of the upper respiratory tract to thermal stimuli, *Am. J. Physiol.*, **115**:181-187 (1936).

56. P. F. Swindle, The architecture of the blood vascular networks in the erectile and secreting lining of the nasal passage, *Ann. Otol.*, **44**:913-932 (1935).

57. Y. Takagi, D. F. Proctor, and S. Evering, Effects of cold air and carbon dioxide on nasal airflow resistance, *Ann. Otol.*, **78**:40-49 (1969).

58. N. G. Toremalm, A heat and moisture exchanger for post-tracheotomy care, *Acta Otolaryngol. (Stockh.)*, **52**:1-12 (1960).

59. T. Unno, Humidification in nasal cavity, *J. Otolaryngol. Jpn.*, **66**:592-610 (1963).

60. P. Webb, Air temperatures in respiratory tracts of resting subjects in cold, *J. Appl. Physiol.*, **4**:378-382 (1951).

61. H. G. Wolff, J. Wolf, W. J. Grace, T. E. Holmes, I. Stevenson, L. Straub, H. Goodell, and P. Seton, Changes in form and function of mucous membranes occurring as part of protective patterns in man during periods of life stress and emotional conflict, *Trans. Assoc. Am. Physicians,* **61**:313-334 (1948).

62. J. Wolfsdorf, D. L. Swift, and M. E. Avery, Mist therapy reconsidered: An evaluation of the respiratory deposition of labeled water aerosols produced by jet and ultrasonic nebulizers, *Pediatrics,* **43**:799-808 (1969).

63. M. H. Wolpoff, Climatic influence on the skeletal nasal aperture, *Am. J. Phys. Anthropol.,* **29**:405-423 (1968).

64. E. Yokoyama and R. Frank, Respiratory uptake of ozone in dogs, *Arch. Environ. Health,* **25**:132-138 (1972).

5

Particle Deposition

R. F. HOUNAM and ARTHUR MORGAN

Atomic Energy Research Establishment
Harwell, Oxfordshire, United Kingdom

I. Introduction

The purpose of respiration is to provide the organism with oxygen and to eliminate carbon dioxide. To supply himself with adequate oxygen man inhales some 10 m^3 (15 kg) of air daily, even when sedentary. Unfortunately, even so-called "fresh air" carries an associated burden of suspended particles and in urban environments this is inevitably increased. Among the contaminants are particles of wide-ranging characteristics and origins which may include spores, pollens, skin scales, bacteria, infected droplets, smuts, smokes, and nuclei from the combustion of coal, oil, and vegetation, fumes emitted by internal combustion engines, and resuspended particles of soil. These vary in size from small nuclei (0.001 μm) to particles of coarse dust and pollen which may reach 100 μm in diameter. While particles of these materials are inhaled inadvertently, others, such as tobacco smoke and aerosols of bronchodilating drugs, may be inhaled intentionally. Typical dust levels in rural and busy urban areas are 0.1 and 1 mg m^{-3}, respectively. In certain industrial environments, however, levels can be considerably higher, and may fluctuate widely over short periods of time.

The fate of the suspended particles in inhaled air is obviously of interest, especially in cases where the contaminants are harmful to a greater or lesser

degree. The first question to be considered is the fraction of the inhaled material
which is deposited in the respiratory tract. Although much information has been
accumulated on the effect of particle characteristics and breathing patterns on
overall deposition, research in this field continues and discrepancies in the published results of experiments with human subjects [19] indicate that some may
be inaccurate for one reason or another. The second and perhaps more important question concerns the sites within the respiratory tract at which deposition
is most likely to occur. For example, the nasal passages, in addition to their
function of warming and humidifying the inspired air, act as a filter for removing
the larger particles of inhaled dust. As a consequence of this, some occupational
diseases (such as the nasal cancer of workers in the furniture and boot and shoe
manufacturing industries) are attributable to the deposition of relatively coarse
dust in the nasal passages. Of the smaller particles which penetrate the nasal
filter, a fraction will deposit in the lower regions of the respiratory tract. Lung
cancer in uranium miners and coal workers' pneumoconiosis are due to the deposition of finer dust in the conducting airways or in the acini, the respiratory units
which together constitute the pulmonary compartment of the lung. Many pathogenic bacteria gain access into the body by inhalation. For example, the common
cold virus and tonsillar infection of a streptococcal sore throat are due to deposition of infected droplets in the upper airways. Deposition in the lower respiratory tract is the route of infection for tuberculosis, inhalation anthrax, and the
classical bacterial and virus pneumonias. It is also in this part of the respiratory
tract that the very fine sulfate nuclei of atmospheric pollution are preferentially
deposited, thus adversely affecting lung function of members of urban communities [14]. Clinical aspects of inhaled particles have been discussed by Muir [58].

   Insight into the processes involved in deposition have followed developments in the techniques of aerosol production and measurement. One of the
most important of these is the production of monodisperse aerosols of non-
hygroscopic particles which has added precision to investigations of the effect
of particle size on deposition. Radioactively labeled particles are also available
which permit the distribution and clearance of deposited particles to be studied
with the aid of external detectors. A century ago, Tyndall demonstrated that
light-scattering particles were removed from air on passage through the lungs.
More recently aerosol photometers, based on conventional or laser light sources
but employing the same principle, have been used to measure deposition by integrating the concentration of respired aerosols continuously at the mouth.
Improvements in experimental techniques have been matched by the development
of lung models in which information on the dimensions and configuration of the
human airways, together with details of the airflow on inspiration and expiration,
have been used to calculate deposition in different compartments of the respiratory tract. These models have reached the state where experimental observations
and theoretical prediction are in broad agreement, at least for particles of regular

shape. In practice, however, dusts tend to be of irregular shape and are frequently aggregated. An extreme case is provided by asbestos fibers which may have aspect ratios of two orders of magnitude or more and it is the prediction of the behavior of such particles that is perhaps in most need of attention at the present time.

In recent years a number of articles dealing with the deposition of inhaled particles have been published. One of these, namely the Report of the Task Group on Lung Dynamics for the International Commission on Radiological Protection [72] formalized experimental data on both deposition and clearance into a model which could be applied generally to calculations of maximum permissible concentrations of radionuclides in air. However, as this involved assessment of much of the published work on particle deposition, it is relevant to conventional as well as to radioactive dusts. Similarly the Technical Report of the Panel on Inhalation Risks from Radioactive Contaminants of the International Atomic Energy Agency [62] deals specifically with the hazards of inhaling radioactive aerosols, but most of the material is of general application. A third source, which approaches the problem more from the viewpoint of occupational hygiene, is the monograph *Pulmonary Deposition and Retention of Inhaled Aerosols* by Hatch and Gross [38].

II. Physical Characteristics of Airborne Particles

Deposition of inhaled particles in the respiratory tract is determined not only by the physical characteristics of the particles themselves, but also by the nature of the airflow in the various regions, which is discussed in the next section. The physical processes of importance include (a) gravitational settling, (b) inertial impaction, (c) diffusion, and (d) interception. Other factors which may be significant are the electrical state of charge, and, because in the moist atmosphere in the respiratory tract water vapor will condense on hygroscopic materials, chemical composition.

Airborne particles exhibit an extraordinary variety of shapes and sizes and a brief outline is included of means adopted to characterize the size distribution and aerodynamic behavior of the particles of dust clouds encountered in practice. Because the aerodynamic behavior of irregular particles is difficult to treat mathematically, it is convenient to limit the discussion to spherical particles settling so that the airflow round them is streamlined. For particles of unit density, this sets an upper limit of 50 μm to the diameter. Because larger particles settle out so rapidly, the possibility of their being inhaled is in any case remote, except perhaps close to their point of origin. Fuller treatments of the physical factors affecting the dynamics of airborne particles have been given by Green and Lane [34] and by Fuchs [30].

A. Gravitational Settling

Perhaps the most obvious characteristic of particles in still air is that they fall
steadily under the influence of gravity. While the weight of a particle acts in a
downward direction, its buoyancy and the resistance of the air act upward. When
the gravitational force equals the sum of the buoyancy and air resistance, the particle continues to fall at a constant velocity known as the terminal settling velocity (V_s). For particles between 1 and 50 μm V_s is given by Stokes's law,

$$V_s = (\rho - \rho_a)\frac{D^2 g}{18\eta}$$

where ρ is the density of the particle, D its diameter, g the acceleration due to
gravity, and ρ_a the density and η viscosity of air. For particles smaller than 1 μm
(a size commensurate with the mean free path of air molecules) which fall more
rapidly than predicted by Stokes's law, a slip factor, known as Cunningham's
correction, must be introduced. In Table 1 are given terminal settling velocities
at 20°C for spherical particles of unit density covering a range of diameters from
0.02 to 20 μm. Values calculated with and without the slip factor are included
to illustrate the range of particle sizes in which its effect becomes significant.
Particles larger than 50 μm generate turbulence when falling freely and have a
lower terminal velocity than predicted.

TABLE 1 Rates of Fall in Air of Spherical Particles of Unit Density
as Given by Stokes's Law and as Modified by the Cunningham Correction
Factor

Diameter of particle (μm)	Uncorrected terminal velocity (cm sec^{-1})	Corrected terminal velocity (cm sec^{-1})
0.02	1.2×10^{-6}	1.39×10^{-5}
0.1	3×10^{-5}	8.6×10^{-5}
0.2	1.2×10^{-4}	2.2×10^{-4}
1.0	3×10^{-3}	3.48×10^{-3}
2.0	1.2×10^{-2}	1.3×10^{-2}
10	3×10^{-1}	3.07×10^{-1}
20	1.2	1.2

B. Inertial Impaction

Particles carried in an airstream tend to impact upon obstructions. This is due to the inertial resistance of the particles to changes in the direction of airflow. Regions of the respiratory tract in which inertial impaction are important are, first, at the entrance to the nose and in the nasal passages where the flow velocity is high and changes direction rapidly, and second, at bifurcations in airways where the flow divides. The probability that a particle moving with an airstream of average velocity $\overline{V}$ in a tube of diameter d will impact at a bifurcation is related to a parameter Stk called the Stokes number

$$Stk = \frac{\rho D^2 \overline{V}}{9\eta d}$$

It will be noted that, as in gravitational settling, the significant parameters of the particle are ρ and D^2. The deposition of aerosols by impaction in models of human lung bifurcations has been discussed by Bell and Friedlander [9] and in bent tubes by Johnston and Muir [43].

C. Diffusion

As particles decrease in size below 1 μm, their motion relative to the air in which they are suspended is increasingly attributable to diffusion. Particles just visible in the optical microscope (0.2 μm) can be seen to be in continual motion due to bombardment with gas molecules. Such movement is termed Brownian motion. For particles of 0.5 μm diameter, the root-mean-square displacement in a second due to Brownian motion is approximately equal to the terminal settling velocity. For particles of smaller sizes therefore, diffusion rather than sedimentation becomes the dominant influence in deposition.

D. Interception

This effect becomes of importance when the size of the particle becomes a significant fraction of the diameter of the airway through which it is passing. As the diameter of the terminal bronchiole is in the region of 0.7 mm interception is unlikely to be important in unconstricted airways for particles of regular shape, because those large enough to show significant interception effects will have already been deposited either by impaction or sedimentation. For particles of extreme shape, however, such as fibers with high aspect ratios, interception in the smaller airways, particularly at bifurcations, can be important. The significance of interception in the deposition of asbestos fibers has been discussed by Timbrell [75].

E. Hygroscopicity

Airborne particles will adsorb moisture to an extent that depends on their affinity for water and on the relative humidity. If the particle is insoluble, or if the relative humidity is below a critical value for soluble particles, adsorbed moisture will not significantly affect the particle size. Above the critical value, however, hygroscopic particles (e.g., tobacco smoke) will grow to reach an equilibrium diameter. On the other hand, petrol engine exhaust particulates, which consist of chainlike aggregates, condense to form smaller regular-shaped particles under conditions of high humidity.

Information on the rate of evaporation and growth of water and saline droplets has been presented by Fuchs [29] and El Golli et al. [25]. At 20°C and 80% relative humidity, a 5-μm water droplet will evaporate completely in about 0.1 sec. Unless the surrounding atmosphere is saturated with water vapor still smaller droplets will evaporate almost instantaneously. The rate of change in the size of droplets is an important practical consideration in the humidification of the respiratory tract by mist therapy as in the treatment of patients with mucoviscidosis.

F. Electric Charge

The possibility that electric charges on inhaled aerosol particles may affect deposition has long been recognized, but quantitative experimental evidence, especially for humans, is sparse. Longley [55] in experiments with rats found deposition in the pulmonary compartment was diminished when the animals were charged and he pointed out that the moist respiratory tract will act as a Faraday cage. Once a particle has entered the respiratory tract, therefore, the only electrical force of importance for deposition is the attraction between the charge on the particle and its image in the collecting surface. Expressions for an image force parameter have been described by Ranz and Wong [66].

The possible effects of charge on deposition in the nose of polystyrene particles of 1 and 4 μm diameter have been investigated by Fry [27]. She found that passing the particles through a krypton-85 deionizer, which effectively equilibrated the charges produced by the method of dispersion, had no significant effect on deposition. This result is in accordance with the calculated effect of mirror-image charges.

G. Aerodynamic Diameters and Shape Factors

For spherical particles of density other than unity, the terminal velocity can be calculated if the diameter is expressed in terms of the diameter of an equivalent unity density sphere (D_e)

$$D_e = \sqrt{\rho D^2}$$

where ρ is the density of the particle and D its actual diameter. In practice airborne particles are rarely spherical, but their aerodynamic behavior can still be expressed in terms of their aerodynamic diameter (D_a). This is the diameter of a sphere of unit density having the same terminal settling velocity and when dealing with deposition this parameter is generally used. Terminal velocities of small particles may be measured with, for example, a Schaefer cell [34] or an aerosol spectrometer such as that of Timbrell [76].

In many cases it is required to predict the aerodynamic behavior of irregular or aggregated particles from measurements of their size with a microscope. To obtain this information, aerosol spectrometers have been used by a number of workers to compare the aerodynamic behavior of typical irregular particles with that of spherical particles of known density. In such work it is common to use the projected area diameter (D_p) of a particle which is the diameter of a circle having the same area as the projected area of the particle. Several shape factors relating D_p to the physical properties of the particle such as mass, surface, and settling velocity have been defined by Cartwright [12]. Reports of measurements by Stöber [70], Timbrell [76], and Kotrappa [44] of the aerodynamic diameters and dimensions of spherical, irregular, aggregated, and fibrous particles have recently been published together in the proceedings of an International Conference on Assessment of Airborne Particles.

In sampling airborne dust clouds to assess inhalation hazards, there is an increasing tendency to measure the mass concentration of aerodynamically sized fractions of the airborne material. The adoption of gravimetric methods avoids the measurement of individual particles under the microscope and the need to derive appropriate shape factors. In the coal mining industries of the United States and the United Kingdom, which give rise to many cases of pneumoconiosis, samplers are used which collect only the fraction of airborne dust which is considered to be respirable (i.e., that which is likely to penetrate to and be deposited in the pulmonary compartment). The selection curve for the respirable fraction of dust recommended by the Dust Panel of the British Medical Research Council [37] and by the 1959 Johannesburg Pneumoconiosis Conference [60] accepts no particles larger than 7.1 μm, 50% of 5 μm, and 100% of very small sizes, all dimensions being in terms of unit density spheres. A second selection curve has been proposed by the U.S. Atomic Energy Commission [54] which accepts no particles greater than 10 μm, 25% of 5 μm, and 100% of 2 μm and smaller particles. The American Conference of Governmental Industrial Hygienists [6] adopts 90% for the acceptance of particles less than 2 μm, the other points remaining unchanged.

H. Parameters of Particle Size Distribution

The aerodynamic diameters of particles in the atmosphere cover a range of five to six orders of magnitude. Particles from a single source, such as an industrial process involving grinding, crushing, or cutting, will have sizes within a range covering two or three decades. Only with special care can monodisperse aerosols (those having particles differing in size by less than a factor of 2) be produced.

There are several ways of defining an aerosol. The size distribution may be measured with the optical or electron microscope or light-scattering spectrometer. Alternatively, the mass distribution may be determined with a gravimetric sampler. While the frequency distributions obtained can be displayed graphically, it is convenient if they can also be expressed mathematically. Particles of monodisperse aerosols conform to normal or Gaussian distributions and can be defined in terms of their mean size and standard deviation. However, many aerosols encountered in practical situations are polydisperse and can be specified by the "log normal" distribution in which frequency is related to the logarithm of particle diameter. The characteristics of the log-normal distribution have been outlined recently by Walton [78]. A distribution is specified by its geometric mean obtained by count, or count median diameter (CMD) and the geometric standard deviation. The geometric mean diameter obtained by weight is also referred to as the mass median diameter (MMD). Once the median diameter is obtained, either by count or by weight, it can then be expressed as equivalent aerodynamic diameter and this gives rise to the terms count median aerodynamic diameter (CMAD) and mass median aerodynamic diameter (MMAD). In defining radioactive aerosols, the term activity median aerodynamic diameter (AMAD) is also used. If the radioactivity is distributed homogeneously in the dust sampled, which is not always the case, the AMAD is identical with the MMAD.

III. The Respiratory Tract

Although this is not the place to present an authoritative description of the respiratory tract, some information on its anatomy is useful in considering regional deposition. Evidence relating to the deposition and clearance of particles summarized by, for example, the Task Group on Lung Dynamics [72], leads to the conclusion that the respiratory tract can be subdivided conveniently into three compartments. These are defined and described briefly. In addition, formalized lung models, used in calculating regional and total deposition are outlined. Finally, a brief account is given of the changing characteristics of the airflow as inhaled air penetrates along the conducting into the terminal airways.

A. The Nasopharyngeal (N-P) Compartment

This extends from the nostrils through the nasal passages, naso- and hypopharynx to the larynx. The nasal passages have a relatively complex structure. After the nostril, the airway cross section diminishes in the vestibule to the anterior narrow constriction which has an oval cross section of only about 0.3 cm^2 (total 0.6 cm^2) in the adult. This constriction is close to the mucocutaneous junction (see Chap. 3). Posterior to this constriction the height and width increase to form the main turbinated passage in which there is only a narrow channel between the turbinates and the septum for the passage of air. The width of this channel is rarely more than 2 mm. Generally the septum is asymmetrical so that the flow through one side of the nose is greater than through the other. The length of the nasal passages from nostril to nasopharynx ranges from 8 to 11 cm in adults and the total surface area of the mucosa from 160 to 180 cm^2 [65]. The nasal passages are lined, except at their anterior end, with a very vascular mucous membrane characterized by ciliated columnar epithelium and scattered mucous glands. The incoming air is warmed and humidified during its passage through the nose.

B. The Tracheobronchial (T-B) Compartment

This consists of the tracheobronchial tree down to and including the terminal bronchioles. The trachea divides at its lower end into two main bronchi which conduct air to the left and right lungs. These subdivide, initially in a dichotomous manner, into airways of decreasing length and diameter of which there are more than 20 generations. By this means, inspired air is distributed to the functional regions of the lung, where gaseous exchange occurs. The trachea is an elastic tube held open by cartilaginous rings which are also present in the bronchi. The ultimate airways of this compartment, the terminal bronchioles, are about 0.7 mm in diameter. The walls of these airways carry ciliated columnar cells which steadily propel a layer of mucinous fluid toward the esophagus.

C. The Pulmonary (P) Compartment

From the terminal bronchioles proceed in succession respiratory bronchioles, alveolar ducts, atria, alveolar sacs, and alveoli. The latter occur as evaginations on the walls of the alveolar sacs, but also less frequently on the walls of the alveolar ducts and respiratory bronchioles. The diameter of an alveolus is typically about 0.15 mm. These structures together comprise the pulmonary compartment which can be regarded as the functional area of the lung, in so far as gaseous exchange is concerned. Its surface consists of nonciliated epithelium with different secretory elements to those present in the tracheobronchial compartment.

D. Formalized Lung Models

Calculations of deposition of inhaled particles in various regions of the respiratory tract require accurate knowledge of the dimensions and morphology of the airways. To simplify calculations it is generally assumed that all airways of the same order have equal lengths and diameters, branch at the same angle and that the airflow through each is the same. Findeisen [26] proposed a model, starting with the trachea, and having nine orders of branching tubes each with increasing numbers of airways of decreasing diameters. Landahl [45] added the mouth and pharynx to this model and increased the number of alveolar sacs. Later [47] he returned to the original number of alveoli but adjusted their volume. Recent advances in morphometry and electron microscopy were used by Weibel [79] to formulate a more accurate model and this has been used extensively in calculations of deposition by, for example, Beeckmans [8]. Davies [18] has recently elaborated on his model of 1961 to give information on the characteristics of airflow in the various subdivisions in addition to physical dimensions. Some of his data are given in Table 2. As he later made slight modifications to his first model, data from the two sources are not necessarily consistent. He also gives information on the distribution of the volume expansion of the lungs for a tidal volume of 600 cm^3.

E. Breathing Patterns and Lung Volumes

A variety of breathing patterns have been used in the course of experiments and calculations of deposition. The Task Group on Lung Dynamics [72] adopted a respiratory cycle of 4-sec duration having an inspiratory portion of 1.74 sec and expiratory portion of 2.06 sec. A pause of 0.2 sec between inspiration and expiration completed the cycle. Tidal volumes of 750, 1450, and 2150 cm^3 (RTPS) were used, the first being associated with a mild to moderate activity state. The higher values were considered appropriate (a) for a relatively sustained work state and (b) a state of greater activity not capable of being well sustained. The data in Table 2 show that the total volume of the lung and airways at resting expiratory level is close to 3 liters. Because of the nature of the airflow it is not possible neatly to allocate the residual (end-reserve) and reserve (end-tidal) volumes to

specific regions of the lungs. At end-expiratory level, reserve air can be found in
the dead space and in the alveolated airways. However, despite their changing
configuration and distribution during breathing, these volumes can be evaluated
and are given by Davies [18] as

Residual volume	1960 cm^3
Reserve volume	1000 cm^3
Functional residual capacity	2960 cm^3

F. Airflow in the Respiratory Tract

In addition to the physical parameters of particle size and density, the other major
factor which affects deposition is the nature of the airflow in the airways. The
purpose of respiration is to effect the transfer of gaseous material from the atmosphere to the functional area of the lung. This is ultimately dependent upon
transfer of gas molecules by diffusion in the terminal airways, but fresh air is provided and spent air is removed by the bulk movement of air during breathing. It
is the bulk movement of air along the upper, conducting, and alveolated airways
that constitutes "air flow." The transport of particles (except for the very smallest) in and out of the lungs is determined by airflow and is not influenced to any
significant extent by the diffusion of air molecules.

Turbulence is an important consideration in the flow of air in tubes. As
the flow increases in a tube with given dimensions, a critical velocity will be
reached at which flow becomes unstable, and turbulence, characterized by large
and irregular velocity fluctuations, will appear. The type of flow can be determined by the use of the dimensionless parameter called the Reynolds number

$$Re = \frac{\rho d\overline{V}}{\eta}$$

where ρ is the density and η the viscosity of the fluid, d is the diameter of and
$\overline{V}$ the flow velocity in the tube. When the Reynolds number is below a critical
value, the flow is normally laminar. When it exceeds this value, the flow becomes
unstable and turbulence will develop. In long cylindrical tubes the critical value
is approximately 2000, and when it exceeds 3000 the flow is usually turbulent.

Some of the characteristics of airflow in the upper airways have been described by Proctor and Swift [65] who drew attention to the fact that, because
of the small cross section, air velocity into the nasal passages through the anterior
narrow constriction must be about 12 times that in any other region of the respiratory tract. This will result in high inertial deposition on the septum and on the

TABLE 2 Formalized Anatomy of the Human Respiratory Tract

Dimensions of airways and characteristics of airflow when breathing flow rate is 20 liters min^{-1}
Columns 1 to 4 from **Ref.** 16; columns 5 to 7 from **Ref.** 18

Region	-1 Number	2 Diameter of lumen (cm)	3 Axial length (cm)	4 Total cross-section of airways (cm^2)	5 Cumulative volume (cm^3)	6 Mean velocity (cm/sec)	7 Reynolds' no. (Re)	Subdivision of lung	Compart-ment
Mouth	1	2	7	3	22	50-100			
Pharynx	1	3	3	7	43				
Trachea	1	1.7	12	2.3	70	145	1670		Tracheobronchial
Main bronchi L, R	2	1.2,1.4	5,2.5	2.67	79.5	125	1100	dead space	
Lobar bronchi	5	0.8	2-4	2.5	86.4	139	740		
Segmental bronchi	18	0.5	6	3.6	103	92	314		

<table>
<tr>
<td>Intrasegmental bronchi</td><td>252</td><td>0.3</td><td>2.5</td><td>17.9</td><td>122</td><td>37</td><td>75</td><td></td><td></td><td rowspan="9">segment</td><td rowspan="4">Tracheobronchial</td>
</tr>
<tr>
<td>Bronchioles</td><td>504</td><td>0.2</td><td>2.0</td><td>16.1</td><td>147</td><td>21</td><td>28</td><td></td><td></td>
</tr>
<tr>
<td>Secondary bronchioles</td><td>3,024</td><td>0.1</td><td>1.5</td><td>24</td><td>177</td><td>14</td><td>9.4</td><td></td><td></td>
</tr>
<tr>
<td>Terminal bronchioles</td><td>12,096</td><td>0.07</td><td>0.5</td><td>47</td><td>199</td><td>7</td><td>3.4</td><td></td><td rowspan="6">secondary lobule</td>
</tr>
<tr>
<td>Respiratory bronchioles</td><td>169,400</td><td>0.05</td><td>0.2</td><td>339</td><td>267</td><td>1</td><td>0.33</td><td rowspan="5">acinus</td><td rowspan="5">Pulmonary</td>
</tr>
<tr>
<td>Alveolar ducts</td><td>847,000</td><td>0.08</td><td>0.125</td><td>4,240</td><td>801</td><td>0.079</td><td>0.042</td>
</tr>
<tr>
<td>Atria</td><td>4,240,000</td><td colspan="2">sphere 0.06 cm diam</td><td>11,900</td><td>1,280</td><td>0.028</td><td>0.01</td>
</tr>
<tr>
<td>Alveolar sacs</td><td>21,200,000</td><td>0.03</td><td>0.05</td><td>14,800</td><td>2,021</td><td>0.222</td><td>0.004</td>
</tr>
<tr>
<td>Alveoli</td><td>530,000,000</td><td colspan="2">sphere 0.015 cm diam</td><td>95,400</td><td>2,960</td><td>0.003</td><td>0.002</td>
</tr>
</table>

anterior ends of the turbinates. Beyond this point the airstream divides into three paths, the main one being directed along the space between the middle meatus and the septum. Ancillary streams pass upward and above the middle turbinate and downward along the floor of the nose. The flow in all these streams is turbulent.

Information on the linear velocities of air flowing through the various conducting airways and the corresponding Reynolds numbers are given in Table 2 from which it can be seen how rapidly the velocity of the inspired air falls as it penetrates the respiratory tract. Incipient turbulence characterizes flow in the larger airways. Changes in direction at bifurcations lead to fluid inertia effects persisting down to a Reynolds number of 1.6 (i.e., above the respiratory bronchioles) but beyond, in the acini, the flow is entirely viscous.

Air entering a tube advances most rapidly along the axis, while the velocity at the walls is zero. Thus the inhaled tidal air advances furthest along the axes of the airways surrounded by an inner sheath of reserve air and an outer sheath of residual air. In the tracheobronchial region, because the flow is not completely streamlined, radial mixing occurs, and there is a nearly uniform particle concentration across each airway. In the alveolated airways, however, where the flow is entirely viscous, no radial mixing occurs and the airborne particles of the tidal, reserve, and residual air are disposed in coaxial envelopes which, at the end of inhalation, have lengths greater than their diameters. In this region, on exhalation, the flow reverses, but higher up in the conducting airways mixing again occurs. From reasoning along these lines, it would appear that the tidal, reserve, and residual air will mix by turbulence, while tidal air is in transit through the conducting airways in the so called "dead space" above the terminal bronchioles. In the alveolated airways, however, mixing will be negligible as far as the inhaled particles are concerned, although gaseous mixing by molecular diffusion will proceed normally.

IV. Total Deposition in the Respiratory Tract

Measurements of total deposition, although relatively simple in principle, require careful attention to experimental detail to obtain reliable results. Measurements of particle concentrations and size are not necessarily straightforward. Breathing cycle parameters require thoughtful selection, control, and measurement and should not be dictated by the limitations of the apparatus. It should also be borne in mind that results obtained with resting are not necessarily applicable to working subjects.

A. Deposition of Particles Less than 0.1 μm Aerodynamic Diameter

No experiments with human subjects appear to have been performed to measure the deposition of freshly formed daughter products of radon-222 and thoron (radon-220) before they become attached to condensation nuclei. These are formed in air as positively charged ions because orbital electrons are stripped from the recoil atom, either by the departing α particle, or by recoil motion. They have diameters of less than 0.001 μm. However, because of the high diffusivity of particles in this size range, it can be assumed that deposition is complete.

The deposition of condensation nuclei was investigated by George and Breslin [31] using particles tagged with radioactive daughter products of radon-222 to facilitate measurement of their concentration in inhaled and exhaled air. Additional measurements were carried out in uranium mines in which high concentrations of radon daughters occurred naturally. Tidal volumes and respiratory rates were measured and the aerosol characterized using a diffusion battery to determine effective diffusion diameters which ranged from 0.005 to 0.08 μm depending upon location in the mine. As would be expected, for particles whose behavior is determined largely by diffusion, total deposition increased from 25 to 50% as the particle size decreased from 0.08 to 0.005 μm. However, as the particle size may have increased in the respiratory tract at the high humidities encountered, the results may be of only qualitative significance. Deposition was found to increase with tidal volume, but was not noticably affected by changes in the rate of breathing. Shanty [68] measured the deposition of 0.08 μm monodisperse particles which correspond to the larger end of the size spectrum of condensation nuclei. He obtained average values of 9 and 16% for mouth and nose breathing subjects, respectively. As the behavior of inhaled nuclei is of importance there is a need for experiments to extend the range downward to monodisperse particles of atomic size.

B. Deposition of Particles Greater Than 0.1 μm Aerodynamic Diameter

Among the experiments carried out for particles in this size range may be mentioned those of Van Wijk and Patterson [77]; Dennis and Sawyer [24]; Brown et al. [11]. To illustrate the relation between deposition and the size of inhaled particles, the collected results of these experiments were presented together by Davies [17] in the form shown in Fig. 1. There is considerable scatter in the results, some of which must be due to factors such as the shape and density of the particles used, differences in tidal and reserve lung volumes and of breathing frequency, nose or mouth breathing, and subject to subject variations. Davies concluded that these considerations could not account for the extreme range of values and that the experimental technique of some workers was faulty. The scatter in the results is greatest for particles below 1 μm in size, but recent work [20,39,59] has shown that for particles in the range 0.2 to 1 μm, deposition in resting, mouth-breathing subjects is fairly constant at 10 to 15%. Results for a similar range of particle sizes (0.2-1.8 μm) have been reported by Giacomelli-

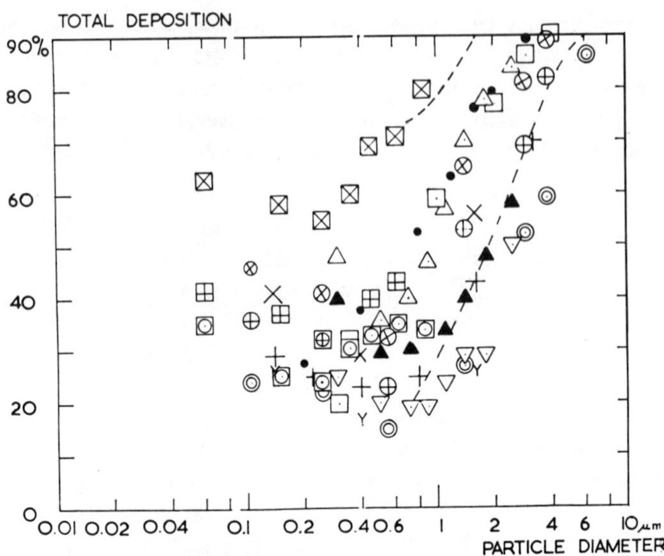

FIGURE 1 Experimental results for the deposition of particles during breathing. Results of a number of authors collected by Davies [17]. (By permission of Pergamon Press Ltd.)

Maltoni et al. [32] who used monodisperse particles of wax. Their values for total deposition showed a minimum of about 20% at 0.5 μm. Above this critical size, deposition increased with tidal volume, but below it was independent. Thus, while there still appears to be some discrepancy in measurements of total deposition at 1 μm, it is apparent that above this size, deposition must increase quite rapidly to become effectively quantitative at 5 or 6 μm.

C. The Effect of Changes in Breathing Pattern on Total Deposition

Breathing patterns may be changed in various ways. The rate of breathing and the tidal volume are the most obvious, but changes in the magnitude of the reserve volume, or in the duration of the pause between inspiration and expiration, are other factors which may affect deposition. In studying the effects of changes in the pattern of respiration on the deposition of inhaled particles, unless all the parameters listed above are controlled, results will not necessarily be reproducible. Dennis [21], using monodisperse particles of stearic acid from 1 to 5 μm in diameter, measured changes in deposition with breathing rate. For sedentary subjects he found minimum values in the region of 15 to 20 breaths/min. When a subject is not performing work, there is a tendency for the minute volume to be constant, so that as the breathing rate increases the tidal volume becomes smaller. During work, however, both the breathing rate and minute volume usually increase together. To investigate this point Dennis [22] performed another series of experiments and found that, under working conditions, the increase in minute volume was generally associated with an increase in percentage deposition. In addition to experiments with monodisperse particles, Dennis also measured deposition in both nose and mouth breathing subjects, using a polydisperse aerosol of spherical particles of calcium carbonate. The inspired and expired air was sampled with conifuges which enabled the relation between deposition and size from 0.75 to 6 μm to be determined in one and the same experiment. The results he obtained are illustrated in Fig. 2 which shows that deposition in nose breathing is generally somewhat greater than in mouth-breathing subjects.

By controlling both the rate of breathing and the tidal volume of resting subjects, Muir and Davies [59] showed that when the breathing rate increased in the range 10-30 breaths/min, percentage deposition of 0.5 μm particles decreased

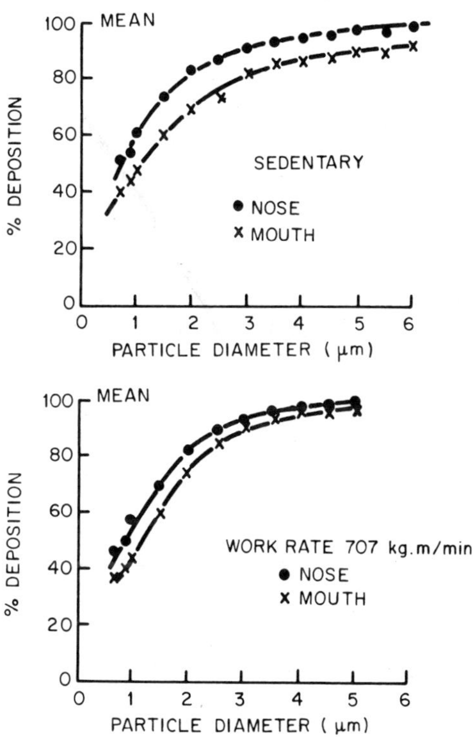

FIGURE 2 Variation in total deposition of calcium carbonate particles with particle size for sedentary and working subjects [22]. (By permission of Unwin Brothers Ltd.)

if the tidal volume was kept constant. Conversely, if the tidal volume was increased at a constant breathing rate, deposition increased. In a further series of experiments with 0.5 μm particles Davies et al. [20] determined the effect of altering the magnitude of the expiratory reserve volume above and below its normal resting value. For three male subjects, breathing at a rate of 16 breaths/min with tidal volumes of 600 cm^3, deposition decreased from 15 to 5% as the expiratory reserve volume was changed from 1200 cm^3 below its normal value to 1200 cm^3 above. This effect is associated with the decrease in the area of the interface between the tidal and reserve air of expanded lungs. An additional factor is the increase in the thickness of the sheaths of residual and reserve air around the tidal air in the lung acini.

Now that information is accumulating on the influence of changes in the breathing pattern, including the degree of inflation of the lungs, on deposition, additional results are required for both sedentary and working populations so that forecasts of deposition can be made for any practical situation.

D. Calculations of Total Deposition

From a knowledge of the aerodynamic behavior of particles and of the morphology of airways, calculations of deposition can be made for specific breathing patterns. Because of the intricate nature of the respiratory passages, uncertainties regarding the nature of the airflow and neglect of factors such as the residual volume, it is prudent to treat the results of such calculations with some reserve. Nevertheless, calculations have yielded results which are in reasonable agreement with experimental observations and have made their own contribution to our understanding of the processes involved in deposition and the distribution of deposited particles.

As a basis for his calculations, Findeisen [26] used the formalized model of the respiratory tract referred to above. Adopting typical values for tidal volume and rate of breathing, he calculated deposition at each junction and in each region for a range of particle diameters. In general, the results obtained were in line with experimentally determined values. Deposition passed through a minimum for particles between 0.1 and 0.3 μm and increased to 100% for particles of 3 μm and larger. Landahl [47] made similar calculations using his model of the respiratory tract already referred to. Lung volumes were modified in conformity with later information. He made calculations for a number of breathing patterns and found that deposition was greater for larger tidal volumes and lower breathing rates. His results also gave minimum values for the total deposition of particles in the range 0.2-0.6 μm. Further calculations along similar lines, but using the concept of intrapulmonary mixing of tidal, reserve, and residual air developed by Altshuler and his coworkers [3,4] were made by Beeckmans [7].

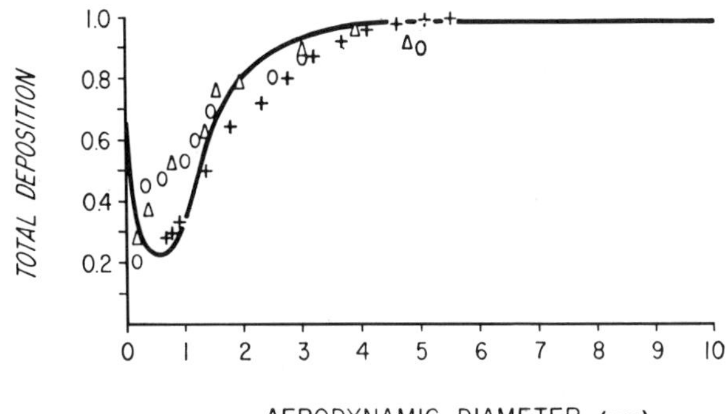

FIGURE 3 Variation of the total deposition of inhaled particles with aerodynamic diameter for nasal breathing. (———) calculated deposition. 15 respirations min^{-1}, tidal volume 750 cm^3. (+) Dennis, 13.3 respirations min^{-1}, tidal volume 720 cm^3. (O) Brown et al., 15 respirations min^{-1}, tidal volume 700 cm^3. (△) Van Wijk and Patterson, 19 respirations min^{-1}, tidal volume 900 cm^3. (From Ref. 72; by permission of Pergamon Press Ltd.)

Later he [8] recalculated deposition using the more elaborate lung model of Weibel, but the curves obtained did not appear to fit the experimental data any better. By this time, computers were available which enabled depositions for a wide range of particle sizes and breathing patterns to be calculated relatively easily.

Further values, based on Findeisen's method, but using Weibel's model of the respiratory tract were computed by the Task Group on Lung Dynamics [72]. Inhalation through the nose was assumed, and allowance made for deposition in this region using an empirical relationship based on Pattle's [63] experimental observations. The results of these calculations will be discussed later, but the variation of total deposition for unit density particles with aerodynamic diameter is shown as the continuous line in Fig. 3. This indicates that in nose breathing subjects, deposition has a minimum value of about 20% at an aerodynamic diameter of 0.5 μm and is essentially complete for particles greater than 5 μm.

Calculations of total deposition have generally shown a continuous variation with aerodynamic diameter for particles less than 1 μm, but results published more recently [39] indicate that for a specific breathing pattern, deposition is constant in the range 0.2-1 μm. In addition, deposition is only about 10% which is lower than for any previously published values. Although it would not be difficult to ensure that calculated deposition agreed with this value for particles of a particular size (say 0.5 μm), the fact that deposition is essentially constant over the range 0.2 to 1 μm is not readily reconciled with methods of calculation used to date. Davies [18] suggests that the deposition of particles in this size range is not primarily dependent upon a combination of the processes of impaction, sedimentation, and diffusion, but to a single process, namely the mechanical mixing of tidal, reserve, and residual air. While deposition in the alveolated region takes place by sedimentation and diffusion from the residual air, the rate-determining process is the transfer of particles from the tidal to the reserve and residual air by turbulent mixing in the larger conducting airways or by incompletely developed turbulence and fluid inertia effects in the smaller ones.

V. Regional Deposition in the Respiratory Tract

The risk to health from the inhalation of toxic or infectious particles is not necessarily in proportion to their total deposition within the respiratory tract. Thus, for a full understanding of the consequences of particle deposition, it is necessary to have information on the sites at which particles of different characteristics are deposited and on the efficiency of deposition at these sites. Particles deposited in the conducting airways are removed quite rapidly by mucociliary clearance unless the ciliated epithelium is damaged for example by exposure to toxic or irritant materials. These particles are generally of less significance therefore than the respirable fraction of the inhaled dust, which is deposited in the pulmonary compartment, and which is retained for much longer periods of time.

A. Deposition in the Nasopharyngeal Compartment

In addition to its air conditioning role, the nose acts as a filter in which most of the larger (>5 μm) and very small (<0.01 μm) particles are removed from the inhaled airstream by inertial impaction and diffusion, respectively. Although this has the effect of reducing the amount of dust reaching the lower regions of the respiratory tract, some diseases of occupational origin result from deposition of specific materials in the nose. For example, ulceration of the nasal septum in chrome workers has been attributed to the deposition of droplets of chromate solution at this site. Evidence has also been presented that nasal cancer in woodworkers in the furniture industry [1] and in operatives in the boot and shoe manufacturing industry [2] is due to deposition of dusts specific to these industries on the septum and middle turbinates.

From the experimental point of view, the nose has the advantage that it can be studied in isolation from the remainder of the respiratory tract. A number of workers have measured deposition by either blowing or sucking aerosols through the nose and mouth of breath-holding subjects. Using this technique Lehman [51] found nasal deposition to be greater in normal than in silicotic subjects. Landahl and Black [48] and Landahl and Tracewell [49] studied the deposition of a variety of materials in the nose, including deposition on the nasal hairs, over a range of flow rates. By separating the particles into sized fractions with an impactor, they demonstrated that aerodynamic diameter (D_a) was important in determining deposition and also that the flow rate through the nose and hygroscopicity of the particles were significant factors. When they plotted the fraction of particles of a given size penetrating the nose against the logarithm of the particle size, they obtained a sigmoid curve which showed that particles greater than 20 μm in diameter were completely removed by the nose and that there was little deposition of 1-μm particles. Dennis and Sawyer [23] carried out similar measurements with particles of lipiodol, stearic acid, and gum arabic, sizing the administered and exhaled aerosol microscopically. They obtained consistent results when penetration was plotted against the logarithm of D_a^2F, where F is the flow rate through the nose. The resulting correlation was still sigmoid in shape and indicated that, for an airflow of 28 liters min^{-1}, deposition was complete for particles exceeding about 6 μm in aerodynamic diameter. Using monodisperse particles of methylene blue ($>$1 μm) produced with a spinning disc aerosol generator, Pattle [63] carried out similar experiments at flow rates of 10-30 liters min^{-1} and found a linear relationship between penetration and the logarithm of D_a^2F. The results of Pattle were used by the Task Group on Lung Dynamics to evaluate deposition in the nasopharyngeal compartment, which they expressed in terms of the empirical equation

$$N = -0.62 + 0.475 \log D_a^2F$$

where N is the fractional deposition and F the flow rate in liters min^{-1}. A further series of experiments, using monodisperse polystyrene particles of various sizes, labeled with bromine-82, was reported by Hounam et al. [41]. They found that, at a given value of D_a^2F, appreciable variations in deposition could be obtained, but the variations were smaller if the difference in pressure across the nose and mouth was used instead of flow rate. This pressure difference is essentially a measure of nasal flow resistance and, for a given flow rate, is known to vary widely between subjects and even in the same subject at different times. Flow resistance is very dependent upon the cross section of the nasal passages which is smallest at the anterior constriction. At this location therefore, the highest air velocities will be encountered and in consequence the major contribution will be made to the total nasal flow resistance. During inhalation the static pressure in the nasal passages is less than ambient resulting in a constricting effect so that the cross section will be somewhat variable.

The nasal deposition of submicrometer particles has been investigated by Hounam [40] using natural condensation nuclei which range in size from 0.001 to 0.1 μm. At a flow rate of 35 liters min^{-1}, deposition was only about 5% but this increased to 20% at 5 liters min^{-1}. George and Breslin [31] using unattached radon daughter products, obtained a value of about 60% for deposition in the nose, a value rather similar to that reported by Black and Hounam [10] for iodine vapor. It appears therefore, that particles of atomic size and molecules of reactive vapors are removed with similar efficiency on passage through the nose.

Landahl [46] published details of a model of the nasal passages which consists, in effect, of a number of inertial separators with different efficiencies arranged in series and parallel. Predictions based on this model are in reasonable agreement with experimental observations. The model also predicts that most deposition occurs in the anterior regions. Hadfield and Macbeth [36] have reported that deposits of wood dust may be observed on the septum and particularly on the anterior ends of the middle turbinates in the noses of woodworkers. Fry and Black [28] studied distribution in the anterior and posterior regions of the nasal passages, following administration of radioactive particles by normal breathing and by forced aspiration. The pattern of deposition was monitored with fixed collimated detectors. They found that on average, over 80% of the deposition occurred in the anterior region and distribution did not appear to be

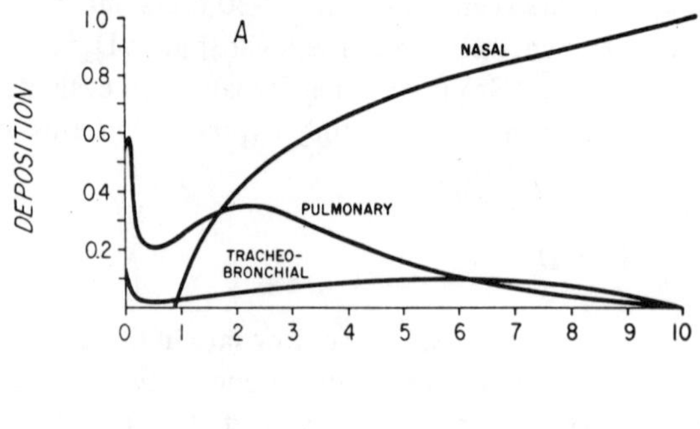

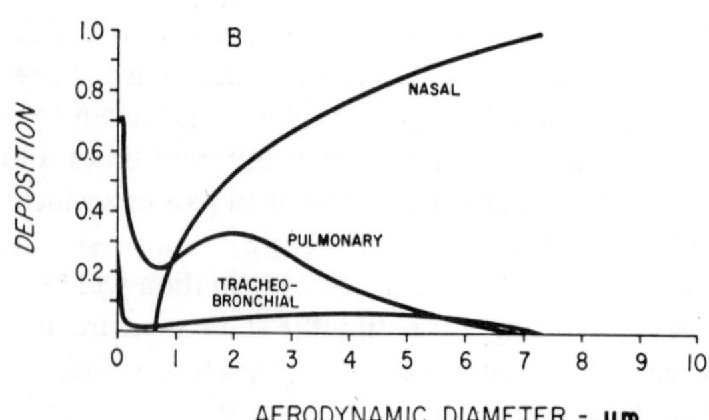

FIGURE 4 Deposition as a function of particle size for 15 respirations min^{-1}. (A) 750 cm^3 tidal volume. (B) 1450 cm^3 tidal volume. (From Ref. 72; by permission of Pergamon Press Ltd.)

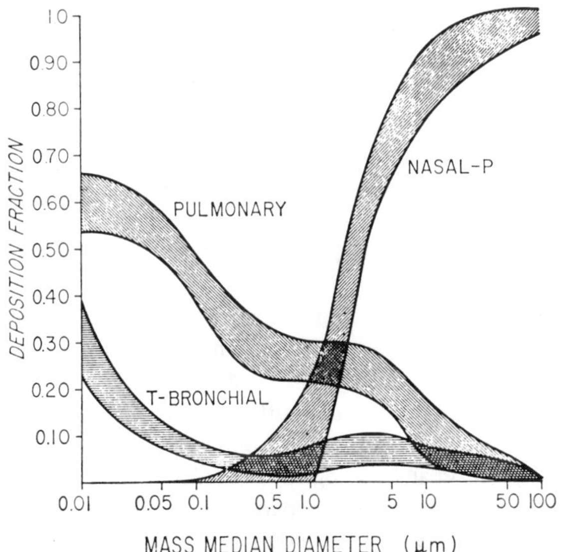

FIGURE 5   Each of the shaded areas (envelopes) indicates the variability of deposition for a given mass median (aerodynamic) diameter in each compartment when the distribution parameter σ_g varies from 1.2 to 4.5 and the tidal volume is 1450 ml. (From Ref. 72; by permission of Pergamon Press Ltd.)

affected either by the size of the particles (2.5-10 μm) or by the method of administration. Lippmann [52] also used radioactive tracer methods to study regional deposition and found that in two out of three subjects deposition was greater in the unciliated anterior region than in the posterior ciliated, but for the other, the situation was reversed.

The relation between deposition in the N-P compartment and particle size proposed by the Task Group is illustrated in Figs. 4 and 5, the latter figure being applicable to some of the heterogeneous aerosols encountered in practice. It will be seen that the nose is a very effective filter for particles of aerodynamic diameter exceeding 8 μm, but the efficiency of deposition drops rapidly and is negligible for particles smaller than $\sim$1 μm diameter. Deposition will increase again for very small particles, less than 0.01 μm diameter, whose behavior is increasingly affected by diffusion as their size diminishes.

The clearance of particles deposited in the nose depends critically on the site of deposition, being rapid from the ciliated surfaces and slow from the anterior unciliated areas. Additional evidence is required on the deposition pattern in the nose for particles greater than 1 μm diam and on the deposition and distribution of very small ($<$0.01 μm diam) homogeneous inhaled particles.

B. Deposition in the Mouth

Although mouth breathing is not uncommon, deposition of inhaled particles in the mouth seems to have attracted relatively little attention. However, it has been assumed intuitively that total deposition will be greater in nose than in mouth-breathing subjects for particles exceeding about 1 μm in size. In almost all experiments using mouth-breathing subjects, the aerosol has been inhaled through a tube held in the mouth, although Dennis [21] showed that this results in lower deposition than would be observed if the subject breathed through the mouth in a normal manner. In more recent investigations, Dennis [22] concluded that deposition during mouth breathing is usually, but not invariably, lower than when breathing through the nose (see Fig. 2). Using values for total deposition obtained during normal mouth breathing and by breathing through a tube, the Task Group on Lung Dynamics derived an empirical relationship between mouth deposition and the logarithm of $D_a{}^2 F$, similar to that describing deposition in the nasopharyngeal compartment. Using radioactive particles, Lippmann et al. [53] measured deposition in the heads of mouth-breathing subjects and concluded that the increase in deposition with log $D_a{}^2 F$ was more rapid than indicated by the calculations of the Task Group. Calculations of deposition in the mouth have been made by Landahl [46] and Beeckmans [7] who simulated the mouth by a tube 7 cm long and 2 cm in diameter. However, as the principal object of these papers was the evaluation of total and pulmonary deposition, losses in the mouth were not reported specifically.

C. Deposition in the Tracheobronchial and Pulmonary Compartments

Deposition of dust in the tracheobronchial compartment is not generally of significance because of its rapid removal, particularly from the larger airways by mucociliary action. However, the elevated incidence of bronchogenic carcinoma in uranium miners is due to irradiation of the respiratory tract epithelium by α-emitting decay products of radon deposited in the tracheobronchial tree. The decay products of radon (radon daughters) behave as though they are of two kinds, characterized by different diffusion coefficients. Generally radium-A is present in ionic form, while radium-B et seq. are attached to condensation nuclei which have an activity median aerodynamic diameter (AMAD) of about 0.1 μm. The free ions, because of their high diffusivity, deposit in the conducting airways and there is negligible penetration to the alveolar region of the lung. In the case of attached daughters, most deposition occurs in the pulmonary compartment and very little in the airways. Deposition in the tracheobronchial compartment is also important in the case of hygroscopic materials. Cigarette smoke, which has a mean size of about 0.2 μm [64] would not normally be expected to deposit in significant amounts in this region. However, if the particles of smoke increase in size to 2 or 3 μm by absorbing moisture while in the lung airways, they will readily deposit on the bronchial epithelium.

Although total deposition and deposition in the nasopharyngeal compartment can be measured directly, it has not been possible to investigate deposition in the other compartments of the respiratory tract until appropriate radioactive tracer methods were developed. It is generally accepted that particles deposited in the conducting airways are cleared by mucociliary action within a day, while those which penetrate to, and are deposited in, deeper nonciliated regions are retained for much longer periods of time. Consequently, measurement of chest radioactivity following the inhalation of particles tagged with a γ-emitting nuclide, enables the relative deposition in the conducting airways and in the pulmonary compartment to be assessed. The Task Group on Lung Dynamics in their model consider that 40% of insoluble particles deposited in the pulmonary compartment are cleared with a half-time of 24 hr, although no mechanism is proposed to account for this. Recent experimental work by Foord (personal communication) in which measurements of chest radioactivity were made following the inhalation of radioactive 2.5-μm particles indicate that the fraction subject to rapid clearance from the pulmonary region is appreciably less than 40%. The first workers to use radioactive tracer techniques for this purpose were Talbot et al. [71] and Wilson and LaMer [80] who used polydisperse droplet aerosols labeled with sodium-24. More recently, Lippmann et al. [53] have used monodisperse particles of ferric oxide labeled with either technetium-99m or gold-198. They found, in mouth-breathing subjects, that of the aerosol deposited below the level of the trachea, more than half was deposited in the pulmonary compartment for particles with aerodynamic diameters less than 4 μm. Above this size, there was a rapid falloff in alveolar deposition, but this was still significant at 10 μm. Thomson and Short [74] using a somewhat similar system, found that in normal subjects, 30-50% of 5-μm particles were deposited beyond the ciliated airways. However, alveolar deposition in nose-breathing subjects will be considerably lower than these values, because of the efficient removal of particles greater than 5 μm in the nasopharyngeal compartment. In this type of measurement it has been assumed that particles deposited in the conducting airways are cleared quite rapidly. While this is true for material deposited in the larger airways, there is evidence that clearance from the smaller bronchioles may not be complete within 1 day. For this reason measurements of chest radioactivity should extend over a period of at least 1 day to ensure that clearance is substantially complete.

Information on the size of dust retained in the pulmonary compartment can be obtained from the analysis of particles isolated from lung tissues of workers exposed occupationally, but, unfortunately, it has not always been possible to obtain information on the size distribution of airborne dust to which these workers were exposed. The literature in this field was reviewed by Davies [17] and Cartwright [13], who published results based on the analysis of coal miners' lungs. He reported that there was a maximum mass frequency at a particle diameter of 1 μm (unit density spheres) with a fall in frequency toward larger sizes. These results indicate that the retention of larger particles is lower than experimental inhalation studies would suggest, but this discrepancy may be due to a size-selective mechanism for particle clearance. There was no clear indication of the size at which cutoff is complete, but at 15 μm the mass frequency was only about 1% of that at 1 μm. Mineral fibers up to 200 μm in length have been found in lung tissues, although the majority are generally shorter than 5 μm [35].

Because most of the experimental measurements of deposition in the tracheobronchial and pulmonary compartments have been made on mouth-breathing subjects, extrapolation to nose-breathing subjects should be made with caution, particularly at larger particle sizes.

D. Calculation of Regional Distribution

The calculation of total deposition in the respiratory tract by various workers has been discussed above. As total deposition is derived by summation of deposition in each region of the respiratory tract, the same references should be consulted for information on regional deposition. Deposition in the various compartments as a function of particle size was calculated by the Task Group on Lung Dynamics and the results are presented in Fig. 4. These values refer to a respiratory rate of 15 breaths min^{-1} and tidal volumes of either 750 cm^3 (Fig. 4A) or 1450 cm^3 (Fig. 4B). Deposition in the tracheobronchial compartment is less than 10% for all particle sizes. In the pulmonary compartment deposition passes through a maximum of about 30% at 2 μm and a minimum at about 0.5 μm. The range of these curves has been extended by Morrow [57] to cover particles of aerodynamic diameter from 0.001 to 1000 μm.

Aerosols encountered in practice are almost without exception polydisperse and must be considered as distributions which generally approximate to log-normal. The Task Group extended their calculations to cover the deposition of aerosols with log-normal distributions and geometric standard deviations ranging from 1.2 to 4.5 at a tidal volume of 1450 cm^3. The results are shown in Fig. 5 which illustrates that the mass deposition of the aerosol occurring in each compartment varies over quite a narrow range, although the particles cover a wide size spectrum.

This conclusion carries the implication that the hazards associated with the inhalation of radioactive aerosols may preferably be estimated from determinations of concentration and AMAD rather than from the concentration of the "respirable fraction" as currently carried out in coal mines in the United Kingdom and the United States. Instruments such as the cascade centripeter described by Hounam and Sherwood [42], which separate airborne dust into sized fractions aerodynamically, facilitate these determinations. Direct evaluation of pulmonary deposition is possible using the sampling head devised by Stevens and Stephenson [69], as its collection efficiency varies with particle size in a manner which closely matches deposition in the pulmonary compartment as illustrated in Fig. 4.

VI. Deposition in Pulmonary Disease

It is to be expected that certain pulmonary diseases or dysfunctions will result in changes in the dimensions of airways which will alter the deposition of inhaled particles. Several conditions such as bronchospasm, inflammation, tumors, and excessive secretions in the lumen can cause a reduction in caliber, or partial obstruction of airways which will, in turn, produce an alteration in flow pattern. For a given flow rate, the velocity in a tube is inversely proportional to the square of its diameter. Thus a reduction in caliber will lead to turbulence in regions of the lung where flow is normally streamlined. This will lead to a greater overall deposition and the regional deposition pattern will be altered. In diseases such as asthma obstruction is more marked during expiration than during inspiration and the transit time of particles will be increased, resulting in enhanced deposition by sedimentation and diffusion. In diseases such as pneumoconiosis, there will be an alteration in breathing pattern and the lung architecture will undergo a progressive change in which the branching angles at bifurcations may be altered, leading to an increase in deposition due to inertial impaction. The deposition of aerosols in pulmonary disease has been discussed in general terms by Goldberg and Lourenco [33].

A number of investigations have been described in which the deposition of inhaled particles in normal subjects has been compared with that in smokers or in patients with chronic obstructive pulmonary disease. Palmes et al. [61] studied the rate of deposition of 0.5-μm particles during breath-holding in normal, bronchitic, and emphysematous subjects. They considered that the rate of deposition should vary inversely with the size of the air spaces in which particles are contained during breath-holding. Although the initial deposition tended to be higher for abnormal subjects, a consistently lower rate of deposition was only demonstrated by subjects with emphysema. Love et al. [56] using 1-μm monodisperse particles, measured total deposition in coal workers with and without simple pneumoconiosis and found no apparent difference between the two groups. Using radioactive aerosols Lippmann et al. [53] studied regional deposition in asthmatic and bronchitic patients and in smokers. Compared with normals, there was enhanced deposition in the tracheobronchial compartment for all three categories with a consequent reduced deposition in the pulmonary compartment. Similar findings were reported by Sanchis et al. [67] who used a radioactive polydisperse droplet aerosol with mass median diameter of 3 μm to investigate regional deposition in smokers, nonsmokers, and patients with chronic obstructive disease. They demonstrated, using a scintillation camera, that in smokers and patients there was enhanced small airway deposition leading to a reduced delivery of aerosol to regions distal to the terminal bronchioles. Results presented by Thomson and Pavia [73] also show that under standard conditions of inhalation 5-μm radioactive tracer particles penetrate deeper into the healthy than into the bronchitic lung.

References

1. E. D. Acheson, R. H. Cowdell, E. Hadfield, and R. G. Macbeth, Nasal cancer in woodworkers in the furniture industry, *Br. Med. J.,* **2**:587-596 (1968).
2. E. D. Acheson, R. H. Cowdell, and B. Jolles, Nasal cancer in the Northhamptonshire boot and shoe industry, *Br. Med. J.,* **1**:385-393 (1970).
3. B. Altshuler, Calculations of regional deposition of aerosol in the respiratory tract, *Bull. Math. Biophys.,* **21**:257-270 (1959).
4. B. Altshuler, E. D. Palmes, L. Yarmus, and N. Nelson, Intrapulmonary mixing of gases studied with aerosols, *J. Appl. Physiol.,* **14**:321-327 (1959).
5. B. Altshuler, L. Yarmus, E. D. Palmes, and N. Nelson, Aerosol deposition in the human respiratory tract, *Arch. Ind. Health,* **15**:292-303 (1957).
6. American Conference of Governmental Industrial Hygienists. Threshold Limit Values for Chemical Substances and Physical Agents in the Workroom Environment with Intended Changes for 1974. Cinncinnati, American Conference of Governmental Industrial Hygienists, 1974.
7. J. M. Beeckmans, The deposition of aerosols in the respiratory tract, *Can. J. Physiol. Pharmacol.,* **43**:157-172 (1965).
8. J. M. Beeckmans, Correction factor for size-selective sampling results, based on a new computed alveolar deposition curve, *Ann. Occup. Hyg.,* **8**:221-231 (1965).
9. K. A. Bell and S. K. Friedlander, Aerosol deposition in models of a human lung bifurcation, *Staub. Reinhalt. Luft.* (in English), **33**:183-187 (1973).
10. A. Black and R. F. Hounam, Penetration of iodine vapour through the nose and mouth and the clearance and metabolism of the deposited iodine, *Ann. Occup. Hyg.,* **11**:209-225 (1968).
11. J. H. Brown, K. M. Cook, F. G. Ney, and T. Hatch, The influence of particle size upon the retention of particulate matter in the human lung, *Am. J. Public Health,* **40**:450-458 (1950).

12. J. Cartwright, Particle shape factors, *Ann. Occup. Hyg.*, 5:163-171 (1962).

13. J. Cartwright, Airborne dust in coal mines: The particle-size-selection characteristics of the lung and the desirable characteristics of dust-sampling instruments. In *Inhaled Particles and Vapours II.* Edited by C. N. Davies. Oxford, Pergamon Press, 1967, pp. 393-405.

14. Community Health and Environmental Surveillance System (CHESS), Health consequences of sulfur oxides. U.S. Environmental Protection Agency. EPA-650/1-74-004, 1974.

15. L. Dautreband, H. Beckmann, and W. Walkenhorst, Licht und elektronen-mikroscopische Untersuchungen zur Abscheidung von Staub in den Atemwegen, *Staub,* 21:141-172 (1961).

16. C. N. Davies, A formalised anatomy of the human respiratory tract. In *Inhaled Particles and Vapours.* Edited by C. N. Davies. Oxford, Pergamon Press, 1961, pp. 82-87.

17. C. N. Davies, Deposition and retention of dust in the human respiratory tract, *Ann. Occup. Hyg.,* 7:169-183 (1964).

18. C. N. Davies, Breathing of half-micron aerosols. II. Interpretation of experimental results, *J. Appl. Physiol.,* 32:601-611 (1972).

19. C. N. Davies, Deposition of inhaled particles in man, *Chem. Ind. (Lond.),* 1974:441-444.

20. C. N. Davies, J. Heyder, and M. C. Subba Ramu, Breathing of half-micron aerosols. I. Experimental, *J. Appl. Physiol.,* 32:591-600 (1972).

21. W. L. Dennis, In Discussion of paper by C. N. Davies, In *Inhaled Particles and Vapours.* Edited by C. N. Davies. Oxford, Pergamon Press, 1961, pp. 88-90.

22. W. L. Dennis, The effect of breathing rate on the deposition of particles in the human respiratory system. In *Inhaled Particles III.* Edited by W. H. Walton. Old Woking, Surrey, England, Unwin, 1971, pp. 91-102.

23. W. L. Dennis and K. F. Sawyer, The retention of particles in the human nose. Porton Technical Paper, No. 95., Chemical Defence Experimental Establishment Porton, Salisbury, England, 1949.

24. W. L. Dennis and K. F. Sawyer, The retention of particles in the human respiratory system. Porton Technical Paper, No. 190., Chemical Defence Experimental Establishment Porton, Salisbury, England, 1950.

25. S. El Golli, J. Bricard, P. Y. Turpin, and C. Treiner, Evaporation of saline droplets, *J. Aerosol Sci.,* 5:273 292 (1974).

26. W. Findeisen, The deposition of small airborne particles in the human lung during inhalation, *Pflügers Arch.,* 236:367-379 (1935).

27. F. A. Fry, Charge distribution on polystyrene aerosols and deposition in the human nose, *J. Aerosol Sci.,* 1:135-146 (1970).

28. F. A. Fry and A. Black, Regional deposition and clearance of particles in the human nose, *J. Aerosol Sci.,* 4:113-124 (1973).

29. N. A. Fuchs, *Evaporation and Droplet Growth in Gaseous Media,* London, Pergamon Press, 1959.

30. N. A. Fuchs, *The Mechanics of Aerosols,* Oxford, Pergamon Press, 1964.

31. A. George and A. J. Breslin, Deposition of radon daughters in humans exposed to uranium mine atmospheres, *Health Phys.*, **17**:115-124 (1969).

32. G. Giacomelli-Maltoni, C. Melandri, V. Prodi, and G. Tarroni, Deposition efficiency of monodisperse particles in human respiratory tract, *Am. Ind. Hyg. Assoc. J.*, **33**:603-610 (1972).

33. I. S. Goldberg and R. V. Lourenco, Deposition of aerosols in pulmonary disease, *Arch. Intern. Med.*, **131**:88-91 (1973).

34. H. L. Green and W. R. Lane, *Particulate Clouds: Dusts, Smokes and Mists*, 2nd ed., London, E. & F. N. Spon, 1964.

35. P. Gross, R. A. Harley, Jr., J. M. G. Davies, and L. J. Cralley, Mineral fiber content of human lungs, *Am. Ind. Hyg. Assoc. J.*, **35**:148-151 (1974).

36. E. H. Hadfield and R. G. Macbeth, Adenocarcinoma of ethmoids in furniture workers, *Ann. Otol., Rhinol., Laryngol.*, **80**:699-703 (1971).

37. R. J. Hamilton and W. H. Walton, The selective sampling of respirable dust. In *Inhaled Particles and Vapours*. Edited by C. N. Davies. Oxford, Pergamon Press, 1961, pp. 465-475.

38. T. Hatch and P. Gross, *Pulmonary Deposition and Retention of Inhaled Aerosols*, New York, Academic Press, 1964.

39. J. Heyder, G. Gebhart, C. R. Heigwer, C. Roth, and W. Stahlhofen, Experimental studies of the total deposition of aerosol particles in the human respiratory tract, *J. Aerosol Sci.*, **4**:191-208 (1973).

40. R. F. Hounam, The deposition of atmospheric condensation nuclei in the nasopharyngeal region of the human respiratory tract, *Health Phys.*, **20**: 219-220 (1971).

41. R. F. Hounam, A. Black, and M. Walsh, The deposition of aerosol particles in the nasopharyngeal region of the human respiratory tract. In *Inhaled Particles III*. Edited by W. H. Walton. Old Woking, Surrey, England, Unwin, 1971, pp. 71-79.

42. R. F. Hounam and R. J. Sherwood, The cascade centripeter: A device for determining the concentration and size distribution of aerosols, *Am. Ind. Hyg. Assoc. J.*, **26**:121-131 (1965).

43. J. R. Johnston and D. C. F. Muir, Inertial deposition of particles in the lung, *J. Aerosol Sci.*, **4**:269-270 (1973).

44. P. Kotrappa, Shape factors for aerosols of coal, UO_2 and ThO_2 in respirable size range. In *Assessment of Airborne Particles*. Edited by T. T. Mercer, P. E. Morrow, and W. Stöber. Springfield, Ill., Charles C Thomas, 1972, pp. 331-358.

45. H. D. Landahl, On the removal of air-borne droplets by the human respiratory tract: I. The lung, *Bull. Math. Biophys.*, **12**:43-56 (1950).

46. H. D. Landahl, On the removal of air-borne droplets by the human respiratory tract: II. The nasal passages, *Bull. Math. Biophys.*, **12**:161-169 (1950).

47. H. D. Landahl, Particle removal by the respiratory system: Note on the removal of airborne particulates by the human respiratory tract with particular reference to the role of diffusion, *Bull. Math. Biophys.*, **25**:29-30 (1963).

48. H. D. Landahl and S. Black, Penetration of air-borne particulates through the human nose, *J. Ind. Hyg. Toxicol.,* **29**:269-277 (1947).

49. H. D. Landahl and T. Tracewell, Penetration of air-borne particulates through the human nose. II, *J. Ind. Hyg. Toxicol.,* **31**:55-59 (1949).

50. H. D. Landahl, T. Tracewell, and W. H. Lassen, On the retention of air-borne particulates in the human lung, *Arch. Ind. Hyg. Occup. Med.,* **3**: 359-366 (1951).

51. G. Lehmann, The dust filtering efficiency of the human nose and its significance in the causation of silicosis, *J. Ind. Hyg.,* **17**:37-40 (1935).

52. M. Lippmann, Deposition and clearance of inhaled particles in the human nose, *Ann. Otol. Rhinol. Laryngol.,* **79**:519-529 (1970).

53. M. Lippmann, R. E. Albert, and H. T. Peterson, Regional deposition of inhaled aerosols in man. In *Inhaled Particles III.* Edited by W. H. Walton. Old Woking, Surrey, England, Unwin, 1971, pp. 105-120.

54. M. Lippmann and W. B. Harris, Size-selective samplers for estimating "respirable" dust concentrations, *Health Phys.,* **8**:155-163 (1962).

55. M. Y. Longley, Pulmonary deposition of dust as affected by electric charges on the body, *Am. Ind. Hyg. Assoc. J.,* **21**:187-194 (1960).

56. R. G. Love, D. C. F. Muir, and K. F. Sweetland, Aerosol deposition in the lungs of coal workers. In *Inhaled Particles III.* Edited by W. H. Walton. Old Woking, Surrey, England, Unwin, 1971, pp. 131-137.

57. P. E. Morrow, Models for the study of particle retention and elimination in the lung. In *Inhalation Carcinogenesis,* USAEC Symposium, Series No. 8, 1970, pp. 103-115.

58. D. C. F. Muir, *Clinical Aspects of Inhaled Particles,* London, Heinemann, 1972.

59. D. C. F. Muir and C. N. Davies, The deposition of 0.5 μ diameter aerosols in the lungs of man, *Ann. Occup. Hyg.,* **10**:161-174 (1967).

60. A. J. Orenstein, Recommendations of the Conference. In *Proceedings of the Pneumoconiosis Conference, Johannesburg, 1959.* Edited by A. J. Orenstein. London, Churchill, 1960.

61. E. D. Palmes, R. M. Goldring, Chiu sen Wang, and B. Altshuler, Effect of chronic obstructive pulmonary disease on rate of deposition of aerosols in the lung during breath-holding. In *Inhaled Particles III.* Edited by W. H. Walton, Old Woking, Surrey, England, Unwin, 1971, pp. 123-129.

62. Panel on Inhalation Risks from Radioactive Contaminants, *Inhalation Risks from Radioactive Contaminants.* Technical Report Series No. 142, Vienna, International Atomic Energy Agency, 1973.

63. R. E. Pattle, Retention of gases and particles in the human nose. In *Inhaled Particles and Vapours.* Edited by C. N. Davies. Oxford, Pergamon Press, 1961, pp. 302-309.

64. J. Porstendörfer and A. Schraub, Concentration and mean particle size of the main and side stream of cigarette smoke, *Staub, Reinhalt. Luft.* (in English), **32**:33-36 (1972).

65. D. F. Proctor and D. L. Swift, The nose—A defence against the atmospheric

environment. In *Inhaled Particles III*. Edited by W. H. Walton. Old
Woking, Surrey, England, Unwin, 1971, pp. 59-69.

66. W. E. Ranz and J. B. Wong, Impaction of dust and smoke particles on surface and body collectors, *Ind. Eng. Chem.*, **44**:1371-1381 (1952).

67. J. Sanchis, M. Dolovich, R. Chalmers, and M. T. Newhouse, Regional dis-
tribution and lung clearance mechanisms in smokers and non-smokers.
In *Inhaled Particles III*. Edited by W. H. Walton. Old Woking, Surrey,
England, Unwin, 1971, pp. 183-188.

68. F. Shanty, The deposition of ultrafine aerosols in human volunteer subjects. Ph.D. Dissertation. Baltimore, Md., Johns Hopkins University,
1974.

69. D. C. Stevens and J. Stephenson, A size selective dust sampler which con-
forms to the lung model proposed by the ICRP Task Group on Lung Dynamics, *J. Aerosol Sci.*, **3**:15-23 (1972).

70. W. Stöber, Dynamic shape factors of nonspherical aerosol particles. In
Assessment of Airborne Particles. Edited by T. T. Mercer, P. E. Morrow, and
W. Stöber. Springfield, Ill., Charles C Thomas, 1972, pp. 249-289.

71. T. R. Talbot, E. H. Quimby and A. L. Barach, A method of determining the
site of retention of aerosols within the respiratory tract of man by the use
of radioactive sodium, *Am. J. Med. Sci.*, **214**:585-592 (1947).

72. Task Group on Lung Dynamics for Committee II of the International
Commission on Radiological Protection, Deposition and retention models,
Health Phys., **12**:173-208 (1966).

73. M. L. Thomson, and D. Pavia, Particle penetration and clearance in the
human lung, *Arch. Environ. Health*, **29**:214-219 (1974).

74. M. L. Thomson and M. D. Short, Mucociliary function in health, chronic
obstructive airway disease, and asbestosis, *J. Appl. Physiol.*, **26**:535-539
(1969).

75. V. Timbrell, The inhalation of fibrous dusts, *Ann. N.Y. Acad. Sci.*, **132**:
255-273 (1965).

76. V. Timbrell, An aerosol spectrometer and its application. In *Assessment
of Airborne Particles*. Edited by T. T. Mercer, P. E. Morrow, and W. Stöber.
Springfield, Ill., Charles C Thomas, 1972, pp. 290-330.

77. A. M. Van Wijk and H. S. Patterson, The percentage of particles of different sizes removed from dust-laden air by breathing, *J. Ind. Hyg. Toxicol.*,
22:31-35 (1940).

78. W. H. Walton, Airborne particles and health. In *Dust Control and Air
Cleaning*. Edited by R. G. Dorman. Oxford, Pergamon Press, 1973,
pp. 1-29.

79. E. R. Weibel, *Morphometry of the Human Lung*, Berlin, Springer-Verlag,
1963.

80. I. B. Wilson and V. K. LaMer, The retention of aerosol particles in the
human respiratory tract as a function of particle radius, *J. Ind. Hyg. Toxicol.*, **30**:265-280 (1948).

6

Uptake of Pollutant Gases by the Respiratory System

MICHAEL S. MORGAN and ROBERT FRANK

University of Washington
Seattle, Washington

I. Introduction

In this chapter, we consider principles governing the absorption and desorption of gases by the respiratory tract. Empirical studies are reviewed with emphasis on sulfur dioxide (SO_2) and ozone (O_3), two important air pollutants. The concentrations of these gases in ambient air are usually low—both are strong irritants—being less than a fraction of a part per million parts of air by volume (ppm); in industrial settings and especially in accidents, their levels may go considerably higher.

To a large degree, the two gases are absorbed by and affect different portions of the respiratory tract. We have ample experimental information about their absorption across the upper airways; less about their absorption across the lower airways and alveoli; and virtually no data on their specific sites of absorption within these regions. Instead, we must rely on inferences drawn from the functional or anatomical changes produced by the gases. It is interesting to note that a number of models exist for predicting the pattern of deposition of inhaled particles within the respiratory system, whereas analogous models for pollutant gases (contrasted with anesthetic gases) are rare. Further on in the chapter, we describe a preliminary version of such a model and indicate some of its potential uses.

II. Physical Principles of Absorption

A. Definitions

Particle-free ambient air behaves in some respects like a simple solution and will be referred to in this way. Thus, an absorbable component of air becomes the "solute," and a contaminant such as SO_2 is an example. The terms used to describe quantitatively the process of solute uptake are now defined with the aid of Fig. 1:

1. *Input* is the airborne delivery of solute to any portion of the respiratory tract.

2. *Absorption* or *uptake* is the transfer of solute from the airstream to the surrounding tissues; desorption is the reverse process.

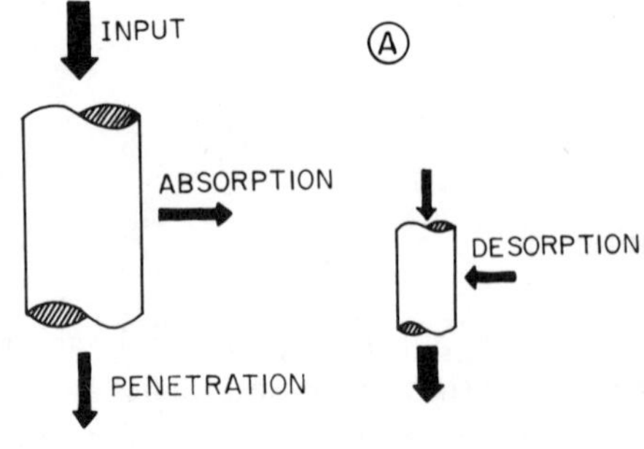

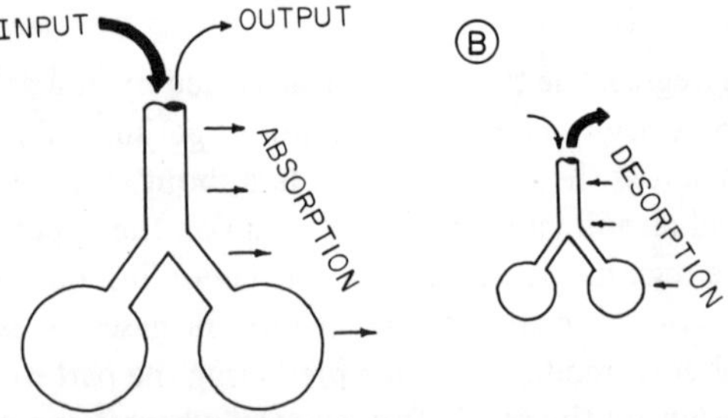

FIGURE 1 Schema of the absorption process, with terminology. (A) A portion of the airway in which some of the inspired gas is absorbed, and the remainder penetrates. Inset shows desorption. (B) The entire respiratory tract, showing absorption and output. Inset indicates desorption.

    3.   *Penetration* is the airborne entry of solute into distal portions of the respiratory tract beyond the upper airways (Fig. 1A).

    4.   *Output* is the elimination of solute via the expired air (Fig. 1B).

Numerical values assigned to these terms may have units of volume, or mass of solute. When rate quantities for each term are considered, the units are volume (or mass)/time. Two additional useful terms are:

    5.   *Fractional uptake* is the ratio of uptake to input, or of the corresponding rates.

    6.   *Fractional penetration* is the ratio of penetration to input, or of the corresponding rates.

The absorption of gaseous solutes as a general physicochemical process is examined next. The importance of absorptive processes to industrial and biological systems has spurred quantitative studies of their kinetics. We can draw upon this knowledge in two ways: first to develop a basic understanding of absorption, permitting us to identify significant factors affecting its rate and extent; second, to adapt the analytic techniques designed originally for industrial processes to the biological system. The techniques provide a means of predicting the pattern of uptake by the respiratory system of gases with different properties, and to suggest how this pattern may be affected by changes in pulmonary mechanical function or ventilation. By way of illustration, the basis for a recent numerical model of respiratory gas absorption will be considered [1].

B. Conservation Principle

The first step in analysis of a transport process such as absorption is the establishment of a mass balance. Based on conservation of mass, the balance states that the difference between the mass input into and output from the respiratory tract must represent the mass which is absorbed. For example, consider an animal breathing air containing a known concentration of solute, S. The mass input rate of S to the respiratory tract is the product of inspiratory flow rate ($\dot{V}_I$) and inspired concentration (C_I). The mass output rate is the same product, with expiratory subscripts. When a steady state exists such that the concentrations are time-independent, the difference between the two rates is the mass rate of uptake ($\dot{m}$):

$$\dot{V}_I C_I - \dot{V}_E C_E = \dot{m} \tag{1}$$

This equation describes the mass balance for the total respiratory tract, as shown in Fig. 1B. It is easily modified to describe the balance for any portion of the airway, as in Fig. 1A. Here, the second term in Eq. (1) becomes the product of inspiratory flow rate, $\dot{V}_I$, and outlet concentration, C_{out}.

The analysis is simplified if the gases are considered to be ideal, at constant temperature and total pressure. The ideal gas law is rearranged to give:

$$C = \frac{P_B M}{RT} \cdot F \tag{2}$$

Where P_B is the barometric pressure, M is the solute molecular weight, R is the gas law constant, T is the absolute temperature, and F is the volume fraction of solute in the gas phase. Substituting this relation into Eq. (1) with some rearrangement gives the balance expression:

$$\dot{V}_I F_I - \dot{V}_E F_E = \dot{V}_S \tag{3}$$

in which $\dot{V}_S = \dot{m}RT/MP_B$ is the rate of solute absorption in volumes/time.

Equation (3) provides a working definition of the rate of uptake, namely $\dot{V}_S$. The equation also gives the upper limit for the rate of uptake: since F_E cannot fall below zero, this maximum rate of uptake, $\dot{V}_{S(max)}$, equals $\dot{V}_I F_I$, the rate at which the solute is delivered to the respiratory tract.

The mass balance principle should be applied whenever gas uptake is considered, regardless of conditions of flow, solute concentration, existence of a steady state,† or portion of the respiratory tract under study. It provides useful definitions of fractional and total uptake, and offers a simple approach to measuring uptake. Rather than attempting to analyze the respiratory tract tissues and blood for the solute, it is easier to determine the flow rate and composition of the gas and to use Eq. (3) to compute uptake.

With the upper limit of rate of solute uptake established, we next examine the factors which influence uptake when it is less than maximal. To do so, we must consider the general principles regarding phase equilibria and kinetics of mass transfer. Whenever the rate of solute uptake is submaximal, it can be shown that equilibrium and kinetic factors determine the observed rate of uptake.

---

†In the unsteady state the mass balance is described by a differential equation containing a term that accounts for changes in concentration with time. The solution to this equation is more complex, but since the conclusions regarding the factors influencing uptake are the same [2], we will consider only the simpler steady state equations.

C. Phase Equilibria

The absorption of a gaseous solute by the respiratory tract involves entry of the solute into some or all tissues of the tract to form a solution. The behavior of dissolved gases in the respiratory tract tissue, and how it influences uptake are considered next.

For a simple solution consisting of one solute in one solvent, there is a unique relation between the concentration of solute and its tension, or pressure, within the solution [3]. In a more complicated system with two or more solutes, each solute exerts a distinct pressure, called the partial pressure. Here, the partial pressure of each solute is not always uniquely related to its concentration, but may depend on interaction with other components of the solution. Many solute-solvent systems show a linear relation between partial pressure and concentration. This is shown in Fig. 2 for nitrogen dissolved in water [4]. The slope of this relation is called the Henry's law constant, H:

$$P_{N_2} = H_{N_2} C_{N_2} \qquad (4)$$

The implication of this relation for gas and water phases that are in contact is illustrated in Fig. 3. Let us assume that initially neither the water nor the gas phase contains nitrogen, and that at time zero sufficient nitrogen is introduced into the gas phase to produce a partial pressure of 100 mmHg of N_2. Nitrogen is subsequently added to or subtracted from this phase to maintain P_{N_2} at 100 mmHg. Initially, the P_{N_2} in the water is zero but increases with time as nitrogen dissolves in the water. When phase equilibrium is established, P_{N_2} is equal in the

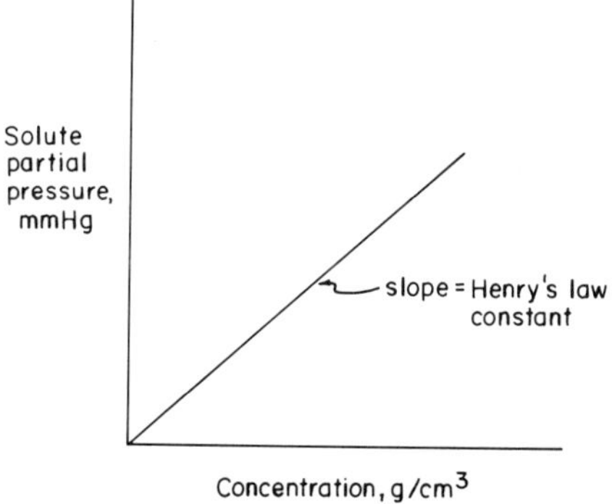

FIGURE 2　The equilibrium curve for a gaseous solute (nitrogen) in water relates partial pressure to mass concentration. The higher the slope, the lower the solubility, since concentration at a given partial pressure is reduced.

(a)   time = 0 :

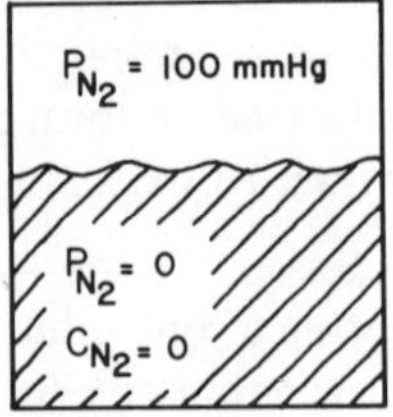

(b)   time >> 0 :

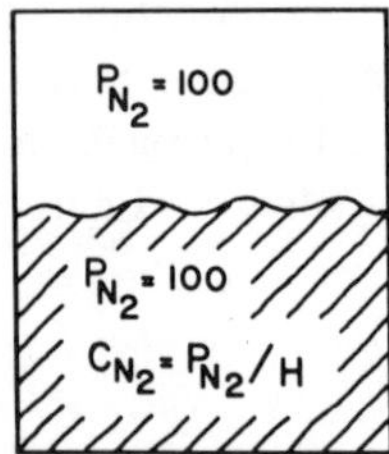

FIGURE 3   Equilibrium is reached when the partial pressures of solute in the two phases are equal. Prior to this time, the driving force for absorption, which is the difference in partial pressures, causes solute to move into the liquid.

the two phases, and the water is saturated with N_2. The concentration of N_2 in the water is given by the Henry's law relation:

$$C_{N_2} = \frac{P_{N_2}}{H_{N_2}} \tag{5}$$

Two points deserve emphasis: One is that the driving force for absorption of the gaseous solute is the difference between its partial pressures in the two phases. Whenever they differ, the system is said to be displaced from equilibrium. On attaining phase equilibrium, the liquid is saturated with solute at the partial pressure in the gas phase, and net transfer ceases. The second point is that the solubility of a gas is its concentration in the liquid phase at equilibrium. From Eq. (5) we see that for any partial pressure in the gas phase, the solubility of a solute is inversely related to the Henry's law constant.

These principles regarding a pure solvent and a chemically inert solute may be applied to respiratory tract absorption if certain complicating factors are recognized. One stems from the complex nature of the fluid which lines the respiratory tract, the first solvent encountered by inspired soluble gases. This fluid contains a variety of molecules and ions [5] capable of altering the equilibrium behavior of a gaseous solute. In general, the solubility of a gas decreases as the concentration of other ions and molecules increases [6]. Also, any chemical reaction between the gaseous solute and a component of the solvent will alter the former's solubility, and will usually result in a nonlinear relation between solute partial pressure and concentration [6]. An example of this effect is seen in the reactions of dissolved sulfur dioxide (SO_2) with water:

$$SO_2 + H_2O \rightleftharpoons H^+ + HSO_3^-$$

$$HSO_3^- \rightleftharpoons H^+ + SO_3^{2-}$$

$$HSO_3^- + \tfrac{1}{2}O_2 \rightleftharpoons H^+ + SO_4^{2-} \tag{6}$$

Here Henry's law applies for the partial pressure and the concentration of solute present only as undissociated SO_2. The undissociated SO_2 is only a fraction of the total SO_2 added to the solution because of the chemical transformations shown in (6). Hence, the partial pressure of SO_2 is less than would be expected if the same amount of gas were added to water with no chemical reactions. This reduction in partial pressure is a general characteristic of reactive gas-solvent systems, so that at low solute concentrations, it is often valid and useful to assume that the partial pressure is zero.

Solubility data are presented in Table 1 for a list of important gases in water. For the chemically reactive gases, a range of solubilities is given. In view of the factor just considered, these data are only approximate when applied to the respiratory tract. Until the composition and chemical reactivity of the fluids which line the tract are known more completely, the use of data based on the behavior of these gases in water remains necessary and justified.

Inspired soluble gases may eventually move beyond the respiratory tract and enter other solvent systems, starting with the circulation. This migration has been shown in studies of anesthetic gases [7]. With some gases, the transport from the mucus or alveolar lining to other fluid compartments occurs quickly; for others, the process is slow and only the respiratory compartment(s) need be considered. The effect of these secondary compartments on the apparent equilibrium depends on the solubility, inspired concentration and reactivity of the gas, so that each set of conditions must be evaluated separately. In general, for gases inspired at low concentrations (100 ppm or less) we may assume that mucus, alveolar liquid, and possibly blood are the most important compartments. This assumption will not hold when the inspired concentrations are much higher, as with anesthetic gases [8].

In summary, one obviously important factor determining the uptake of a gaseous solute by the respiratory tract is its solubility in the fluid lining. A

TABLE 1 Aqueous Solubilities of Selected Gases

Gas	Temp. ($^{\circ}$C)	Solubility (g/100 cm^3-mmHg)	Ref.
Acetaldehyde	25	2.04×10^{-2}	30
Acetic acid	25	6.26×10^{-2}	30
Acetone	25	4.93×10^{-3}	30
Acetylene	30	1.29×10^{-4}	4
Ammonia	30	8.69×10^{-2} at 11.5 mmHg 9.09×10^{-2} at 110 mmHg 5.56×10^{-2} at 719 mmHg	4
Carbon dioxide	30	1.73×10^{-4}	4
Carbon monoxide	30	3.29×10^{-6}	4
Chlorine	30	8.48×10^{-3} at 5 mmHg 1.57×10^{-3} at 100 mmHg 7.73×10^{-4} at 750 mmHg	4
Chloroform	25	1.83×10^{-3}	30
Formaldehyde	25	5.57×10^{-2}	30
Hydrochloric acid	30	1.74×10^{3} at 0.005 mmHg 5.00×10^{2} at 0.5 mmHg 3.93×10^{1} at 9.9 mmHg	4
Hydrogen sulfide	30	4.08×10^{-4}	4
Methane	30	2.61×10^{-6}	4
Nitric oxide	30	7.08×10^{-6}	4
Nitrogen	30	2.22×10^{-6}	4
Nitrous oxide	30	1.24×10^{-4}	4
Oxygen	30	4.93×10^{-6}	4
Ozone	25	4.93×10^{-5}	56
Phenol	25	1.72	30
Sulfur dioxide	30	2.12×10^{-2} at 4.7 mmHg 1.34×10^{-2} at 52 mmHg 1.09×10^{-2} at 688 mmHg	4

second factor affecting uptake is the duration of contact between the gas and the fluid lining. With adequate time, the difference in partial pressure (driving force) between these two phases will disappear so that no further uptake can occur, and equilibrium is established. If the residence time is too short for equilibration, the rate of transport of the solute from the gas to liquid phase will control uptake. Finally, equilibration may be delayed or even precluded by the buffering capacity of mucus or the transfer of the solute to the circulation and thence to other absorbent tissues.

D. Kinetics of Mass Transport

In the preceding discussion of equilibrium relationships, a motionless gas phase containing a solute and in contact with the liquid phase of the respiratory tract served as a model. For that model the net driving force acting on the solute is its partial pressure gradient. When either phase is in motion, the solute molecules in that phase are also carried along; that is, they have an additional bulk or convective component of motion. Figure 4 shows schematically a section of the

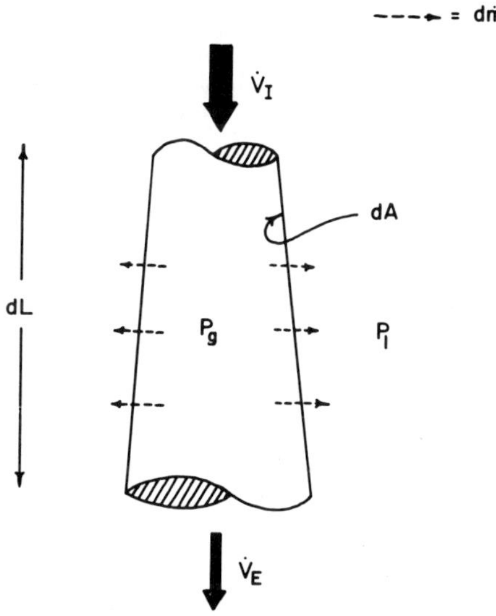

FIGURE 4 The specific nomenclature for analysis of absorption includes the surface area, A, the partial pressures, P_g and P_1, and the mass transfer rate, ṁ. For integrating over the complete tract to compute total absorption, A and ṁ are treated as differentials, dA and dṁ. The airway is shown as a truncated cone to indicate that cylindrical geometry need not be assumed; arbitrary flow tract geometry is permitted in this analysis.

airway through which air containing an absorbable solute is flowing. Note the competing influences of (a) the concentration gradient tending to move solute toward the wall; and (b) the axial bulk motion tending to move solute parallel to the airstream. Here, in contrast to the static situation, not all of the solute entering a section of the airway may have time to equilibrate by radial movement to the walls before, owing to axial bulk motion, it leaves the section. During gas flow, the rate of radial transport of a solute may become an important factor limiting its uptake. Next, we consider the factors influencing this competition between axial and radial transport, and how they may apply to respiratory tract absorption.

Mass transport under conditions of bulk fluid motion, that is, convective mass transport, can be described mathematically using the terminology of Fig. 4. The purpose is to express the mass of solute transported per unit time, $\dot{m}$, across the contact surface. At each point in the flow tract, this amount is proportional to the area of contact, A, and the local driving force, $P_g - P_1$ [9], expressed by the equation:

$$\dot{m} = r(P_g - P_1)A \tag{7}$$

where P_g and P_1 are the partial pressures of the solute in the gas phase and absorbent liquid phase (e.g., mucus, alveolar liquid, or blood); the proportionality factor, r, is the mass transfer coefficient; it is equivalent to the local uptake coefficient of Aharonson et al. [10]. Rearrangement of Eq. (7) indicates that r is a conductance relating the mass flux to the driving force:

$$\text{mass flux} = \text{conductance} \times \text{driving force}$$

$$\dot{m}/A = r \times (P_g - P_1) \tag{8}$$

This relation describes the local mass flux. If P_g and P_1 vary over the length of the flow tract, Eq. (8) must be evaluated at intermediate smaller sections, and the results summed (integrated) to obtain the total flux. For each small section of the tract (Fig. 4), Eq. (7) is rewritten in differential form:

$$dm = \frac{\dot{V}M}{RT} dP_g = r(P_g - P_1)\,dA \tag{9}$$

in which the ideal gas law, Eq. (2), has been used to convert $d\dot{m}$ into a term containing the flow rate and solute partial pressure. Integration of Eq. (9) provides a relation between the change in P_g over the entire respiratory tract and the total surface area, flow rate, and mass transfer coefficient. Numerous techniques have been described for accomplishing this integration in a variety of systems and experimental conditions [9-11]. All show that the change in P_g from inlet to outlet, from which the overall $\dot{m}$ is calculated, is influenced by the following factors:

1. $(\dot{V}/\dot{Q})$ (HM/RT): The first half of this product, $\dot{V}/\dot{Q}$, is the effective ventilation/perfusion ratio of the absorption tract; the second half, HM/RT, is the modified Henry's law constant for the solute-solvent system. This product is inversely related to uptake: as the value increases, the driving force for transfer of solute to the liquid or tissue phase is reduced [9], and absorption decreases. This occurs under circumstances of high gas flow rate, low liquid flow rate, and low solubility of the solute (high H).

2. A/L: This is the total surface area per unit length of the absorbing region; if this value is high, as for the nose [12], absorption is enhanced.

3. r: The value of the mass transfer coefficient (conductance) affects the rate of uptake directly. An increase in r leads to a proportionate increase in mass transfer. To predict the uptake of a solute under different physical conditions, we must account for the factors influencing r.

To identify what determines the steady state value of r as defined by Eq. (7), it is convenient to divide the total driving force $P_g - P_1$ into a series of components for each of the different phases [13]. Figure 5 shows a longitudinal section of the respiratory tract with the transfer zones identified. For the gas zone, the driving force is the difference between the solute partial pressure in the gas phase, P_g, and the value at the interface, P_0. Defining a mass transfer coefficient for this zone, r_g, analogous to Eq. (7), we have:

$$\dot{m}_g = r_g A(P_g - P_0) \tag{10}$$

Similarly for the tissue-liquid zone we may write:

$$\dot{m}_1 = r_1 A(P_0 - P_1) \tag{11}$$

where P_1 is the solute partial pressure in the mucus, alveolar lining fluid, interstitial fluid, or blood. The value of r_1 pertains to the transfer zone shown in Fig. 5. For steady-state conditions, Eqs. (10) and (11) may be combined to give:

$$\dot{m}_g = \dot{m}_1 = \dot{m} = r A(P_g - P_1) \tag{12}$$

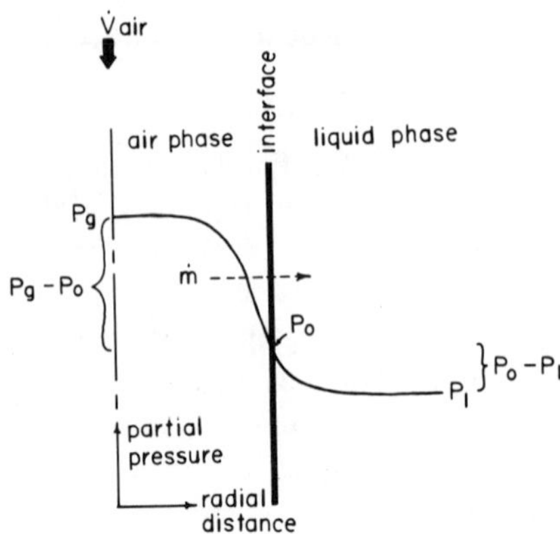

FIGURE 5 Longitudinal section of a short length of airway and surrounding tissues, with the airway center line at the left. Here the partial pressure is plotted against radial distance from the center for the case of absorption. The partial pressure at the interface, P_0, is not measured but serves merely to define the gas and liquid mass transfer conductances (see text).

where the overall conductance, r, is given by:

$$\frac{1}{r} = \frac{1}{r_g} + \frac{1}{r_1} \tag{13}$$

Equation (13) shows that if one zonal r is much smaller than the other, it will control the value of the overall coefficient and the other component may be neglected. In this circumstance, the "resistance" of either the gas or liquid phase is said to control the absorption rate.

Values of r_g and r_1 have not been reported for the respiratory tract, nor have sufficiently detailed measurements of absorption been made to permit calculation of these coefficients. We know of no studies which include measurements of blood levels or perfusion rate for the absorptive region involved. There is, however, considerable information for a wide range of nonbiological absorptive systems which may be used to infer how r will behave for the respiratory tract.

E. Theory

Occasionally, r can be calculated theoretically for a specific zone. The calculation involves finding the solution to the differential equations governing fluid mass transfer [2], and can be done only for simple, well-defined geometric structures and conditions of flow. One such condition is that of fully developed, laminar flow in a cylindrical tube [2] for which r for the fluid is given as:

$$r = B \frac{D}{d} \tag{14}$$

Here, d is the tube diameter, D is the solute diffusivity, and B is a constant dependent on the nature of the mass transfer at the boundary. While these conditions of flow and geometry are seldom encountered in the trachea and major airways [14], and never in the upper airways, Eq. (14) does provide a relation which can be modified empirically to fit the geometry and flow characteristics of portions of the respiratory tract. For the liquid phase, predicting the value of r_1 is more difficult. The complexity of both the geometry of the capillary bed surrounding the airways and the dynamics of blood flow discourage attempts to calculate r_1. As an approximation, however, we can assume that the initial liquid resistance to be encountered is that of a stagnant layer of mucus. Fick's law for diffusion then gives the mass transfer conductance [13] as:

$$r_1 = \frac{D}{xH} \tag{15}$$

where x is the thickness of the stagnant layer. While it is known that mucus is not stagnant, its velocity is sufficiently low [15] relative to that of the gas phase to justify such an approximation. If blood, the other moving liquid phase, presents a significant resistance, its coefficient will be of a form similar to Eq. (14), as modified by Henry's law constant and by (as yet) unknown empirical factors:

$$r_1 = \frac{BD}{dH} \tag{16}$$

F. Experimental Data

Studies of specially designed absorbers [9] have shown that r_1 and r_g depend on (a) the nature of the fluid motion at and near the phase boundary, (b) the thickness of the fluid over which the driving force exists, and (c) the diffusivity of the solute in the fluid. Factors (a) and (b) are strongly dependent upon flow rate and the geometry of the flow tract; for the upper airways, the geometry is not easily characterized and we can apply the experimental results in only an approximate way; the lower airways are more suitable for applying experimental values. Data are reported as dimensionless correlations, the general form of which is [13]:

$$\frac{rd}{D}\frac{RT}{M} = \lambda(Re)^a(Sc)^b\left(\frac{d}{L}\right)^c \tag{17}$$

where d is the diameter (or appropriate linear dimension) of the flow tract, D is the solute diffusivity in the fluid, and L is the length of the tract. Re is the Reynolds number, defined as:

$$Re = \frac{du}{\nu} \tag{18}$$

and Sc is the Schmidt number:

$$Sc = \frac{\nu}{D} \tag{19}$$

where u is the mean fluid velocity, and ν is the kinematic viscosity of the fluid. The constants λ, a, b, and c depend on whether flow is laminar or turbulent [9], and whether flow is within tubes or around solid objects [4]. In all cases, the exponents are positive and less than unity, so that r increases with increasing fluid velocity by reason of its dependence on the Reynolds number; dependence on d is variable. Perry [4] presents a complete listing of mass transfer correlations for a variety of flow conditions.

With respect to the situation of complex or unknown surface geometry, as for the nose, experiments on absorption in beds of granular solids have led to a useful simplification. When the surface area, A, in Eq. (7) is unknown, the measured mass transfer coefficients are often reported as a combined group, ra, where a is the (unknown) surface area.

The product ra must be averaged over the entire absorptive region when Eq. (9) is integrated. Aharonson et al. [10] have reported their results for nasal absorption in this fashion, calling the integrated average coefficient R:

$$R = \overline{ra}\,\frac{RT}{MP_B} \tag{20}$$

where $\overline{ra}$ is the average value of ra.

In summary, the primary determinants of the rate of solute uptake are the gas and (effective) liquid flow rates, solubility of the solute, surface area, and mass transfer conductance. The last factor depends chiefly on the ratios D/d or D/x for gas and liquid phases, but is modified significantly by flow, fluid property, and geometric parameters. For the respiratory tract, only approximate values of the liquid phase, gas phase, and overall conductances can be calculated from empirical data, owing to the complex and often unknown structural geometry and fluid dynamics.

The terminology of gas uptake defined at the outset can now be combined with the physical principles to form useful quantitative relations:

$$\text{Rate of input} = \frac{\dot{V}_I P_I}{P_B}$$

$$\text{Rate of uptake} = \frac{\dot{V}_I (P_I - P_E)}{P_B}$$

$$\text{Rate of penetration} = \frac{\dot{V}_I P_E}{P_B}$$

$$\text{Rate of output} = \frac{\dot{V}_E P_E}{P_B} \tag{21}$$

Each expression has units of volume/time. The subscripts I and E refer to inlet and outlet values, or inspired and expired values, respectively. The fractional quantities are

$$\text{Fractional uptake} = (P_I - P_E)P_I$$

$$\text{Fractional penetration} = \frac{P_E}{P_I} \tag{22}$$

Finally, the rate of uptake is related to the average mass transfer coefficient and average driving force by integration of Eq. (9):

$$\frac{\dot{V}_I (P_I - P_E)}{P_B} = R\overline{\Delta P} \tag{23}$$

where $\overline{\Delta P}$ is the value of $P_g - P_1$ averaged over the absorptive zone [9].

III. Upper Airways

A. Structural and Functional Attributes

The nasal passages are superbly designed to heat and humidify inspired air, to entrap airborne particles, and to scrub soluble gases: they are irregular, tortuous, and compressed mediolaterally. In adults, the section extending from the bony nasal opening to the ends of the turbinates is estimated to have a surface area of about 160 cm^2, compared to a volume of only about 20 ml [12]. The relation between applied pressure and flow across the nose suggests that turbulence which would be expected to enhance gas absorption is present at virtually all rates of gas flow [16,17].

Nasal blood flow and blood volume, the amount and physicochemical attributes of nasal mucus, and the temperature and patency of the passages all interact, in a manner not explicitly known, to influence gas uptake. The blood vessels have features conducive to rapid changes in their permeability and conductance. The superficial capillaries, especially those of the lateral nasal walls, are large with fenestrated basement membranes [18]. In the arterioles, there is no internal elastic membrane separating the smooth muscle layer from the porous basement membrane of the endothelium; apparently, agents can reach this musculature directly from blood [18]. Blood flows from the capillaries into a sinusoidal network which is capable of swelling like erectile tissue [19]. The sinusoids empty into venules which, like the capillaries, have porous endothelial basement membranes.

Blood flow is governed principally by nerve reflexes and chemical agents. The blood vessels have both cholinergic and adrenergic fibers [18]. Cholinergic stimulation causes vasodilatation and diminished patency of the nasal passages; adrenergic stimulation acts reciprocally [20]. The adrenergic innervation is concentrated in the anterior vessels, while cholinergic innervation is mostly distal.

A distinctive alignment of the blood vessels probably acts to enhance the absorptive and air-conditioning functions of the nose. The arterioles run in two parallel rows in a posterior-to-anterior direction [19]. One row lies directly under the mucosa; the second is in proximity to the mucous and serous glands. Brain [21] has likened the movements of nasal mucus and blood to a "countercurrent" system, said to be analogous in principle to engineering scrubbing towers: the mucus under the impetus of the cilia (and perhaps gravity) flows toward the pharynx, the same direction taken by inspired air, while the blood flows in the opposite direction toward the nasal opening. This arrangement is shown schematically in Fig. 6.

Mucous production is increased by cholinergic stimulation and decreased by adrenergic stimulation. Thus, acetylcholine may be expected to cause distinctive changes within the nose in the flow and volume of blood, the volume and composition of mucus, the patency of the passageways (affecting principally the turbinates), and, consequently, the degree of turbulence at a specified rate of gas flow. There is little explicit information about the effects of these events on the absorptive capacity of the nose for soluble gases.

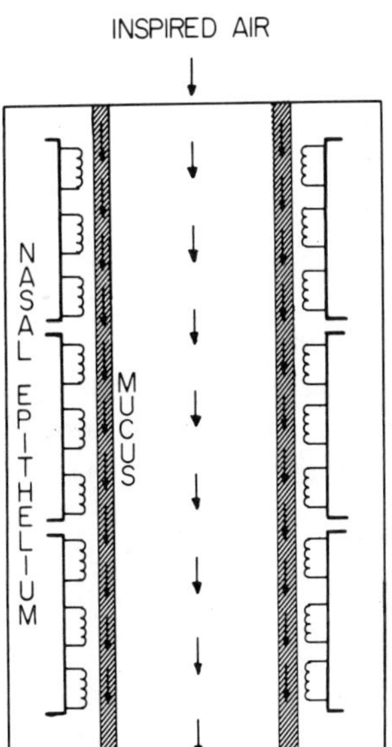

FIGURE 6 A schematic view of the nose. The dark lines in the nasal epithelium represent arteries and veins. The smaller curving lines represent capillaries. (Reprinted from Ref. 21, p. 537, by courtesy of the Annals Publishing Co.)

B. Methods of Measuring Uptake

Although it is customary to refer to gas absorption by the "upper airways" or the "nose," in many experiments the test gas also traverses a portion of either the mouth (studies on human subjects) or the larynx and uppermost trachea (studies on animals). These extra segments may absorb some gas [22-24] so that the absorptive rate attributed directly to the nose or upper airways is probably slightly overestimated.

In studies of nasal uptake involving human subjects, a sampling tube is customarily placed in the mouth [25] or oropharynx [26]. In the first instance, the remainder of the throat, extending as far as the pharynx, becomes an extension of the sampling tube. The subject may breathe quietly, the sampling occurring in synchrony with inhalation [25], or may hold his breath (glottis closed) while the gas is pulled through the oral tube and nose [26].

In animals, the upper airways may be isolated surgically [10,24,27,28]. The cranial end of the trachea is cannulated close to the larynx; gas is either drawn through the laryngeal cannula or pumped through the nose or mouth (mask or cannula). Gas flow is steady and unidirectional rather than to-and-fro. The isolated segment may be flushed with humidified air periodically to prevent drying of the mucosal surfaces. Generally, the lower airways are also cannulated and may be studied separately for gas uptake. While the constant unidirectional flow provides a simplified and useful means of analysis, it may pose problems of interpretation. Ordinarily, cyclic flow permits desorption from the mucosal surfaces during expiration when the average partial pressure of the airborne pollutant in expired air is relatively low. Consequently, equilibration between the gas and liquid phases may proceed at contrasting rates for the two patterns of flow.

Radioactive $^{35}SO_2$ has provided a high degree of precision in the measurement of uptake by the upper airways [28]. The same radionuclide has been used to detect the excretion of minute amounts of SO_2 by the lung [29]. In 1955, LaBelle et al. [30] hypothesized that the site of uptake of a soluble gas such as SO_2 would be shifted toward the lower airways by the addition of a droplet aerosol, a factor that might contribute to the exaggerated biological response seen when the two agents are inhaled together. Radionuclides would prove useful for testing the hypothesis.

C. Factors Influencing Absorption

1. Flow Rate

Aharonson et al. [10] have shown that the efficiency of the nose for scrubbing soluble foreign gases improves with increasing rates of airflow. Their mathematical analysis, referred to earlier, utilized an uptake coefficient defined as the flux or amount of gas diffusing through the nasal walls per unit area, per unit time, and per unit difference in partial pressure between air and blood. A steady state between the gas and tissue compartments is assumed, and any back-pressure of the solute in the blood (and mucus) is neglected.† While the authors consider a number of factors which might alter uptake coefficients locally (r), including the properties of the gas, configuration of the nasal passages, type of gas flow, volume of mucus and rate of perfusion, there is no attempt to assess local coefficients within the nose. Instead, they calculate a single overall coefficient, R, knowing the rate of airflow ($\dot{V}$) and partial pressures of the foreign gas entering (P_{in}) and leaving (P_{out}) the nose:

---

†The assumption is reasonable when low concentrations of gas are administered, or if the gas decomposes upon contact with the tissue, as does O_3. With higher concentrations ranging up to several percent or more, which is generally true for anesthetic gases, significant back-pressures do develop [8].

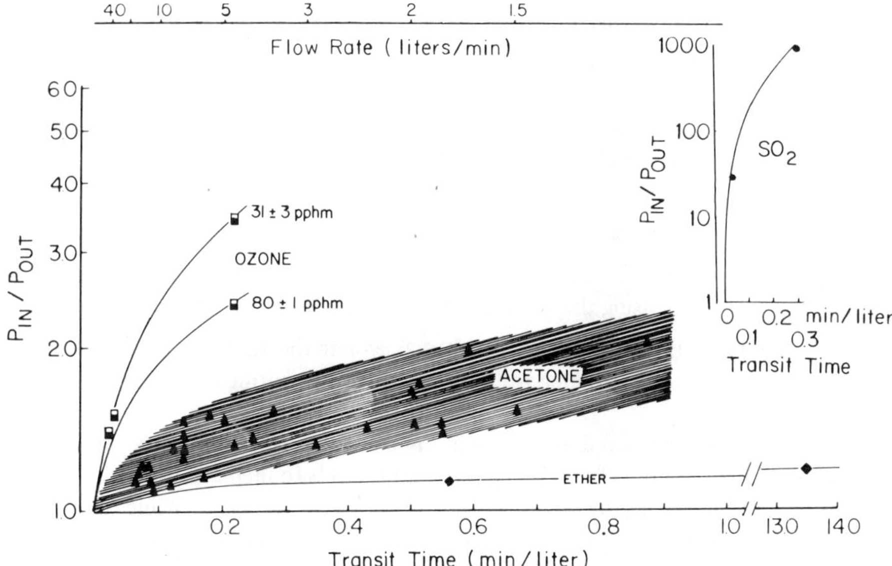

FIGURE 7 Effect of airflow rate on acetone, ether, ozone, and SO_2 removal by the nose. Data for ozone were taken from Ref. 31 and data for SO_2 were taken from Ref. 21. P_{in}/P_{out} = pressure entering the nose relative to pressure leaving the nose. Shaded area includes experimental points for acetone. (Reprinted from Ref. 10, p. 656, by courtesy of the American Physiological Society.)

$$R = \frac{\dot{V}}{P_B} \ln\left(\frac{P_{in}}{P_{out}}\right) \tag{24}$$

where P_B is the barometric pressure.

Aharonsen et al. apply this analysis to their own empirical observations made with ether and acetone, and to data obtained in other investigations using O_3 [31] and $^{35}SO_2$ [21,24]. The results are shown in Fig. 7. The coefficient R is proportional to the slope of the tangent drawn to the curves for the individual gases. In each instance, R is maximal for the shortest transit time, or, conversely, the highest flow rate.

The basis for the improved efficiency of uptake with increasing flow rate is conjectural. A number of possibilities may be enumerated:

1. Changes in the amount and/or distribution of turbulent flow in the passageways might increase contact with the mucosal surfaces.

2.   At higher flows, additional pathways within the nasal cavity might be recruited, thereby increasing the area for absorption.

3.   Higher flows might cause secondary changes in blood perfusion and mucus production which narrow the passageways and enhance air-liquid contact.

4.   As part of the response suggested in 3, the buffering capacity of nasal mucus may increase.

5.   Higher flows may lower temperature in the nasal cavity, increasing the solubility of inhaled gases.

6.   Higher gas velocities may attenuate the air-liquid boundary film, reducing resistance to diffusion through the film.

Despite this improved efficiency in uptake, the percentage of gas penetrating beyond the nose (fractional penetration) tends to increase as flow increases. The point is illustrated in Table 2 for both O_3 and SO_2. While the rise in R implies protection of the lower airways against exposure to the pollutant, the increase in fractional penetration has the opposite implication. The sensitivity of this relation of penetration to flow rate varies among gases, and is quite different for the nose and mouth, as will be described.

2 Concentration

The relation between the concentration of a solute gas and its fractional absorption by the upper airways may differ radically among the pollutants. For SO_2 which is highly soluble, absorption is approximately complete and independent of concentration over a wide range [24]. To be more precise, its fractional absorption does go up slightly with increasing concentration.† The change, however, is minuscule: in dogs, the nasal passages absorb about 99.9% of 1 ppm of $^{35}SO_2$, about 99.99% of 10 ppm, and over 99.99% of both 25 and 50 ppm. With O_3, a less soluble gas, the absorptive rate appears to be inversely related to concentration [31,33]; for one of these studies [31], the experimental prep-

---

†A conflicting conclusion was drawn by Strandberg [32] from studies on anesthetized rabbits breathing spontaneously. He measured the tracheal concentrations of SO_2 at "maximum" inspiration and expiration during exposures to 0.05–700 ppm of the gas. At maximum inspiration, the mean absorptive rate between the nares and trachea rose from about 5% at the lowest concentrations to over 90% at the highest concentrations (see Fig. 9 in which results are expressed reciprocally as percentage penetration). A conductometric method was used in analyzing SO_2. The large variance in results and uncertainties over the method of sampling raise questions about the validity of the findings.

TABLE 2   Nasal Penetration, %[a]

| | O_3 | | |
liters/min	26–34 pphm	78–80 pphm	$^{35}SO_2$ (1 ppm)
3.5–6.5	28.3	40.8	0.1
35–45	63.1	73.3	3.2

[a]Dogs, unidirectional flow; measurements made 15–20 min after onset of exposure to O_3, and during first 5 min of exposure to $^{35}SO_2$, pphm = parts per hundred million by volume; ppm = parts per million by volume. The calculated increase in exposure per minute for a 10-fold increase in flow rate (3.5 to 35 liters/min) is about 20-fold for O_3, and 320-fold for $^{35}SO_2$. (Taken from Refs. 24 and 31.)

aration was virtually identical with that used for $^{35}SO_2$. The inverse relation holds whether O_3 is administered by nose or mouth (Fig. 8). The sensitivity of this relation between concentration and absorptive rate is influenced by flow rate (Fig. 8).

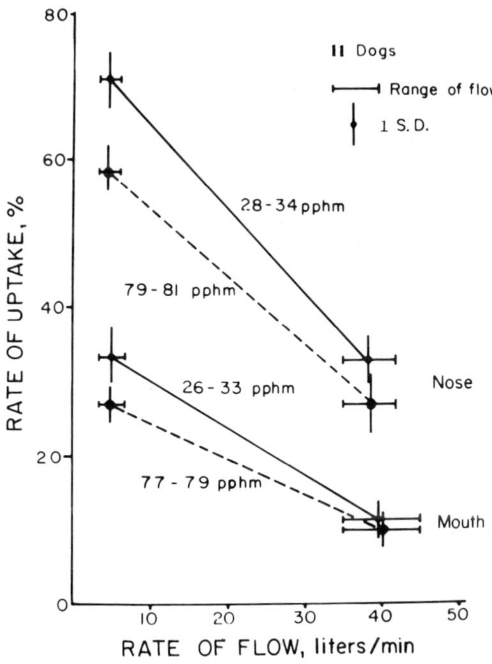

FIGURE 8   Percentage change in the rate of nasal uptake of O_3 under conditions of increased or decreased transnasal pressure (at constant flow), the latter produced by aerosols of histamine or phenylephrine. (Reprinted from Ref. 31, p. 136, by courtesy of HELDREF Publications, Inc.)

3. Duration of Exposure

The attributes of the respiratory tract which help determine absorption of a gas
during a single breath or series of breaths may be modified if exposure is prolonged. Depending on the chemical reactivity of the irritant gas, on the dosage,
and on the responsiveness of the organism, there may be structural, functional,
or biochemical changes having the potential for altering uptake. Some of the
changes may be the consequences of inflammation or injury; others may be
adaptive.

The net nasopharyngeal uptake of 25 ppm of SO_2 by human subjects is
apparently unchanged after 4 to 6 hours [26]. This sustained performance may
be due in part to the partial desorption of the gas that attends each expiration,
a process to be described subsequently. However, the same study [26] does
suggest that portions of the nasal passages may in time become saturated with
SO_2, and that the principal site(s) of uptake may therefore shift distally. The
evidence for this shift is indirect: the authors show with the use of radioactive
particles placed on the inferior turbinate that there is a slowing of mucous flow
during exposure which starts anteriorly and then migrates to the deeper portions of the nose. The effect is less notable at concentrations lower than 25 ppm.

No consistent changes have been noted in the nasopharyngeal uptake of
O_3 administered continuously (unidirectional flow) to dogs for 30 min in concentrations under 1 ppm [34]. However, after chronic exposures lasting 18
months, there is an increase in fractional penetration of the gas as measured in
the upper trachea. This increase in penetration above that found in control
animals is of the order of several percent of the inhaled concentration (1-3 ppm),
or up to about 0.2 ppm. It is worth noting that a Mast O_3 Analyzer, relied on
in this study, is ill-suited for measuring the rapid changes in concentration that
occur during spontaneous breathing.

4. Airway Caliber

Narrowing of the nasal passages would be expected to enhance absorption by
increasing the likelihood for contact between the inspired solute gas and mucosal
surfaces. Accordingly, at any specified flow there should be an increase in the
absorptive rate of the solute as nasal flow resistance increases.

It would be difficult to demonstrate such a relation for SO_2, a gas that is
almost completely taken up by the nose under normal circumstances. Attempts
to influence the uptake of O_3 in the isolated upper airways of dogs, with either
a decongestant, phenylephrine, or a congestant, histamine, have given equivocal
results [31]. The decongestant reduced the transnasal pressure drop by 15 to

43% at two constant flow rates without causing a measurable change in uptake. Histamine aerosol increased transnasal pressure drop by 12 to 270%, and, while the uptake of O_3 rose in 6 out of 7 trials, the change never exceeded 8% of the control values.

D.   Local Effects of Uptake: Reflex Responses

In serving as an absorptive surface for irritant gases, the upper airways are subjected to functional and structural changes. Reference has already been made to changes that may implicate blood perfusion, secretion of mucus, and mucous flow rate [26]. SO_2 may increase nasal flow resistance [26,35], influence the outcome of infection of the upper airways [36], and cause edema, necrosis, and desquamation of the nasal lining [36]. To produce this severe tissue damage, the dosage must be far in excess of those ordinarily experienced in urban or occupational settings.

Reflex changes in blood pressure, pulse rate, ventilatory pattern, and tracheobronchial smooth muscle tone may be evoked by stimulation or irritation of the nasopharyngeal mucosa. The question of whether bronchoconstriction is evoked (sometimes referred to as the nasobronchial nasopulmonary reflex) has been the subject of long-standing interest. The receptor for this reflex would be contained in the trigeminal nerve, the efferent limb in the vagal nerve. Electrical, mechanical, thermal, and chemical stimuli, including irritant gases, have been applied directly to the nasal mucosa in attempts to elicit the reflex. The results have not been notable for consistency. Early studies purporting to demonstrate it have been criticized by Nadel and Widdicombe [37] on the basis that the methods used did not permit separation of elastic and flow resistive pressures nor account for changes in lung volume. The latter investigators could not produce any change in respiratory flow resistance in anesthetized cats with either chemical or physical irritation of the nasal mucosa. Widdicombe has since participated in a study on cats in which stimulation of the nasal mucosa and epipharynx produced bronchodilation [38]. And more recently others [39] have been able to show an increase in pulmonary flow resistance in dogs after applying ice water to the nasal vestibule (the water then flowed out of the dependent nares), an effect that was dissipated within four to five breaths.

How prevalent or important the reflex may be in healthy subjects is unknown. The reflex has been suggested as a basis for the change in the maximal ventilatory performance seen in volunteers exposed to SO_2 by nose [26], the rationale being that penetration of SO_2 beyond the upper airways is negligible. There is informal evidence for existence of the reflex in asthmatic patients [40].

E. Nose and Mouth Compared

Thus far our focus has been on the absorptive capacity of the nose. The mouth
by comparison serves this function less efficiently. (Presumably, the performance
of the mouth is strongly influenced by the position of the lips, teeth, and tongue,
a factor that should be considered in the experimental design.) To illustrate, the
fractional uptake of $^{35}SO_2$ by the mouth falls significantly as the concentration
of the gas is increased, while nasal uptake improves slightly [24]. Moreover, the
absorptive capacity of the mouth is keenly sensitive to flow rate. If 1 ppm of
$^{35}SO_2$ is administered orally to dogs at 3.5 liters/min for 4-5 min, over 99% of
the gas is removed locally. However, if for the same concentration flow is increased tenfold, uptake falls to about 33% within minutes (Fig. 9). Obviously,
such a limited capacity for scrubbing gases has important implications for the
lower airways and alveolar regions. It may be said that any circumstance dis-
posing to mouth breathing is likely to increase exposure of the lung to pollutants.
During exercise, once mouth breathing becomes obligatory, the total amount of
pollutant reaching the tracheobronchial system, whether it be a soluble gas or

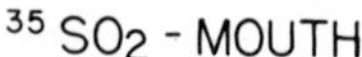

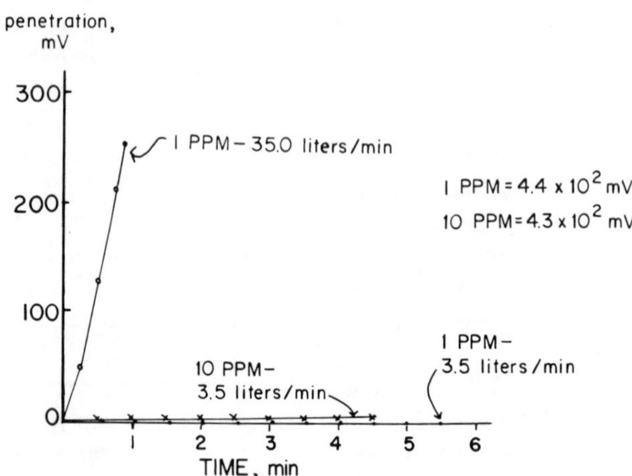

FIGURE 9 $^{35}SO_2$ leaving the segment was monitored by means of an ion
chamber coupled to a vibrating reed electrometer; the voltage output (millivolts)
of the electrometer was proportional to the concentration of $^{35}SO_2$ in the ion
chamber. Outlet concentration (penetration) is expressed as percentage of inlet
concentration. The mouth permitted considerably more penetration by $^{35}SO_2$
at 1 ppm X 35 liters/min than at 10 ppm X 3.5 liters/min (same dosage). (Reprinted in part from Ref. 24, p. 318 and Ref. 31, p. 137, by courtesy of
HELDREF Publications, Inc.)

aerosol, will go up for two reasons: (a) minute ventilation increases; and (b) fractional penetration by the pollutant increases. In the experiment alluded to with $^{35}SO_2$ (1 ppm at both 3.5 and 35 liters/min), the tenfold increase in minute volume was associated with a 1300-fold increase in the amount of $^{35}SO_2$ reaching the upper trachea. Two observations point to the importance of these factors: the functional changes caused by low concentrations of O_3 at rest are exaggerated during exercise [41]; there is evidence suggesting that the performance of track athletes is impaired by photochemical smog [42].

IV. Lower Airways and Parenchyma

A. Modeling

We considered earlier how knowledge of the mass transfer coefficient R would permit the calculation of gas uptake anywhere within the respiratory tract under specified conditions of flow. A theoretical prediction of R for the upper airways is extremely difficult owing to their complex morphometry; empirical determinations are a virtual necessity. For the airspaces distal to the larynx, the lower airways, we have better information on morphometry and flow behavior [14,43, 44] and an analysis of solute uptake is practicable. We now outline briefly an initial attempt to model uptake by the lower airways and parenchyma. Some shortcomings and potential uses of the model are described, followed by a brief review of empirical measurements.

As mentioned in the discussion of factors influencing R in the upper airways, the approach to the theoretical prediction of uptake is to establish differential equations for mass transport in the airways and solve them for the specific conditions of flow. McJilton et al. [1] using a modification of Weibel's morphometry of the lungs, and applying the physical principles of fluid flow and diffusion, derived a mathematical model of uptake that should give results corresponding to those based on empirically derived values for R. The model is based on a differential equation relating axial diffusion, convection, and radial absorption to depletion of solute in the gas phase. A sinusoidal breathing pattern was adopted rather than steady flow. In this way, the model accounts for physiological flow characteristics that are not treated adequately by the mass transfer coefficient (R) method. The model, admittedly preliminary, does have limitations so that the solution to the equation is only approximate. The limitations include: flux across the gas-liquid boundary is assumed to depend solely on solubility; there is no consideration of chemical reactions that might occur in the mucus layer; the dimensions of the airways [44] are assumed to be fixed during the breathing cycle; radial diffusion is assumed to be essentially instantaneous. In

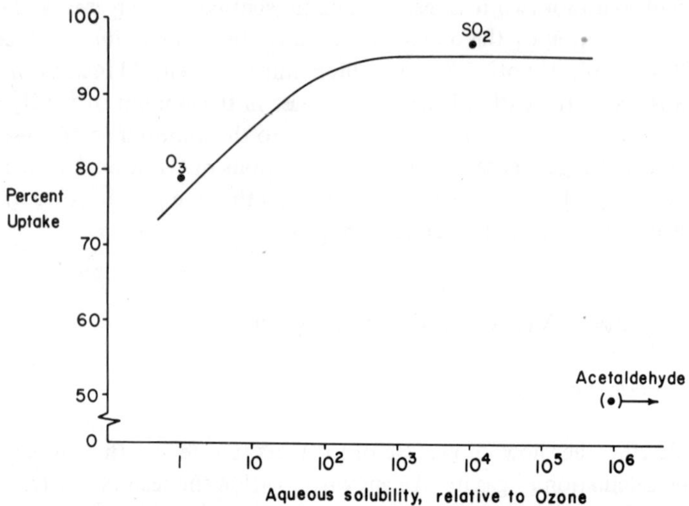

FIGURE 10 Lower airways: predicted fractional uptake in percent is related to the aqueous solubility relative to that of ozone. The calculations are for a tidal volume of 500 ml and a 2-sec breathing period. Data points shown are for ozone [31], sulfur dioxide [45], and acetaldehyde [46]. The last solute has very high solubility, but its observed uptake was much lower than the model prediction. (From Ref. 1.)

addition, the partial pressure of the solute beyond the mucus in the surrounding tissue and blood is considered to be negligible and is ignored. Despite these shortcomings, the model is an encouraging first step toward an important form of analysis.

The absorption of O_3 and SO_2 by the lower airways and parenchyma predicted by the model may be summarized as follows: (a) O_3, the less soluble gas, is absorbed over the entire length of the system, while SO_2 is absorbed in the proximal regions; (b) total absorption (Fig. 10) decreases with lower solubility; (c) increasing tidal volume is associated with increasing total uptake when inspiratory time is held constant. The authors have also developed the concept of local dose, defined as the mass of solute absorbed per unit area of airway surface, and calculated the dose of the two gases for the entire length of the system. Such a model could be useful for predicting the variations in local uptake and tissue dosage that might attend abnormalities in the distribution of ventilation. Insofar as patients with pulmonary disease may respond differently than healthy subjects to pollutants one contributing factor may be a grossly uneven exposure of the tissues. Figure 11 is an example of a predicted change in local dosage arising from a respiratory mechanical defect.

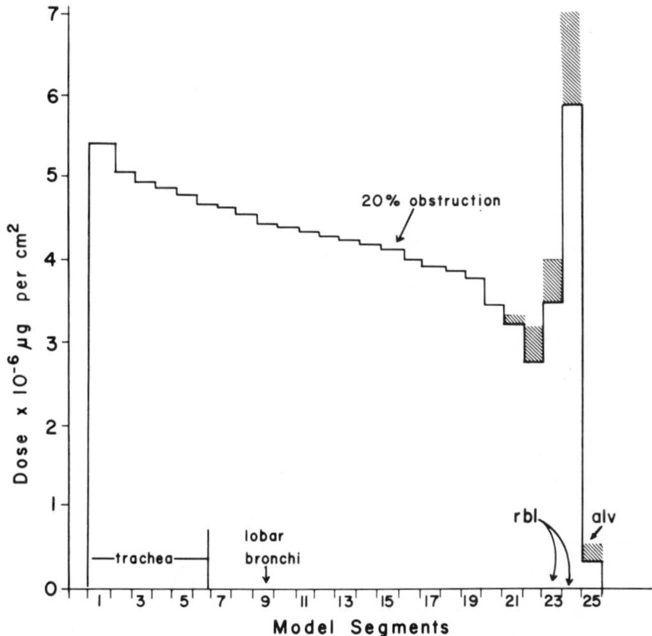

FIGURE 11   To simulate a pulmonary mechanical defect, one out of every five conducting airways is occluded at the level of the 15th model segment; the latter corresponds to Weibel's eighth generation of airways [44]. O_3 is administered before and after. The heavy black line is the predicted distribution of dose before occlusion. The shaded area indicates the increase in dose to the normal model segments beyond the occlusion. Note that the biggest predicted increase in dose occurs in the respiratory bronchioles. Rbl = respiratory bronchioles, alv = alveolar region. (From Ref. 1.)

B. Experimental Data

Reported measurements of lower airway uptake are few, perhaps owing to the many technical problems. In intact preparations it is virtually impossible to determine the sites of absorption, to measure how much solute reaches the alveolar region, or to distinguish between what is absorbed during inspiration and expiration. Sampling with tubes may interfere with patterns of flow; if the sampling rate in an airway should exceed the local flow rate, air will be drawn from other parallel pathways or from levels beyond the sampling site. Intermittent sampling of small amounts is required, a considerable difficulty for most analytical techniques. One alternative to direct sampling is the use of radiotracer techniques, using short-lived isotopes when feasible to keep radiation exposure low.

Measurements of uptake by the total respiratory tract (the difference between inspired and expired concentrations measured at the mouth or nose) may be combined with those of the upper airways alone to solve for lower tract absorption. The measurements of total uptake can be made more informative if the different sources of expired air are analyzed separately, that is, dead space, alveolar compartment, and the mixture of the two.

The absorption of O_3 by the lower respiratory tract has been measured in tracheostomized, mechanically ventilated dogs [31]. (The lining of the pump was coated with Teflon.) The results, indicating a fractional uptake of 0.80 to 0.87, agree well with predictions based on the model of McJilton et al. [1]. With an increase in ventilatory rate from 20 to 30 breaths/min and with tidal volume and lung volume held constant, the average uptake fell only 4%.

Ichioka [45] used a series of glass tubes lined with different absorbing liquids in attempting to model uptake of SO_2 and NO_2 by the lower airways. The uptake of SO_2 across the tubes averaged 98.6% in physiological saline and 99.8% in bovine albumin solution. The latter solvent was proposed as an analog of respiratory mucus. Uptake of the "less soluble" NO_2 averaged 47% with either solvent. Flow was constant and unidirectional. The variation of flow rate over the range of 0.5 to 3 liters/min influenced the uptake of NO_2 but not of SO_2.

Reports are available on the uptake of a miscellany of gases other than SO_2, O_3, and NH_3 by the respiratory tract and its subdivisions. These include acetaldehyde [46,47], acetone [47], ammonia [48,49], fluorinated hydrocarbons [50], iodine vapors [49], war gases [51], and anesthetic gases [52]; mathematical modeling of the uptake and distribution of anesthetic gases is also available [52].

C. Desorption and Excretion

With the probable exception of O_3 which decomposes rapidly upon contact with liquids, pollutant gases may be expected to desorb whenever their partial pressure in air is less than in the adjacent tissue lining. In the case of SO_2, the movement from liquid to air reflects two distinct processes. The first is a simple reversal of the previous absorptive process. Gas desorbed from mucosal surfaces may be excreted in expiration or be carried deeper into the airways if fresh air is inspired. Excretion reduces the body burden. In human subjects, about 15% of the SO_2 absorbed by the nasal passages is excreted in this way [25]; the process continues for some time following the end of exposure. Desorption occurring during inspiration acts to move the gas into more distal airways and could be the basis for functional responses that are either delayed or prolonged.

The second process by which SO_2 may reenter the gas phase is round-about; that is, some of the SO_2 previously absorbed in the nasal passages enters the circulation, apparently in physical solution, and is thereupon released into the lungs as a gas [29]. How much SO_2 reenters the gas phase from the pulmonary circulation is not known. In dogs, the expired $^{35}SO_2$ collected through a sampling tube placed in the lower trachea represented less than 1% of the concentration administered to the isolated upper airways. It is possible that a significant fraction of any SO_2 that may be excreted into alveoli is absorbed by the airways while en route to the trachea during expiration.

Whether this form of "exposure from below" affects smooth muscle tone is not known. Such indirect phenomenon could provide a basis for the changes in maximal ventilatory performance seen by Andersen et al. [26] in subjects exposed to 1 and 5 ppm of SO_2 by nose; the authors noted that very little of the inspired SO_2 would be expected to penetrate beyond the nose and suggested instead that a nasobronchial reflex might be the mediator of the functional changes.

D. Excretion of Ammonia

According to a recent hypothesis, desorption into the lungs of another gas, ammonia (NH_3), may mitigate the effects of inhaled acid sulfate aerosols [53]. In the region of St. Louis, Missouri, sulfuric acid (H_2SO_4)aerosol has been found in concentrations up to 10-25 $\mu g/m^3$ [54]. The pH of this aerosol is less than 1. The addition of about 10 parts per billion by volume (ppb) of NH_3 will neutralize these concentrations of H_2SO_4 and raise pH to about 5 [53]. If it is assumed that the ammonium ion normally present in blood is in equilibrium with alveolar gas, the calculated alveolar NH_3 will then be 7 ppb or greater; preliminary analyses of expired air from healthy subjects confirm the calculation [53]. Such levels of NH_3, by raising the pH of inhaled H_2SO_4 at the alveolar level, may afford protection to the lungs. The hypothesis has yet to be tested. Using an improved technique, Larson et al. (*Science,* in press) have recently measured expired ammonia concentrations ranging from 30 to 300 ppb in healthy subjects. At these levels NH_3 altered the molecular form of an inhaled acid aerosol.

E. Influence of Aerosol on Absorptive Site of Gas

Conceivably, the site of uptake of a gas within the respiratory tract might be influenced by aerosol droplets that are present in the inspired air. Atmospheric particles which are hygroscopic and submicronic in size [54,55] are of concern to public health. The hypothesis that a pollutant gas might be absorbed by these droplets, rendering the droplets (and gas) more irritating to the lungs, is time-honored [30]. "Synergism" is the term often applied to the exaggerated functional impairment produced by this interaction. One of the mixtures most commonly used to study synergism is SO_2 and sodium chloride aerosol.

Whether the synergism is due to changes in the site of absorption of SO_2, that is, increased delivery of the gas by the aerosol to the lower airways, or to the bisulfite, sulfite, or hydrogen ions that are formed in solution [see Eq. (6)] is not known. If it is assumed that 1 ppm of SO_2 and 1 mg/m^3 of liquid NaCl aerosol are mixed at a temperature of 20°C, then at equilibrium only about 1% of the gas will have been absorbed by the liquid (personal communication, Neil Horike and Timothy Larson, University of Washington) for transport to another site. Apparently, this small amount of interaction is effective.

Following inhalation of the mixture, the partial pressure of SO_2 in the gas phase must fall as absorption by the surrounding mucosa proceeds. Whether there is ensuing desorption of SO_2 from the airborne droplets is not known. It is theoretically possible for the SO_2 to be released from the droplet in the lower airways and to provoke a biological response, without the droplet itself striking the respiratory lining. Synergism appears to be predicated on the absorption of only a small fraction of the SO_2 in the ambient mixture, and it is possible that a piggyback process of this type, were it to occur, might have significance.

References

1. C. E. McJilton, J. Thielke, and R. Frank, Ozone uptake model for the respiratory system. Presented at the American Industrial Hygiene Association Conference, San Francisco, May, 1972. A33:20, 1972. (Abstract)
2. W. M. Kays, *Convective Heat and Mass Transfer,* New York, McGraw-Hill, 1966, pp. 53-77.
3. K. G. Denbigh, *Principles of Chemical Equilibrium,* 3rd ed., Cambridge, University Press, 1971, pp. 221-231.
4. J. H. Perry, *Chemical Engineers Handbook,* 3rd ed., New York, McGraw-Hill, 1950, pp. 668-693.
5. A. D. Barton, J. L. Powers, M. Lopata, and R. V. Lourenço, Biochemical characteristics affecting the consistency of bronchial secretions, *Chest,* 63:59S-61S (1973).
6. V. M. Ramm, *Absorption of Gases,* Jerusalem, IPST press, 1968, pp. 27-30.
7. E. I. Eger, *Anesthetic Uptake and Action.* Baltimore, Williams and Wilkins, 1974, pp. 77-120.
8. E. M. Papper and R. J. Kitz, *Uptake and Distribution of Anesthetic Agents.* New York, McGraw-Hill, 1963, pp. 5-39.
9. W. L. McCabe and J. C. Smith, *Unit Operations of Chemical Engineering,* 2nd ed., New York, McGraw-Hill, 1967, pp. 648-674.
10. E. F. Aharonson, H. Menkes, G. Gurtner, D. L. Swift, and D. F. Proctor,

Effect of respiratory airflow rate on removal of soluble vapors by the nose, *J. Appl. Physiol.*, **37**:654-657 (1974).

11. B. A. Hills and G. M. Hughes, A dimensional analysis for oxygen transfer in the fish gill, *Resp. Physiol.*, **9**:126-140 (1970).

12. D. F. Proctor, Physiology of the upper airway. In *Handbook of Physiology*, Vol. 1, Section 3: Respiration. Edited by W. O. Fenn and H. Rahn. Baltimore, Waverly Press, 1965, pp. 309-345.

13. B. A. Hills, *Gas Transfer in the Lung,* Cambridge, University Press, 1974, pp. 55-82.

14. D. E. Olsen, G. A. Dart, and G. F. Filley, Pressure drop and fluid flow regime of air inspired into the human lung, *J. Appl. Physiol.*, **28**:482-494 (1970).

15. A. Bouhuys, *Breathing: Physiology, Environment and Lung Disease.* New York, Grune and Stratton, 1974, pp. 385-415.

16. F. E. Speizer and N. R. Frank, A technique for measuring nasal and pulmonary flow resistance simultaneously, *J. Appl. Physiol.*, **19**:176-178 (1964).

17. B. G. Ferris, Jr., J. Mead, and L. H. Opie, Partitioning of respiratory flow resistance in man, *J. Appl. Physiol.*, **19**:653-658 (1964).

18. N. Cauna, Electron microscopy of the nasal vascular bed and its nerve supply, *Ann. Otol. Rhinol. Laryngol.*, **79**:443-450 (1970).

19. F. N. Ritter, The vasculature of the nose, *Ann. Otol. Rhinol. Laryngol.*, **79**:468-474 (1970).

20. R. T. Jackson, Pharmacologic responsiveness of the nasal mucosa, *Ann. Otol. Rhinol. Laryngol.*, **79**:461-467 (1970).

21. J. D. Brain, The uptake of inhaled gases by the nose, *Ann. Otol. Rhinol. Laryngol.*, **79**:529-539 (1970).

22. T. Dalhamn and L. Strandberg, Acute effect of sulphur dioxide on the rate of ciliary beat in the trachea of rabbit, in vivo and in vitro, with studies on the absorptional capacity of the nasal cavity, *Int. J. Air Water Pollution,* **4**:154-167 (1961).

23. T. Dalhamn, M.-L. Edfors, and R. Rylander, Mouth absorption of various compounds in cigarette smoke, *Arch. Environ. Health,* **16**:831-835 (1968).

24. N. R. Frank, R. E. Yoder, J. D. Brain, and E. Yokoyama, SO_2(^{35}S-labeled) absorption by the nose and mouth under conditions of varying concentration and flow, *Arch. Environ. Health,* **18**:315-322 (1969).

25. F. E. Speizer and N. R. Frank, The uptake and release of SO_2 by mouth and by nose, *Arch. Environ. Health,* **12**:725-728 (1966).

26. I. Andersen, G. R. Lundqvist, P. L. Jensen, and D. F. Proctor, Human response to controlled levels of sulfur dioxide, *Arch. Environ. Health,* **28**: 31-39 (1974).

27. G. R. Cameron, J. H. Gaddum, and R. H. D. Short, The absorption of war gases by the nose, *J. Pathol. Bacteriol.*, **58**:449-455 (1946).

28. T. Dalhamn, A. Rosengren, and R. Rylander, Nasal absorption of organic matter in animal experiments. I. Experimental technique and tobacco smoke absorption, *Arch. Environ. Health,* **22**:554-556 (1971).

29. N. R. Frank, R. E. Yoder, E. Yokoyama, and F. E. Speizer, The diffusion of $^{35}SO_2$ from tissue fluids into the lungs following exposures of dogs to $^{35}SO_2$, *Health Phys.*, **13**:31-38 (1967).

30. C. W. La Belle, J. E. Long, and E. E. Christofano, Synergistic effects of serosols, *Arch. Ind. Health*, **11**:297-304 (1955).

31. E. Yokoyama and R. Frank, Respiratory uptake of ozone in dogs, *Arch. Environ. Health*, **25**:132-138 (1972).

32. L. G. Strandberg, SO_2 absorption in the respiratory tract, *Arch. Environ. Health*, **9**:160-166 (1964).

33. T. R. Baughan, Jr., L. F. Jennelle, and T. R. Lewis, Long-term exposure to low levels of air pollutants. Effects on pulmonary function in the beagle, *Arch. Environ. Health*, **19**:45-50 (1969).

34. W. J. Moorman, J. J. Chmiel, J. F. Stara, and T. R. Lewis, Comparative decomposition of ozone in the nasopharynx of beagles, *Arch. Environ. Health*, **26**:153-155 (1973).

35. F. E. Speizer and N. R. Frank, A comparison of changes in pulmonary flow resistance in healthy volunteers acutely exposed to SO_2 by mouth and by nose, *Br. J. Ind. Med.*, **23**:75-79 (1966).

36. W. E. Giddens, Jr. and G. A. Fairchild, Effects of sulfur dioxide on the nasal mucosa of mice, *Arch. Environ. Health*, **25**:166-173 (1972).

37. J. A. Nadel and J. G. Widdicombe, Reflex effects of upper airway irritation on total lung resistance and blood pressure, *J. Appl. Physiol.*, **17**:861-865 (1962).

38. Z. Tomori and J. G. Widdicombe, Muscular bronchomotor and cardio-vascular reflexes elicited by mechanical stimulation of the respiratory tract, *J. Physiol., (Lond.)*, **200**:25-49 (1969).

39. J. H. Whicker and E. B. Kern, The nasopulmonary reflex in the awake animal, *Ann. Otol. Rhinol. Laryngol.*, **82**:355-358 (1973).

40. S. Ingelstedt, Discussion of Panel on Physiologic Relationships of Nose and Lungs. Symposium on the pathology of the nose and adjoining cavities, Washington, D.C., 1-3 December 1969. Quoted by R. Frank. The effects of inhaled pollutants on nasal and pulmonary flow-resistance, *Ann. Otol. Rhinol. Laryngol.*, **79**:540-546 (1970).

41. D. V. Bates, G. M. Bell, C. D. Burnham, M. Hazucha, J. Mantha, L. D. Pengelly, and F. Silverman, Short-term effects of ozone on the lung, *J. Appl. Physiol.*, **32**:176-181 (1972).

42. W. S. Wayne, P. F. Wehrle, and R. E. Carroll, Oxidant air pollution and athletic performance, *JAMA*, **199**:901-904 (1967).

43. K. Horsfield, G. Dart, D. E. Olsen, G. F. Filley, and G. Cumming, Models of the human bronchial tree, *J. Appl. Physiol.*, **31**:207-217 (1971).

44. E. R. Weibel, *Morphometry of the Human Lung*, New York, Academic Press, 1963, p. 139.

45. M. Ichioka, Model experiments on absorbability of the airway mucous membranes of SO_2 and NO_2 gases, *Bull. Tokyo Med. Dent. Univ.*, **19**:361-375 (1972).

46. J. L. Egle, Retention of inhaled acetaldehyde in the dog, *Arch. Environ. Health,* **24**:354-357 (1972).

47. T. Dalhamn, M.-L. Edfors, and R. Rylander, Mouth absorption of various compounds in cigarette smoke, *Arch. Environ. Health,* **16**:831-835 (1968).

48. T. Dalhamn and J. Sjöholm, Studies on SO_2, NO_2 and NH_3: Effect on ciliary activity in rabbit trachea of single in vitro exposure and resorption in rabbit nasal cavity, *Acta Physiol. Scand.,* **58**:287-291 (1963).

49. R. E. Pattle, The retention of gases and particles in the human nose. In *Inhaled Particles and Vapours,* Vol. 1. Edited by C. N. Davies. New York, Pergamon Press, 1961, pp. 302-309.

50. A. Morgan, A. Black, M. Walsh, and D. R. Belcher, The absorption and retention of inhaled fluorinated hydrocarbon vapours, *Int. J. Appl. Rad. Isotopes,* **23**:285-291 (1972).

51. G. R. Cameron, J. H. Gaddum, and R. H. D. Short, The absorption of war gases by the nose, *J. Pathol. Bacteriol.,* **58**:449-455 (1946).

52. E. I. Eger, *Anesthetic Uptake and Action,* Baltimore, Williams and Wilkins, 1974, p. 122.

53. D. Covert, R. Frank, T. Larson, and N. Horike. Chemistry of sulfate/ sulfite-containing aerosols: The effects of ammonia and water vapor. Presented at Symposium on Sulfur Pollution and Research Approaches, Duke University Medical Center, Durham, N.C., May, 1975, sponsored by EPA and Duke University.

54. R. J. Charlson, A. H. Vanderpol, D. S. Covert, A. P. Waggoner, and N. C. Ahlquist, Sulfuric acid-ammonium sulfate aerosol: Optical detection in the St. Louis region, *Science,* **184**:156-158 (1974).

55. D. F. S. Natusch and J. R. Wallace, Urban aerosol toxicity: The influence of particle size, *Science,* **186**:695-699 (1974).

56. R. C. Weast (ed.), *Handbook of Chemistry and Physics,* 45th ed., Cleveland, Chemical Rubber Company, 1964, p. B-200.

Unit Two

CILIA AND AIRWAYS SECRETIONS

The airways are protected to some extent by the upper respiratory tract; they, in turn, contribute to the protection of the alveolar parenchyma. The ciliary escalator of the airway epithelium is the primary mechanism for clearing airway surfaces of foreign materials. This clearance function is evident when bronchoscopy is used to directly observe particles on the mucus blanket moving mouthward. Yet the origin and control of the periciliary liquid layer is unclear. Excess fluid "drowns" the cilia and means that sticky secretions do not move; too little means that cilia may become entangled.

How immunoglobulins, normally present in the airway fluids, contribute to maintaining the normal sterility of the lung is not clear. Nor is it known to what extent diminished immunoglobulins allow the lung's defenses to be breached so that infection can occur. A large population of bacteria is normally found on the surface of gut and skin. However, detection of organisms in the airways of the lungs marks a stage in the breakdown of its defenses.

A variety of cell types have been identified in the airway epithelium, several only recently; in the rat there are 10: eight epithelial and two mesenchymal. Some of them can still be considered "cells in search of a function." A few of the cell types are found elsewhere, such as the brush and Kulchitzky cells, while others seem peculiar to the lung, like the Clara cell. Although structure may suggest a role, physiological studies have not correlated these cells with identified functions of the lung. It has been postulated that once the Clara and serous cells are differentiated, they are separate cell types. Experimental studies suggest the contrary; an appropriate stimulus may cause each to change to a goblet cell, indicating close relationships among the three.

In the following chapters control of mucus secretion is discussed. Drugs can change the quantity of mucus or alter its properties, but their effects are only beginning to be studied clinically. Sputum is a mixture of so many components that the effect of a drug may not be apparent from analysis of total sputum. Organ culture of human bronchial glands has contributed significantly to our understanding of the control of mucus secretion. Chemical analysis of the mucus collected in organ culture may be expected to extend it further.

The histochemical study of secretion, both within the cell and after discharge, contributes to our understanding of the types of acid glycoprotein incorporated into the final macromolecules of mucus. Chemical analyses of the glycoprotein of bronchial mucus by various specialists have given different results. Before biochemical analysis, mucus must be pretreated, whether it is from whole sputum or aspirated directly from the bronchial lumen. Histochemical studies reveal that a wider variety of molecular types are produced than the few identified by biochemical studies, and certain types suggested by histochemistry have not been identified biochemically. Recently, by a method of separation that introduces minimal destruction of the mucus macromolecule, one type has been identified which evidently represents a polymer of all histochemical types. Histochemical and biochemical results can be reconciled and the seeming contradiction of the earlier biochemical studies explained by the methods of preparation used. For this reason, the section discussing the chemical constituents of sputum deals with methods in some detail, albeit in a general way.

The rheological studies described here have mainly investigated viscosity and elasticity, since techniques are available for measuring them accurately. While other characteristics may prove to be as or more important clinically, it seems best to concentrate on these two until the chemical basis for them has been established. In view of the fact that sputum is a complex mixture, it is surprising, but encouraging that certain features are so consistent. This justifies the study of whole sputum as well as its constituents.

Lynne M. Reid

7

The Respiratory Mucous Membrane

PETER K. JEFFERY

Chelsea College
University of London
London, England

LYNNE M. REID

The Children's Hospital Medical Center
Boston, Massachusetts

I. Introduction

The lining of the respiratory tract can be regarded as part of the surface of the body, albeit a part that is invaginated in a complex way within the body. Although the gut is similar in that its lining epithelium is in continuity with the body surface, the way the ambient air moves frequently and freely over the lung lining makes it more susceptible to the gaseous environment. In man the volume of air entering the nares each day can be calculated to be of the order of 10,000 to 15,000 liters.

The lining epithelium of most of the airways is thinner than the surface skin. It appears pink because of underlying blood vessels and velvety because of surface irregularity due to various cell types and their arrangement. It is moist because of overlying tissue fluid which is constantly renewed. Streaks of mucus may be seen on its surface, when challenged by infection or irritant and rarely under normal conditions. In disease, more mucus is secreted and the wall becomes "soggy" and edematous.

While air moves in two directions, the mucus moves toward the cranium, together with any foreign material that is impacted on the airway wall. The movement of mucus in the main airways of the lower respiratory tract is against the forces of gravity in upright man, although at the periphery there are some airways that drain downward (i.e., with gravity). Ciliary movement acts similarly in the upper respiratory tract. Mucus formed in the sinuses drains to the nares, much of this also against gravity.

A. Upper Respiratory Tract

The larynx is considered to be a boundary between the upper and lower respiratory tracts. Thus the upper tract is here taken to extend from the external nares to the larynx.

The upper respiratory tract consists of two nasal cavities lying on either side of the nasal septum. The superior part of each is entirely surrounded by bone; the posterior opening is into the naso- and oropharynx. Each nasal cavity is wider below than above and wider centrally than at either its anterior or posterior ends. On the lateral wall of each are three conchae, which run roughly horizontally, and, under each, a meatus.

Following the oropharynx is the larynx, a valve protecting the lungs and the organ of phonation. An extraordinary degree of control is maintained over it, the size of its opening, and the tautness of its vocal cords.

The upper respiratory tract is linked to a system of sinuses, those air boxes within the skull which offer resonance to the voice and lighten the skull, and also to each of two continuous Eustachian tubes which are responsible for maintaining air in the middle ear. Air may move in and out of these sinuses and liquid and mucus formed here must also drain to the nasal cavities. Table 1 indicates the position and names of the accessory tubes of the upper respiratory tract.

B. Lower Respiratory Tract

The lower respiratory tract is of great complexity, the pattern of airway branching being described as the "bronchial tree." The larynx opens into a single tube, the windpipe or trachea. The trachea then divides into two main bronchi which, in man, give rise to five lobar bronchi, three on the right and two on the left. After entering the hilum of the lung, the lobar bronchi give rise to 20 segmental bronchi and their intrasegmental branches [1]. Airways throughout the lung are of varying width and length.

In man the larger airways are supported by cartilage which directly influences their function. In the trachea and main extrapulmonary bronchi the cartilage is in C-shaped plates opening posteriorly. Bundles of muscle fibers,

TABLE 1 Sinuses and Other Accessory Tubes Opening into the Nose and Oropharynx

Inferior meatus	Nasolacrimal canal
Middle meatus	Maxillary sinus—hiatus semilunaris, posterior part
	Anterior ethmoid into infundibulum
	Frontal sinus—anterior part middle meatus
	Middle ethmoidal—near bulla
Superior meatus	Posterior ethmoidal
Above superior meatus	Sphenoidal sinus
Nasopharynx	Eustachian tube and mastoid

present only in the gap, interlace and attach laterally to the posterior ends of the cartilage. Elastic fibers are scattered irregularly in the anterior wall, but posteriorly they collect into dense bands running the length of the airway. In the intrapulmonary airway they fan out and are seen under the lining epithelium around its entire circumference. Where present, cartilage and muscle surround the intrapulmonary airway, this muscle coat is virtually suspended from the last plates of cartilage. The arrangement of airway muscle is geodesic [2] so that as it contracts it also shortens.

The designation of intrasegmental airways by size is unsatisfactory, since their size varies with the phase of inspiration or expiration and the state of muscle contraction. An airway must be designated by structure and by its position along a pathway (i.e., generation or order of division). The following definitions are based on the presence of cartilage: *Bronchi* are those air tubes lying proximal to the last plate of cartilage along an airway. The proximal five generations can be considered "large bronchi" as cartilage is so abundant in their walls and is included in any cross section. Between the 5th and 15th generation, there are "small bronchi" as the cartilage is so sparse that many cross sections will not include cartilage. *Bronchioli* are those airways distal to the last plate of cartilage and proximal to the alveolar region. A "respiratory bronchiolus" is an airway where the alveoli open into the lumen; a "terminal bronchiolus" is that airway immediately proximal to this. Ciliated epithelium covers the nonalveolated wall of the respiratory bronchiole; the ciliary escalator begins in this airway. The *acinus* includes all the lung distal to a terminal bronchiolus and up to eight generations of respiratory bronchioli.

II.   The Mucous Membrane

The mucous membrane consists of epithelium and elastic lamina, that of the upper respiratory tract being in continuity with the lower.

But for the anterior nares (where it is stratified squamous and keratinized), much of the epithelium of the upper respiratory tract is ciliated, pseudostratified, columunar, and with goblet cells. In the human nasopharynx the area lined by stratified squamous epithelium is predominant with 60% of its anterior aspect and between 80-90% of its posterior aspect lined by such epithelium (see Ref. 3).

The lining of the pharyngeal tonsil and that of both lateral walls shows a pattern alternating between patches of ciliated and squamous epithelia. In the larynx itself stratified squamous epithelium is found over the anterior surface of the epiglottis, the upper half of its posterior surface, the upper part of the aryepiglottic folds, the vocal cords, and a few patches above the glottis. Elsewhere it is ciliated, pseudostratified, columunar and contains goblet cells. There are differences in epithelial thickness and goblet cell number in different parts of the upper respiratory tract. For example, the epithelium lining the paranasal sinuses is thinner and contains fewer goblet cells than that of the nasal cavities themselves.

The epithelium of the lower respiratory tract becomes progressively thinner toward the periphery of the lung and is continuous with the flattened epithelium of the alveolar duct and alveolus. Proximally, in the trachea and main bronchi, the epithelium appears stratified but, since all the cells touch the basement membrane, it is actually pseudostratified. In man it loses its pseudostratified nature some distance after entering the lung, thereafter becoming a simple, single layer of cells. In other species it becomes simple in type earlier on. In the rat the transition is at the hilum of the lung. In the lower respiratory tract of man, goblet cells are numerous in the extrapulmonary airway but sparse in bronchioli of less than 1 mm in diameter [4,5]. Here also there is a species difference: whereas goblet cells may be numerous in the trachea and bronchi of the guinea pig and cat, they are few in the rat, mouse, hamster, and rabbit. Unlike most other species they are abundant in the bronchioli of the cat [6].

The submucosa consists of glands, with both serous and mucous cell types, which are the major contributors to the mucus found in the respiratory tract. In the upper respiratory tract, racemose glands are most numerous around the upper end of the pharynx and Eustachian tube, epiglottis, and at the margins of the aryepiglottic folds and saccules. Glands are relatively few in number in the thin walls of the paranasal sinuses. In the human lower respiratory tract the submucosa consists of gland and cartilage, the two occurring together, the gland mainly internal to cartilage lying between it and epithelium but also between the plates of cartilage (Fig. 1). Glands are most numerous in the trachea pro-

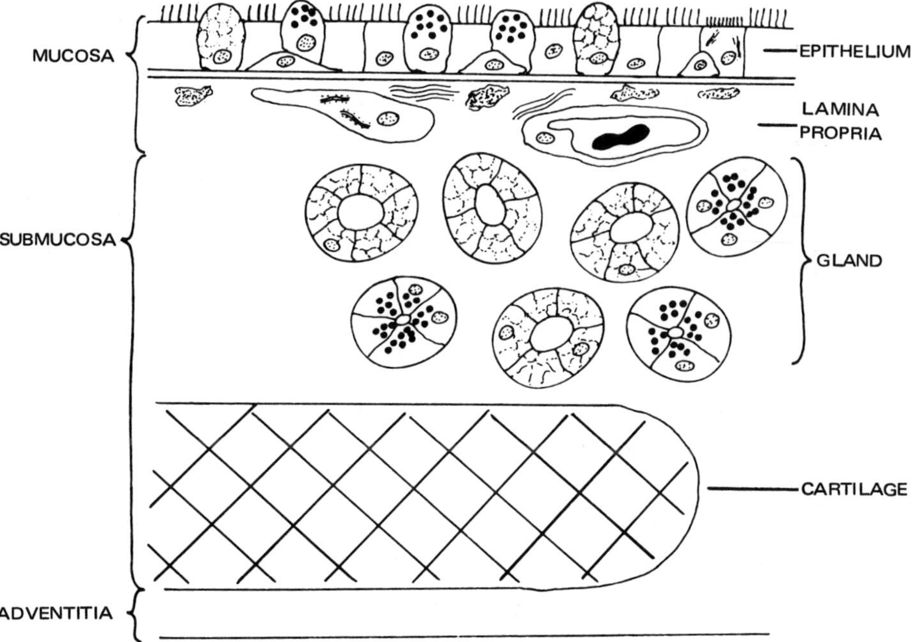

FIGURE 1   Diagrammatic representation of airway wall.

gressively decreasing distally until, along with cartilage, they are absent from airways smaller than 1 mm in diameter. Here there is a species difference. In the rat, cartilage is absent from any intrapulmonary airway and gland is concentrated centrally in the upper trachea. In the rabbit, gland is absent while cartilage extends well into the lung. In the cat, both cartilage and gland extend into small intrapulmonary bronchi.

A.   Epithelial Cell Types

There are a great many similarities between the cells of the upper and lower respiratory tract epithelium and also between the respiratory epithelia of different species. In each the epithelium is comprised of a number of distinct cell types which are to be here considered. For the sake of clarity the following descriptions will relate to mammalian species studied and, where marked and relevant, differences will be emphasized. With regard to epithelial ultrastructure much more is known about the lower than the upper respiratory tract, so most of the following will relate to the ciliated epithelium of the lower. The specialized epithelium of the olfactory region will not be dealt with here.

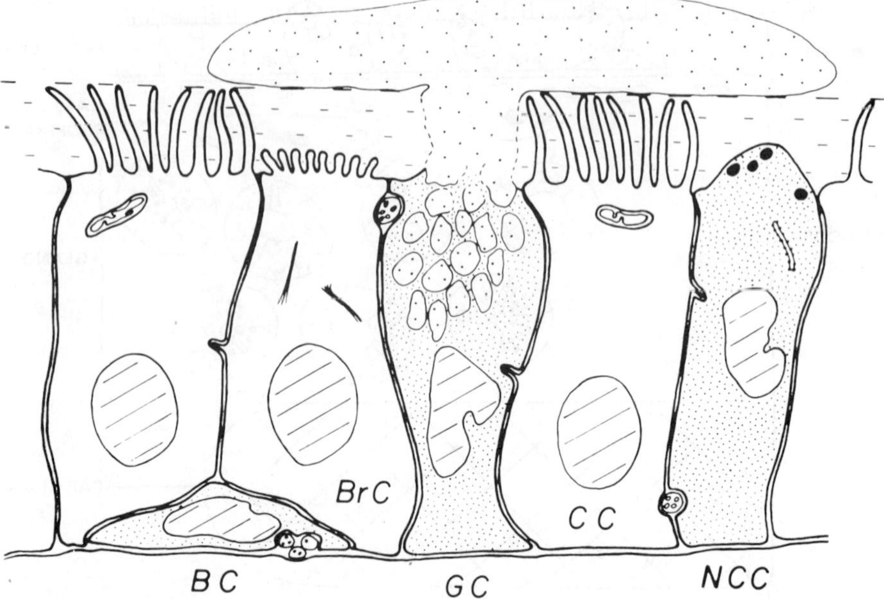

FIGURE 2 The ultrastructure of airway epithelium represented diagrammatically. Cilia beat in a fluid layer of low viscosity above which move flakes of mucus. Ciliated (CC), goblet (GC), nonciliated "serous" (NCC), brush (BrC), and basal cells (BC) are shown as are nerves penetrating the epithelium.

The epithelium is thought to be covered by a fluid layer of low viscosity in which the cilia beat (Fig. 2). Scanning electron microscopy has shown that above this "periciliary" layer lie discrete flakes or droplets of mucus [7,8], which are moved cranially by the tips of the cilia, and are not a continuous mucus blanket as was originally thought [9,10].

Early light microscopic studies [4,11-14] of the lining epithelium of man and mammals showed that the epithelium consisted of at least three cell types: ciliated, goblet (mucus-secreting), and basal cells (Fig. 3). A fourth cell, said to produce a nonmucoid secretion and now known as the "Clara cell" was reported in the terminal bronchioli of man and rabbit [15,16].

Now, by electron microscopy, eight epithelial and one migratory cell have been found in the human and, in the lower respiratory tract, an additional migratory cell in the rat and dog [17,18]. The eight epithelial cells can be divided on the basis of their position within the epithelium, on the presence of cilia, and the type of their secretory granules, if any (Fig. 4). Each cell has other distinguishing features, such as electron density of cytoplasm and type of endoplasmic reticulum; these are summarized in Tables 2 and 3.

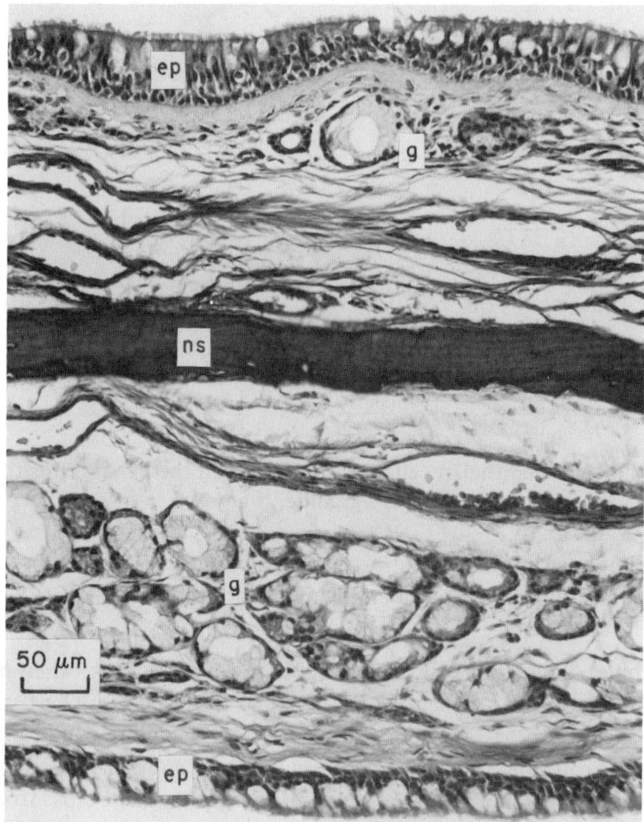

FIGURE 3 Photomicrograph of the epithelium (ep) lining the nasal cavities.
The fringe of cilia can be clearly seen, goblet cells do not take up the stain and
basal cells do not reach the airway lumen. Gland (g) is present beneath the
epithelium. The nasal septum (ns) divides the mucosa of the two cavities.
Hematoxylin and eosin: ×190.

1. *Ciliated Cell*

In man and other mammals the ciliated cell has electron-lucent cytoplasm and a
nucleus whose membrane is smooth in outline [17,19-23] . Ribosomes are
usually not associated with endoplasmic reticulum and occur singly or, more
often, in groups as polysomes. From each cell 200 cilia, and numerous microvilli between them, project into the airway lumen [19,23] . The cilia beat at
about 1000 times per minute, the effective stroke being more or less cranial
in direction. The cells are grouped into fields, each ciliated field being separated
by areas of nonciliated cells. All the cilia in one field beat in the same direction,

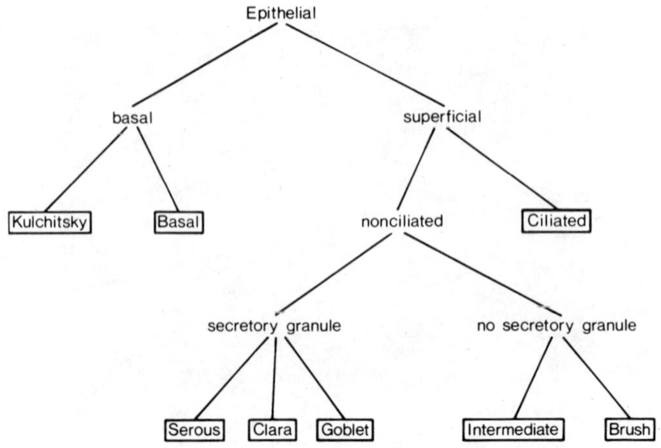

FIGURE 4   The eight epithelial cell types classified on the basis of position, presence of cilia, and type of secretory granule, if any.

whereas those of adjacent fields beat together in a slightly different direction [24], the resultant movement of mucus being toward the throat where it is swallowed.

The tubules, which drive the cilium by sliding each over the other, have a characteristic spatial arrangement which is similar in all species, whether metazoan or acellular [25]. However, the rootlet of the cilium, which anchors the cilium to the cell shows a species variation in length and striation. In mammals the rootlet is short and in man and cat, it shows prominent cross-striations which are absent in rabbit and mouse.

The tip of each cilium is significant, since this is the part of the cilium which will make contact with mucus on the surface of the periciliary layer. Sleigh [25] describes the cilium as "clawing" the mucus forward, and recently [17] electron microscopy has shown that in man, mouse, and rat there are small clawlike projections at the tip (Fig. 5).

The studies of Greenwood and Holland [26] in rabbit and Jeffery and Reid [17] in rat have shown that cilia decrease in length progressively toward the periphery of the lung and that, at least in rat, they are present well into the respiratory bronchiolar region. At the periphery the proportion of ciliated cells is greater than proximally, and the cilia have been reported to beat faster.

2. Serous Cell

The epithelial serous cell has only recently been described in man and rat and is one of three types of secretory cell found in airway epithelium. In man it is

TABLE 2 Distinguishing Ultrastructural Features of Cells Within the Epithelium[a]

Cell	Granule and density	Endoplasmic reticulum	Cytoplasmic density	Projections
Epithelial				
				Luminal
Ciliated	–	R	L	Cilia + filiform
Serous	+ D	R	D	Filiform
Clara	+ D	S	L	Blunt
Goblet	+ L	R	D	Filiform
Intermediate	–	R	L or D	Filiform
Brush	–	R	Intermediate	Microvillus brush
				Intercellular
Basal	–	R*	D	Blunt
Kulchitsky	Neurosecretory	R	L or D	Elongated
Mesenchymal				
Globule leukocyte	+ D	R*	Intermediate	Filiform
Lymphocyte	–	R*	L	Blunt

[a]*Symbols:*
 * in small amount   S smooth
 + present   L Lucent
 – absent   D dense
 R rough

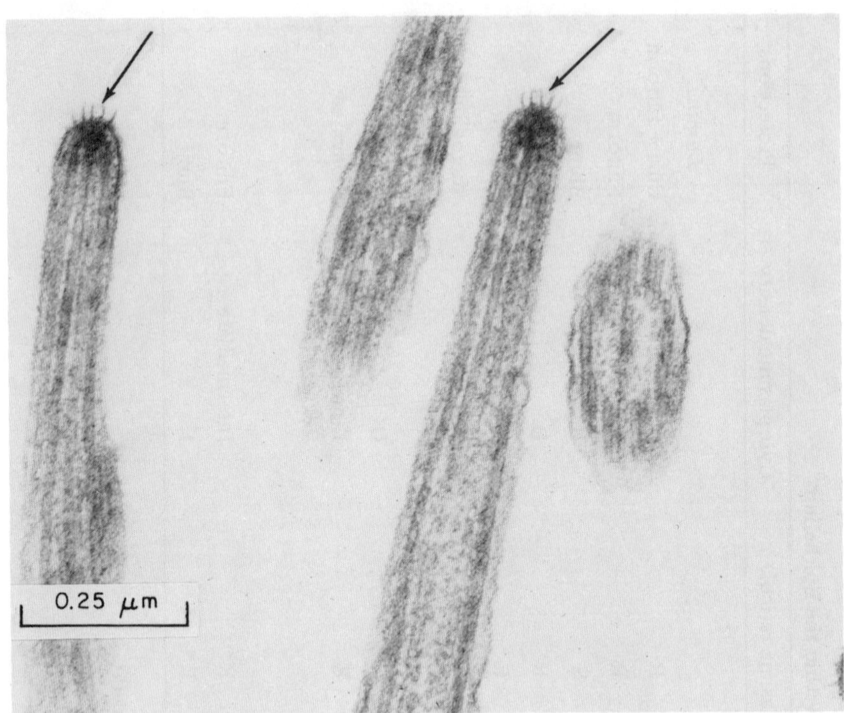

FIGURE 5 An electron micrograph of the tips of two cilia, taken from rat, which show small clawlike projections (arrows). Glutaraldehyde and osmium: lead citrate + uranyl acetate, ×120,000.

frequently found during fetal life; little is known of its fate postnatally (Fig. 6). In the specific pathogen-free (SPF) rat the serous cell is the most frequent type of secretory cell found proximally, that is in the extrapulmonary airway (see Table 3). The cell has electron-dense cytoplasm, a nucleus with irregular outline and a number of electron-dense secretory granules each 600 nm in diameter, discrete, and enclosed by membrane (Fig. 7). These granules are not osmiophilic and contain much protein probably in the form of neutral glycoprotein, since by light microscopy they stain pink with PAS in combined Alcian blue (pH 2.6)/periodic acid-Schiff staining. In this they resemble the serous cells of

---

FIGURE 6 Electronmicrograph of an epithelial serous cell in the trachea of a human fetus of 16 weeks gestation. The cell has many electron-dense secretory granules (arrow), a well-developed Golgi apparatus (Go) and an abundance of glycogen (gly). Glutaraldehyde and osmium:lead citrate + uranyl acetate. ×6720.

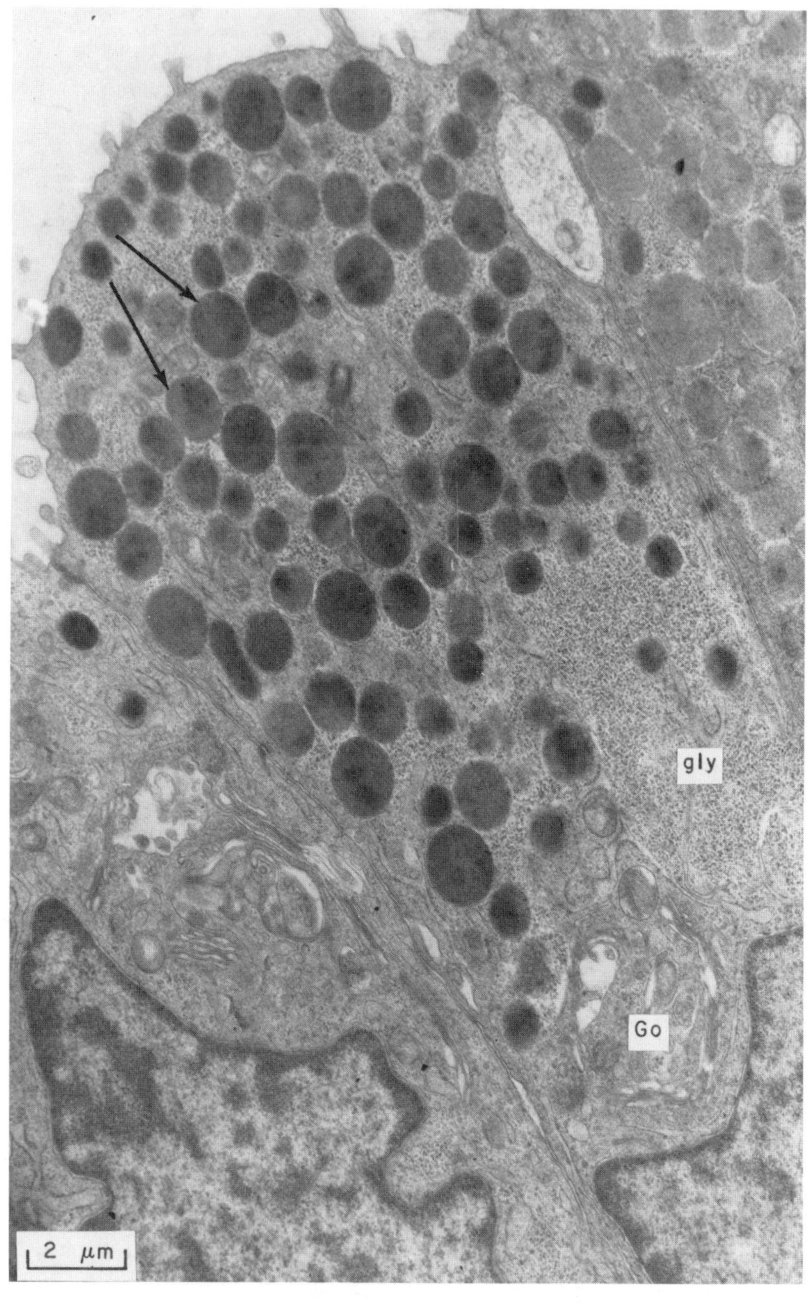

gly
Go
2 μm

TABLE 3 Mean Percentage of Each Cell Type at Five Airway Levels (Rat)

| | Epithelial cell | | | | | Migratory |
| | | Nonciliated granule | | | | |
	Ciliated	Dense	Lucent	None	Basal	% epithelial cells
I Trachea—upper	17	20	<1	27	36	14
II Trachea—lower	33	27	<1	13	27	8
III Bronchus—extrapulmonary	35	21	<1	16	27	4
IV Bronchus—intrapulmonary	53	20	<1	12	14	4
V Bronchiolus	65	20	<1	15	<1	1

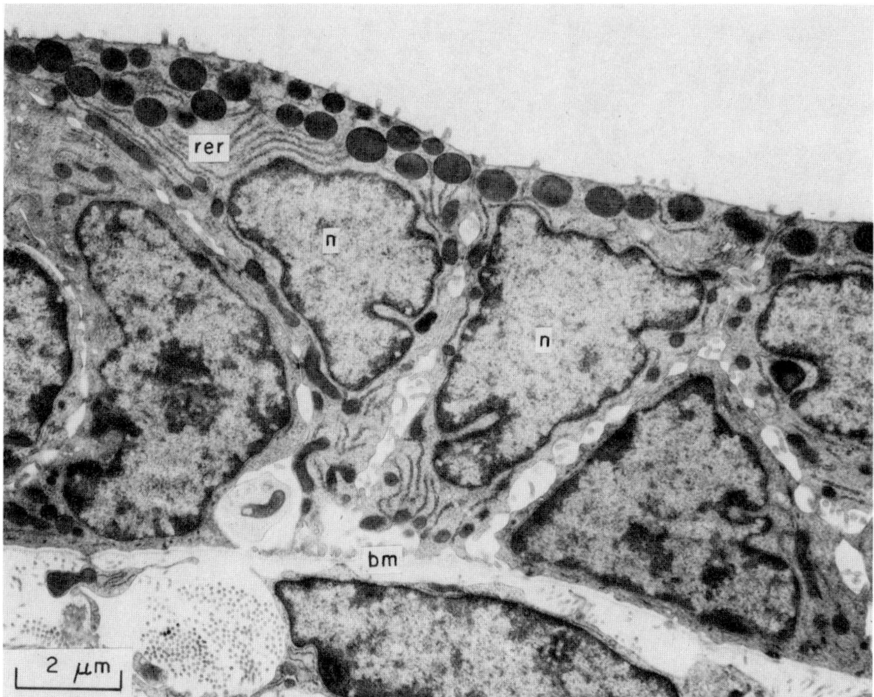

FIGURE 7   Epithelial serous cells (sc) in rat trachea, similar to those of the human fetus each with discrete electron-dense granules, electron-dense cytoplasms, and irregular nucleus (n).   Rough endoplasmic reticulum (rer) and basement membrane (bm).   Glutaraldehyde and osmium: lead citrate + uranyl acetate. X9000.

the human and rat submucosal glands.   Recent experiments [27] have indicated that this cell may transform into a goblet cell following irritation (see Sect. III).

3. Goblet Cell

Owing to their chalice-like shape because of their mucus content, the mucus-producing cells of the human lung have been named goblet cells [12].   Nevertheless their shape varies according to the degree of distention caused by the mucus they contain.   The term "goblet cell" is sometimes extended to include cells which contain small amounts of secretion.   In this chapter the term is restricted to cells distended by their mucus, goblet in shape, and probably producing acidic glycoprotein.   In the human airway these cells all produce sialomucin or a mixture of sulfo- and sialomucin.

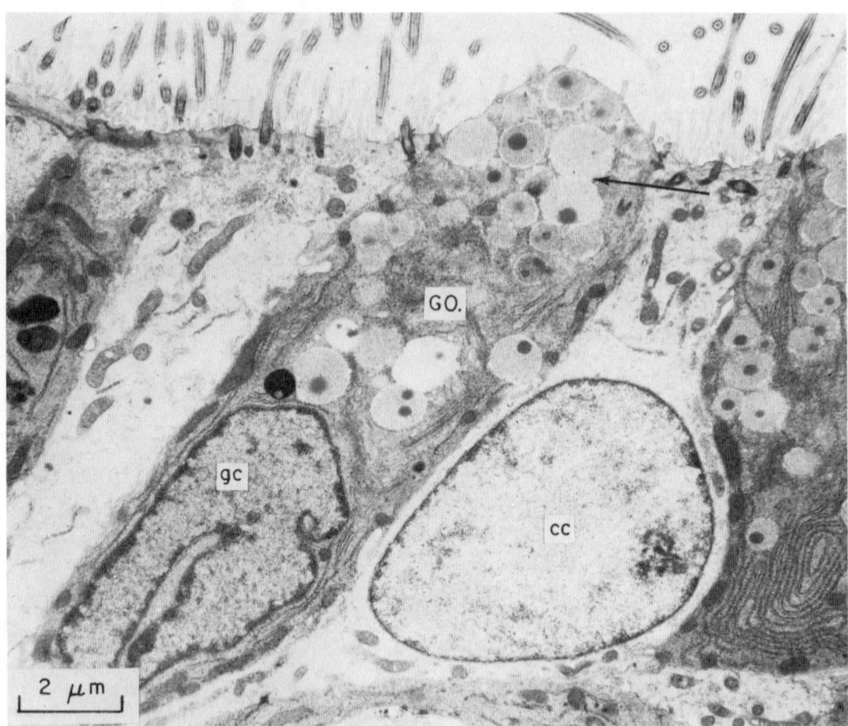

FIGURE 8   An electronmicrograph of a goblet cell (gc) in rat airway epithelium. The secretory granules are large, electron-lucent and tend to fuse (arrow): the cytoplasm is electron-dense with a Golgi apparatus (Go) and rough endoplasmic reticulum. Adjacent ciliated cell (cc). Glutaraldehyde and osmium:lead citrate + uranyl acetate. X9000.

By electron microscopy the appearance of goblet cells in all species is similar [17,19,22,23]. The cell is distended by electron-lucent secretory granules. The granules are larger than those of the serous cell and sometimes have an electron-dense core (Fig. 8). Each granule usually has an incomplete membrane and there is fusion between them. The cytoplasm is electron-dense and, like the serous cell, the nucleus irregular in outline.

4. Clara Cell

In 1937 Max Clara [15] published his findings on the structure, determined by light microscopy, of certain nonciliated cells found particularly in the terminal bronchiolus of man and rabbit. The apex of each cell projected into the lumen and contained a secretion which did not stain for mucus. The cell, found in many species, is commonly called the "Clara cell." By electron microscopy and in keeping with the other two secretory cells, the nucleus is irregular in outline but, in contrast, the cytoplasm is not electron-dense and is similar to that of adjacent cells (Fig. 9). As with the serous cell the secretory granule of the Clara cell is electron-dense and small, although very often it is irregular and angular in outline. One of the ultrastructural features of this cell is the abundance of endoplasmic reticulum which, in contrast to that of the other two types of secretory cell, is smooth and without ribosomes (Fig. 10). This structural difference suggests that the cell, though secretory, produces a different secretion from those of the goblet and serous cells. The nature of this secretion is still obscure (see below).

By electron microscopy, Clara cells have been described in a number of mammalian species including human, rat, rabbit, mouse, and pig [17,28-39]. The amount of smooth endoplasmic reticulum varies between species, being greatest in the rat and mouse and least in human. Secretory granules are usually homogeneous and electron-dense but in the human appear mottled [30]. In the rabbit they appear to be absent altogether [34]. The nonciliated bronchiolar cell, originally described in the mouse [35,28], has unusual mitochondria in that they are deficient in cristae and have encircling whorls of smooth endoplasmic reticulum. This does not appear to be the case in any other species. There is variation in electron density of cytoplasm both between species and in the same species [37]. Although the original light microscope observations led Clara [15] to suggest that this cell was confined to terminal bronchioli, electron microscopy has shown cells with features of Clara cells as far proximally as the hilum in rat and the trachea and nasal mucosa in the mouse [17,40,41], although, in each species, they appear to be most frequent in the region of the terminal bronchiolus.

The nature of the secretion produced by the Clara cell is the subject of controversy. Histochemical studies have shown that the granules are PAS-positive, indicating a neutral glycoprotein, and digested by pepsin, indicating the presence of protein [42,43]. Others [36] have demonstrated bound lipids and phosphoglycerides in them. By electron microscopic autoradiography, Niden [44] and Etherton and Conning [38] demonstrated in mice that within 5 minutes of giving an intraperitonal injection of labeled precursor of dipalmitoyl lecithin, radioactivity was apparent over the Clara cell and its granules. The significance of this is that dipalmitoyl lecithin is a constituent of surfactant, a substance known to be present in the alveolar region and whose function is to reduce alveolar and perhaps bronchiolar surface tension. This led to the suggestion that Clara cells were responsible for production of surfactant, but the weight of evidence is toward the alveolar type II pneumonocyte and not the Clara cell as responsible for the production of alveolar sur-

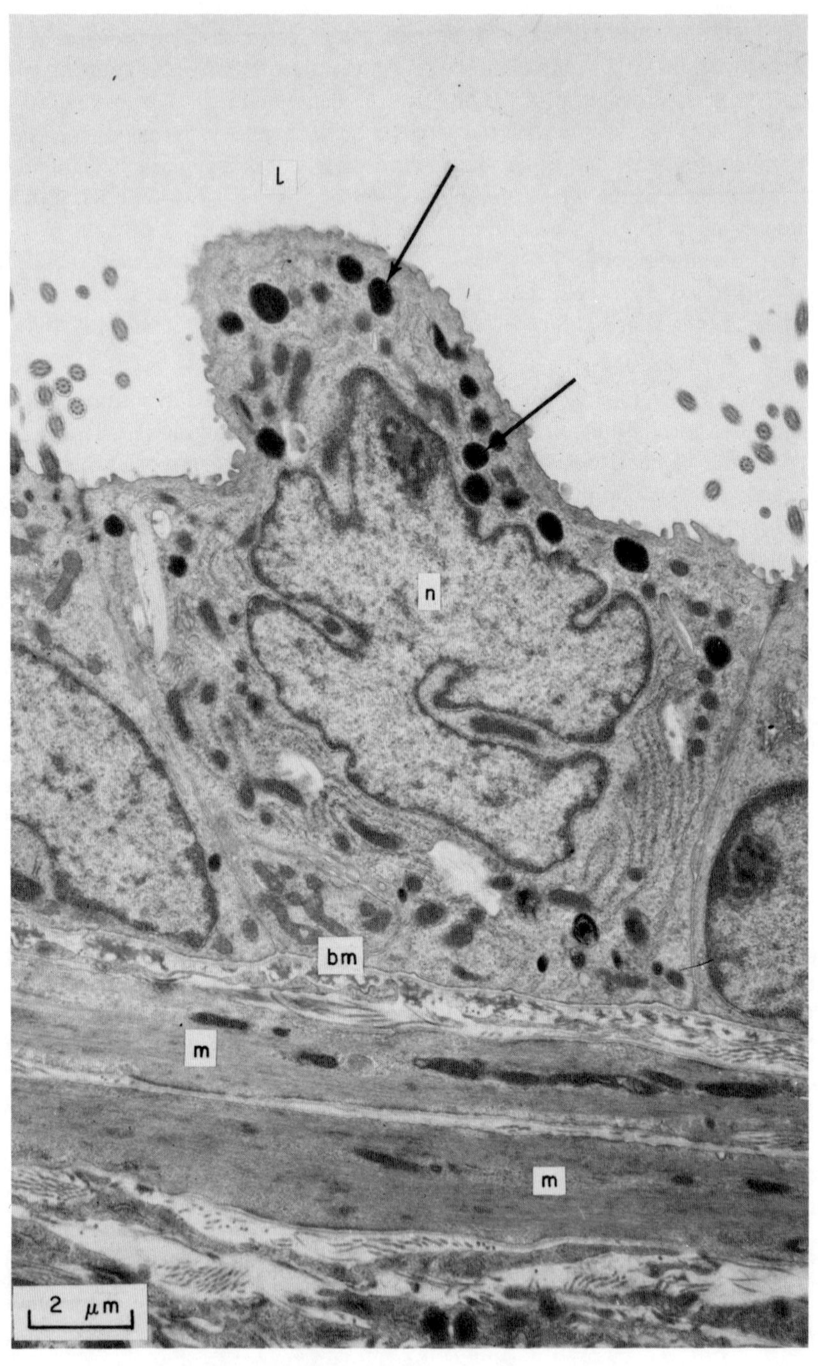
L
n
bm
m
m
2 μm

factant (see Ref. 45). Recently Petrik and Collet [46] showed that tritiated choline, a specific precursor of the dipalmitoyl-phosphatidylcholine active fraction of surfactant is not incorporated into mouse Clara cell granules; tritiated leucine, galactose, and acetate are, which suggests synthesis and secretion of protein, carbohydrate, and possibly cholesterol. Indeed the Clara cell does have ultrastructural features in common with cells which produce certain steroids, for example, testicular interstitial cells, whose secretion is based on cholesterol [47]. The true nature of the Clara cell secretion (and indeed whether it produces only one secretion) remains to be determined: recent experiments indicate these cells may transform and produce mucus (see Sect. III).

5. Other Nonciliated Cells

Four of the nonciliated epithelial cell types lack any secretory granule. The intermediate and brush cells are superficial in position and form part of the surface; the basal and Kulchitsky cells are basally situated and do not add to the surface.

a. Intermediate Cell

"Intermediate" cells have been described in the respiratory epithelium of a wide variety of species [4,14]. In pseudostratified epithelium, such as that of man, rabbit, and pig, they are elongated vertically, often spindle-shaped, and usually but not always reach the airway lumen [19,21,22]. Where the epithelium is more simple in type (such as in rat) the intermediate cell is more cuboidal and reaches the surface [17,23].

These cells are called intermediate because they lack either secretory granule or cilia and their apical processes are few and short. In this respect they are "undifferentiated" and capable of transformation into any of the epithelial cell types. The electron density of the cytoplasm is variable, presumably electron-dense if it is to form a secretory or electron-lucent if it is to form a ciliated cell. The cell is found both in proximal and distal airways.

FIGURE 9 A Clara cell in rat bronchiolus. The nucleus (n) is irregular, the cell apex bulges into the airway lumen (l), the cytoplasm is not electron-dense but similar to that of adjacent ciliated cells. Secretory granules (arrows) are small and electron-dense. Basement membrane (bm) and muscle (m). Glutaraldehyde and osmium:lead citrate + uranyl acetate. X7500.

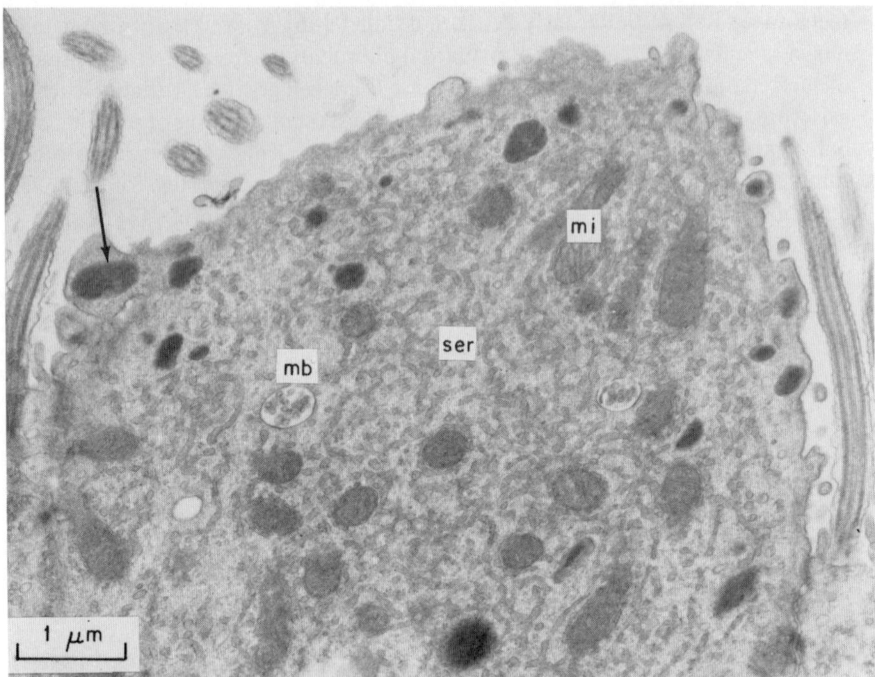

FIGURE 10   An electron micrograph of the apex of a Clara cell showing the
abundance of smooth endoplasmic reticulum (ser) present as short profiles.
Mitochondrion (mi), multivesicular body (mg), and secretory granule (arrow).
Glutaraldehyde and osmium:lead citrate and uranyl acetate. ×18,750.

b.   Brush Cell

While lacking secretory granules the brush cell is distinct from the intermediate
in having a pronounced brush border of microvilli, about 2 μm thick, vesicles
and fibrils in the cytoplasm and, often, glycogen (Fig. 11). While reported to
be present in man [28,48], morphologically it bears little resemblance to the
brush cell as described in rat [17,23,49] and pig [39]. A brush cell has been
found in rat alveolus [50], nasal mucosa [51], gut [52], and bile duct [53].
The function of the cell is unknown, but it is likely to be one of absorption,
perhaps of fluid from the periciliary layer.

c.   Basal Cell

Of the two basally situated cells the basal cell is the most frequently found, the
Kulchitsky being rare. Early electron microscopic studies [23] suggested that
the basally situated cells were blood-derived white cells, but it is now clear that,
while some are lymphocytes (see below), most are truly epithelial (endodermal

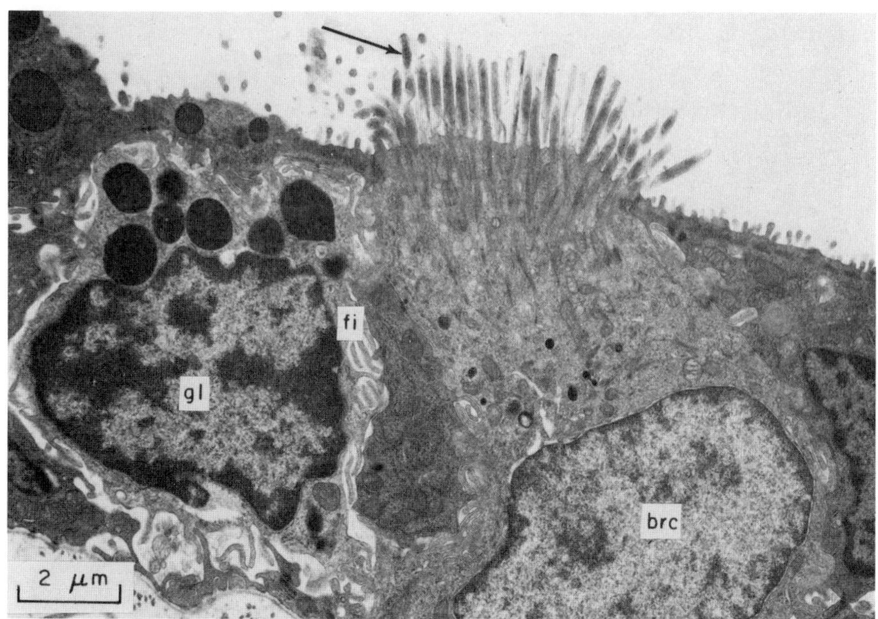

FIGURE 11 A brush cell (brc) and globule leukocyte (gl) in rat epithelium.
The brush cell has a pronounced border of microvilli (arrow), with many vesicles
and fibrils in its cytoplasm; the globule leukocyte has a nucleus with much peripheral chromatin, large electron-dense granules, and filiform projections (fi) from its
surface. Glutaraldehyde and osmium:lead citrate and uranyl acetate. ×9000.

in origin). The basal cell is likened to cells of the germinal layer of the epidermis
which have sparse electron-dense cytoplasm and many filaments called "tono-
filaments." Like them the basal cell is the stem cell from which the overlying
superficial cells develop by maturation [54], and it is in this basal layer that
mitotic figures are most often, though not exclusively, seen [54,55]. In the
bronchiolus, basal cells are infrequent and here the single layer of cells must
both divide and differentiate. Generally, differentiation is not compatible with
division, the more differentiated the cell (e.g., nerve cell) the less it is likely to
be able to divide. This is true of respiratory epithelium also. Here differentiation
is not so advanced and specialized as it would be in the nerve cell, but then
mitotic figures are not very frequent, the turnover of the entire epithelium being
somewhat slow (45-200 days, depending on whether it is a central or distal airway) [55]. Irritation of the epithelium has profound effect on epithelial turnover and on which cells divide (see below).

d. Kulchitsky Cell

Bronchial cells resembling the "Kulchitsky" or enterochromaffin cells of the gut
have recently been found in human airway as well as those of other species [29,

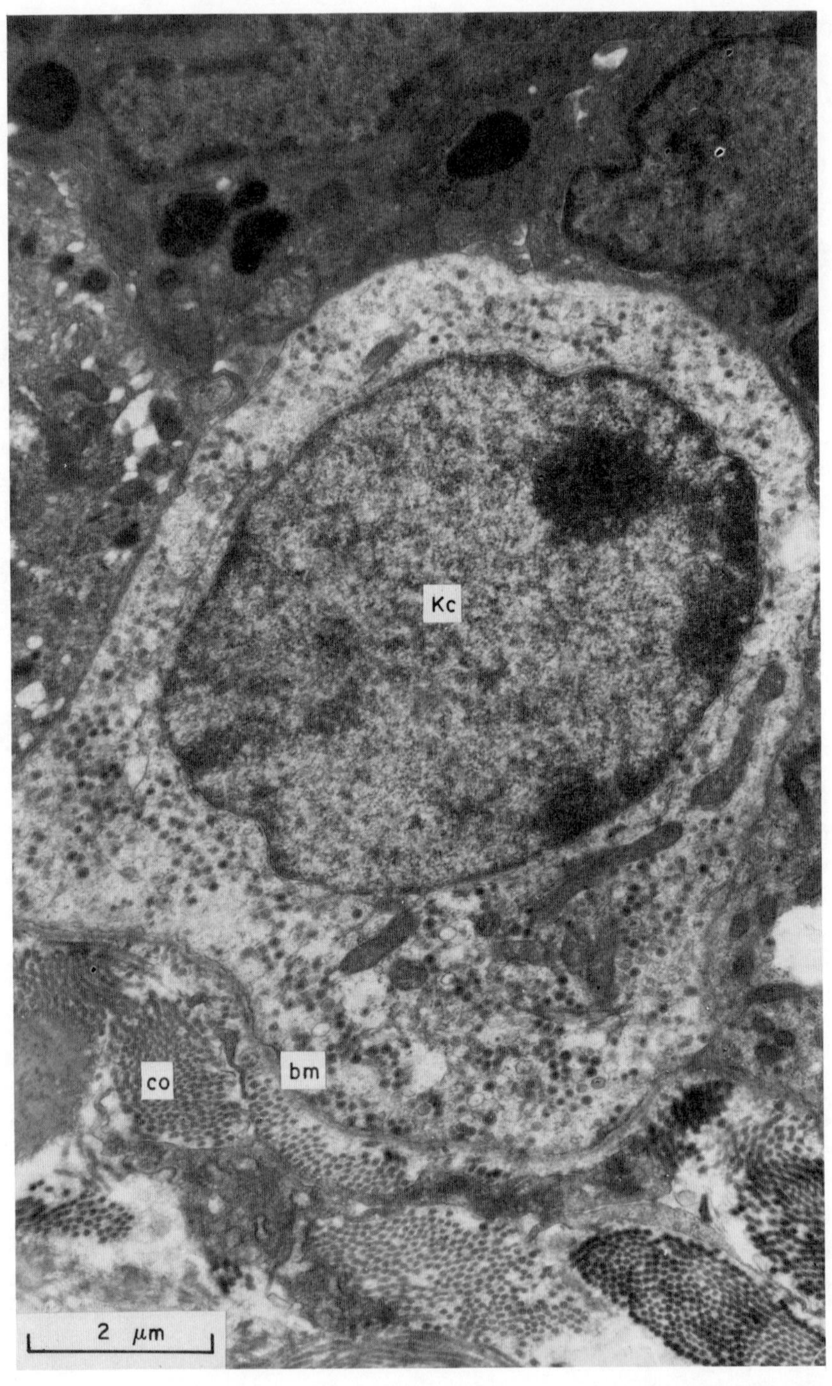
Kc
co
bm
2 μm

30,56-62]. The cell has many synonyms, including AFG (argyrophil, fluorescent, and granulated) and the Feyrter cell.

By electron microscopy the bronchial Kulchitsky cell contains large numbers of dense-cored vesicles distributed throughout its electron-lucent cytoplasm (Fig. 12). Fibrils and lipid bodies may or may not be present. The cell does not usually reach the airway lumen, although morphologically related cells called neuroepithelial bodies do [63].

The bronchial Kulchitsky cell is found both in the adult human and with greater frequency in the prematurely born infant [29,30,60]. The cell has also been found in fetal and 3-day-old rat [59,64].

It is generally agreed that the bronchial Kulchitsky cell should be included in the diffuse system of APUD (amine precursor uptake and decarboxylation) cells [65-67]. There is evidence of its amine uptake and storage [68] and of discharge of secretory granules, shown to contain serotonin, in response to hypoxia [64,69]. The role of this cell and its special secretion in respiratory epithelium is still unclear. If it is part of the APUD system then it may not be endodermal, as are the other epithelial cells, but rather ectodermal and have closer links with the nervous system (see Ref. 1, Chap. 3 for further details of the origin of the bronchial Kulchitsky cell).

6. The Migratory Cells

Two migratory cells are found normally in airway epithelium. These cells do not show the specialized zones of cell contact and adhesion (i.e., desmosomes and tight junctions) shown by other cells of the epithelium. In the human, only the lymphocyte has been found, in the rat the lymphocyte and globule leukocyte as well.

a. Lymphocyte

By light microscopy lymphocytes have long since been described in airway epithelium [16,70,71]. Ultrastructurally they have been found in the human fetus (see Ref. 1, Chap. 3), rat fetus, and adult rat [17]. It is likely that they are long-lived and remain in the epithelium for long periods of time. Their function is unknown. The lack of organelles suggests that they are in an unstimulated state and not actively producing immunoglobulins (Fig. 13).

FIGURE 12 A bronchial "Kulchitsky" cell (Kc) at the base of the epithelium lining the rat trachea. Its cytoplasm is electron-lucent and contains numerous dense-cored vesicles. Basement membrane (bm) and collagen (co). Glutaraldehyde and osmium:lead citrate and uranyl acetate. X12,500.

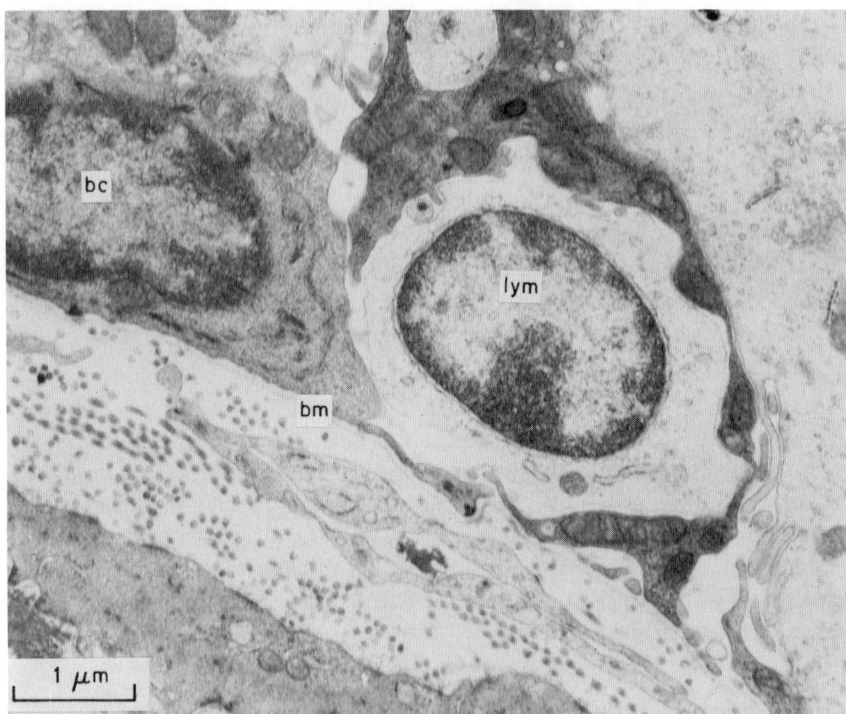

FIGURE 13 An electron micrograph showing a lymphocyte (lym) in rat airway epithelium. The lymphocyte has sparse cytoplasm which is electron-lucent and contains virtually no organelles. Adjacent basal cell (bc) and basement membrane (bm). Glutaraldehyde and osmium:lead citrate + uranyl acetate. ×20,000.

b. Globule Leukocyte

The globule leukocyte, identified light microscopically by its markedly eosinophilic granules, is present in mucous membranes throughout the body in a wide variety of species [72]. In the human it has been identified in gut mucosa [73] but as yet not in airway epithelium.

Ultrastructurally the cell is similar to a mast cell, but the granules or "globules" are larger. The granules are unlike those of the "eosinophil" in that they are rounded and homogeneously electron-dense (see Fig. 11). It is a characteristic of the granule that it loses its electron density when fixed by osmium alone [17,18]. Its function is unknown, although there is some evidence to suggest it has an immunological role during parasitic (nematode) infections [74-77].

B. Intraepithelial Nerves

Nerve fibers have now been shown to penetrate the basement membrane of the epithelium of both the upper and lower respiratory tract and of submucosal glands, each nerve ending deep within the epithelium or gland in close association with epithelial cells.

With the electron microscope intraepithelial nerves have been identified in the upper and lower airways of man, rat, cat, rabbit, chicken, and goose and a number of other species [19,58,62,63,78-81]. In man and rat, nerves have been shown to penetrate the submucosal gland also [56,82,83].

In each species the nerves are unmyelinated, have no Schwann cell covering or basement membrane and lie in very close apposition—15 nm—to adjacent cells. They are recognized by their electron-lucent cytoplasm, characteristically small, elongated mitochondria and neurotubules (Fig. 14). In man, cat, rabbit, and mouse, they are typically devoid of any neurosecretory vesicle and sometimes have accumulations of mitochondria: these features have led to the suggestion

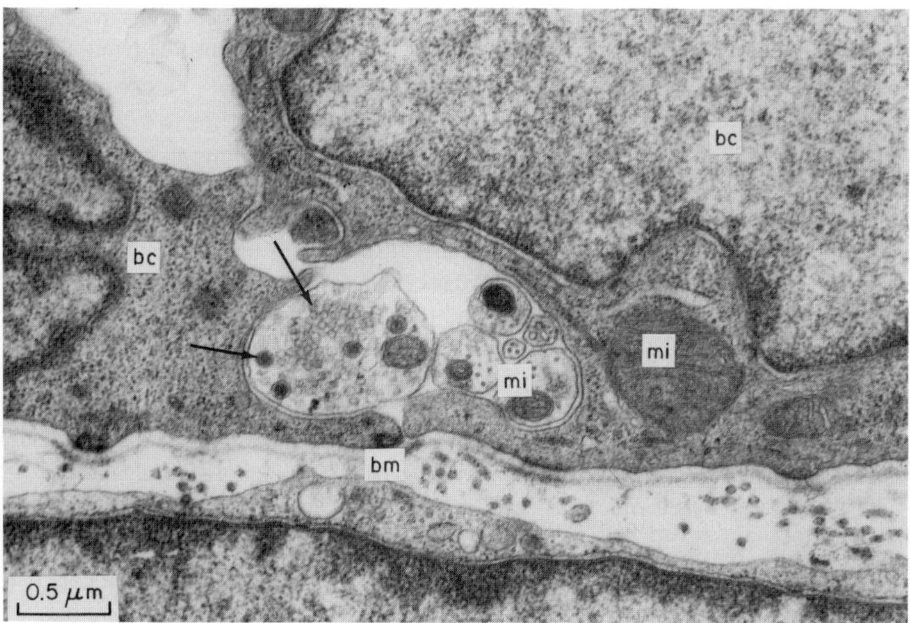

FIGURE 14 A group of six intraepithelial nerves just superficial to the epithelial basement membrane (bm) and enclosed by the processes of two basal cells (bc). All the axons have electron-lucent cytoplasm and neurotubules; the largest has neurosecretory vesicles of two types (arrows). Mitochondria (mi) are typically small in diameter when compared to those of adjacent epithelial cells. Glutaraldehyde and osmium: lead citrate + uranyl acetate. ×30,000.

that they are sensory. In rat, chicken, and goose, many axons (up to 50%) also have collections of vesicles and if the predictions of Katz [84], Bennet [85], and others are correct, then these vesicles indicate that the axons are motor. It may then be that in some way they regulate epithelial function. More experimental work is needed to confirm these suggestions, but there is already physiological evidence from the cat, goose, and rat that drugs which mimic the actions of the autonomic nervous system, and others such as the prostaglandins, stimulate the nerves supplying the epithelium and affect either the ciliary rate or the mucous output or both [86,87]. Fewer observations have been made on the nerves supplying the cells of the submucosal gland, but some axon terminals do appear to have vesicles and physiological data suggest that both sympathetic and parasympathetic nerves are present, that they are motor and regulate mucus discharge [86,88].

C. Epithelial Cell Turnover

Henle [89] was one of the first anatomists to mention the presence of mitotic figures in the tracheobronchial epithelium, claiming that they were scattered through the tracheal epithelium and that under normal conditions, the renewal of this epithelium was slow.

It was not until 1951 that Bertalanffy [90], using colchicine to arrest mitosis, estimated the turnover time of the rat tracheal epithelium to be 48 days. Using autoradiography and tritiated thymidine in mice, Spencer and Shorter [91] reported the epithelial turnover in small bronchi and respiratory bronchioli as 7-10 days, while Gottesberge and Koburg [93] reported longer turnover times: 18 days in large bronchi and 60 days in small bronchi. Shorter and colleagues [93,94] studied mice and rats and found in adult Sprague-Dawley rats, weighing between 150 and 160 g, no difference in epithelium turnover time between 3- and 10-week-old male animals. In Blenkinsopp's result [54] the turnover time in 330-g rats tended to be slower than in those of 200 g. Different airway levels were compared by Shorter et al. [94] and Blenkinsopp [54] who reported a shorter turnover time in the trachea than in more peripheral airways and by Bertalanffy [95] who recorded the reverse.

Wells [96] studied the kinetics of cell preliferation in the tracheobronchial epithelium of rats with and without chronic respiratory diseases. Studying upper and lower tracheal levels in 5-week-old animals (their sex is not stated) he found that the turnover time is slower in those with less disease. He reported finding cells in mitosis in both the basal and superficial layer.

Recently Bolduc and Reid [55] studied cell turnover in the rat lung, at five different airway levels in both male and female rats and at three ages. The pattern of cell turnover was established with the light microscope after adminis-

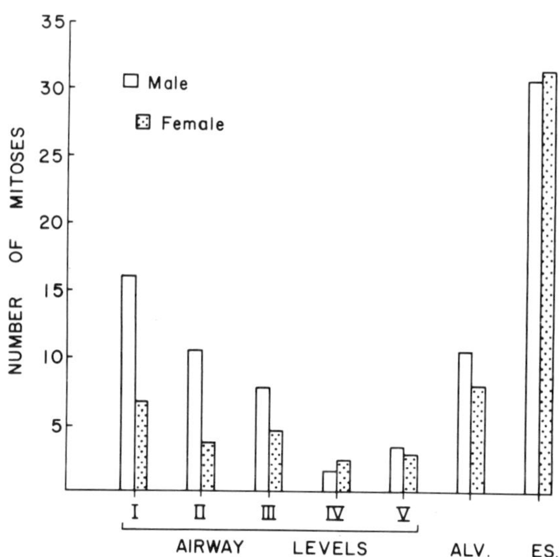

FIGURE 15 The "Mitotic Index" in a group of 33-day-old rats, males and females. Upper, mid, and lower trachea, (levels I–III, respectively), main bronchus
(level IV) and a peripheral bronchiole (V). The alveolar region and esophagus
are included for control purposes.

tration of colchicine, or tritiated thymidine. The position of the dividing nucleus
within the epithelium was also determined. The animals were specific pathogen-
free. Three tracheal levels, a main bronchus, and a peripheral intrapulmonary
airway were analyzed. At each airway site the Mitotic Index was estimated,
i.e., the number of cells in mitosis per 1000 epithelial cells (Fig. 15).

In the young group (age 33 days; mean weight: male 133.5 g, female
114.5 g) the Mitotic Index decreased progressively from trachea to axial pathway. It was significantly higher in the trachea of the male than the female, yet
the animals were gaining weight at the same rate. In the axial and lateral bronchi
the proportion of superficial cells increases, being about a quarter in the axial
bronchus and a third to a half more distally in the lateral bronchiolus. Up to
half of the dividing nuclei were near the basement membrane, rather fewer in
the male than in the female.

In the second group (age 47 days; mean weight: male 241 gm female
199 g) in both the male and female animals, the trend is the same but less
marked.

In contrast, in the group with the oldest rats (age 105 days; mean weight:
male 555 g, female 334 g) no significant difference was seen between any tracheal
level and the other airways in either the male or female animals.

The concentration of cells was significantly greater in the trachea than in the intrapulmonary airways. It was greater in the male than in the female and in old than in the younger animals. But it should be stated that where cell concentration is high a given Mitotic Index represents a faster replacement per unit of airway than where concentration is low. For example, in the peripheral airway where cells are less concentrated, a low Mitotic Index will suffice to give the same turnover time per unit area of airway wall as a higher one in an extrapulmonary airway where cell concentration is greatest.

D. Tracheobronchial Submucosal Gland

Since the early descriptions of mucus-producing glands in airway submucosa [14,97] much has been learned of their histochemistry, anatomical arrangement, and ultrastructure.

Using hematoxylin and eosin stains two cell types have been identified in submucosal glands of both the upper and lower respiratory tracts: mucous and serous, the latter often described as forming demilunes around mucous acini. After a combination of Alcian blue (pH 2.6) and periodic acid-Schiff stains, mucous cells stain blue for acid glycoprotein, while most of the serous cells stain pink. Details of their histochemistry are considered later.

1. Reconstruction of Gland

Recently a new duct region of the gland has been found which does not produce mucus and, for the first time in the human, four distinct regions of the gland are recognized: (a) the ciliated duct, (b) collecting duct, (c) mucous tubules, and (d) serous tubules [98]. The relationships of these regions to each other have now been determined by a graphic reconstruction of serial sections of a gland taken from a normal human bronchus (Fig. 16): similar results have been obtained from 19 other specimens. Details of the cell and duct dimensions are given in Table 4.

a. Ciliated Duct

Dipping into the airway wall and continuous with its lumen and epithelial lining is a nonbranching duct with ciliated epithelium similar to that of the airway: this is the ciliated duct.

b. Collecting Duct

In the "collecting duct" the ciliated epithelium is replaced by one with unusually tall nonciliated cells packed with mitochondria and with markedly eosinophilic

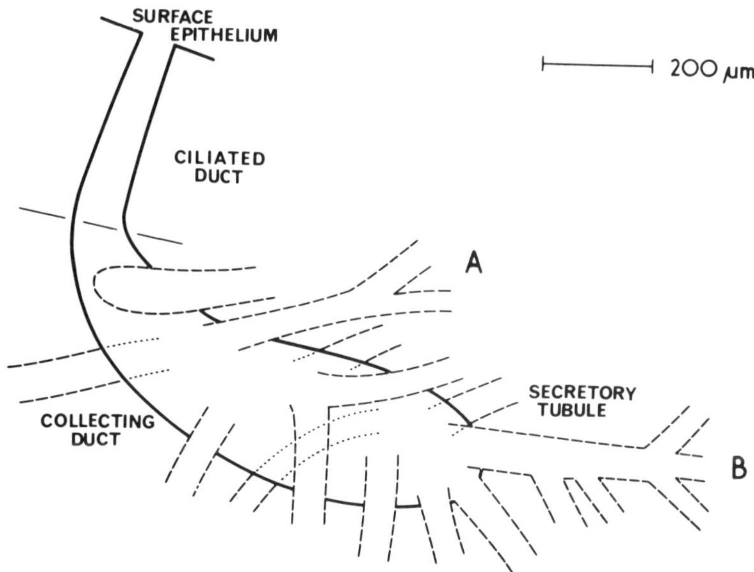

FIGURE 16 A graphic reconstruction of serial sections of a submucosal gland from a normal human bronchus. Secretory tubules A and B are reconstructed in Fig. 17.

TABLE 4 Dimensions (μm) of Ducts and Secretory Tubules in Graphic Reconstruction of Human Submucosal Gland from Main Bronchus

	Length	External diameter	Height of epithelium
Ciliated duct	350	65-90	20-30
Collecting duct	800	90-250	60-70
Mucous tubule	500	50-75	25-30
Serous tubule	180	50-70	20-30

cytoplasm [99]. The duct is about 1 mm in length and runs obliquely to the surface epithelium. Branching of the duct has been seen. From its junction with the ciliated duct the collecting duct increases in external diameter to a maximum of 250 μm. Due to its thick epithelial lining the widest point of the lumen is 100 μm and this narrows toward its distal end. The transition from duct epithelium to that of a secretory tubule is abrupt, the junction of the two types of epithelia usually coinciding with the point of branching. Blood capillaries appear to be more numerous around the collecting duct than around the secretory tubules.

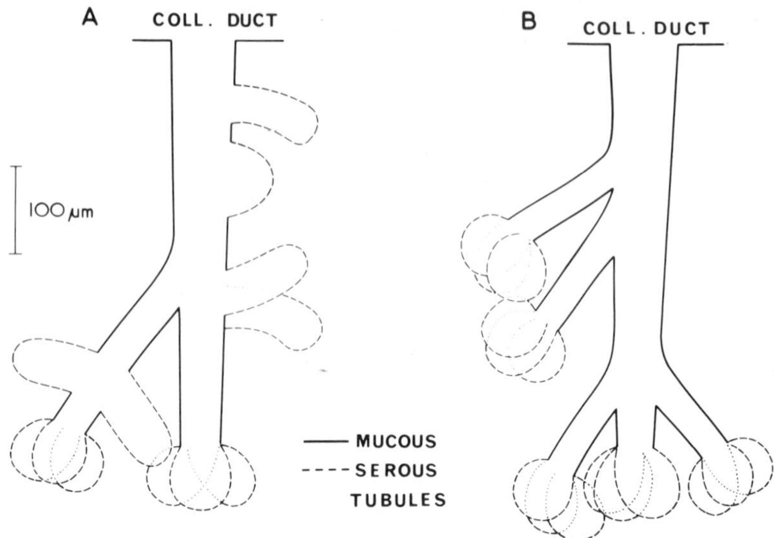

FIGURE 17   Reconstructions of secretory tubules A and B showing their origin with the collecting duct, mucous, and more distally placed serous tubules.

c. Mucous Tubules

The branches arising directly from the collecting duct are usually mucous: that is, they are lined by mucous cells. In the reconstructed gland there were 13 major branches at least 50 μm long and with a distinct lumen: Fig. 17 shows two which were reconstructed. Invariably they subdivide in a random manner and usually several times. The mucous cells are filled with a confluent mass of secretion that compresses the nucleus against the base of the cell.

d. Serous Tubules

Most serous tubules are located at the ends of the mucous tubules, either singly or in groups: some arise like "buds" from the lateral walls (Fig. 17). A serous tubule may be as short as 50 μm, while the longest seen was 180 μm. Their lumen is small and occasionally contains secretion.

For each gland the lumens of all its tubules and ducts are continuous and their anatomical arrangement ensures that secretion of the serous cell always passes over that of the mucous. Both secretions then pass into the collecting duct (where ionic adjustments may be made) and through it and the ciliated duct onto the surface lining epithelium of the airway.

2. Gland Ultrastructure

The ultrastructural features of the human submucosal gland [56,99] and of the opossum also [100] have been recently reported. In the normal human gland five cell types have been found: (a) mucous, (b) serous, (c) collecting duct, (d) myoepithelial, and (e) "clear" cells.

a. Mucous Cell

The mucous cell is usually packed with secretory granules of moderate electron density. The size range of granules is between 300 and 1800 nm, the smallest being central and near the Golgi apparatus. The largest granules are near the periphery of the cell where they appear to lose their limiting membrane and fuse. In this latter respect they resemble the goblet cell of the surface epithelium.

b. Serous Cell

The serous cell has numerous secretory granules usually concentrated toward the apex of the cell. Although they are packed densely, they do not distend the cell as in the mucous cell. Individual granules are usually spherical and, even at the cell apex, completely surrounded by a membrane (Fig. 18). While the majority of serous granules are electron-dense, a few pale granules are seen near the Golgi apparatus. The majority of granules are between 300 and 1000 nm in diameter, but smaller granules of 100 nm are found, and occasionally larger granules up to 1800 nm. Where there are cell projections into the acinar lumen or intercellular canaliculi, numerous vesicles are seen nearby containing material with an electron density similar to that of the secretory granule.

An apical dilatation of the intercellular space represents the intercellular canaliculus seen with the light microscope. At the polar extremities of this space tight junctions are found. Some intercellular canaliculi project into the cell cytoplasm to a depth equal to one-third of the cell diameter. Secretory granules are found in the cytoplasm surrounding canaliculi (Fig. 19).

The total number of granules varies from cell to cell even within the same section. The size distribution of the granules suggests that within an acinus the cells are often in the same phase of the secretory cycle. Although in some cells no granules below 300 nm were seen, in others they were the predominant size. When granules are numerous, those larger than 1300 nm do not predominate. A single cell does not show the full size range of granules, but within each cell the granules are grouped over a limited size range. A cell seems never to be emptied completely of its granules. Although granules may reach 1800 nm in size it is thought they are secreted at sizes ranging from 1000 to 1500 nm [99].

　　　　　　　　　　　　　　　　　　P. K. Jeffery and L. M. Reid

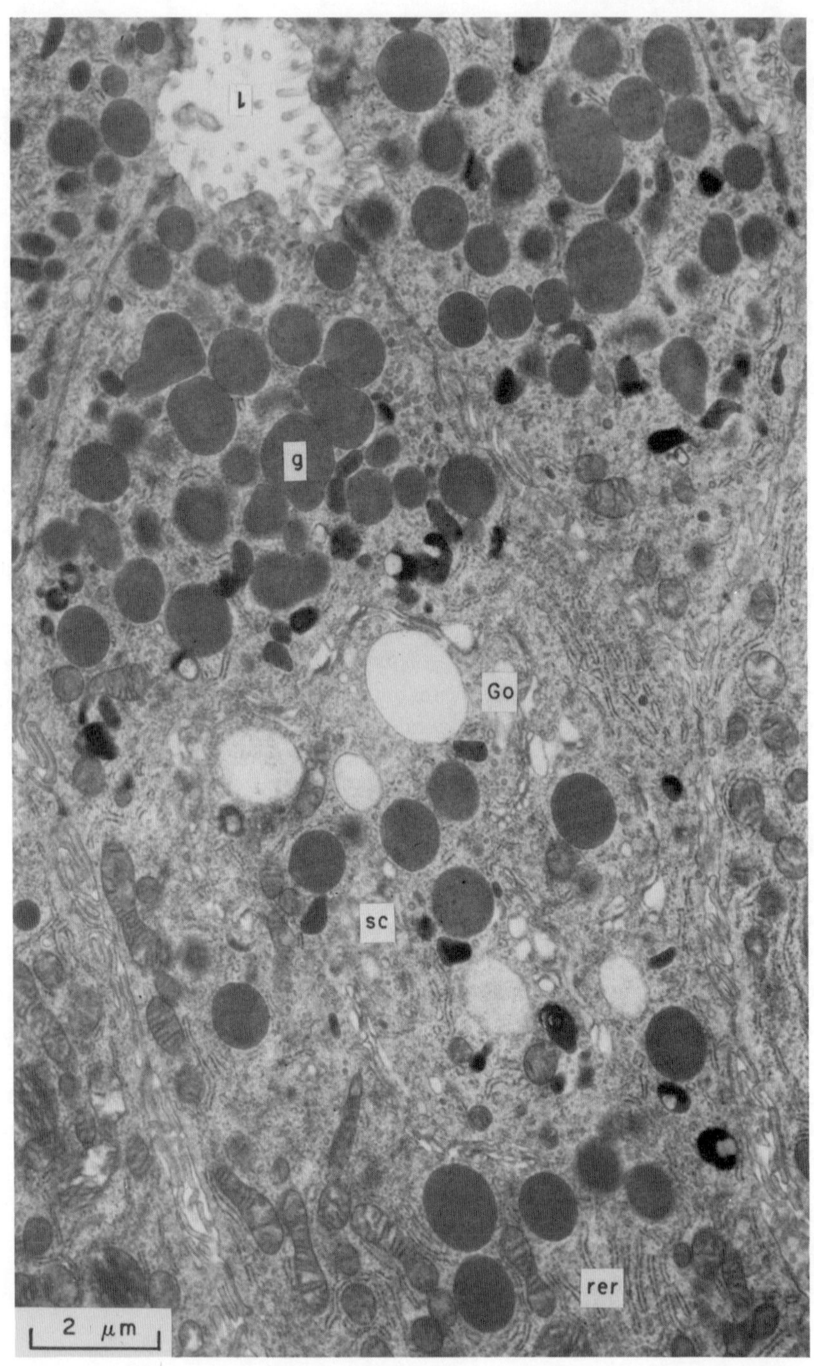

c. Collecting Duct Cell

The cells of the collecting duct are striking in their ultrastructure (Fig. 20). The
nucleus is ovoid, centrally placed, and with a prominent nucleolus. These cells
are 60–80 μm high and striking because of their densely and uniformly packed
mitochondria, mainly orientated to run parallel with the lateral cell membrane.
In the supranuclear region of the cell a well-developed, discrete Golgi apparatus
is found, containing basket-shaped groups of lamellae, numerous vesicles measuring 40–60 nm in diameter with only an occasional vacuole. Osmiophilic granules, measuring 150–400 nm in diameter, are seen in this area and, in the majority of cells, lipochondria. Transitional forms between the osmiophilic granules
and lipochondria are seen. There is little endoplasmic reticulum but free ribosomes are numerous. Around the whole cell circumference, immediately under
the cell membrane, lies a narrow zone free from mitochondria and with a fib-
rillar appearance. The apex of the cell is flat with a few cytoplasmic extensions.
The base lacks the striations seen in its salivary counterpart.

The cells are found directly against mucous cells but not against serous
cells, and myoepithelial cells are found beneath the duct cells also.

d. Myoepithelial Cell

Myoepithelial cells are found beneath the serous, mucous, and collecting duct
cells. They contain numerous fibrils running parallel to the basement membrane. These are thought to be contractile, their action probably contributing
to the passage of secretion along the secretory tubules and collecting duct.
The acinar basement membrane is narrow, and not well defined. The secretory
cell membrane is crenated at junctional regions of myoepithelial cells.

e. Lymphocyte ("Clear Cell")

In most sections about half of the acini show at least one "clear cell" (Fig. 21).
These cells are small; they have not been seen to reach the lumen but are
found between the cells and the basement membrane associated with the serous,
mucous, and especially the collecting duct cells. They are identified by their
sparse cytoplasm, almost free of organelles; in any one section only a few mitochondria are seen and a few cisternae of rough endoplasmic reticulum. The

---

FIGURE 18   An electron micrograph of part of a human serous acinus showing
the electron-dense secretory granules (g) of a serous cell (sc) accumulating in a
region adjacent to the acinar lumen (l). The serous cell has a prominent Golgi
apparatus (Go) and rough endoplasmic reticulum (rer). Glutaraldehyde and
osmium: lead citrate + uranyl acetate. X9000.

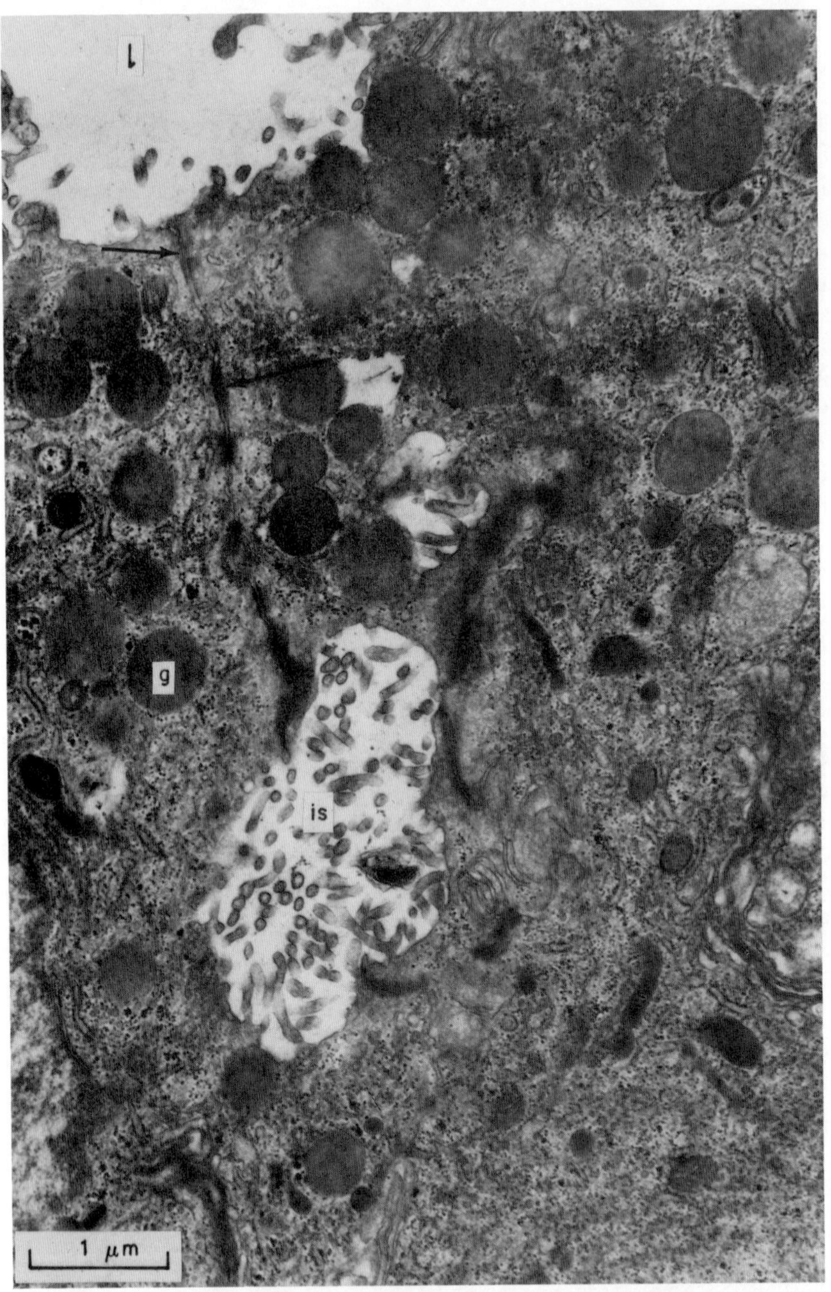

FIGURE 19 A canaliculus seen by electron microscopy as a dilated intercellular space (is) with filiform cell projections. Secretory granules (g) of two serous cells lie nearby. Acinar lumen (l) and cell junctions (arrows). Glutaraldehyde and osmium:lead citrate and uranyl acetate. ×18,000.

224

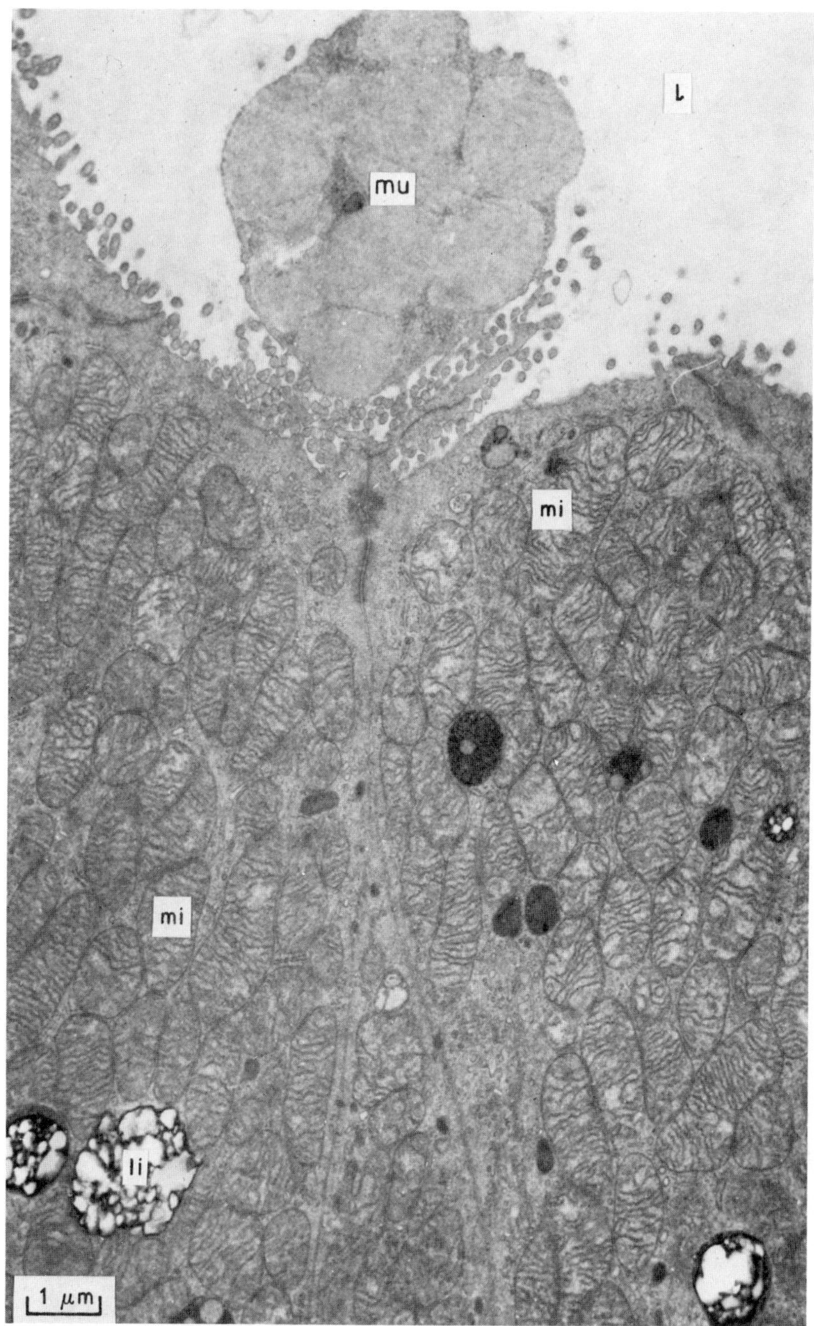

FIGURE 20   An electron micrograph of collecting duct cells in human submucosal gland. Each cell is packed with mitochondria (mi). Lipochondria (li) may occasionally be present also. Mucus (mu) is present in the duct lumen (l). Glutaraldehyde and osmium:lead citrate and uranyl acetate. ×10,000.

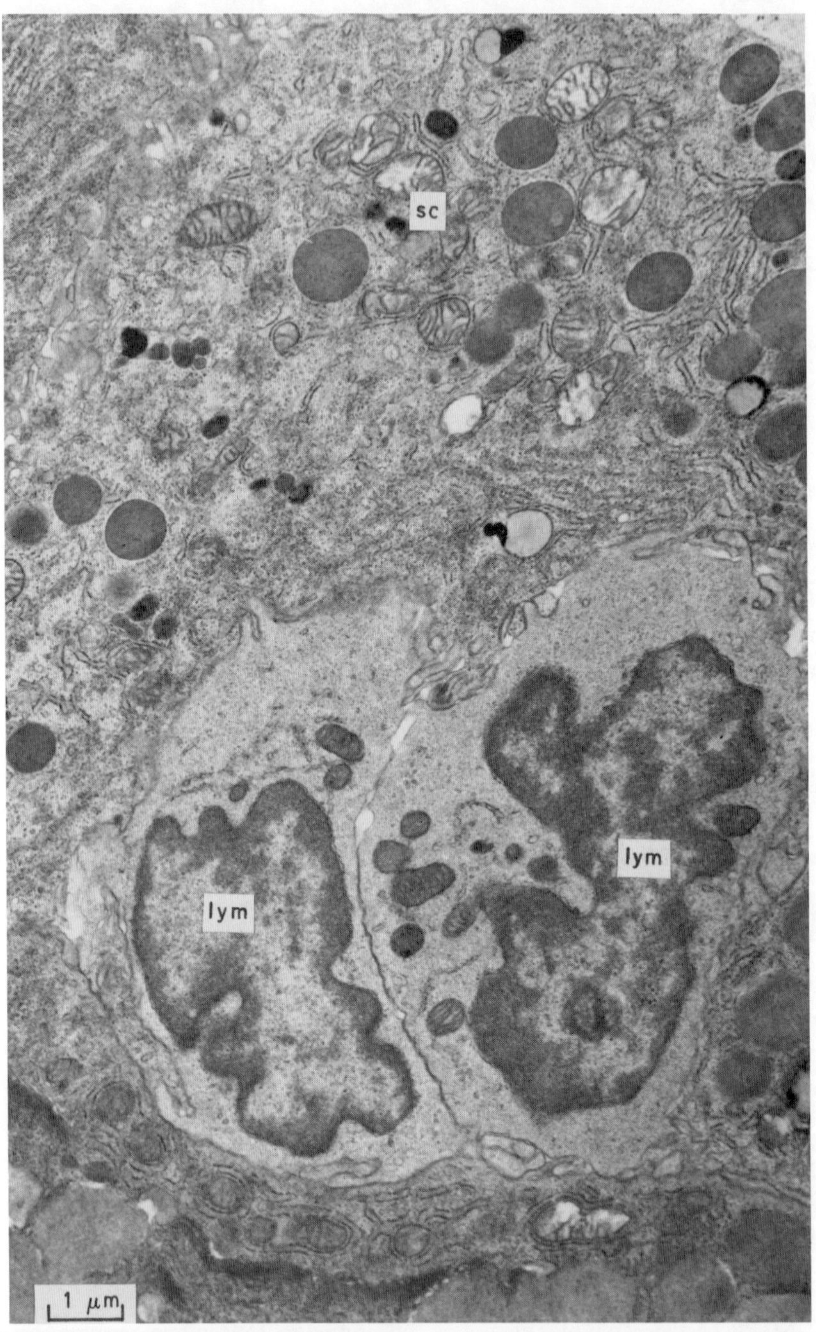

FIGURE 21 Two lymphocytes (lym) in human submucosal gland, each with sparse cytoplasm almost free of organelles. Serous cell (sc) and mucous cell (mc). Glutaraldehyde and osmium:lead citrate and uranyl acetate. ×10,000.

cytoplasm includes numerous free ribosomes and polysomes. This cell would seem to be a lymphocyte and as such resembles that found in the surface epithelium lining airways.

Occasionally a mast cell may be found penetrating the acinar basement membrane. Bensch and his colleagues [56] have described a Kulchitsky cell and its elongated processes at this site also. Bundles of unmyelinated axons are present between acini and, within the basement membrane, endings which may either be nerve or a Kulchitsky cell process are found.

E. Histochemical Types of Glycoprotein

Both neutral and acid glycoproteins have been demonstrated in surface epithelium and submucosal gland. Of the various stains available the periodic acid-Schiff and Alcian blue techniques (henceforth referred to as PAS and AB, respectively) are much used. The PAS technique demonstrates all epithelial glycoproteins by staining them pink [101,102]. Alcian blue was first introduced as a specific stain for the demonstration of acid glycoprotein in 1950 [103] and, used at pH 2.6, stains both sialic acid and sulfate radicals bright blue. If a combination of AB/PAS staining is used [104], the blue AB staining demonstrates the presence of acid radicals, the remaining pink PAS staining, neutral glycoprotein. Used in conjunction with the AB/PAS technique, neuraminidase (i.e., sialidase) digestion demonstrates the presence of digestible (susceptible) sialic acid by the loss of AB staining in the section [105]: acid hydrolysis removes all sialic acid but leaves sulfate radicals which stain blue [106]. The presence of sulfate radicals as the acidic component can be shown by the uptake of radioactive sulfur [107,108] or specific stains for sulfate, although the latter stain only a proportion of the total area containing sulfomucin. Recently the control of AB staining, by varying its pH between 0.5 and 2.6, has been found to be satisfactorily selective and distinguishes between sialo- and sulfomucin [109,110].

1. Surface Epithelium

In the human and using the combined AB/PAS technique there is little neutral glycoprotein in the airway lining epithelium. Goblet cells have mostly acid glycoprotein, are either heliotrope or blue, and contain both sialomucin and sulfomucin [108,111].

In the mouse the AB staining is due to the presence of sialidase-susceptible sialomucin which is present at all levels of its respiratory tract [112,113]. In the rat there is a regional distribution; peripheral airways have sialidase-susceptible sialomucin, the trachea has sialidase-resistant sialomucin and sulfomucin, while hilar bronchi show a mixture. Furthermore in any goblet cell there may be a range of color indicating a mixture of glycoproteins within a single cell [114]. In the normal SPF rat a large proportion of the glycoprotein is neutral [115]. However, in the dog there is little neutral glycoprotein. After AB/PAS, goblet cells stain uniformly blue-purple and uptake of radioactive sulfate by them, as well as specific stains, indicate they contain sulfomucin [116,117].

2. *Submucosal Gland*

In some submucosal glands, as in the mouse larynx, all the cells contain sialomucin, which is susceptible to digestion by sialidase. In the human gland the variety of intracellular acid glycoprotein types is greater. Furthermore certain combinations of acid glycoprotein are found within cells: the serous and mucous cell each has a characteristic pattern.

a. Mucous Cell

The mucous cells of the tracheobronchial glands of man have been shown to produce four groups of acid glycoprotein: (a) sialomucin susceptible to sialidase; (2) sialomucin resistant to sialidase; (3) sulfate identified by AB staining after acid hydrolysis (which removes all the sialomucin) and by uptake of radioactive sulfate but not staining with specific stains for sulfate; and (4) sulfate which stains after hydrolysis, takes up radioactive sulfate and which stains with the specific stains for it [118]. A single cell may produce mucin with one or more of the four acidic groups just described but only in the following combinations: 1; 1 and 2; 2 and 3; 2, 3 and 4; 3 and 4; or 4.

b. Serous Cell

A histochemical and autoradiographic investigation (after incorporation of $^{35}S[^3H]$ glucose) of the serous cells of the human bronchial tree shows that the intracellular secretion includes a PAS-staining carbohydrate component associated with two acidic groups, sialic acid and sulfate.

Quantitative autoradiography shows that 70-80% of the carbohydrate, labeled with radioactive glucose, is lost after 1 hour of mild acid hydrolysis suggesting that a large proportion of the carbohydrate component consists of sialic acid. The histochemical reactions of the two acid groups in the serous cell, sulfate radical and carboxyl group of sialic acid, have different patterns of histochemical behavior than the sulfate and sialic acid found in the adjacent mucous cell; those of the former fail to fit the available histochemical classification of acid glycoproteins.

Comparison of the uptake of radioactive glucose and sulfate between adjacent mucous and serous cells show that the latter have a higher degree of sulfation per unit of radioactive glucose uptake [119].

III. Irritation

Bronchial epithelium may be altered, experimentally, to produce excess mucus by various agents including irritant gases, drugs, and bacteria. These agents act by stimulating existing secretory cells to synthesize and discharge at a faster rate and also by inducing an increase in the number of goblet cells.

Of the irritant gases, formalin, ammonia, chlorine, sulfur, and nitrous dioxide and tobacco smoke have all been shown to produce goblet cell hyperplasia in the experimental animal [120-128]. Drugs have not been investigated as extensively as irritants. Isoprenaline sulfate, when given to rats subcutaneously, causes marked goblet cell hyperplasia in both extra- and intrapulmonary airways, and so does pilocarpine nitrate [129]. Experimentally, infection with *Mycoplasma hyorhinis* has been shown to produce goblet cell hyperplasia in the pig [130]. With most of the above, the submucosal gland hypertrophies with goblet cell hyperplasia.

The above agents not only increase goblet cell number but also the number of cells dividing, i.e., the Mitotic Index. Concurrent with this there is probably an effect on cell differentiation so that there is a shift to the goblet cell.

The origin of "new" goblet cells has recently been analyzed by light and electron microscopy. Histochemical studies [115] suggested that the goblet cell, distended with secretion, starts as a cell with few secretory granules, that consist of neutral glycoprotein (Alcian blue-negative/PAS-positive). This cell then changes to one producing acid glycoprotein (Alcian blue-positive) and becomes distended by this type of secretion. The goblet cell distended with acid glycoprotein is typical of the airway responding to irritation.

Recent ultrastructural studies support this pattern of conversion [131]. Specific pathogen free rats, which have few goblet cells in their extrapulmonary airways and none in bronchioli of less than 0.4 mm diameter, have been exposed for a maximum of 6 weeks to air containing either sulfur dioxide or tobacco smoke. At intervals during the exposure, animals have been killed and their airways examined for goblet and other epithelial cells.

1. Irritant Aerosols

At a dose of 400 ppm most SO_2 is absorbed in the nasal mucosa and less than 10 ppm reaches the trachea [132]. Due to its high solubility, the trachea and

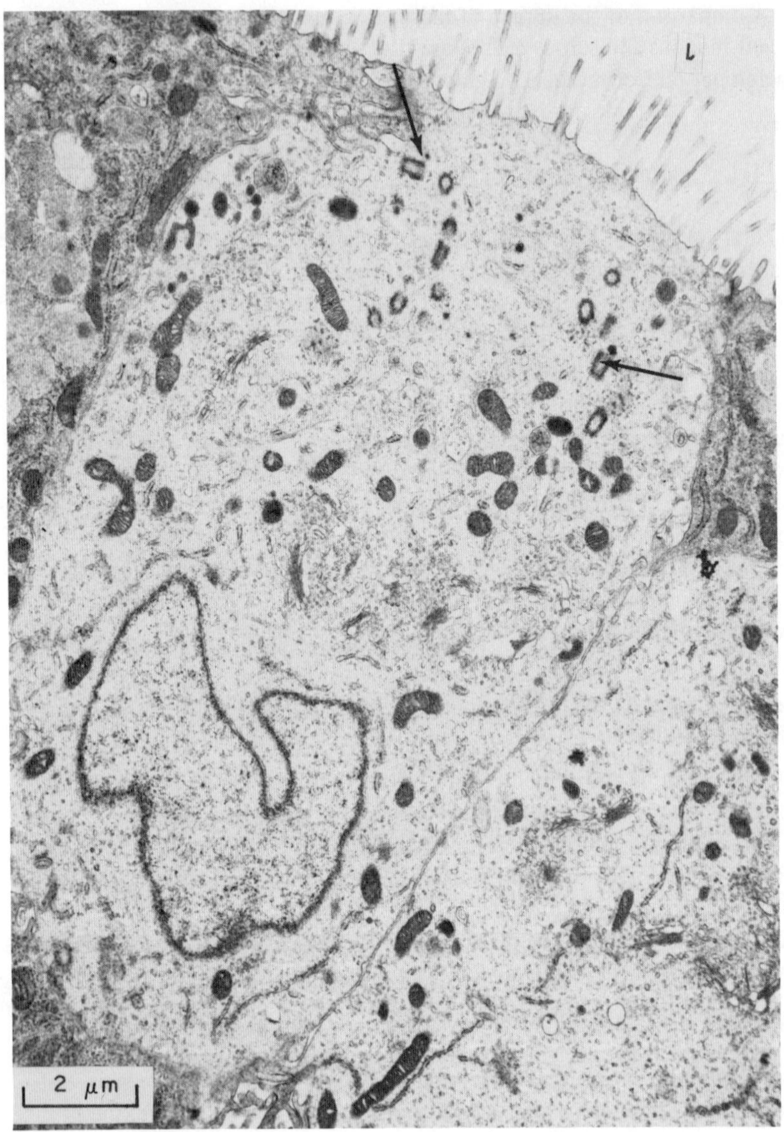

FIGURE 22 Rat trachea after 3 weeks exposure to sulfur dioxide. This area
has lost its cilia and a cell is shown with many centrioles at its apex (arrows).
These indicate ciliogensis is in progress. Microvilli project into the airway
lumen (l). Glutaraldehyde and osmium:lead citrate and uranyl acetate.
X8,000.

main bronchus are severely affected with loss of cilia and epithelial ulceration: some areas show healing, others squamous cell metaplasia, and yet others an intact epithelium with goblet cell hyperplasia (Fig. 22). In the first week the Mitotic Index is raised and, although it then falls somewhat, it is still consistently higher than the control levels (Fig. 23). In the small bronchiolus, at the periphery of the lung, the concentration of SO_2 is relatively less and so the damage is less severe. Direct observation shows that at no time is there epithelial ulceration and the Mitotic Index is not significantly increased: yet by 3 weeks, goblet cells appear and as exposure continues they increase in number [123]. These new goblet cells must represent transformation from existing bronchiolar cells, and electron microscopy has shown cells intermediate in appearance to a Clara and a goblet cell (Figs. 24 and 25). It appears that under these conditions of irritation the granules of the Clara cells change, from being highly protein, into acid glycoprotein granules that become confluent; that is, the Clara cell becomes a typical goblet cell (Fig. 26).

In airways of rats exposed to an atmosphere of tobacco smoke (25 cigarettes per day) goblet cells also increase but, the bronchiolar region is less affected than by 400 ppm SO_2. The trachea shows a significantly higher Mitotic Index than the controls, but ulceration is not marked [127]. At no time does the main extrapulmonary bronchus show ulceration, although the epithelium is significantly thickened (Fig. 27), not because of stratified epithelium, but rather from cell hypertrophy with a change in cell shape from cuboidal to tall columnar [131]. Since the increase in cell division causes more cells to be

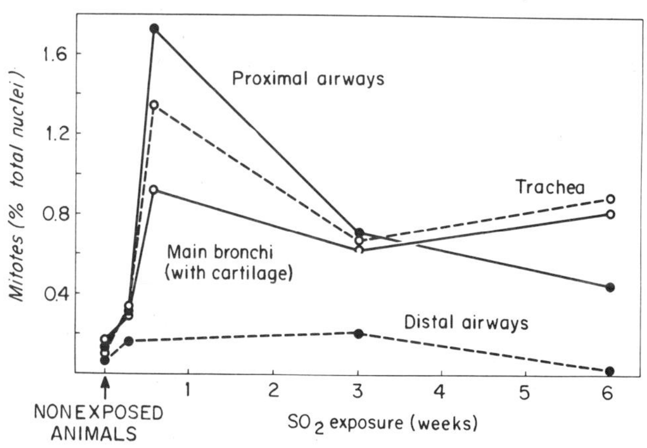

FIGURE 23 Mitotic count (4-hr period) after exposure to sulfur dioxide for up to 6 weeks.

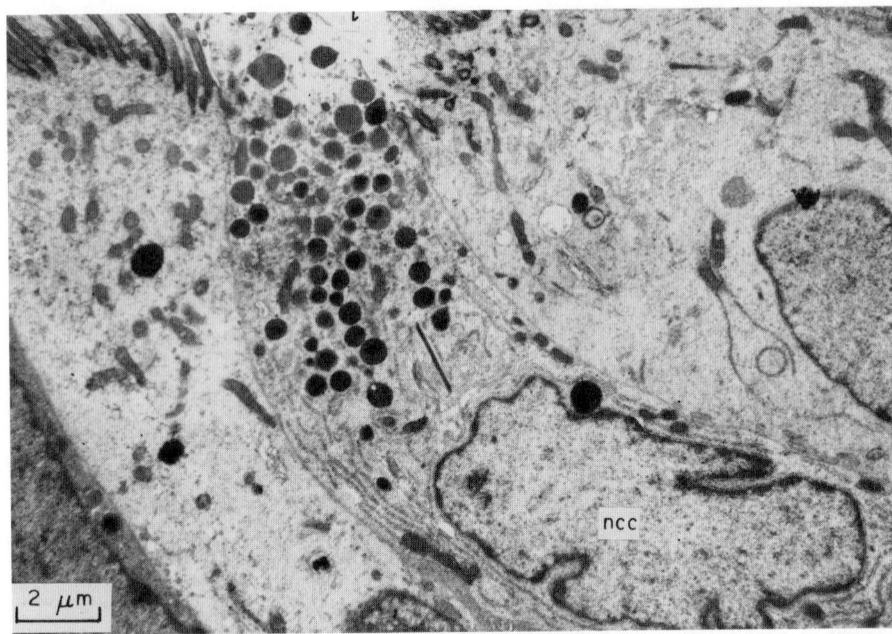

FIGURE 24 Rat bronchiolus after exposure to sulfur dioxide for 3 weeks. Within the enclosed epithelium nonciliated cells (ncc) are present which resemble Clara cells but have larger numbers of secretory granules and little smooth endoplasmic reticulum. Airway lumen (l). Glutaraldehyde and osmium:lead citrate and uranyl acetate. ×7500.

formed than are lost by epithelial sloughing, the cell concentration in unit length of epithelium is significantly increased; i.e., more cells are crowded into the same length of epithelium. Under these conditions of irritation goblet cells develop at the expense of both the serous and intermediate cell and stages of transformation from a serous to a goblet cell can be seen (Fig. 28). Tobacco smoke is unlike SO^2, in that cilia are not structurally damaged and indeed the number of ciliated cells increases: it does not decrease as was formerly thought. That ciliated cells lose their cilia and develop secretory granules has not been supported in these ultrastructural studies.

2. Drugs

Although acting via the bloodstream (the drugs are given subcutaneously) isoprenaline sulfate, a sympathomimetic drug, has similar effects to the aerosolized irritants SO_2 and tobacco smoke. The Mitotic Index is increased and goblet cells appear in peripheral bronchioli where they are normally absent. After this drug,

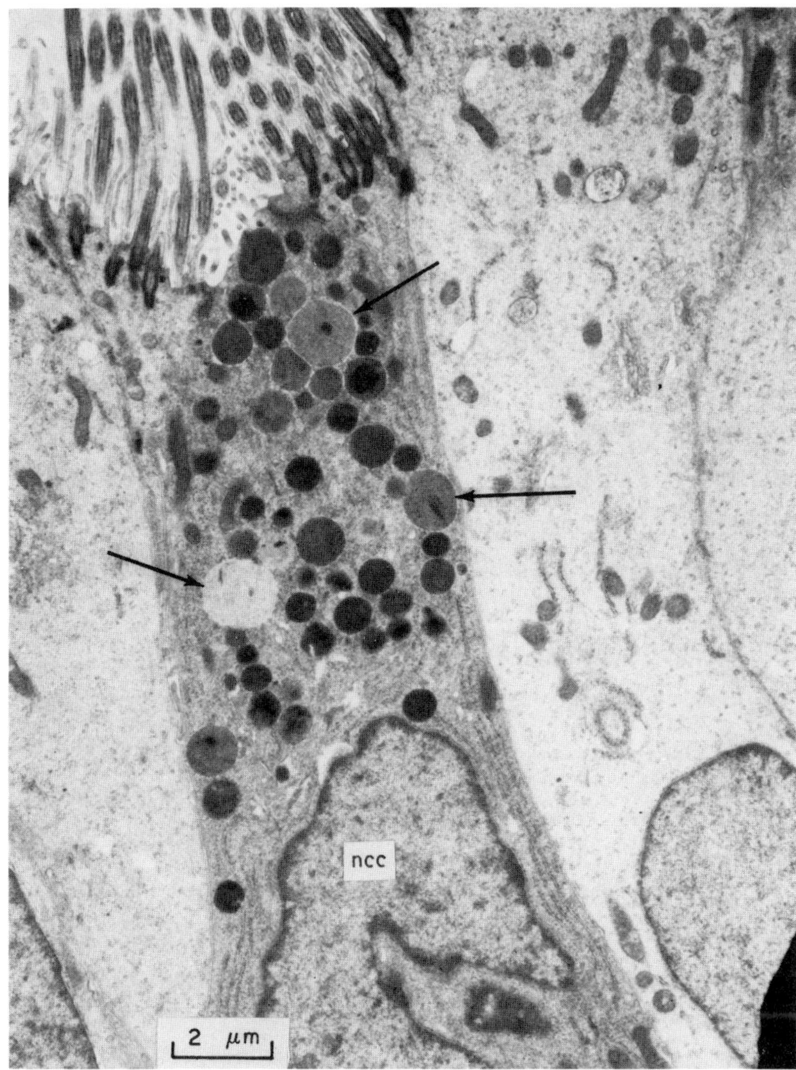

FIGURE 25 Rat bronchiolus after 3 weeks exposure to sulfur dioxide. A nonciliated cell (ncc) intermediate to that shown in Fig. 24 and a goblet cell is present. The majority of granules are electron-dense, but three larger and more electron-lucent granules are developing (arrows). Glutaraldehyde and osmium:lead citrate and uranyl acetate. ×7500.

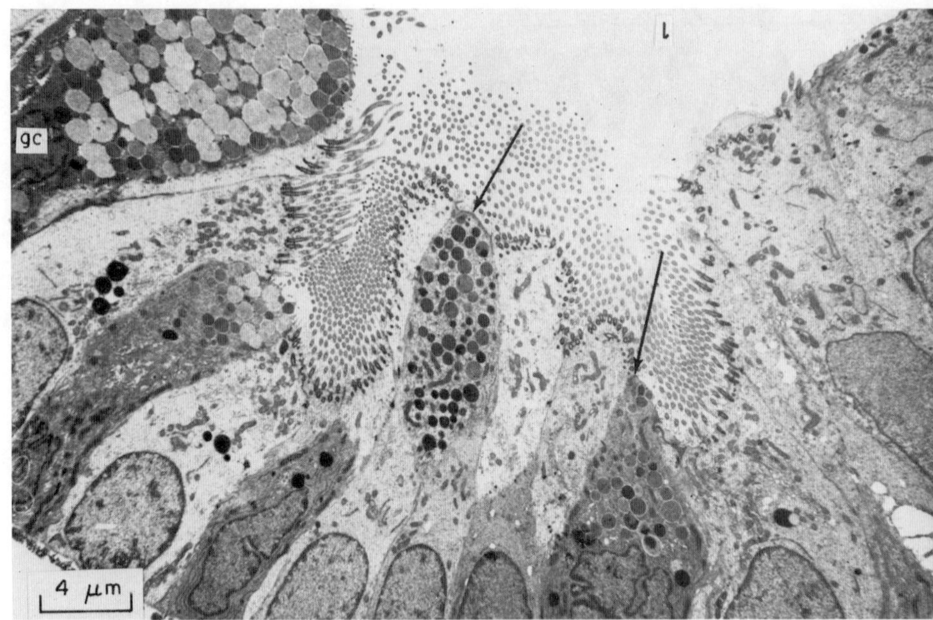

FIGURE 26   A rat bronchiolus after prolonged exposure to sulfur dioxide. Goblet cells (gc) distended mainly by electron-lucent granules appear. Nonciliated cells intermediate to the Clara and goblet cell are seen (arrows). Airway lumen (l). Glutaraldehyde and osmium:lead citrate and uranyl acetate. ×3750.

"differentiated" goblet cells have been seen in division supporting the concept that cell differentiation does not necessarily exclude division. This effect is perhaps related to stimulation of β_2 receptors and the cyclic AMP system [133]. If the experiment is repeated using Salbutamol, a β_1 stimulant, there is no goblet cell hyperplasia [134].

3. Infection

Infection is not a prerequisite for goblet cell hyperplasia, and in chronic bronchitis it is now thought that goblet cell hyperplasia and submucosal gland hypertrophy precede subsequent bouts of infection. Recently Jones and colleagues [130] have shown, for the first time, that gland hypertrophy can be produced by infection.

Enzootic pneumonia was produced, experimentally, in the piglet by nasal instillation of *Mycoplasma hyorhinis*. The bronchial gland hypertrophied, mucous and serous cells both increasing in number and size. The total glycoprotein con-

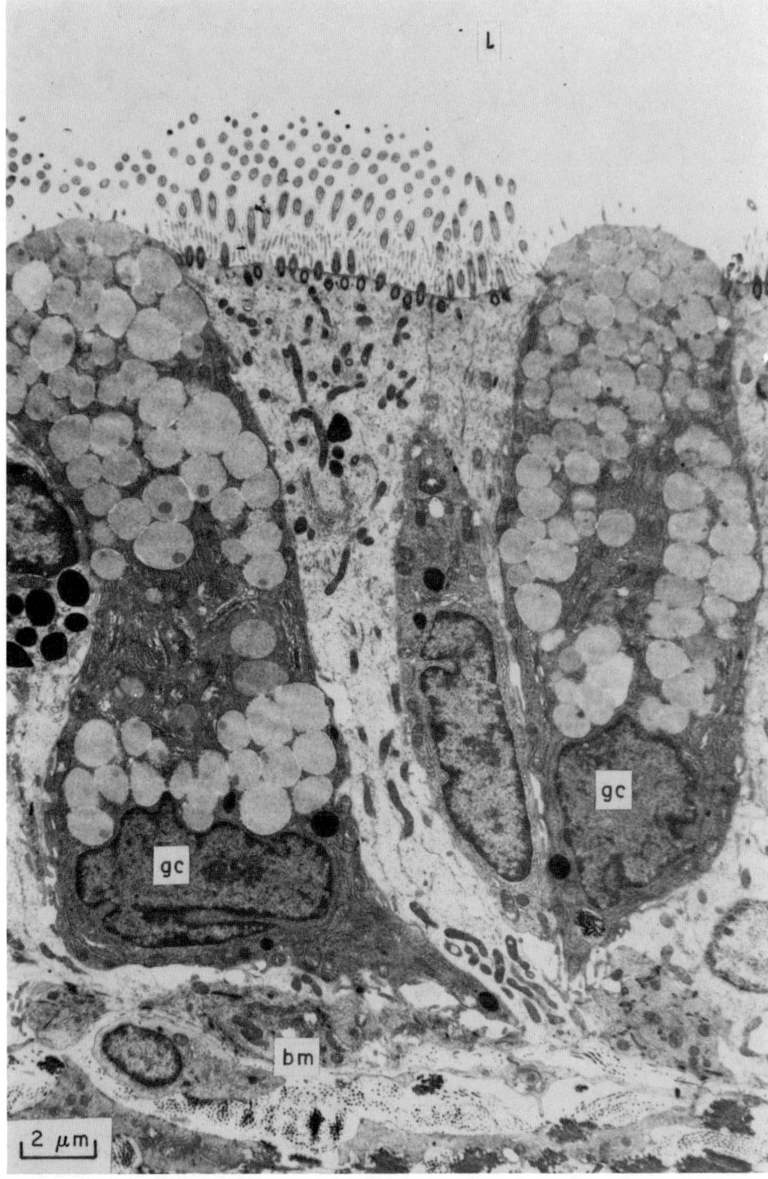

FIGURE 27 Rat extrapulmonary bronchus after exposure to tobacco smoke for 2 weeks. The epithelium is thickened due to cell hypertrophy and many goblet cells (gc) are present filled with electron-lucent granules. Basement membrane (bm) and lumen (l). Glutaraldehyde and osmium:lead citrate and uranyl acetate. ×5250.

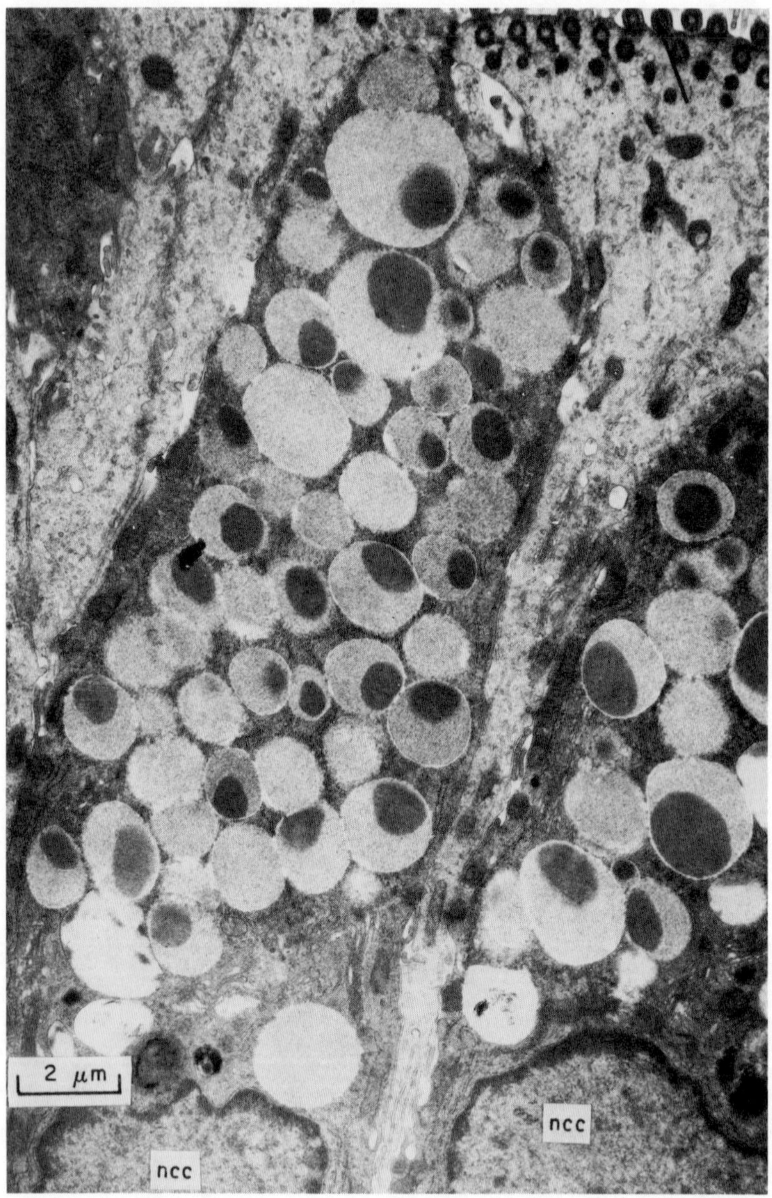

FIGURE 28   A nonciliated cell (ncc) intermediate to a serous and goblet cell in bronchial epithelium of a rat exposed to tobacco smoke for 2 weeks. The secretory granules have large cores of similar electron density to the serous cell granule, while the outer halo has the electron density of granules found in the goblet cell. Airway lumen (l). Glutaraldehyde and osmium: lead citrate and uranyl acetate. ×7500.

tent of the gland increased and acid glycoprotein increased, relative to neutral, in both the mucous and serous cell. In the mucous cell, sialidase-susceptible sialomucin and sulfomucin increased at the expense of the sialidase-resistant sialomucin, in the serous cell, the sialidase-resistant sialomucin and sulfomucin was increased. The thickness of surface epithelium was increased, but the goblet cells were depleted of their mucus-content and therefore difficult to identify after stains for intracellular glycoprotein.

4. Antiinflammatory Drugs

Antiinflammatory drugs such as the salicylates, indomethacin, and steroids may protect against some of the changes caused by irritants. For example, recently phenylmethyloxadiazole (PMO), when added (2% by weight) to the tobacco of cigarettes, has been shown to protect against some of the effects of tobacco smoke just described [127] ; (a) There is complete protection against goblet cell hyperplasia but none against the change of mucus from neutral to acid, (b) no protection against submucosal gland hypertrophy, and (c) partial protection against the increase in cell division and epithelial thickness.

The respiratory and alimentary tracts have common features, but the former is more susceptible to the constant replacement of its gaseous environment with included pollutants. Both the upper and lower respiratory tracts are, in the main, ciliated with mucus-producing goblet and gland cells, mucus and cilia acting together to remove entangled particulate and gaseous pollutants to the pharynx where they are normally swallowed. The lining epithelium to airways is complex; up to 10 cell types are currently recognized, each with its particular role. It has been shown that nerves penetrate the epithelium and, as well as being sensory, probably regulate epithelial function also. More is being learned about the replacement of cells lost through sloughing and their differentiation. The structure of the cells of the submucosal gland and their spatial arrangement within the gland has been determined and a new collecting duct region has been recognized. There are many types of epithelial glycoprotein and alterations of their relative proportions may be significant in disease. It is now recognized that, in the experimental animal, irritant aerosols, certain drugs, and (most recently) infection, produce submucosal gland hypertrophy. These are the anatomical hallmarks of chronic bronchitis and our understanding of these changes contributes to our knowledge of the pathogenesis of airway hypersecretory disease.

Acknowledgments

We are grateful to the Cystic Fibrosis Research Trust for supporting Dr. P. K. Jeffery as a Research Fellow and many of the studies presented here.

References

1. E. A. Boyden, *Segmental Anatomy of the Lung*, New York, McGraw-Hill, 1955.
2. W. S. Miller, *The Lung*, 2nd ed., Springfield, Ill., C. C Thomas, 1947.
3. I. Friedmann, L. Michaels, J. Gerwat, and E. S. Bird, The microscopic anatomy of the nasopharyngeal tonsil by light and electron microscopy, *ORL*, **34**:195-209 (1972).
4. W. S. Miller, The epithelium of the lower respiratory tract. In *Special Cytology*, 2nd ed. Edited by E. V. Cowdry. New York, Hafner, 1932, pp. 133-150.
5. L. Reid, Pathology of chronic bronchitis, *Lancet*, **1**:275-279 (1954).
6. R. de Haller, Development of mucus-secreting elements. In *The Anatomy of the Developing Lung*. Edited by J. Emery. London, Heinemann, 1969, pp. 94-115.
7. J. Iravani and A. van As, Mucus transport in the tracheobronchial tree of normal and bronchitic rats, *J. Pathol.*, **106**:81-93 (1972).
8. A. van As and J. D. Wessels, Scanning electron microscopy of rat tracheal epithelium. In *Proc. S. African Electron Microscopy Society*, Johannesburg, pp. 43-44, 1972.
9. A. C. Hilding, On cigarette smoking, bronchial carcinoma and ciliary action. II. Experimental study on the filtering action of cows' lungs, the deposition of tar in the bronchial tree and removal by ciliary action, *N. Engl. J. Med.*, **254**:1155-1160 (1956).
10. A. M. Lucas and L. C. Douglas, Principles underlying ciliary activity in the respiratory tract. II. Mucous clearance in man, monkey, and other mammals, *Arch. Otolaryngol.*, **20**:518 (1934).
11. J. E. Purkinje and G. Valentine [Cited by Sharpey (1836)], Cilia. In *The Cyclopaedia of Anatomy and Physiology*, Vol. 1. Edited by R. B. Todd. London, Sherwood, Gilbert and Piper, 1834, pp. 606-638.
12. F. E. Schulze, The lungs: 1. The lungs of mammals. In *Human and Comparative Histology*. Edited by S. Stricker; translated by H. Power. London, New Sydenham Soc., 1972, pp. 49-68.
13. A. Kolliker, Special histology of the lungs. In *Manual of Human Histology*, Vol. 2. Edited and translated by G. Busk and T. Huxley. London, Sydenham Soc., 1953, pp. 160-180.
14. C. Frankenhaeuser, Untersuchungen uber den Bau der Tracheo-Bronchial-Schleimhaut, Thesis, St. Petersburg. (Buchdruckerei der Kaiser-Lichen Akademie der Wissenschaften), 1879.
15. M. Clara, Zur Histobiologie des Bronchalepithels, *Z. Mikrosk. Anat. Forsch.*, **41**:321-347 (1937).
16. H. von Hayek, *The Human Lung*. Translated by V. Krahl. New York, Hafner, 1960.
17. P. K. Jeffery and L. Reid, New features of rat airway epithelium a quantitative and electron-microscopic study, *J. Anat.*, **120**:295-320 (1975).
18. J. M. Frasca, O. Auerbach, V. R. Parks, and J. D. Jamieson, Electron

microscopic observations of the bronchial epithelium of dogs. I. Control dogs, *Exp. Mol. Pathol.*, **9**:363-379 (1968).

19. J. Rhodin, Ultrastructure and function of the human trachel mucosa, *Am. Rev. Respir. Dis.*, **93**:1-15 (1966).

20. J. M. Frasca, O. Auerbach, and V. R. Parks, Electron microscopic observations of bronchial epithelium, *Exp. Mol. Pathol.*, **6**:261-273 (1967).

21. V. Konradova, The ultrastructure of the tracheal epithelium in rabbit, *Folia Morphol.*, **14**:210-214 (1966).

22. A. Baskerville, Ultrastructural studies of the normal pulmonary tissue of the pig, *Res. Vet. Sci.*, **11**:150-155 (1970).

23. J. Rhodin and T. Dalhamn, Electron microscopy of the tracheal ciliated mucosa in rat, *Z. Zellforsch. Mikrosk. Anat. Mikrosk. Anat.*, **44**:345-412 (1956).

24. A. van As and I. Webster, The organization of ciliary activity and mucus transport in pulmonary airways, *S. Afr. Med. J.*, **46**:347-350 (1972).

25. M. A. Sleigh, Metachronism of cilia of metazoa. In *Cilia and Flagella*, London, Pergamon Press, 1974.

26. M. F. Greenwood and P. Holland, The mammalian respiratory tract surface: A scanning electron microscopic study, *Lab. Invest.*, **27**:296-304 (1972).

27. P. K. Jeffery and L. Reid (in preparation).

28. J. Rhodin, *An Atlas of Ultrastructure*, Philadelphia, Pa., Saunders, 1963, pp. 82-86.

29. R. C. Rosan and J. M. Lauweryns, Mucosal cells of the small bronchioles of prematurely born human infants (600–700g), *Beitr. Pathol.*, **147**:145-174 (1972).

30. F. Basset, J. Poirier, M. le Crom, and J. Turiaf, Etude ultrastructurale de l'epithelium bronchiolaire humain, *Z. Zellforsch. Mikrosk. Anat. Mikrosk. Anat.*, **116**:425-442 (1971).

31. F. B. Askin and C. Kuhn, The cellular origin of pulmonary surfactant, *Lab. Invest.*, **25**:260-268 (1971).

32. P. Smith, D. Heath, and H. Moosavi, The Clara cell, *Thorax*, **29**:147-163 (1974).

33. E. Cutz and P. E. Conen, Ultrastructure and cytochemistry of Clara cells, *Am. J. Pathol.*, **62**:127-142 (1971).

34. N. S. Wang, R. V. Kotas, M. E. Avery, and W. M. Thurlbeck, Accelerated appearance of osmiophilic bodies in fetal lungs following steroid injection, *J. Appl. Physiol.*, **30**:362-365 (1971).

35. H. E. Karrer, Electron microscopic study of bronchiolar epithelium of normal mouse lung. Preliminary report, *Exp. Cell Res.*, **10**:237-241 (1956).

36. A. H. Niden and E. Yamada, Some observations on the fine structure and function of the nonciliated bronchiolar cells. In *Proc. 6th Int. Congr. Electron Microscopy, Kyoto*, p. 599 (1966).

37. J. M. Lauweryns, M. Cokelaere, and L. Boussauw, L'ultrastructure de l' epithelium bronchique et bronchiolaire de la souris, *Bull. Assoc. Anatomistes (Nancy)*, **146**:548-560 (1971).

38. J. E. Etherton, D. M. Conning, and B. Corrin, Autoradiographical and morphological evidence for apocrine secretion of dipalmitoyl lecithin in the terminal bronchiole of the mouse lung, *Am. J. Anat.*, **138**:11-36 (1973).

39. A. Baskerville, Ultrastructure of the bronchial epithelium of the pig, *Zentralbl. Veterinarmed.*, **17**:796-802 (1970).

40. M. M. Hansell and R. L. Moretti, Ultrastructure of the mouse tracheal epithelium, *J. Morphol.*, **128**:159-170 (1969).

41. D. H. Mutalionis and H. F. Parks, Ultrastructural morphology of the normal nasal respiratory epithelium of the mouse, *Anat. Rec.*, **176**:65-84 (1973).

42. A. Azzopardi and W. M. Thurlbeck, The histochemistry of the nonciliated bronchiolar epithelial cell, *Am. Rev. Respir. Dis.*, **99**:516-525 (1969).

43. E. Cutz and P. E. Conen, Ultrastructure and cytochemistry of Clara cells, *Am. J. Pathol.*, **62**:127-142 (1971).

44. A. H. Niden, Bronchiolar and large alveolar cell in pulmonary phospholipid metabolism, *Science*, **158**:1323-1324 (1967).

45. B. Meyrick and L. Reid, Electron microscopic aspects of surfactant secretion, *Proc. R. Soc. Med.*, **66**:386-387 (1973).

46. P. Petrik and A. Collet, Quantitative electron microscopic autoradiography of in vivo incorporation of ^{3}H-choline, ^{3}H-leucine, ^{3}H-acetate and ^{3}H-galactose in non-ciliated bronchiolar (Clara) cells of mice, *Am. J. Anat.*, **139**:519-534 (1974).

47. A. K. Christensen, The fine structures of testicular interstitial cells in guinea pigs, *J. Cell Biol.*, **26**:911-936 (1965).

48. J. H. L. Watson and G. L. Brinkman, Electron microscopy of the epithelial cells of normal and bronchitic human bronchus, *Am. Rev. Respir. Dis.*, **90**:851-866 (1964).

49. L. Luciano, E. Reale, and H. Ruska, Uber eine "chemoreceptive" Sinneszelle in der Trachea der Ratte, *Z. Zellforsch. Mikrosk. Anat. Mikrosk. Anat.*, **85**:350-375 (1968).

50. B. Meyrick and L. Reid, The alveolar brush cell in rat lung—a third pneumonocyte, *J. Ultrastruct. Res.*, **23**:71-80 (1968).

51. H. Brettschneider, Electronmicroscopic studies of the nasal mucosa, *Anat. Anz.*, **105**:194-204 (1958).

52. D. B. Silva, The fine structure of multivesicular cells with large microvilli in the epithelium of the mouse colon, *J. Ultrastruct. Res.*, **16**:693-705 (1966).

53. D. J. Riches, Histochemical and ultrastructural observations on the mucus-secreting cells of the guinea-pig common bile duct, *J. Anat.*, **106**:409-410 (1969).

54. W. K. Blenkinsopp, Proliferation of respiratory tract epithelium in the rat, *Exp. Cell Res.*, **46**:144-154 (1967).

55. P. Bolduc and L. Reid, Epithelial turnover of rat bronchial and alveolar lining, *Am. J. Respir. Dis.*, In press (1977).

56. K. G. Bensch, A. B. Gordon, and L. R. Miller, Studies on the bronchial

counterpart of the Kultschitsky (argentaffin) cell and innervation of bronchial glands, *J. Ultrastruct. Res.*, **12**:668-686 (1965).

57. J. T. Gmelich, K. G. Bensch, and A. A. Liebow, Cells of Kultschitsky type in bronchioles and their relation to carcinoid tumour, *Lab. Invest.*, **17**:88-98 (1968).

58. P. K. Jeffery and L. Reid, Intra-epithelial nerves in normal rat airways: A quantitative electron microscopic study, *J. Anat.*, **114**:35-45 (1973).

59. E. Cutz, W. Chan, V. Wong, and P. E. Conen, Endocrine cells in rat fetal lungs: Ultrastructural and histochemical study, *Lab. Invest.*, **30**: 458-464 (1974).

60. J. A. Terzakis, S. C. Sommers, and B. Anderson, Neuro-secretory appearing cells of human segmental bronchi, *Lab. Invest.*, **26**:127-132 (1972).

61. E. Hage, Endocrine cells in the bronchial mucosa of human foetuses, *Acta Pathol. Microbiol. Scand. [A]*, **80**:225-234 (1972).

62. J. M. Lauweryns, J. C. Peuskens, and M. Cokelaere, Argyrophil, fluorescent and granulated (peptide and amine producing ?) AFG cells in human infant bronchial mucosa. Light and electron microscopic studies, *Life Sci.*, **9**:1417-1429 (1970).

63. J. M. Lauweryns, M. Cokelaere, and P. Theunynck, Neuroepithelial bodies in the respiratory mucosa of various mammals, *Z. Zellforsch. Mikrosk. Anat. Mikrosp. Anat.*, **135**:569-592 (1972).

64. H. Moosavi, P. Smith, and D. Heath, The Feyrter cell in hypoxia, *Thorax*, **28**:729-741 (1973).

65. A. G. E. Pearse, The cytochemistry and ultrastructure of polypeptide hormone-producing cells of the APUD series and the embryonic, physiologic and pathologic implications of the concept, *J. Histochem. Cytochem.*, **17**:303-313 (1969).

66. A. G. E. Pearse and J. M. Polak, Cytochemical evidence for the neural crest origin of mammalian C cells, *Histochemie*, **27**:96-102 (1971).

67. E. Hage, Electron microscopic identification of several types of endocrine cells in bronchial epithelium of human foetuses, *Z. Zellforsch. Mikrosk. Anat. Mikrosk. Anat.*, **141**:401-412 (1973).

68. L. E. Ericson, R. Hakanson, B. Larson, Ch. Owman, and F. Sundler, Fluorescence and electron microscopy of amine-storing enterochromaffin-like cells in tracheal epithelium of mouse, *Z. Zellforsch. Mikrosk. Anat. Mikrosk. Anat.*, **124**:532-545 (1972).

69. J. M. Lauweryns and M. Cokelaere, Hypoxia-sensitive neuro-epithelial bodies, intrapulmonary secretory neuro-receptors modulated by the CNS, *Z. Zellforsch. Mikrosk. Anat., Mikrosk. Anat.*, **145**:521-540 (1973).

70. W. R. Flemming, Studien uber Regeneration der Gewebe, *Arch. Mikrosk. Anat.*, **24**:50-91 (1885).

71. W. Andrew and M. R. Burns, Leucocytes in the tracheal epithelium of the mouse, *J. Morphol.*, **81**:317-341 (1947).

72. J. F. Kent, The origin, fate, and cytochemistry of the globule leucocyte of the sheep, *Anat. Rec.*, **112**:91-116 (1952).

73. W. Bloom and D. W. Fawcett, *Textbook of Histology,* 8th ed., Philadelphia, Pa., Saunders, 19 .

74. A. B. Dawson, Modified lymph nodes from dogs with a known history of irradiation including a note on "globule" leucocyte formation, *Anat. Rec.,* **36**:1-29 (1927).

75. C. Dobson, Immunofluorescent staining of globule leucocytes in the colon of the sheep, *Nature,* **211**:875 (1966).

76. P. Whur, Relation of globule leucocytes to gastrointestinal nematodes in sheep and *Nippostrongylus brasiliensis* and *Hymenolepsis nana* infections in rats, *J. Comp. Pathol.,* **76**:57-65 (1966).

77. P. Whur and M. Gracie, Histochemical comparison of mast cell and globule leucocyte granules in the rat, *Experientia,* **23**:655-657 (1967).

78. R. D. Cook and A. S. King, A neurite-receptor complex in the avian lung; electron microscopical observations, *Experientia,* **25**:1162-1164 (1969).

79. C. Walsh and J. McLelland, Intra-epithelial axons in avian trachea. *Z. Zellforsch. Mikrosk. Anat. Mikrosk. Anat.,* **147**:209-217 (1974).

80. P. K. Jeffery and J. G. Widdicombe (In Preparation).

81. N. Cauna, K. H. Hinderer, and R. T. Wentges, Sensory Receptor organs of the human nasal respiratory mucosa, *Am. J. Anat.,* **124**:187-210 (1969).

82. B. Meyrick and L. Reid, Ultrastructure of cells in the human bronchial submucosal glands, *J. Anat.,* **107**:281-299 (1970).

83. P. K. Jeffery, Unpublished.

84. J. del Castillo and B. Katz, Biophysical aspects of neuro-muscular transmission, *Progr. Biophys. Biophys. Chem.,* **6**:122-170 (1956).

85. E. D. P. de Robertis and H. S. Bennett, Some features of the submicroscopic morphology of synapses in frog and earthworm, *J. Biophys. Biochem. Cytol.,* **1**:47-58 (1955).

86. J. A. Gallagher, P. W. Kent, M. Passatore, R. J. Phipps, and P. S. Richardson, The composition of mucus and the nervous control of its sevretion in the cat, *Proc. Roy. Soc. (Lond.) B.,* **192**:49-76 (1975).

87. J. Iravani and G. M. Melville, Mucociliary activity in the respiratory tract as influenced by prostaglandin E_1, *Respiration,* **32**:305-315 (1975).

88. H. Florey, H. M. Carleton, and A. Q. Wells, Mucus secretion in the trachea, *Br. J. Exp. Pathol.,* **13**:269-284 (1932).

89. W. Bowman, Mucous Membranes. In *Cyclopaedia of Anatomy and Physiology,* Vol. 3. pp. 484-506.

90. F. D. Bertalanffy, The mitotic activity and renewal of the lung. M.Sc. Thesis, McGill University, Montreal, 1951.

91. H. Spencer and R. G. Shorter, Cell turnover in pulmonary tissues, *Nature,* **194**:880 (1962).

92. A. M. Gottesberge and E. Koburg, Autoradiographic investigations of cell formation in the respiratory tract, Eustachian tube, middle ear and external auditory canal, *Acta Otolaryngol. (Stockh.),* **56**:353 (1963).

93. R. G. Shorter, J. L. Titus, and M. B. Divertis, Cell turnover in the respiratory tract, *Dis. Chest,* **46**:138 (1964).

94. R. G. Shorter, J. L. Titus, and M. B. Divertis, Cytodynamics in the respiratory tract of the rat, *Thorax,* **21**:32 (1966).

95. F. D. Bertalanffy and C. Lau, Cell renewal, *Int. Rev. Cytol.,* **13**:357-366 (1962).

96. A. B. Wells, The kinetics of cell proliferation in the tracheobronchial epithelia of rats with and without chronic respiratory disease, *Cell Tissue Kinet.,* **3**:185-206 (1970).

97. W. Bowman, Mucous membranes. In *Cyclopaedia of Anatomy and Physiology,* Vol. 3. 1847, pp. 484-506.

98. B. Meyrick, J. Sturgess, and L. Reid, Reconstruction of the duct system and secretory tubules of the human bronchial submucosal gland, *Thorax,* **24**:729-736 (1969).

99. B. Meyrick and L. Reid, Ultrastructure of cells in the human bronchial submucosal glands, *J. Anat.,* **107**:281-299 (1974).

100. S. P. Sorokin, On the cytology and cytochemistry of the opossum's bronchial glands, *Am. J. Anat.,* **117**:311-318 (1965).

101. J. F. A. McManus, Histochemical demonstration of mucin after periodic acid, *Nature,* **158**:202 (1946).

102. R. D. Hotchkiss, A microchemical reaction resulting in the staining of polysaccharide structures in fixed tissue preparations, *Arch. Biochem.,* **16**:131-141 (1948).

103. H. F. Steedman, Alcian Blue 8GS; a new stain for mucin, *Q. J. Micros. Sci.,* **91**:477-479 (1950).

104. R. W. Mowry, Alcian blue techniques for the histochemical study of acidic carbohydrates, *J. Histochem. Cytochem.,* **4**:407 (1956).

105. S. S. Spicer and L. Warren, The histochemistry of sialic acid containing mucoproteins, *J. Histochem. Cytochem.,* **8**:135-137 (1960).

106. M. G. Butler, A new method of hydrolysis for the identification of acid glycoprotein, *Med. Lab. Tech.,* **30**:105-112 (1973).

107. M. A. Jennings and H. W. Florey, Autoradiographic observations on the mucous cells of the stomach and intestine, *Q. J. Exp. Physiol.,* **41**:131-152 (1956).

108. D. Lamb, Intracellular development of secretion of mucus in the normal and morbid bronchial tree. Ph.D. Thesis, University of London, 1969.

109. R. Jones and L. Reid, The effect of pH on Alcian blue staining of epithelial glycoproteins. I. Sialomucins and sulphomucins (singly or in simple combinations), *Hist. J.,* **5**:9-18 (1973).

110. R. Jones and L. Reid, The effect of pH on Alcian blue staining of epithelial glycoproteins. II. Human bronchial submucosal gland, *Hist. J.,* **5**:19-27 (1973).

111. C. McCarthy and L. Reid, Intracellular mucopolysaccharides in the normal human bronchial tree, *J. Exp. Physiol.,* **49**:86-94 (1964).

112. R. C. Curran and J. S. Kennedy, The distribution of the sulphated mucopolysaccharides in the mouse, *J. Pathol. Bacteriol.,* **70**:449-457 (1955).

113. L. Reid, C. McCarthy, J. Duvenci, and R. A. Gibbons, Intracellular mucus in the human bronchial tree, *Nature,* **195**:715-716 (1962).

114. C. McCarthy and L. Reid, Acid mucopolysaccharide in the bronchial tree in the mouse and rat, *Q. J. Exp. Physiol.,* **49**:81-84 (1964).

115. R. Jones, P. Bolduc, and L. Reid, Goblet cell glycoprotein and tracheal gland hypertrophy in rat airways: The effect of tobacco smoke with or without the anti-inflammatory agent phenylmethyloxadiazole, *Br. J. Exp. Pathol.,* **54**:229-239 (1973).

116. S. S. Spicer, L. W. Chakrin, J. R. Wardell, and W. Kendrick, Histochemistry of mucosubstances in the canine and human respiratory tract, *Lab. Invest.,* **25**:483-490 (1971).

117. L. W. Chakrin, A. P. Baker, S. S. Spicer, J. R. Wardell, N. de Sanetis, and C. Driss, Synthesis and secretion of macromolecules by canine trachea, *Am. Rev. Res. Dis.,* **105**:368-381 (1972).

118. D. Lamb and L. Reid, Quantitative distribution of various types of acid glycoprotein in mucous cells of human bronchi, *Hist. J.,* **4**:91-102 (1972).

119. D. Lamb and L. Reid, Histochemical and autoradiographic investigation of the serous cells of the human bronchial gland, *J. Pathol.,* **100**:127-138 (1970).

120. H. Florey, H. M. Carleton, and A. Q. Wells, Mucus secretion in the trachea, *Br. J. Exp. Pathol.,* **13**:269-284 (1932).

121. P. C. Elmes and D. Bell, The effects of chlorine gas on the lungs of rats with spontaneous pulmonary disease, *J. Pathol. Bacteriol.,* **86**:317-327 (1963).

122. L. Reid, An experimental study of hypersecretion of mucus in the bronchial tree, *Br. J. Exp. Pathol.,* **44**:437-445 (1963).

123. D. Lamb and L. Reid, Mitotic rates, goblet cell increase and histochemical changes in mucus in rat bronchial epithelium during exposure to SO_2, *J. Pathol. Bacteriol.,* **96**:97-111 (1968).

124. A. Freeman and G. B. Haydon, Emphysema after low level exposure to NO_2, *Arch. Environ. Health,* **8**:125-128 (1964).

125. G. Freeman, S. C. Crane, N. J. Furiosi, R. J. Stephens, M. J. Evans, and W. D. Moore, Covert reduction in ventilatory surface in rats during prolonged exposure to subacute nitrogen dioxide, *Am. Rev. Respir. Dis.,* **106**:563-579 (1972).

126. D. Lamb and L. Reid, Goblet cell increase in rat bronchial epithelium after exposure to cigarette and cigar tobacco smoke, *Br. Med. J.,* **1**:33-35 (1969).

127. R. Jones, P. Bolduc, and L. Reid, Protection of rat bronchial epithelium against tobacco smoke, *Br. Med. J.,* **2**:142-144 (1972).

128. S. S. Spicer, L. W. Chakrin, and J. R. Wardell, Effect of chronic sulfur dioxide inhalation on the carbohydrate histochemistry and histology of the canine respiratory tract, *Am. Rev. Respir. Dis.,* **110**:13-24 (1974).

129. J. Sturgess and L. Reid, The effect of isoprenaline and pilocarpine on (a) bronchial mucus-secreting tissue and (b) pancreas, salivary glands, heart, thymus, liver and spleen, *Br. J. Exp. Pathol.,* **54**:388-403 (1973).

130. R. Jones, A. Baskerville, and L. Reid, Histochemical identification of glycoproteins in pig bronchial epithelium. (a) Normal (b) Hypertrophied from enzootic pneumonia, *J. Pathol.*, **116**:1-11 (1975).

131. P. K. Jeffery and L. Reid. (In preparation).

132. T. Dalhamn and L. Strandberg, Acute effects of SO_2 on rate of ciliary beat in rabbit trachea (vivo and vitro) and absorption capacity of nasal cavity, *Int. J. Air Water Pollution*, **4**:154 (1960).

133. T. Barka, Stimulation of protein and ribonucleic acid synthesis in rat submaxillary gland by isoproterenol, *Lab. Invest.*, **18**:38-41 (1968).

134. P. K. Jeffery and L. Reid. (In preparation).

8

The Nature and Action of Respiratory Tract Cilia

MICHAEL A. SLEIGH

University of Southampton
Southampton, England

This chapter is written by a biologist interested in the functioning of cilia from a comparative point of view. The structure, chemistry, and functioning are considered in such a way as to relate what is known about the cilia of the mammalian respiratory system with knowledge of cilia in general, so that the reader can gain not only a general appreciation of our present understanding about cilia but also get a feeling for those features that are of particular significance in the functioning of the mucociliary system of mammals, especially man.

I. The General Characteristics of Cilia

Cilia are cellular projections that show intrinsic motility. They contain cytoplasmic fibrils whose active motion causes changes in shape of the cilium, which result in propulsion of the fluid that overlies the cell. The fibrils of the cilium are surrounded by cytoplasm and enclosed by the membrane of the cell. Certain cells of almost all animals and many plants bear cilia, but they can only exist where the cell surface is covered with fluid because of their naked cell membrane.

The characteristic function of cilia in active propulsion of fluid is of physiological value in mammals in the respiratory and reproductive systems where transport of mucus and other secretions and of gametes is achieved by activity of cilia. In other animals, particularly invertebrates, cilia perform important functions in the propulsion of water for purposes of locomotion, collection of food, ventilation of respiratory surfaces, transport of excretory products, and circulation of body fluids as well as generalized functions concerned with cleansing surfaces by clearance of mucus and entrapped particles. In addition, ciliated cells have been incorporated into many types of sense organs, where they commonly form part of the sensory cells themselves, e.g., in the retina and auditory and olfactory epithelia; such cilia are sometimes nonmotile.

A cilium has a diameter of about 0.25 μm and commonly a length of 5–50 μm, the length varying according to position and function. Cilia with specialized functions may be compounded together to form much bigger structures; generally these compound cilia are formed by the loose adherence of large numbers of normal complete cilia, each with its own membrane, but giant cilia constructed of numerous bundles of ciliary fibrils within a common membrane also occur. The largest compound cilia contain thousands, or even hundreds of thousands, of component cilia, and extend to lengths of a millimeter or more. Compound and simple cilia show similar patterns of movement and appear to function in a basically similar way. Flagella, including sperm tails, have an identical basic structure to cilia, but show modified patterns of movement and sometimes possess additional structural elements.

The majority of cilia display a pattern of movement that can be broken down into two phases, one in which the cilium is extended and moves more quickly, the other in which the cilium is bent and moves more slowly; because the former movement produces the main flow of fluid over the ciliated cell, it is referred to as the effective stroke, while the slower motion of the curved cilium is the recovery stroke. The beat of a cilium may be planar, but more frequently it does not take place completely in one plane and the precise pattern of movement in three dimensions is not easy to analyze. It is concluded that the effective stroke normally takes place approximately in one plane, but that the recovery stroke does not necessarily occur in the same plane, because in many cases the recovering cilium moves back close to the cell surface at one side of the plane of the effective stroke. Often the deviation from a plane is rather minor and ciliary beat patterns are then described as if they are planar, or as projections upon the plane of the effective stroke. An example of a pattern of ciliary beating of a characteristic type is shown in Fig. 1, together with information on the dimensions and rate of movement of the organelle, which is actually a compound cilium with about 25 components.

The main part of our knowledge of the movement and biochemistry of cilia and flagella is derived from work on invertebrate animals, notably ciliated and flagellated Protozoa, mollusks, worms, and echinoderms. These have been

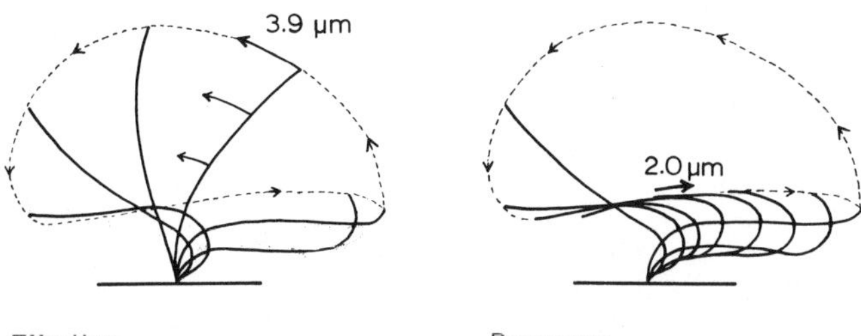

FIGURE 1 A sequence of profiles illustrating the cycle of beating of a compound cilium 32 μm long. The duration of the cycle is about 60 msec and the profiles represent the position and shape of the cilium at intervals of 6 msec. The numbers given on the drawing indicate the distance moved by the ciliary tip in one msec at that stage of the cycle. (Simplified from Rikmenspoel and Sleigh [50].)

used because of convenience of access, size, or ease of isolation in quantity. There is no reason to doubt that the basic mechanisms of biochemistry and motility are the same in mammalian cilia as in these other types, but the pattern of movement and the mechanism of control are more specialized features that vary from animal to animal and from tissue to tissue.

II. Cilia in the Mammalian Respiratory Tract

A. Distribution and Form of the Cilia

The presence of cilia in the respiratory tract of mammals was recognized as early as 1835 by Purkinje and Valentin. Ciliated epithelial cells are now known to occur in all parts of the respiratory tract except the alveolar sacs and the anterior part of the nasal cavity. Because of the relative ease of access, the activity of cilia in the nasal cavity, the trachea and the larger bronchi has been observed more commonly than that in the smaller bronchioles, but sufficient observations of all areas have now been made to show the distribution and principal functions of the cilia. Studies on many mammals using light microscopy have been supported in recent years by transmission electron micrographs and most dramatically illustrated by scanning electron microscopy (Fig. 2).

The following information on the distribution of cilia in the respiratory tract of the mouse is derived from a scanning electron microscope study by Greenwood and Holland [21]. In the nasal cavity the ciliated areas of the nasal septum, turbinates, nasal fossa, and nasopharynx possess numerous cells bearing cilia 5-6 μm long interspersed with cells bearing microvilli. The cilia drive secreted mucus downward and backward to the oropharynx. The posterior part of

FIGURE 2   Scanning electron micrograph of ciliated epithelium of the hamster trachea. Cells bearing cilia that are about 6 μm long are interspersed with cilia bearing microvilli ($\times$2000). (Unpublished micrograph by Ellen R. Dirksen.)

the superior nasal fossa is clothed with long slender olfactory cilia that form a network of intertwined strands over the surface of the olfactory epithelium; it is believed that these long cilia form the primary sensory structures of the epithelium. Ciliated epithelial cells occur at all levels in the trachea, bronchi, and bronchioles, right down to the respiratory bronchioles, the final airways of the system, but are not found in the alveoli themselves. At all levels the ciliated cells, which occur singly or more commonly in groups, are interspersed with polyhedral cells that are clothed with microvilli. The cilia are generally about 5 μm long, and those occasional cells that have much shorter cilia are supposed to be in the process of regeneration or replacement. In the smaller bronchioles the ciliated cells are less frequent and their cilia shorter.

The results of the study agree well with those of Karrer [30] and Hansell and Moretti [22] from transmission electron micrographs of mouse respiratory epithelia, except that the scanning micrographs give rather little information on the distribution of secretory cells, which seem to be sparse or modified in mice. Studies on human respiratory cilia show no obvious differences [51]. Histological observations and transmission electron micrographs of other mammals indicate that mucus-secreting goblet cells are abundant in epithelia of the trachea and larger bronchial airways, but become less frequent in the smaller bronchioles, disappearing at the level of the terminal bronchioles. The respiratory bronchioles, which are about 0.5 mm in diameter, lack mucus-secreting cells but possess other secretory cells that are believed to produce a serous secretion; this watery fluid was shown by Gil and Weibel [18], from electron microscope studies, to be coated at the liquid-air interface by an osmiophilic film that they believe to be of the same nature as the surfactant film that covers the liquid lining layer of the alveoli, but is thicker in the bronchioles than in the alveoli.

B. Activity and Functions of the Cilia

Bronchiolar cilia, and even those in the larger bronchi, are extraordinarily difficult to study, indeed, no study of the functioning of any mammalian lung cilia that can reliably be regarded as reflecting the normal situation in every respect is known to have been made, for it is likely that the activity of the cilia and/or the amount and nature of secretory activity is disturbed by preparative procedures.

This difficulty may be illustrated, for example, by the variation reported for mucociliary clearance rates. The rate of transport of mucus by cat trachea was recorded as between 0.25 mm/min (a misprint for 0.25 mm/sec?) and 14 mm/min in in vivo experiments by Carson et al. [9], while the in vitro observations of Kensler and Battista [31] gave values of between 30 and 45 mm/min. In contrast, Dalhamn [10] found that the mean rate of flow of mucus in 5 in vivo rat trachea preparations was 12.9 mm/min, and that 7 min after removal of these trachea the mean rate of flow had fallen to 6.9 mm/min; in the same experiment the rate of ciliary beat fell only very little, from 21.2 to 20.3 Hz.

In spite of uncertainties about actual values, a general picture of the functioning of the system is emerging, although there is still an urgent need for studies that separate ciliary activity and features of mucous secretion. One of the most exciting recent developments is the technique used by Iravani for observing ciliary activity and mucous clearance in rat lung airways isolated with minimum disturbance [25,28].

The direction of ciliary beating is always toward the oropharynx, whether in nasal or thoracic parts of the respiratory tract (except in some disease situations). Measurements of the frequency of beating of tracheal cilia generally fall close to those recorded from rats by Dalhamn [10] of about 22 Hz at a body temperature of 37°C. Iravani [25] reported that the frequency of ciliary beating in rat preparations increased from about 7 Hz in the peripheral airways to about 18 Hz in a main lobar bronchus; he also observed that the rate of beat increased with exposure to fluid and that he did not find as much mucus as other workers had reported. Ciliary metachronism was confined to groups of contiguous ciliated cells, isolated among nonciliated cells, which are more frequent in the peripheral airways; the precise form and degree of coordination that occurs in this metachronal activity is not clear. No changes in the potential difference across the membrane of the ciliated cell could be recorded during ciliary beating and metachronal activity [26].

Extensive studies of the rate of mucous flow in the rat trachea gave an average of 13.5 mm/min [10], and, corresponding to the gradient of ciliary beating, Iravani and van As [28] found that the rate of mucous transport in rat lung preparations increased from 0.4 mm/min in the smallest bronchioles to 11.5 mm/min in the lobar bronchus. A similar gradient of mucous flow was found in dogs, where mucociliary clearance rates rose from 1.6 mm/min in the distal bronchi to 4 mm/min in segmental bronchi and to 12.6 mm/min in the trachea [2], and a comparable acceleration of flow was found in human lungs by Morrow et al. [44].

C. Some Problems of Mucous Transport

In the observations of Iravani and van As [28] mucus was seen to travel as "droplets" less than 4 μm in diameter, 10-70 μm "flakes," or aggregates of these called "plaques," the degree of aggregation increasing progressively from smaller to larger airways. This compares with the more conventional view of a carpet or blanket of mucus described in many earlier studies and supported, for example, by observations of Dalhamn [10] on rapidly frozen epithelia which show the mucous secretion of healthy rats as a layer of rather consistent thickness (5 μm); although Dalhamn's report did not state what part of the respiratory epithelium was observed, it is probable that it came from one of the larger airways. The ciliated cells must be covered with a layer of fluid several micrometers thick in order for the cilia to function, so uncertainty remains about the quantity and character of the bronchiolar mucus and the relative proportion of serous fluid—is it possible, perhaps, that Iravani's preparative technique results in modified or reduced mucus production?

Uncertainty about the quantity and character of normal mucus and about any variation in the rate or pattern of movement of the cilia means that the true interrelationships of ciliary activity and mucous transport remain obscure. For example, it is not clear how one should interpret Dalhamn's [10] observations that in some cases mucous transport ceased while the cilia were still active, and that in rats exposed to SO_2 reduced ciliary activity was accompanied by an accumulation of mucus. Similarly, what is the true significance of the observation of Iravani and van As [28] on rats with chronic bronchitis, where the amount of mucus present was increased, larger flakes were present in smaller bronchioles, and the mucous transport was more rapid than in normal animals?

What is responsible for the changes in the rate of mucous transport—is it changes in the nature or quantity of the mucus, changes in the activity of the cilia or changes in the contribution of both cilia and mucus? Changes in the viscosity of the mucus or in the proportion of mucus to serous fluid could influence the freedom of movement of the cilia and produce the same apparent result as a change in the intrinsic rhythm of the ciliary beat. Only in rather dramatic cases is it possible to be certain that failure of the ciliary mechanism is involved. Thus, some of the bronchitic rats studied by Iravani and van As showed areas in which the direction of ciliary beating had become disorganized and sometimes even reversed, often only in local patches, suggesting that the regeneration of ciliated cells had been disturbed under these conditions; this disorganization of ciliary beating resulted in local accumulation of mucus. In rats suffering from acute bronchitis, the overall mucus transport was sluggish and patches of inactive cilia were seen, suggesting that ciliary failure may be the primary cause of feeble mucous transport, but sticky mucus could still be responsible; ciliary inactivity was also reported by Iravani and Melville [27] in rats exposed to cigarette smoke.

The picture that emerges from these studies is that there is a steadily increasing density of ciliation and an increasing rate of ciliary beating from the alveolar margins to the cranial end of the trachea. In the finest bronchioles the fluid lining lacks mucus, and consists of a watery serous fluid with surfactant limiting layer continuous with the thinner fluid lining of the alveoli. In the wider bronchioles the addition of mucous secretions to this fluid lining provides first isolated droplets of mucus floating on the fluid surface and being wafted slowly forward by the rather scattered cilia, and later larger rafts of mucus form by the aggregation of droplets and are carried more quickly by the more numerous and faster cilia of larger airways. Finally, in the trachea, the mucus has converged to a more or less continuous sheet with the serous layer almost unseen beneath the layer of mucus.

It was pointed out by Asmundsson and Kilburn [2] that the mucous layer converges about 2000-fold as it is moved from the smaller bronchioles to the trachea, and that absorption of fluid must occur during the passage of the mucus. However, if the ciliary activity is much more efficient at propelling the surface mucus component than it is in moving the serous fluid beneath—because the mucus feels only the forward effective strokes, while the serous fluid feels both forward effective and backward recovery strokes (Sect. VIII)—then the aggregation of isolated droplets into rafts and then into a thicker sheet could accommodate most if not all of the material that converges along the progressively accelerating ciliary escalator, without the need for extensive fluid absorption.

D. Ciliary Function in the Nasal Cavity

A comparable system for the clearance of mucus occurs in the nasal cavity, where mucous secretions are again transported to the pharynx by cilia. Lucas [35] has described detailed studies of ciliary distribution and function in the nasal cavity of the rhesus monkey, and Lucas and Douglas [36] provided a comparison of the nasal cavity of man, the monkey, and some other mammals. Ciliary currents progressively increase in rate from the anterior toward the pharynx, transport of mucus being fastest near the floor of the nasal cavity at about 6.6 mm/min (at room temperature). Hilding [23] found that the rate of transport of india ink droplets on the nasal mucosa of living human subjects reached 4-6 mm/min in more posterior nasal regions. Here again the rate of ciliary activity may be related to the extent of the fluid layer over the cilia, beating being restricted where the fluid zone is too shallow.

III. The Structure of Cilia

A. Parts of the Cilium

This general description of ciliary structure is based on the best information available, irrespective of the organism on which the cilium is borne. Most features are quite homogeneous in different organisms, and mention will be made of variants found in mammalian respiratory cilia where possible.

The cellular projection that is the more obvious part of the cilium arises, developmentally and positionally, from a basal structure in the cell cortex, the basal body, and this in turn may be the site of attachment of extensive "rootlet" fibers that ramify in the cell and provide anchorage for the motile cilium. There are therefore three main structural regions to consider: the ciliary shaft, the basal body and the ciliary roots. The organization of the ciliary shaft structures has been reviewed in detail by Warner [66], and information on the structure and function of the basal bodies has been discussed by Pitelka [48].

B. Components of the Ciliary Shaft

The ciliary shaft is composed of an array of longitudinal fibrils, called the axoneme, embedded in a cytoplasmic matrix and enclosed in a membrane that is continuous with the surface membrane of the rest of the cell. The diameter of the axoneme is consistently about 0.2 μm and the total diameter of the cilium at the membrane is about 0.25 μm. The axoneme has a highly characteristic structure, known as the "9 + 2" bundle, but this "label" conveys little of the modern view of the complexity of the axoneme because not only are each of the nine outer fibrils double, but they are associated with important transverse components. The main elements of structure of the axoneme are shown diagrammatically in Fig. 3 and in the form of electron micrographs in Fig. 4.

The longitudinal components of the axoneme are protein fibrils of a type called microtubules that occurs widely in cells, though their arrangement and attachments are specialized in cilia. The two central members of the complex are separate single microtubules about 24 nm in diameter, while each member of the outer ring of nine consists of one entire (A) microtubule and one partial (B) microtubule attached laterally to make the second component of a doublet that measures about 30 nm in the radial direction and about 45 nm in the tangential direction (Fig. 3f).

Structures known as arms project laterally from the A microtubule of each doublet toward the B microtubule of the adjacent doublet; pairs of arms are borne at intervals of about 17 nm along the length of each A microtubule, and each arm is curved, but the two members of each pair are different in that the outer arm is longer, with an extra segment at its base. Under certain conditions it appears that the distal ends of the arms may attach to the adjacent B subfibril, but most electron micrographs show the distal ends projecting freely.

Some treatments, notably chemical removal of the arms, reveal that there is a second type of connection between the peripheral doublets; these "nexin" or circumferential links appear as straplike strands, up to 30 nm wide, arranged as the rungs of a ladder between the A microtubules of adjacent doublets, or between the A microtubule of one doublet and the B component of the next.

The A microtubules also bear structures about 36 nm long that project toward the center of the ciliary axis and are known as spokes or radial links because they appear to connect the peripheral doublets with the central elements of the axoneme. The radial links are borne along the A microtubule at intervals of about 32 nm, but frequently the links occur in groups of two or three, or in more complex patterns that vary in different organisms (see Fig. 4c, for example). The main part of each radial link is a slender shaft which appears to terminate in a dilated head, although the details of structure vary and in no case are really well known.

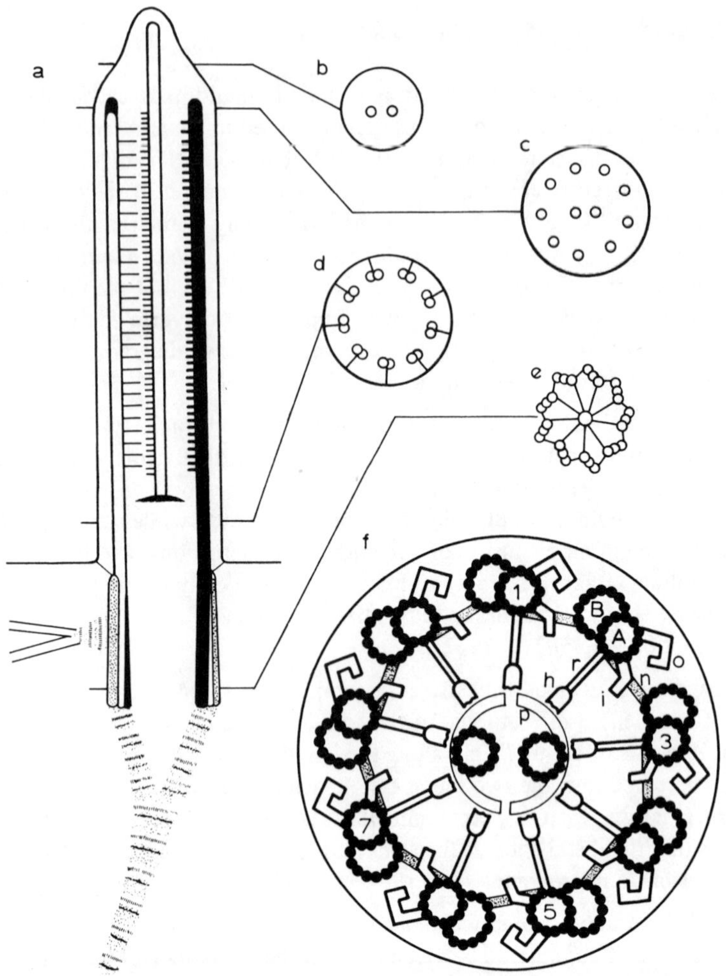

FIGURE 3 Details of the structure of a cilium than can be seen in longitudinal and transverse sections. (a) A longitudinal section through a ciliary shaft, basal body and root; the peripheral A microtubules are shaded, the B and central microtubules are unshaded and the C microtubules (in the basal body) are stippled; at the left are shown those shaft features seen in a radial plane—the projections on the central microtubule and the radial links on the A microtubule, and at the right the dynein arms are shown on the A microtubule as they would appear in a more tangential section; the basal body carries a basal foot, with attached microtubules at the left, and a striated root extending down into the cell below. (b-e) Transverse sections to show the arrangement of microtubules and connecting links at the indicated levels. (f) Detailed diagram of components seen in a transverse section of the main part of the shaft. In (f) the central microtubules carry lateral projections (p), the peripheral doublets are numbered clockwise (as seen

The single central microtubules also bear lateral projections in the form of curved rods 18 nm long, that are attached at 16 nm intervals along the length of each microtubule at either side in such a fashion that they appear to form an incomplete central sheath. The heads of the radial links appear to make connection with these projections from the central fibrils.

At the ciliary tip the peripheral doublets become single microtubules by the termination of the B subfibril, and no lateral arms or radial links are found distal to this level. The single microtubules decrease in number toward the tip, but, although it is known that the central microtubules are sometimes the longest, there is no known general rule about the order in which the fibrils terminate.

C. Structures at the Ciliary Base

At the ciliary base the axoneme continues into the basal body in a modified form. The central fibrils terminate near the level of the cell surface, where there is often a dense granule or partial septum in which the proximal ends of the central microtubules are embedded (Fig. 3a). There are often additional interconnections between the peripheral doublets at this level, and sometimes also links between the peripheral doublets and the ciliary membrane; Linck [34] found that the fibrils of the axoneme remain linked together in this region after chemical disruption of the distal part of the array into separate divergent fibrils.

The peripheral fibrils extend throughout the length of the basal body, but somewhat below the level of the cell surface each doublet becomes a triplet by the addition of another incomplete (C) microtubule attached alongside the B microtubule. The axis of each triplet becomes twisted, with the A microtubule nearest the center, and generally the triplets are interconnected by slender filaments in various configurations, notably near the inner end of the basal body where a pattern of "hub and spokes" is frequently seen (Fig. 3e). Fine filaments or sheets of material are generally found to connect the distal end of the basal body, near the outer ends of the C microtubules, with the cell membrane around the base of the ciliary shaft.

Pitelka [48] describes two types of basal body which differ in the form of the transition zone between the distal end of the basal body and the proximal

---

FIGURE 3 (continued)
from the basal end of the cilium, with dynein arms directed clockwise) and composed of A and B microtubules; adjacent doublets are connected by nexin links (n), and the A microtubules bear outer (o) and inner (i) dynein arms extending toward the B microtubule of the adjacent doublet and radial links (r) composed of a shaft and a dilated head (h) which lies close to the central complex

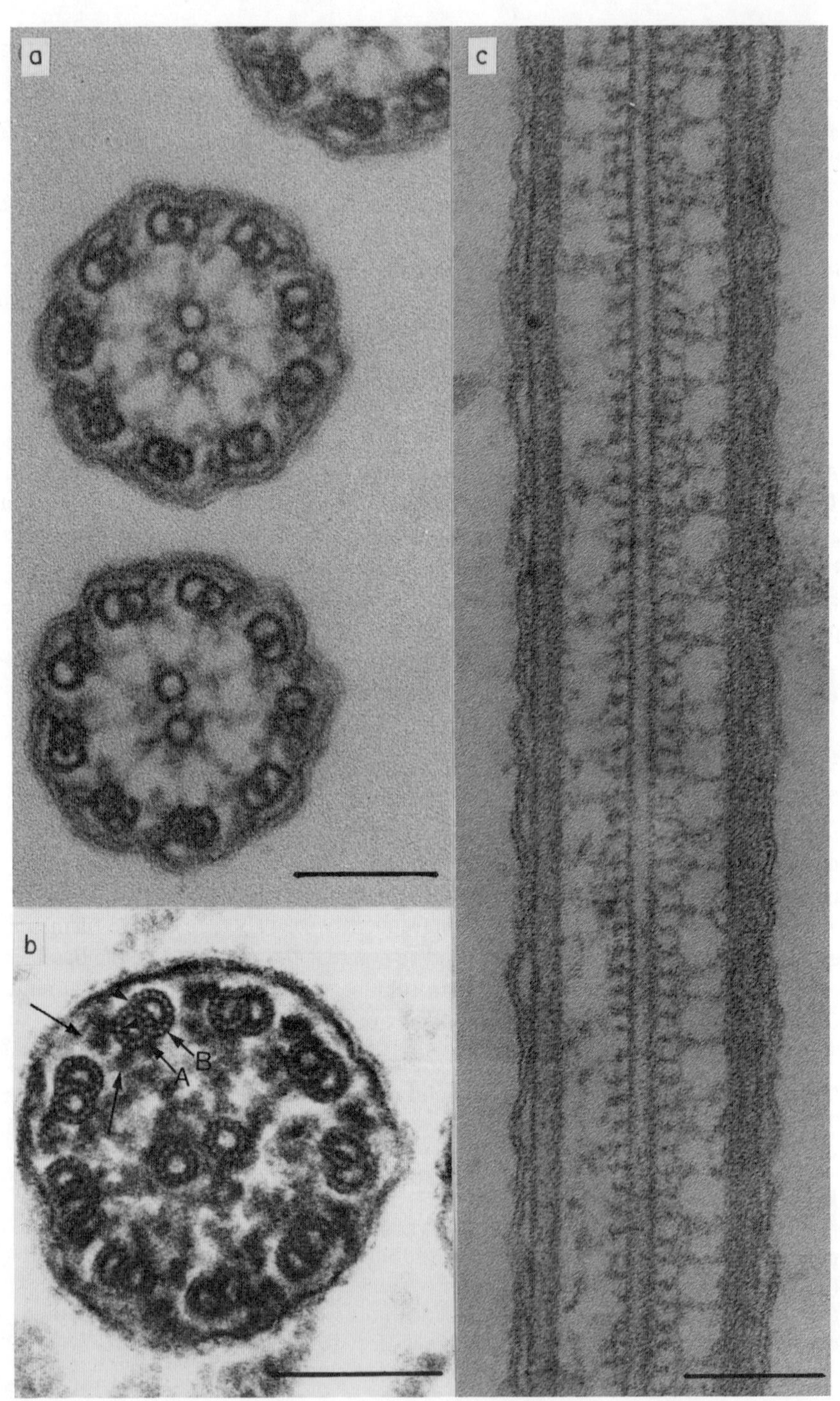

end of the ciliary shaft; the form of her type II basal body is shown in Fig. 3, and
type I basal bodies differ in that the termination of the central fibrils at the axoneme granule or septum is level with the general cell surface, near to the distal ends
of the C microtubules. The significance of these differences is not known, but
type I basal bodies are usual in birds, mammals and many Protozoa, while type
II basal bodies are common in invertebrates and lower vertebrates, and are seen
in human tracheal ciliated cells [51].

The structure of the main part of the basal body is identical with the structure of the centriole that is concerned with the organization of the mitotic spindle
at nuclear division. Both structures are involved in the development of organelles constructed of microtubules. In many types of ciliated cells, including some
cells of vertebrates, products of the replication of "nuclear" centrioles migrate
to the cell surface and act as the organizing centers for ciliary axonemes. It is
likely, however, that in evolutionary terms centrioles were derived from basal
bodies, since in some unicellular forms the flagellar basal bodies appear to act as
the poles of the nuclear spindle at the same time as providing anchorage for the
flagella; one might conclude that primitively the flagellar root fibrils were made
use of to orientate the migration of chromosomes, and that later the functions
of the centriole in providing the nuclear spindle became separated from the function of providing the flagellar axoneme and root fibrils.

D. Ciliary Roots

Although the ciliary membrane around the ciliary base and the filaments connecting it with the basal body may help to maintain the position of the cilium at
a particular point on the cell surface, the main anchorage of the axoneme is normally provided by "root" fibers attached to the basal body. The fibers are pre-
sumed to be proteinaceous and are usually either microtubular or aggregates of
filaments showing a marked cross-striation; frequently dense amorphous or fila-
mentous material surrounds the attachment of these fibers to the basal body.

FIGURE 4 Transmission electron micrographs of sections of cilia showing the
arrangement of the longitudinal microtubules and transverse components. (a)
Transverse sections of gill cilia of the mollusk *Elliptio*, in which dynein arms and
radial links may be seen. (b) Transverse section of a cilium from the chick
trachea stained with tannic acid to reveal the subunit structure of the microtubules—the central and peripheral A microtubules have 13 subunits in their
circumference, while the B microtubules have 11 more subunits. (c) Longitudinal
section of a gill cilium of *Elliptio* showing the arrangement of radial spokes and
central projections. In each micrograph the bar represents a length of 0.1 μm.
(a) and (c) are taken from Warner and Satir [69, pp. 39 and 41], by courtesy of
The Rockefeller University Press, and (b) is taken from Kalnins and Wassmann
[29] published by the Microscopical Society of Canada.

The ciliary basal bodies of higher animals frequently carry a lateral "basal foot" that projects as a short-banded cone from near the midregion of the basal body (Fig. 3a). Microtubules are commonly attached to the tip of the basal foot and extend away into the superficial cytoplasm of the cell. The basal foot of the cilium is found at that side of the basal body toward which the cilium bends in its effective stroke, and microtubules attached to it are ideally placed to resist forces developed in the effective stroke. This positional relationship is found in cilia of mechanoreceptor cells in the accousticovestibular system of vertebrates (reviewed by Barber [3]) as well as in propulsive cilia [17].

Larger striated roots generally extend from the inner end of the basal body toward the cell nucleus. These roots appear to be composed of bundles of fine longitudinal filaments whose structural subunits are precisely aligned to form a complex pattern of cross-striations with a major periodicity of 60-70 nm (sometimes as low as 14 nm or as high as 90 nm), with several clear subbands. Striated roots vary in size and form; sometimes they are massive and extend deep into the cell, in other examples shorter branching roots may diverge to make contact with nearby structures like mitochondria or the lateral cell membrane.

It is assumed that roots attached to basal bodies have the primary function of anchorage of the base of the axoneme, such that the motile efficiency of the cilium can be maximized. The relatively feeble mechanical activity of individual mammalian respiratory cilia may be reflected by the fact that the striated roots attached to the basal bodies appear to be slender and rather short [51]. However, the presence of large striated roots on the basal bodies of such cilia as those in the rod cells of the vertebrate retina suggests that mechanical anchorage may not be their only function.

E. Form and Function of Ciliated Cells

Ciliated cells may carry any number of cilia between one and many thousands, indeed some ciliated protozoan cells possess several different types of cilia with different functions and quite different patterns of activity. The cilia of cells of multicellular animals are usually concerned with only a single function and usually, but not invariably, the ciliated cell carries many similar cilia. Ciliated cells of the respiratory tract bear about 200 cilia packed at a density of about 8 to the square micrometer at the apex of the cell (Fig. 5). The basal ends of the ciliated cells taper as they approach the basement membrane of the epithelium. The nuclei are found toward the lower, tapered, ends of the ciliated cells, Golgi bodies are common in the central regions of the cells, and mitochondria are more abundant in the apical region among the ciliary rootlets than in other parts of the cell. Profiles of nerves have been seen among the basal ends of the ciliated and mucus-secreting cells in electron micrographs [51], but no synaptic

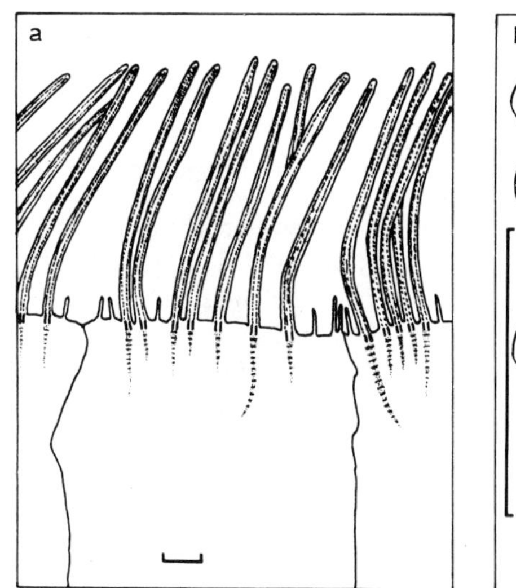
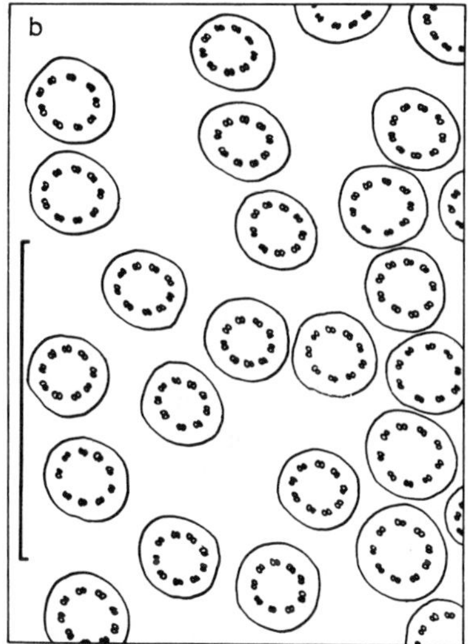

FIGURE 5 Tracings from electron micrographs by Rhodin [51] of human tracheal cilia (a) in longitudinal section and (b) in transverse section to indicate the length and spacing of the cilia. The bars each represent a length of 1 μm.

contacts were found and the function of the nerves is not known—it is probable that they are sensory rather than motor, but nerves of both types could be present.

IV. Biochemical Characteristics of Ciliary Components

A. Sources and Methods

Cilia or flagella from a variety of organisms have been separated from their cell bodies, purified, and subjected to fractionation; subsequently their components have been studied biochemically to discover their structure and enzymatic properties. The cilia of ciliated protozoans, mollusk gills, or sea urchin embryos, or the flagella of flagellate protozoans or sea urchin sperms have been used in most of these studies because these sources provide abundant, relatively clean organelles.

The separation and isolation of the major structural proteins by such workers as Gibbons [15,16], Mohri and colleagues [40,43], and Stephens and colleagues [58,60] was followed by intense activity in this field, and by the isolation of some of the minor components. A comprehensive review of the techniques of isolation and fractionation of cilia and of the results of subsequent biochemical investigations has recently been written by Stephens [59]; only the main results and implications of this work will be considered here, and specific reference will be made only to more recent studies.

B. The Microtubule Protein—Tubulin

The main longitudinal structural elements of the axoneme are the microtubules, composed of a type of protein, named *tubulin* by Mohri [41], which can be isolated either as a monomer with a molecular weight of about 60,000 daltons or as a dimer of about twice this size. High-resolution electron micrographs of negatively stained preparations of disrupted and spread microtubules [68] and of sectioned microtubules stained with tannic acid (see Fig. 4b and Ref. 64) show that the microtubule wall is made up of 13 longitudinal protofilaments, each of which is composed of a linear array of globular units about 4 nm in diameter. Other studies have indicated an 8-nm periodicity along the length of the protofilaments. A globular protein of diameter about 4 nm could have a molecular weight of about 60,000 daltons, but Warner and Meza [67] suggest that the 8-nm unit represents a dimer in which two 1.7 nm diam strands (the two monomers) are joined together in a figure-of-eight. Most analyses of micro-tubule (tubulin) proteins agree that all microtubules contain two types of tubulin, and it is suggested that the tubulin dimer is a heterodimer containing one mole-cule of each type. Each dimer of tubulin is associated with two molecules of guanosine nucleotide, one GTP and one GDP, one tightly bound and one ex-changeable.

Tubulins from each of the central microtubules and from peripheral A and B microtubules can be isolated because they show minor differences in solubility in ionic solutions; these tubulins show almost identical amino acid composition with minor differences as would be expected from the different solubility prop-erties as well as the small differences in structure and in diverse specific attach-ments of the different microtubules.

C. The ATPase Protein—Dynein

The other major protein component of the axoneme is an ATPase protein with a molecular weight of about 540,000 daltons which can be removed from de-membranated axonemes by dialysis at low ionic strength in the presence of EDTA, a technique pioneered by Gibbons [16]. Electron microscope prepa-rations of axonemes after removal of the ATPase protein show that the arms have been removed from the A microtubules of peripheral doublets; addition of the purified ATPase protein and magnesium ions to these axoneme prepara-tions resulted in reappearance of the arms in the normal positions on the doub-let fibrils. Gibbons and Rowe [17] named this protein *dynein* (force protein), since they believed the ATPase was in some way connected with the production of motive force in the axoneme.

The dynein enzyme has a high specificity for ATP as substrate, other nucleoside triphosphates being hydrolyzed at much lower rates, usually well below one-tenth of the rate of ATP hydrolysis. The ATPase activity of dynein requires a divalent cation; normally Mg-ATP^{2-} is the preferred substrate for dynein, and, although calcium, manganese, iron, cobalt, and nickel all give some activation of dynein ATPase, only with magnesium is the ATPase activity coupled to motility. The movement of cilia in association with dynein Mg-ATPase continues in the presence of EGTA, so that it is assumed that calcium is not necessary for this activity, although it may modify it (see Sect. VII). It is thought that the dynein interacts with other components of the axoneme during normal mechanochemically coupled activity, because the behavior of the enzyme in relation to pH and monovalent ions (particularly K^{+}) is quite complex.

Two rows of dynein arms are attached to each A microtubule; the longer arms of the outer row are sometimes more easily removed by ionic solutions, but axonemes perform normal (though slower) movements without the outer arms [13], and there is no substantial evidence for any difference in properties between arms of the two rows. Evidence will be presented in Sect. V which indicates that bridging links can be formed though the dynein arms between the A microtubule of one doublet and the B microtubule of the adjacent doublet, and that ciliary bending depends upon active sliding between adjacent doublets in association with ATP-powered mechanochemical activity involving the dynein arms.

D. Properties of the Tubulin/Dynein System

Comparisons have frequently been made between the tubulin/dynein system of cilia and the actin/myosin system found in muscle, but the details of structure and functioning are very different.

Actin and tubulin have somewhat similar amino acid compositions, but the peptides of which the two proteins are composed show few similarities. Tubulin aggregates as a heterodimer to form cylindrical fibrils and binds guanosine nucleotides, while actin monomers aggregate to form solid filaments of two helically twisted strands and bind adenosine nucleotides.

Myosin molecules possess ATPase properties analogous to those of dynein, but are commonly aggregated to form complete myosin filaments which form temporary bridges through the heads of myosin molecules to actin filaments, while dynein appears to be more or less monomeric, permanently associated with one tubulin microtubule and forming temporary bridges to another micro-

tubule. The myosin ATPase is generally less specific in its nucleotide substrate
and requires both Ca^{2+} and Mg^{2+} for mechanochemically coupled activity, while
dynein is more specific for ATP as a substrate and requires only Mg^{2+}. Other
differences and similarities in properties and enzyme activities, including the
functioning of control proteins, are already suspected and will certainly be
clarified in the near future.

The actin/myosin and tubulin/dynein systems have evolved in parallel to
perform different types of functions. Actin/myosin mechanisms are found not
only in all types of muscle, but also in ameboid cells and probably in the surface
layers of animal cells undergoing division, morphogenetic shape changes or
phagocytosis; the degree of aggregation of the myosin component and the functioning of the control proteins are known to vary. Tubulin/dynein mechanisms
are probably nearly as widespread and again show variations in the details of
the relationship between the participant proteins; they are present in some non-
ciliary motile organelles in protozoan cells and perhaps in some cells that show
cytoplasmic flow, notably nerve axons and some pigment cells. Although the
nuclear mitotic spindle is also microtubular, the mechanism of chromosome
transport probably does not involve dynein.

E. Other Components of Cilia

The identification and study of the properties of minor protein components of
the ciliary axoneme has not yet progressed very far [59]. Two identified protein components associated with the axoneme are adenylate kinase (MW 35,000
daltons), responsible for converting ADP to ATP + AMP, and the nexin protein
of the interdoublet links (MW ~150,000 daltons). Other components have
tentatively been identified as proteins of the radial link complex, and there is a
hint from some studies that an ATPase protein may be incorporated somewhere
in this complex [42,59]. There is evidence for the presence of an arginine kinase
(or perhaps creatine kinase in some animal groups) in the ciliary matrix [70]
and a variety of other enzyme proteins will probably be discovered there.

Isolated scraps of evidence concerning the ciliary membrane suggest that
its structure and properties will not be very different from those of other cell
membranes. There is no evidence of ultrastructural specialization other than
the particle arrays near the ciliary base revealed in freeze-fracture studies [19].
Isolation procedures have generally concentrated on the protein components
and discarded the lipids, so that very little seems to be known about the latter.
Gibbons [16] reported that about 22% of the total ciliary protein appeared to
be in the membrane, and that the membrane fraction possessed ATPase activity
that could be activated by either Mg^{2+} or Ca^{2+}. Stephens [59] observed that
membrane fractions appear to consist principally of a single protein of 200,000-

240,000 daltons molecular weight, although in one study two proteins of molecular weights 115,000 daltons and 60,000 daltons were found that could be fragments of the larger molecules; the relationship of these proteins to the ATPase enzyme is not clear.

V. The Mechanism of Ciliary Bending

A. The Site of Activities that Cause Bending

It is clear that the elements responsible for the characteristic bending movements of a cilium are present within the ciliary axoneme. Cilia that have been removed from their parent cell above the level of the basal body, and subsequently demembranated, are still able to undergo bending cycles when provided with an ATP energy supply and an appropriate ionic environment.

Earlier explanations of the mechanism of this bending were largely based on the belief that the peripheral fibrils of the cilium could actively contract [54], but current hypotheses almost without exception assume that bending is caused by active sliding of the peripheral doublets relative to one another. The first substantial evidence for the sliding microtubule model was obtained by Satir [52], and he has recently reconsidered the extent of support for the model [53]. One attraction of the sliding microtubule model is the parallel with the longer-established sliding fibril theory of muscle contraction.

B. The Sliding Microtubule Hypothesis

The sliding microtubule hypothesis assumes that the length of the ciliary microtubules remains constant and that when the microtubule doublets are caused to slide relative to one another the cilium bends because of crosslinking connections between the doublets. If microtubules of constant length slide against one another, Satir argued in 1965 [52], then, since all microtubules are anchored together in the basal body, the arrangement of the tips of the microtubules should vary with the bending of the cilium, and these tips should be displaced in one direction in the effective stroke and in the opposite direction in the recovery stroke (Fig. 6). Satir demonstrated that such displacements do occur in cilia of mollusk gills and found that the extent of sliding also provided quantitative support for the hypothesis. Agreement with the theoretical model was not quite perfect because the plane of maximum sliding did not lie in the axis of symmetry —this could result from the three-dimensional beat of these cilia or from some other asymmetry of the active sliding process. This work was not regarded as fully conclusive because it was based upon measurements made upon electron micrographs of fixed preparations.

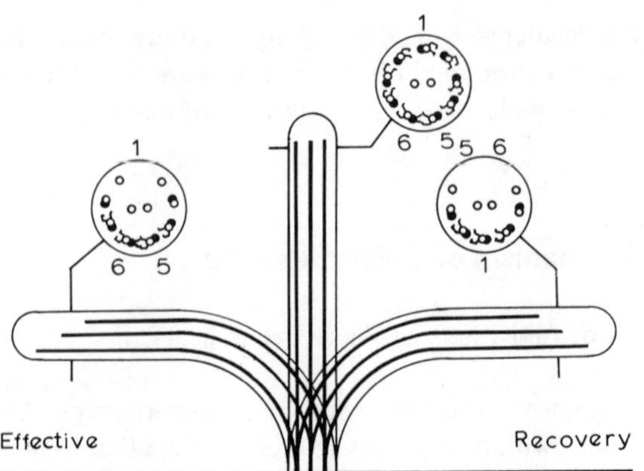

FIGURE 6 Diagram to illustrate the sliding microtubule hypothesis of Satir (see text). In the central part the microtubules of the straight cilium all terminate at about the same level (B microtubules shaded for identification), but at either side it is found that the microtubules end at different levels, those which run at the inside of the bend, and lie closest to the cell surface in cross sections, extend furthest at the ciliary tip. It can be seen that different microtubules appear at the tips in cross sections at the right and at the left because it is possible to make definite identification of microtubule doublets 5 and 6, since they are connected by a "bridge" structure in the mollusk cilia that were being studied.

Strong support was provided by studies on isolated axonemes of sea urchin sperm flagella by Summers and Gibbons [62]. Axonemes detached from their sperm heads and treated with detergent to remove their membranes may be activated with a Mg-ATP^{2-} medium to perform normal bending movements. When similar naked axonemes were briefly digested with a low concentration of trypsin to disrupt the radial and circumferential (nexin) links and then activated with the Mg-ATP^{2-} medium, the axonemes lengthened as groups of doublets slid along one another without the normal restraints of linkage elements, and eventually some doublets separated completely from one another. The increase in length of the axonemes could be seen in the light microscope, and the change in arrangement of the doublets was confirmed by electron microscopy.

C. The Role of Dynein in Sliding of Microtubules

The most likely explanation for this ATP-induced active sliding of the doublet microtubules appears to be that the dynein arms generate a shearing force between adjacent doublets; the ATPase properties of the dynein molecule and the similarity of the ionic requirements of the dynein ATPase and the sliding activity support this view. Any participation of the radial link is not likely, since these structures seem to be more sensitive to digestion by trypsin than the dynein arms [63].

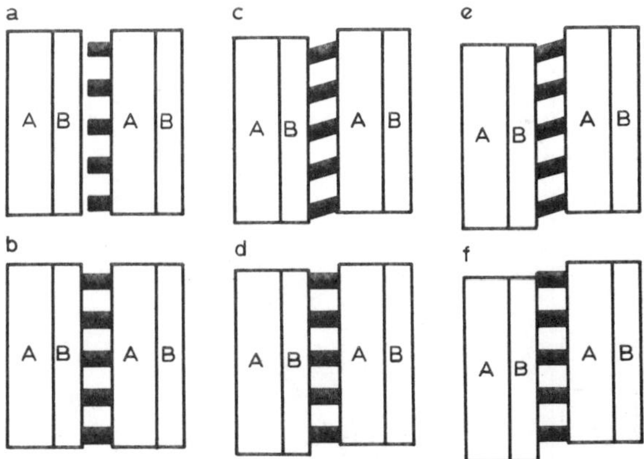

FIGURE 7   Diagram to illustrate a possible way in which one microtubule doublet is moved by the action of the dynein arms attached to another doublet. In (a) the arms on microtubule A are free; in (b) they attach distally to the adjacent B microtubule; in (c) the dynein molecules are assumed to undergo a change in configuration to create a shearing force between the doublets; the dynein arms then release their contact with the B microtubule, return to their original shape and reattach distally (d); further cycles of configuration change (e) and subsequent release and reattachment (f) can result in repeated shearing movements.

It was therefore suggested that the dynein molecules anchored to the A microtubule of one doublet are capable of cyclically changing shape as they bind and hydrolyze ATP in a reaction sequence that is coordinated with the formation and breakage of linkages with binding sites on the B microtubule of the adjacent doublet. Changes in shape or angular orientation of the dynein molecules in the period while they are attached distally to the B microtubule can generate the required shearing force; one can envisage one doublet moving along an adjacent one as its dynein arms make contact, swing and release (Fig. 7), in much the same way as a millipede moves along a twig.

There is some evidence from the work of Summers and Gibbons [63] on partially digested axonemes that when ATP is not available the dynein arms are linked with the adjacent B microtubule (a situation analogous to the rigor state in muscle), and that addition of ATP causes the generation of an active shearing force and sliding between the doublets. Evidence was mentioned in Sect. IV that normal, though slower, bending movements could be performed by axonemes from which the outer rows of arms had been removed.

D.   The Role of Radial Links

The radial links and nexin bridges are not thought to take part directly in the production of forces of active sliding, but the radial links, at least, are believed

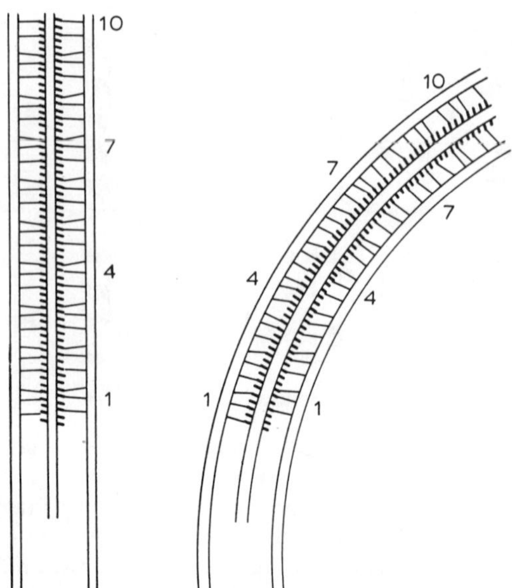

FIGURE 8   Changes in the orientation of radial links and their connections to the projections on the central microtubules during bending of a cilium, according to data given by Warner and Satir [69].

to play an important role in the formation of bends by limiting and regulating the sliding activities of the doublet microtubules, and they may also be responsible for the propagation of bending along the axoneme.

The amount of sliding that would be expected to accompany a bend of normal radius in a cilium must either greatly stretch the nexin and radial links or cause release and reattachment. There is evidence that the radial link heads detach from and reattach to sites on the central complex during the passage of a bending wave. Warner and Satir [69] have found that in electron micrographs of longitudinal sections of straight axonemes the radial links normally project perpendicularly from the doublets and the attachments to the central complex form a regular pattern (as in Fig. 4c), but in bent axonemes the links are not all perpendicular to the doublets, some of them being tilted in the direction of sliding in association with irregularities in the spacing of attachments of the link heads at the central complex (Fig. 8).

The suggestion has been made in Sect. IV that the link head may have ATPase activity, and it is possible that bond formation between the link head and the central complex may be an active process responsible for regulating the extent of ciliary bending and the rate of propagation of bending waves.

E.   Energy Supply for Sliding of Microtubules

Mechanochemical activity of the dynein molecules appears to require ATP as a
specific energy source. Observations that severed flagella can only beat for a
limited period [6,20] suggest that the ATP supply available within the organelle
is limited and that continued motility is dependent upon ATP diffusing into the
flagellum from the cell body (normally from the mitochondria). Calculations
have shown that the rate of diffusion of ATP within a flagellum is adequate to
maintain the ATP concentration above the level required for dynein activity,
even at the tip of the organelle [7,47]. The presence of adenylate kinase and
arginine kinase or creatine kinase (see Sect. IV) could provide a means of
mobilizing a short-term reserve of high-energy phosphate and overcome irregularities in the energy supply or demand.

F.   Summary of Present Understanding

In brief, the current view is that ciliary bending results from active sliding between the peripheral doublet fibrils, caused by ATP-powered changes in shape
of dynein molecules that can form links between these doublets. Other elements
of the ciliary axoneme, notably the radial links, play a part in resisting sliding
and limiting the amount of bending in a manner that induces propagation of
the region of bending along the cilium. It is still not clear how these activities
are integrated into the whole cycle of ciliary bending, and whether, for example,
the dynein arms at one side of the axoneme bend it to one side in the effective
stroke—or indeed whether the same arms can cause bending in both directions.
It is likely that evidence concerning these problems will become available in the
next few years.

VI.   The Coordination of Cilia

A.   The Phenomenon and Patterns of Metachronal Coordination

The early microscopists observed that an active ciliated epithelium was recognizable by the progression of waves of movement across the ciliated surface. The
mechanism by which the movement of cilia is coordinated to produce these
waves has been the subject of much speculation, and only recently has it been
explained in a reasonably convincing manner.

Where an epithelium shows such coordination, the waves that move across
the field of cilia can be seen to contain cilia at different stages of their cycle of
beating. Lines parallel with the wave crests contain cilia that are all at the same

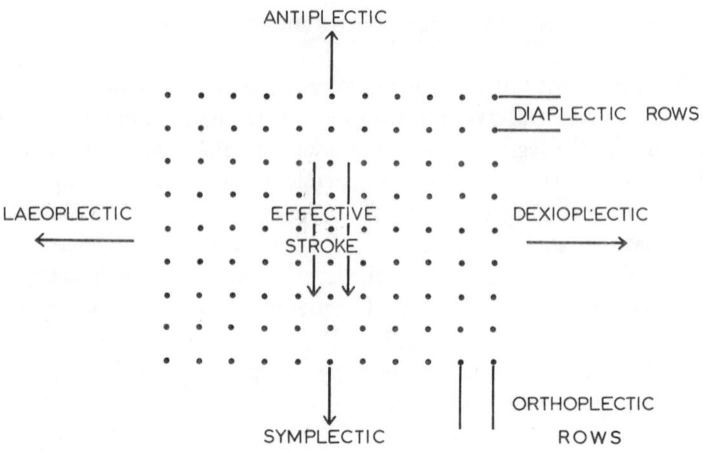

FIGURE 9   The nomenclature of the arrangement of ciliary rows and the metachronal wave patterns in relation to the direction of the effective stroke of the cilia, according to Knight-Jones [32].

stage of their ciliary beat—they are in synchrony. Lines on the epithelium at an angle to the wave crests contain cilia at different phases of their beat—the cilia show metachrony. In lines perpendicular to the wave crests the phase difference between adjacent cilia is maximal, and such a line was referred to by Machemer [37] as the mainline of metachrony.

The direction of the effective stroke of the ciliary beat, and hence the direction of propulsion of fluid by the cilia, normally bears a consistent relation to the direction of progression of the metachronal waves over a particular epithelium, but different systems show diverse relationships between the direction of effective strokes and wave propagation. The principal patterns of metachronism were recognized and named by Knight-Jones [32]. Rows of cilia that lie parallel to the plane of the effective stroke are referred to as orthoplectic rows, and rows perpendicular to this are diaplectic rows (Fig. 9). Waves that are propagated along orthoplectic rows in the same direction as the effective stroke show symplectic metachrony and waves propagated in the direction opposite to the effective stroke show antiplectic metachrony. Waves that pass along diaplectic rows are involved in dexioplectic metachronism when the effective stroke moves toward the right when the observer looks in the direction of wave propagation, and in laeoplectic metachronism when the effective stroke is toward the left of the direction in which the waves travel.

B.   The Mechanism of Metachronal Coordination

The different patterns of metachronism reflect different relationships between the cilia that form the waves. A close examination of the structure of the waves

and of the forms of interaction between the cilia involved has led to the belief
that the formation and propagation of metachronal waves is due to hydrodynamic forces between the moving cilia; these result from the sorts of viscosity-
dominated interactions between the fluid and the cilia discussed in Sect. VIII
(see also Ref. 56).

In the protozoan *Opalina* the closely packed cilia move rather slowly
throughout their beat: when the cilia of a stationary *Opalina* begin to beat, they
move with almost random timing (though consistent direction), but within a few
beats adjacent cilia become more nearly synchronous and regular waves appear
and spread to form a coherent pattern over the whole surface. The stable metachronal pattern is one where the viscodynamic interaction between cilia has been
reduced to a minimum by averaging of frequency and interciliary constraints
that induce formation of coordinated waves.

In other examples where the cilia beat more quickly the metachronal
waves appear almost instantaneously because of the strong viscous coupling between the cilia. Diaplectic metachronism is found in most ciliary systems that
propel water (e.g., Fig. 10a), and in these cases it is found that the direction of
propagation of the metachronal waves corresponds with the dominant viscous
coupling of a recovery stroke in which the cilium swings to one side or the other
[55]; the strongest interciliary coupling occurs during the effective stroke, and
this serves to synchronize the beat of the cilia in the orthoplectic direction.

The principal result of metachronism is to minimize interference between
adjacent cilia—the viscous coupling between adjacent synchronous or nearly
synchronous cilia reduces energy dissipation. Metachronal systems also achieve
greater efficiency because a considerable number of cilia move synchronously
in their effective strokes and influence a larger body of fluid than a single cilium
could move, then the immediately adjacent cilia perform their propulsive stroke
and carry the fluid further along, and this is repeated with more groups of cilia
to maintain a continuity of flow. Wave patterns are adaptive, and can be seen
to have exploited various devices to increase the effectiveness of the cilia in
propulsion of fluids, particularly with regard to the maintenance of a continuous
flow by means that effectively reduce the influence of the nonpropulsive recovery phase [5].

C. Metachronism and Mucous Transport

Metachronism is not nearly as well developed in ciliated epithelia that transport
mucus as it is in systems where water is propelled (Fig. 10). The feeble development or absence of pluricellular metachronism in mucociliary systems may be
assumed to have a functional basis in that arrhythmic cilia are probably better
suited for mucous transport. Where the function of cilia is to transport mucus,
which, at least occasionally, may be thick, particle-laden or sticky, then there

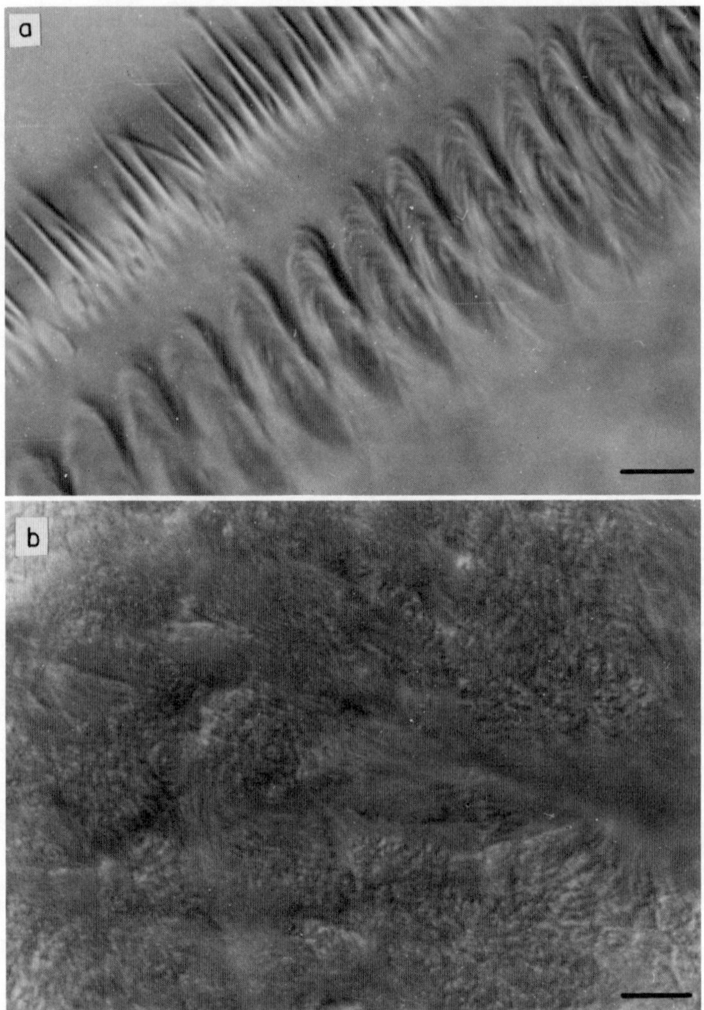

FIGURE 10 Light micrographs illustrating the contrast between (a) the prominent and well-defined metachronal waves of a ciliary tract that propels water (the lateral ciliated cells of *Mytilus* gill filaments), and (b) the erratic metachronal activity seen on a mucus-transporting epithelium (frog palate). The bar in each case indicates a length of 10 μm.

are probably real mechanical advantages in the cilia being short and densely aggregated together, so that the mucus can be supported by the cilia without the cilia being bent over and rendered incapable of beating movements. It is also likely that, as suggested in Sect. VIII, the motion of the tip of a short cilium is better suited to the penetration of and release from a layer of mucus.

Mucus-transporting cilia are short, and the height of the effective stroke is only a little above that of the recovery stroke, but the cilia are very closely packed and the viscous interaction between them must be very large indeed, especially in the orthoplectic direction, so that adjacent cilia on the same cell must always be nearly in phase. However, a rather small discontinuity at the cell boundaries would be sufficient to interrupt the coordination of these very short cilia and prevent the formation of extensive metachronal waves. This may be yet another advantage of short cilia, for the absence of coherent pluricellular metachronal waves may permit the more random motion of groups of many clawlike ciliary tips to move any overlying mucous layer forward most effectively without significant propulsion of the lower serous layer.

When observing a ciliated epithelium that is transporting mucus it is easy to determine the direction of the effective stroke of the cilia from the direction of mucous transport, but the more visible phase of the ciliary beat (because it is slower) is the recovery stroke which moves in the opposite direction to the flow of mucus, so that one has the illusion of active motion, interpreted as more or less well-developed waves of ciliary movement, moving in the opposite direction to the effective stroke. This may explain some of the reports in the literature describing antiplectic metachrony in mucociliary systems, and it appears, for example, that the metachronism of cilia on the palate epithelium of the frog is normally antiplectic. However, there have been almost as many reports describing symplectic metachrony, and films of rat bronchiolar cilia taken by Iravani (personal communication) clearly show metachronal waves of long wavelength that pass in the same direction as the propulsion of mucus. The metachronism of cilia that transport mucus is certainly irregular, and it is possible that the degree of coupling between cilia and the character of the metachronism could vary with the frequency of beating.

VII. Control of Cilia by the Organism

A. Forms and Mechanisms of Ciliary Control

The diversity of ciliary function has led to a variety of requirements for control by the organism. Clearly, where cilia are used for locomotion, then the organism, be it a single-celled animal like *Paramecium* or the multicelled larva of an oyster, will find it necessary to regulate the activity of the locomotive cilia. In *Paramecium* the main form of control is the reversal of ciliary beating brought about by an ionic mechanism dependent upon a change in potential difference across the surface membrane of the cell [46], and associated with a less obvious ability to regulate the rate of ciliary beating [38]; these permit the animal to respond to environmental stimuli. The more complex mollusk larva responds to some stimuli by a long-lasting inhibition of the locomotor cilia, mediated through the larval nervous system, that allows the larva to drop away from the source of the stimulus, and it responds to other stimuli by an increased rate of ciliary beating that is mediated by the hormone serotonin (5-hydroxytryptamine) (see Ref. 1).

Where cilia are used to maintain a water current through a gill cavity for respiration, and often also for feeding, ciliary control may involve one or more of ciliary arrest, ciliary reversal, acceleration of beating, and deceleration of beating. Thus, in an ascidian, where muscular action can close the pharyngeal cavity and can cause squirting of water from the pharynx to remove unwanted foreign particles, the only form of control of gill cilia identified by Mackie et al. [39] was a ciliary arrest mediated by pharyngeal nerves and coincident with muscular closure of the pharyngeal entrances. In larvacean tunicates particles may be removed from the pharynx by reversed beating of the gill cilia under nervous control [12]. The current-producing cilia of the gill of the bivalve *Mytilus* show not only brief periods of ciliary arrest, but also acceleration of beating mediated by serotonin and deceleration of beating mediated by dopamine—all three types of response can be obtained by appropriate stimulation of the nerve supply to the gill (see Ref. 1); reversed beating is not known to occur here.

It is significant that in systems as diverse as ciliary reversal of *Paramecium* [46], arrest of the pharyngeal cilia of the ascidian *Corella* [39], and activation of the oviduct cilia of the amphibian *Necturus* [45], the control exerted within the cell seems to be mediated by changes in the concentration of calcium ions; in the amphibian the concentration of Ca^{2+} influences the energy supply to the cilia, but there is no evidence of a similar effect in the other cases.

B. Control of Cilia of Vertebrates

The requirements for control of cilia in vertebrates appear to be much more restricted. The uses of cilia in vertebrates do not appear to demand controlled reversal of beating because they function to move mucus, particles, spermatozoa or ova in a constant direction.

There is clear evidence (reviewed by Aiello, Ref. 1) for control of activity of the cilia on the palate of the frog; these cilia are normally quiescent but can be stimulated to beat by mechanical stimulation and by stimulation of the motor nerves supplying the palate, which are supposed to act by releasing acetylcholine.

Evidence of ciliary control from studies on mammals is not conclusive, partly because in most studies ciliary activity has been measured in terms of particle transport, which can vary with the quantity and character of mucous secretion as well as with the rate of ciliary beating. Florey et al, [11], Wardell et al. [65], and Sturgess and Reid [61] are among those who have suggested that mucous glands are under nervous control. Particle transport by rabbit tracheal cilia was found by Kordik et al. [33] to be slightly modified by drugs that act on cholinergic systems, but Aiello [1] concluded from a survey of the available information that the cilia of the mammalian respiratory tract are not under nervous control. There remains the possibility that direct mechanical stimulation and hormonal agents which modify cell metabolism are able to influence the activity of these mammalian cilia.

The following conclusions about the regulation of ciliary activity in the human respiratory tract can be derived from present evidence. The cilia develop in a particular morphological pattern in which the basal feet of all of the cilia have a similar orientation and all the cilia beat in a fixed direction with their effective strokes propelling fluid toward the oropharynx (although this may be disrupted by disease, e.g., chronic bronchitis; see Ref. 28); change in the direction of beating is not necessary. The cilia are assumed to beat at a characteristic rate dependent upon the metabolic state of the ciliated cells, and apparently not under the direct control of the nervous system, although hormonal regulation of the cell metabolism may prove to influence ciliary rate. The extent to which mechanical stimulation of respiratory cilia can influence the rate of beat is not known, but the lower rate of beating characteristic of cilia in smaller bronchioles in comparison with those in the larger bronchi and trachea could be a reflection of the amount of direct mechanical stimulation of the cilia. Precise control of the rate of beating of respiratory cilia does not seem to be necessary; indeed, it is difficult to identify reasons for regulation of ciliary activity that could not be satisfied by a direct response of the cilia to the degree of mechanical stimulation.

VIII. Fluid Propulsion and Ciliary Activity

A. Some Hydrodynamic Considerations

The primary function of cilia is to propel fluids. Most of the devices for propelling fluids with which we are familiar, like fish fins, bird wings, propellers of aircraft and ships, depend primarily upon inertial forces, while forces associated with the viscosity of the fluid in which the motion takes place are of little importance. The contrary is true of ciliary systems, where viscous forces dominate, as can be illustrated by the fact that if the cilia cease moving the fluid they propel comes to rest immediately. A measure of the relative importance of inertial forces and viscous forces in a system is given by the Reynolds number

$$\text{Re} = \frac{\text{inertial forces}}{\text{viscous forces}}$$

This nondimensional number can be estimated in a particular system from the expression

$$\text{Re} = \frac{\text{fluid density} \times \text{size} \times \text{speed}}{\text{fluid viscosity}}$$

In the motion of a ship or a larger fish moving in water, large size and high speed result in a high Reynolds number (often exceeding 10^3), indicating relatively insignificant viscous forces, whereas in the motion of cilia in water the very small size and relatively low speeds give a Reynolds number of very much less than 1 (often 10^{-3}, or less), indicating the relative insignificance of inertial components.

When considering the motion of fluid over a ciliated cell, attention must be paid both to the interaction between the cilium and the fluid and to the interaction between the cell surface and the fluid [4,5,56]. Some features of such interaction can be illustrated by consideration of what happens in the fluid around a moving body. If a small sphere is allowed to fall in water under the influence of gravity, the water immediately around the surface of the sphere is carried with the sphere at the same velocity—it obeys the "no-slip" rule; water further from the sphere is influenced less by its movement and is therefore carried more slowly, and so on until eventually one reaches the limit of the zone of influence of the moving sphere. The extent of the zone of influence of the moving body will increase if one or more of the following increase: the viscosity of the fluid, the size of the body, or its speed of motion. The shape of the body is also important, for a slender cylinder moving along its long axis carries only about half the amount of fluid that it influences if it moves perpendicular to its long axis at the same speed.

B. Fluid Movement Around a Cilium

No-slip layers and shear zones of decreasing velocity will surround a cilium moving in a fluid and will also cover the surface of a cell over which fluid is propelled. The fact that the fluid tends to stick to the cell surface means that when a cilium attached to a cell attempts to propel fluid over the cell, the fluid near the ciliary tip is easier to move than that near the cell surface; in fact, one can define an approximately conical zone around a cilium within which the influence of the cilium dominates and outside which the influence of the cell surface dominates— at any point on the cilium the radius of the conical zone is about equal to half the height of that point above the cell surface. The extent of influence of the cilium on the fluid in front of and behind the cilium in the plane of beat is about twice that perpendicular to this plane, so that the shape of the zone of influence

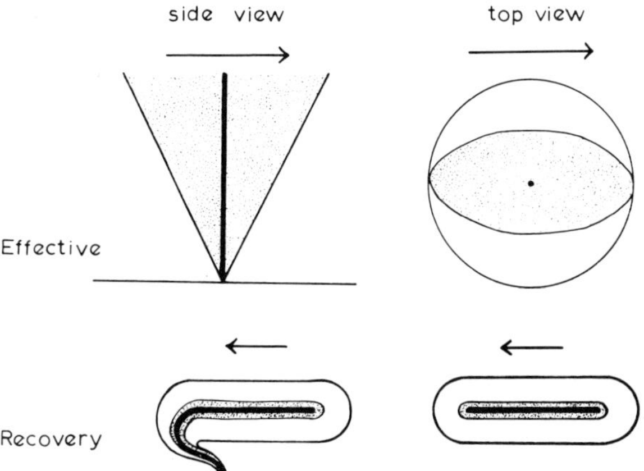

FIGURE 11 A diagram to illustrate the relative amounts of water moved in the effective and recovery strokes of a ciliary beat. Around the cilium in each case is an outer limit showing the extent of the zone of influence of the cilium according to the height of that part of the cilium above the cell surface, while the smaller stippled area indicates the approximate magnitude of the zone of transported water, taking into account the speed and orientation of the cilium as well as its height above the cell surface.

is a cone of elliptical cross section, with the longer axis of the ellipse in the plane of beat (Fig. 11).

Interactions between the fluid and the cilium and between the fluid and the cell surface are both of importance in determining the propulsive action of the cilium, and their relative importance will be different at different parts of the ciliary cycle. In the effective stroke the cilium is extended to its full length, and therefore carries a conical zone of fluid of maximum height, it moves with a high velocity, with the result that it carries a large fluid volume, and it moves perpendicular to its long axis, which maximizes the fluid volume that can be influenced. The ciliary tip moves fastest and is furthest from the cell surface, so that its influence on fluid propulsion is greatest. In the recovery stroke the cilium is bent nearer to the cell surface and moves more slowly with much of its length being drawn through the water along its long axis. Each of these features, height, speed, and orientation, contributes to reduced fluid transport in this phase of the beat (Fig. 11).

A larger fluid volume is thus moved forward quickly in the effective stroke, and a smaller volume is moved more slowly backward in the recovery stroke, fluid carried forward round the cilium in the effective stroke being "scraped off" the cilium in the recovery stroke because of the close proximity to the cell surface. Moreover, a large part of the forward propulsion occurs near the ciliary

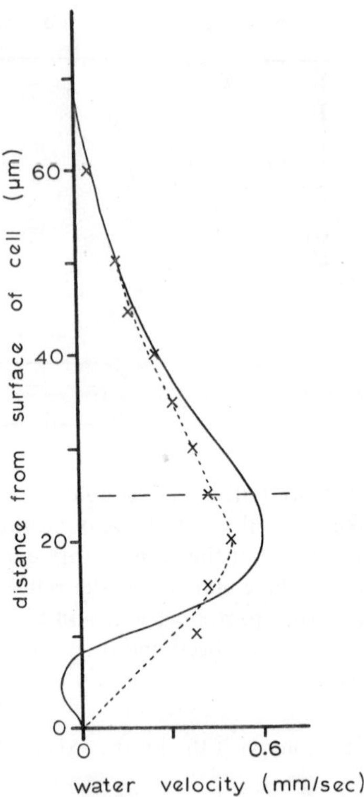

FIGURE 12   An example of the fluid velocity profile through a group of cilia that are propelling water (compound membranelle cilia of the protozoon *Stentor*). The cilia are about 25 μm long. The values plotted (x) on the dotted curve are observed values, while the continuous line indicates an "expected" curve. (Observed data from Sleigh and Aiello [57].)

tip in a zone that is not influenced by the recovery stroke; one can therefore separate the fluid into two layers, a lower region nearer the cell surface in which the fluid will oscillate and show little net forward flow because it is carried forward by effective strokes and backward by recovery strokes, and an upper layer which feels only the effective strokes and shows the maximum forward flow. This can be illustrated in terms of a fluid-velocity profile through a ciliated surface (Fig. 12). Many of these features of the propulsion of fluids by cilia have been discussed in more detail by Blake and Sleigh [4,5] and Sleigh [56].

C. Effects of Ciliary Size and Pattern of Beating

The forward flow produced by cilia can be enhanced in a variety of ways. If the difference in height of the cilium above the cell in effective and recovery strokes is increased, the importance of backflow caused by the recovery stroke can be minimized; one way of achieving this is to lower the profile of the recovery stroke, usually by the cilia moving to one side, close to the cell surface, in this phase of their beat. Since the maximal flow of water can be attained near the level of the ciliary tips, it is clear that an increase in ciliary length will provide a greater volume flow because of the greater distance above the cell surface, and can provide higher flow velocities if the frequency of beating can be maintained. However, longer cilia or higher frequency mean increased viscous resistance to movement, which has two consequences, first that a greater energy supply is required and second that the cilia lose propulsive efficiency if the cilia bend backward because their intrinsic stiffness is inadequate to withstand the increased resistance. A solution to this problem is to combine several cilia together into a compound structure in which the cilia can be longer and/or beat faster because of increased stiffness and greater energetic resources. The formation of close metachronal waves can achieve a similar result. Some comparative features of the rate of propulsion of fluids by cilia can be seen from Table 1; consideration of the way in which fluids are propelled by cilia can explain the significance of many variants of beat pattern and coordination.

D. Propulsion of Mucus

Evidence concerning the way in which mucus is propelled by cilia is sparse. We have seen that when water is propelled over a ciliated epithelium an outer zone that feels only the effective stroke is moved forward, while beneath it is a layer that moves to and fro in each cycle of beating because both effective and recovery strokes act upon it. In the case of mucous transport this two-zone system can be exploited by the presence of a more fluid, periciliary serous layer and an upper zone that includes mucus as islands, strings or sometimes as more continuous areas of mucus. If the regions of mucus are penetrated by the effective strokes and remain above the recovery strokes, the higher viscosity of the mucus will increase the effective zone of influence of the cilia in their effective stroke and accentuate the difference between the motion of the two layers. The fact that the mucus has considerable elasticity means that a continuity of propulsive strokes of many cilia over an extensive area will be required to achieve movement and overcome the tendency for elastic recoil; much of the work done against elasticity can be recovered if continuous ciliary activity is maintained. The viscoelastic character of mucus is also important because it makes possible the movement by cilia of thick bodies of mucus as well as thin layers—unless the mucus becomes too thick and sticky.

These features can explain why the speed of mucous transport comes closer to the estimated tip velocity of the cilia in their effective stroke than does the velocity of waterflow (see Table 1), and give support to the view that the

TABLE 1 Some Examples of Parameters of Ciliary Propulsion of Fluids

Organelle source	Ciliary length (μm)	Frequency (Hz)	Estimated tip velocity of cilium (mm/sec)	Mean fluid velocity (mm/sec)[a]		Comments[b]
Mytilus gill, lateral cilia	15	20	4.0	1.2	(w)	Close metachronal waves [57]
Stentor membranelle	25	30	4.4	0.8	(w)	Single row of compound cilia [57]
Pleurobrachia comb plate	600	5	50	6.0	(w)	Single row of large compound cilia;
	600	12	70	17	(w)	Re nearly 1 [57]
Rat trachea	5	22	0.5	0.2-0.25	(m)	[10]
Rat bronchus	5	18	0.45	0.2	(m)	[10]
Rat bronchiole	5 (or less)	7	0.15	0.007	(m)	Sparse cilia [28]

[a]m, mucus; w, water.
[b]Numbers in brackets are references

ciliary tips actually penetrate the mucus in their effective stroke, rather than remaining in the serous layer beneath and carrying the mucus by shearing forces between the two layers, as some authors have claimed. The forces acting on the cilia are probably not very different in "aquatic" situations where the mucus is completely immersed in water and in "terrestrial" situations where the mucus forms a surface with air, although these forces will change if the character of the mucus alters significantly.

E. Ciliary Beating, Metachronism, and Mucous Transport

We could go some way toward understanding the mechanism of fluid propulsion in mucociliary systems if we knew the form of beat of the cilia. Attempts to analyze the pattern of ciliary beating by study of rapidly fixed respiratory epithelia in the scanning electron microscope have not so far given meaningful information, and high-speed cinephotography is difficult with such epithelia because of problems of illumination and orientation.

The absence of published evidence may be partly a result of preconceptions about what one should see. Theoretical consideration of the form of beat that such short cilia might be expected to perform may throw some light on the problem.

An important influence will be exerted on the cycle of beating by the intrinsic stiffness of the cilium, which effectively limits the amount of bending. Integrity of the axoneme and some crosslinking of its components are essential both for the performance of the cycle of bending and to provide the stiffness required to exert a propulsive force on the fluid. Active bending is opposed by the stiffness of the longitudinal components of the axoneme and the more the cilium bends the greater the proportion of work that is done against internal resistance rather than against the fluid outside.

It seems likely that the extreme lower limit of the radius of curvature that can be achieved by active bending in a ciliary axoneme must lie between 1.0 and 1.5 μm; figures of simple cilia have been published that show a radius of about 2 μm, Brokaw [8] found values of 4-5 μm in flagellar sperm tails (though a value of 1.23 μm was observed under extreme experimental conditions in *Chaetopterus* sperm), and Rikmenspoel and Sleigh [50] found minimum values of 2.5-3.5 μm in two compound cilia.

Figure 13 shows a reconstruction of a possible form of beat for a cilium 5 μm long performing effective and recovery strokes of a standard form within the constraints imposed by a minimum radius of curvature of 1.25 μm in any bend. It is assumed that the beat is planar because very closely spaced cilia would otherwise be expected to interfere strongly with one another in a lateral direction and probably form prominent diaplectic metachronal waves. An obvious

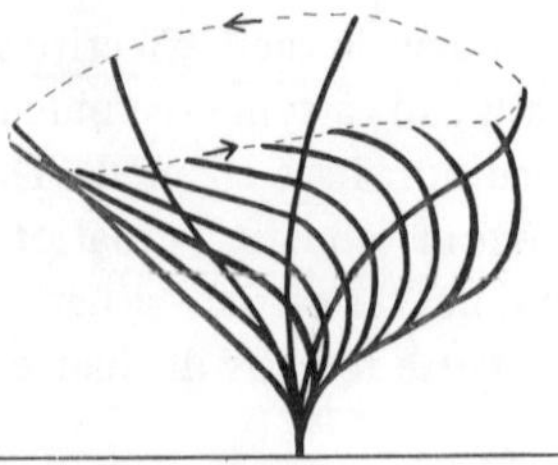

FIGURE 13 Diagrammatic construction of a possible beat pattern for a cilium 5 μm long that shows a minimal radius of bending of 1.25 μm. Many of the shapes seen here are also visible in Fig. 2.

difference between this cycle of beating and that of a longer cilium is in the relative heights of the two strokes, as can be seen by comparison of Fig. 13 with Fig. 1.

Whatever the exact form and extent of the beat of these short cilia, with this radius of curvature the recovery stroke must have a height of well over 3 μm, so that the difference between the height of the effective stroke and that of the recovery stroke cannot exceed about 2 μm, and will probably be considerably less than this. Cilia of this length would not be very efficient in propulsion of water, but in a two-layer system of mucus and serous fluid they could show high efficiency. The serous fluid layer would be about 3-4 μm deep, and the tips of the cilia would penetrate only 1 μm or so into the islands or patches of mucus (Fig. 14). A beat of the form shown in Fig. 13 would be rather suitable for the movement of mucus because the tips of the cilia should easily penetrate the mucous layer at the start of their stroke and "claw" it forward during a short, sharp, propulsive stroke, recalling the action of the "flick" stroke discussed by Hilding [24]. It is hoped that accurate details of the pattern of beating of respiratory cilia will be available soon, and that reliable data on the relationship of mucus and serous layers be obtained, so that the relation of the movement of mucus to the beating of cilia can be understood more clearly.

IX. Conclusions

Research on the structure, biochemistry, and functioning of cilia has now reached an advanced stage. We can state with some confidence that we understand the structural relationships of the various components and have some promising hypotheses about how they interact in causing bending of cilia. We also have sufficient information on the coordination and control of cilia to give a reasonably comprehensive description of these phenomena, but we do not yet understand in most cases how controlling agents act on the molecular mechanism of

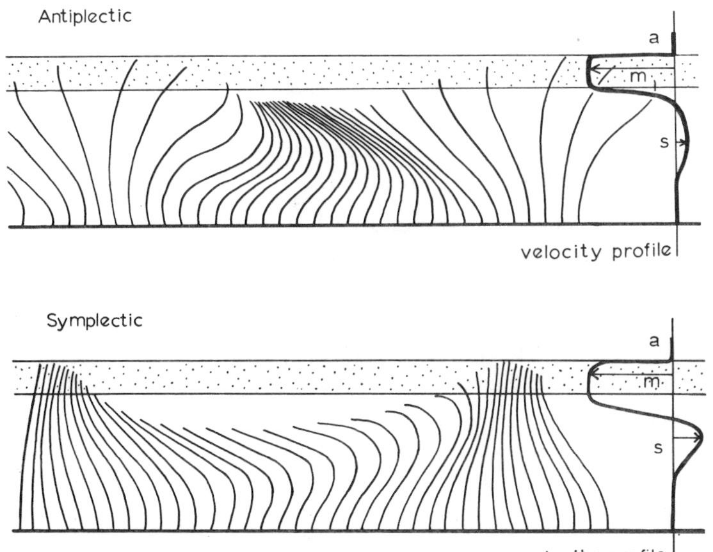

FIGURE 14 Constructions of metachronal waves using profiles similar to those in Fig. 13 for antiplectic waves (above) and symplectic waves (below), which are being used in the propulsion of mucus (m) over a layer of serous fluid (s) by a ciliated surface lining an air cavity (a). In both cases the effective stroke of the beat is toward the left; metachronal waves also move toward the left in the symplectic system but move toward the right in the antiplectic system. In life these waves, especially the symplectic ones, would probably contain more cilia, closely packed, but the relative heights of effective and recovery strokes are thought realistic. Approximate velocity profiles expected from these waves are shown.

the cilia. Knowledge on ciliary functioning has been gained from new studies on the movement of cilia, and an improved understanding about the propulsive action of cilia has come from a recent consideration of the fluid-dynamic principles of ciliary movement.

Very little of this information or understanding has come from work on mammalian respiratory cilia. There has been some ultrastructural work that confirms that respiratory cilia have a conventionsl structure, and there can be very little doubt that their internal mechanism is the same as that of other cilia. In terms of functioning, our knowledge of respiratory cilia is rudimentary—they are small and difficult to study, but, until we know more about their movement, any features of coordination, and the relationship of these to the movement of surrounding fluids, we shall not achieve a substantially improved understanding of the mucociliary transport mechanism.

The easiest measure of the activity of respiratory cilia is obtained by studying the rate of particle transport, but this is the result of two components, the activity of the cilia and the character of the mucus; a deficiency of mucociliary transport could result from a deficiency of mucus or cilia or both—direct observations of the cilia may sometimes make it possible to decide which, but other supporting evidence will usually be required.

Studies of the control of respiratory cilia suggest that nerves are not involved, but the adaptation of new techniques like that pioneered by Iravani should make it possible to study whether nervous stimulation or hormones can influence the mucociliary transport system and indicate whether this occurs through effects on the cilia or mucous secretion or both. Such techniques could also provide an understanding of the extent to which the cilia respond to mechanical stimulation.

References

1. E. Aiello, Control of ciliary activity in Metazoa. In *Cilia and Flagella*. Edited by M. A. Sleigh. London and New York, Academic Press, 1974, pp. 353-376.
2. T. Asmundsson and K. H. Kilburn, Mucociliary clearance rates at various levels in dog lungs, *Am. Rev. Respir. Dis.*, **102**:388-397 (1970).
3. V. C. Barber, Cilia in sense organs. In *Cilia and Flagella*. Edited by M. A. Sleigh. London and New York, Academic Press, 1974, pp. 403-433.
4. J. R. Blake and M. A. Sleigh, Mechanics of ciliary locomotion, *Biol. Rev.*, **49**:85-125 (1974).
5. J. R. Blake and M. A. Sleigh, Hydromechanical aspects of ciliary propulsion. In *Swimming and Flying in Nature*. Edited by T. Y. Wu, C. J. Brokaw and C. Brennen. New York and London, Plenum Press, 1975, Vol. 1, pp. 185-209.
6. C. J. Brokaw, Non-sinusoidal bending waves of sperm flagella, *J. Exp. Biol.*, **43**:155-169 (1965).
7. C. J. Brokaw, Mechanics and energetics of cilia, *Am. Rev. Respir. Dis. [Suppl.]*, **93**:32-40 (1966).
8. C. J. Brokaw, Effects of increased viscosity on the movements of some invertebrate spermatozoa, *J. Exp. Biol.*, **45**:113-139 (1966).
9. S. Carson, R. Goldhamer, and R. Carpenter, Mucus transport in the respiratory tract, *Am. Rev. Respir. Dis. [Suppl.]*, **93**:86-92 (1966).
10. T. Dalhamn, Mucous flow and ciliary activity in the trachea of healthy rats and rats exposed to respiratory irritant gases, *Acta Physiol. Scand. [Suppl. 123]*, **36**:1-163 (1956).
11. H. W. Florey, H. M. Carleton, and A. O. Wells, Mucus secretion in the trachea, *Br. J. Exp. Pathol.*, **13**:269-284 (1932).
12. C. P. Galt and G. O. Mackie, Electrical correlates of ciliary reversal in *Oikopleura*, *J. Exp. Biol.*, **55**:205-212 (1971).

13.   B. H. Gibbons and I. R. Gibbons, Effect of partial extraction of dynein arms on the movement of triton-extracted sea urchin sperm, *J. Cell Biol.*, **55**:84a (1972).

14.   I. R. Gibbons, The relationship between the fine structure and direction of beat in gill cilia of a lamellibranch mollusc, *J. Biophys. Biochem. Cytol.*, **11**:179-205 (1961).

15.   I. R. Gibbons, Studies on the protein components of cilia from *Tetrahymena pyriformis*, *Proc. Natl. Acad. Sci. USA*, **50**:1002-1010 (1963).

16.   I. R. Gibbons, Chemical dissection of cilia, *Arch. Biol. (Liege)*, **76**:317-352 (1965).

17.   I. R. Gibbons and A. J. Rowe, Dynein: A protein with adenosine triphosphatase activity from cilia, *Science*, **149**:424-426 (1965).

18.   J. Gil and E. R. Weibel, Extracellular lining of bronchioles after perfusion-fixation of rat lungs for electron microscopy, *Anat. Rec.*, **169**:185-200 (1971).

19.   N. B. Gilula and P. Satir, The ciliary necklace. A ciliary membrane specialization, *J. Cell Biol.*, **53**:494-509 (1972).

20.   S. F. Goldstein, M. E. J. Holwill, and N. R. Silvester, The effects of laser microbeam irradiation of the flagellum of *Crithidia (Strigomonas) Oncopelti*, *J. Exp. Biol.*, **53**:401-409 (1970).

21.   M. F. Greenwood and P. Holland, The mammalian respiratory tract surface. A scanning electron microscopic study, *Lab. Invest.*, **27**:296-304 (1972).

22.   M. M. Hansell and R. L. Moretti, Ultrastructure of the mouse tracheal epithelium, *J. Morphol.*, **128**:159-170 (1969).

23.   A. Hilding, The physiology of drainage of nasal mucus. I. The flow of the mucus currents through the drainage system of the nasal mucosa and its relation to ciliary activity, *Arch. Otolaryngol.*, **15**:92-100 (1932).

24.   A. C. Hilding, Phagocytosis, mucous flow and ciliary action, *Arch. Environ. Health*, **6**:61-73 (1963).

25.   J. Iravani, Flimmerbewegung in den intrapulmonalen Luftwegene der Ratte, *Pflügers Arch.*, **297**:221-237 (1967).

26.   J. Iravani and C. Leymann, Koordination der Flimmerbewegung in Bronchialepithel der Ratte, *Pflügers Arch.*, **305**:199-209 (1969).

27.   J. Iravani and G. N. Melville, Long term effect of cigarette smoke on mucociliary function in animals, *Respiration*, **31**:358-366 (1974).

28.   J. Iravani and A. van As, Mucus transport in the tracheobronchial tree of normal and bronchitic rats, *J. Pathol.*, **106**:81-93 (1972).

29.   V. I. Kalnins and M. Wassmann, Subunits in microtubules of procentrioles, centrioles and cilia, *Proc. 1st Annu. Mtg. Microsc. Soc. Canada, Toronto,* 1974, pp. 38-39.

30.   K. E. Karrer, Electron microscopic study of bronchiolar epithelium of normal mouse lung, *Exp. Cell. Res.*, **10**:237-241 (1956).

31.   C. J. Kensler and S. P. Battista, Chemical and physical factors affecting mammalian ciliary activity, *Am. Rev. Respir. Dis. [Suppl.]*, **93**:93-102 (1966).

32.   E. W. Knight-Jones, Relations between metachronism and the direction of ciliary beat in Metazoa, *Q. J. Microsc. Sci.*, **95**:503-521 (1954).

33.   P. Kordik, E. Bülbring, and J. H. Burn, Ciliary movement and acetyl choline, *Br. J. Pharmacol. Chemother.*, **7**:67-79 (1952).

34.   R. W. Linck, Comparative ultrastructure and biochemistry of cilia and flagella from the Lamellibranch mollusc *Aequipecten irradians*. Ph.D. Thesis, Boston, Mass., Brandeis University (quoted from Stephens, Ref. 59).

35.   A. M. Lucas, The nasal cavity and direction of fluid by ciliary movement in *Macacus rhesus* (Desm.), *Am. J. Anat.*, **50**:141-171 (1932).

36.   A. M. Lucas and L. C. Douglas, Principles underlying ciliary activity in the respiratory tract II. A comparison of nasal clearance in man, monkey and other mammals, *Arch. Otolaryngol.*, **20**:518-541 (1934).

37.   H. Machemer, Ciliary activity and metachronism in Protozoa. In *Cilia and Flagella*. Edited by M. A. Sleigh. London and New York, Academic Press, 1974, pp. 199-286.

38.   H. Machemer, Frequency and directional responses of cilia to membrane potential changes in *Paramecium, J. Comp. Physiol.*, **92**:293-316 (1974).

39.   G. O. Mackie, D. H. Paul, C. M. Singla, M. A. Sleigh, and D. E. Williams, Branchial innervation and ciliary control in the ascidian *Corella, Proc. R. Soc. Lond. B*, **187**:1-35 (1974).

40.   H. Mohri, Extraction of ATPase from sea urchin and fish sperm tails, *Biol. Bull. Mar. Biol. Lab., Woods Hole*, **127**:381 (1964).

41.   H. Mohri, Amino-acid composition of "Tubulin" constituting microtubules of sperm flagella, *Nature*, **217**:1053-1054 (1968).

42.   H. Mohri, Personal communication, 1974.

43.   H. Mohri, S. Murakami, and K. Maruyama, On the protein constituting 9 + 2 fibers of the tail of sea-urchin spermatozoa, *J. Biochem.*, **61**:518-519 (1967).

44.   P. E. Morrow, F. R. Gibb, and K. M. Gazioglu, A study of particulate clearance from the human lungs, *Am. Rev. Respir. Dis.*, **96**:1209-1221 (1967).

45.   A. Murakami and R. Eckert, Cilia: Activation coupled to mechanical stimulation by calcium influx, *Science*, **175**:1375-1377 (1972).

46.   Y. Naitoh and R. Eckert, The control of ciliary activity in Protozoa. In *Cilia and Flagella*. Edited by M. A. Sleigh. London and New York, Academic Press, 1974, pp. 305-352.

47.   A. C. Nevo and R. Rikmenspoel, Diffusion of ATP in sperm flagella, *J. Theor. Biol.*, **26**:11-18 (1970).

48.   D. R. Pitelka, Basal bodies and root structures. In *Cilia and Flagella*. Edited by M. A. Sleigh. London and New York, Academic Press, 1974, pp. 437-469.

49.   J. E. Purkinje and G. Valentin, Discovery of the existence of continual vibratory movements produced by cilia, in Amphibia, Birds and Mammiferous Animals, *Dublin J. Med. Chem. Sci.*, **7**:279-284 (1835).

50.   R. Rikmenspoel and M. A. Sleigh, Bending moments and elastic constants in cilia, *J. Theor. Biol.*, **28**:81-100 (1970).

51.   J. A. G. Rhodin, Ultrastructure and function of the human tracheal mucosa, *Am. Rev. Respir. Dis. [Suppl.]*, **93**:1-15 (1966).

52.   P. Satir, Studies on cilia. II. Examination of the distal region of the ciliary shaft and the role of the filaments in motility, *J. Cell Biol.*, **26**:805-834 (1965).

53.   P. Satir, The present status of the sliding microtubule model of ciliary motion. In *Cilia and Flagella*. Edited by M. A. Sleigh. London and New York, Academic Press, 1974, pp. 131-142.

54.   M. A. Sleigh, *The Biology of Cilia and Flagella*, Oxford, Pergamon Press, 1962.

55.   M. A. Sleigh, Metachronism of cilia of Metazoa. In *Cilia and Flagella*. Edited by M. A. Sleigh. London and New York, Academic Press, 1974, pp. 287-304.

56.   M. A. Sleigh, Fluid propulsion by cilia and the physiology of ciliary systems. In *Perspectives in Experimental Biology, Vol. 1, Zoology*. Edited by P. Spencer Davies. Oxford, Pergamon Press, 1976, pp. 125-134.

57.   M. A. Sleigh and E. Aiello, The movement of water by cilia, *Acta Protozool.*, **11**:265-278 (1972).

58.   R. E. Stephens, On the structural protein of flagella outer fibers, *J. Mol. Biol.*, **32**:271-283 (1968).

59.   R. E. Stephens, Enzymatic and structural proteins of the axoneme. In *Cilia and Flagella*. Edited by M. A. Sleigh. London and New York, Academic Press, 1974, pp. 39-76.

60.   R. E. Stevens, F. L. Renaud, and I. R. Gibbons, Guanine nucleotides associated with the protein of the outer fibers of flagella and cilia, *Science*, **156**:1606-1608 (1967).

61.   J. Sturgess and L. Reid, An organ culture study of the effect of drugs on the secretory activity of the human bronchial submucosal gland, *Clin. Sci.*, **43**:533-543 (1972).

62.   K. E. Summers and I. R. Gibbons, Adenosine triphosphate-induced sliding of tubules in trypsin-treated flagella of sea-urchin sperm, *Proc. Natl. Acad. Sci. USA*, **68**:3092-3096 (1971).

63.   K. E. Summers and I. R. Gibbons, Effects of trypsin digestion on flagellar structures and their relationship to motility, *J. Cell. Biol.*, **58**:618-629 (1973).

64.   L. G. Tilney, J. Bryan, D. J. Bush, K. Fujiwara, M. S. Mooseker, D. B. Murphy, and D. H. Snyder, Microtubules: evidence for 13 protofilaments, *J. Cell Biol.*, **59**:267-275 (1973).

65.   J. Wardell, L. Charkin, and B. Payne, The canine tracheal pouch: A model for use in respiratory mucus research, *Am. Rev. Respir. Dis.*, **101**:741-754 (1970).

66.   F. D. Warner, The fine structure of the ciliary and flagella axoneme. In *Cilia and Flagella*. Edited by M. A. Sleigh. London and New York, Academic Press, 1974, pp. 11-37.

67.   F. D. Warner and I. Meza, Configuration of flagellar microtubule subunits, *J. Cell Sci.*, **15**:495-511 (1974).

68.   F. D. Warner and P. Satir, The substructure of ciliary microtubules, *J. Cell Sci.*, **12**:313-326 (1973).

69.   F. D. Warner and P. Satir, The structural basis of ciliary bend formation.
      Radial spoke positional changes accompanying microtubule sliding, *J. Cell
      Biol.*, **63**:35-63 (1974).
70.   D. C. Watts and L. H. Bannister, Location of arginine kinase in the cilia of
      Tetrahymena pyriformis, *Nature*, **226**:450-451 (1970).

9

Airway Secretion:
Source, Biochemical and Rheological Properties

MARIA TERESA LOPEZ-VIDRIERO and INDRAJIT DAS

Brompton Hospital
Cardiothoracic Institute
London, England

LYNNE M. REID

The Children's Hospital Medical Center
Boston, Massachusetts

I. Introduction

Sputum is airway secretion contaminated by other secretions, particularly saliva, but even bronchial fluid aspirated directly from the airway is a mixture. In spite of this, measurement of certain properties is sufficiently consistent and reproducible to justify study of the features of whole sputum. It does call for the application of a variety of techniques, the results of each helping in the interpretation of the others [1-4]. For example, an understanding of the histochemical types of intracellular secretion helps in interpreting the results of biochemical analysis.

The normal must be compared with disease. This calls for the same quantitative analysis of the histochemical and structural features of the airway surface epithelium and submucosal glands in states of hypersecretion as in the normal, particularly since it is only in these conditions that enough "airway secretion" can be collected for biochemical and rheological study, either as sputum or aspirated bronchoscopically [1]. Because sputum is a mixture (Fig. 1), biochemical studies need to be carried out on total sputum to identify its various constituents. But this is not enough, since it is the interaction of the various molecules which establishes its behavior within the airways.

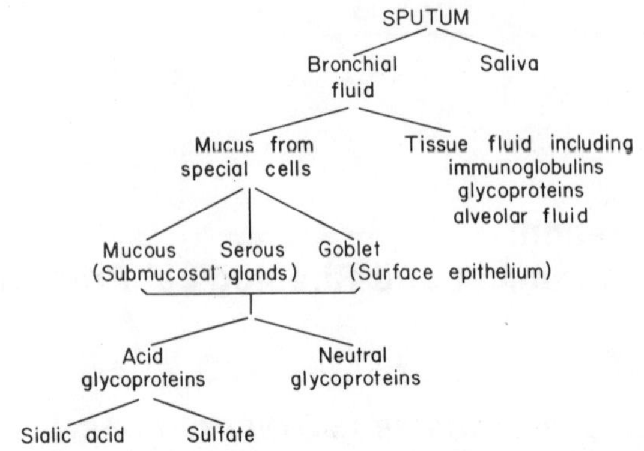

FIGURE 1   The constituents of sputum are a mixture.

The features of airway disease can be analyzed using animal models; hypersecretory states have been produced in a range of species by a variety of irritants —in the rat by exposure to sulfur dioxide, nitrous oxide, or tobacco smoke or by the administration of drugs, (isoproterenol or pilocarpine); in the pig by infection with *Mycoplasma hyorhinis* or isoproterenol; in sheep by tobacco smoke [5-9].

A. "Normal" Sputum

In the normal person any bronchial secretion that reaches the larynx is swallowed; none becomes available for examination. Through bronchoscopy it is known that the airway lining is glistening and moist, but no fluid may be collected. If an infection occurs or an irritant is inhaled, sputum may be expectorated. Recently we have administered prostaglandin F_{2a} to "normal" subjects, inducing expectoration of bronchial secretion [10]. This is an acute stimulus since expectoration occurs within 20-30 min of administering the drug.

Sputum production represents an increase in secretion above the normal level. How much of an increase, or what proportion is coughed up, is not known. Chronic production of sputum is associated with a measurable degree of submucosal gland hypertrophy as in chronic bronchitis, bronchiectasis, cystic fibrosis, and, sometimes, in asthma. Thus "normal sputum" is a contradiction in terms, and for this reason we compared sputum from various diseases, or mucoid sputum with purulent, rather than to "normal" secretion. Our histochemical studies indicate that mucoid chronic bronchitic sputum represents secretion from bronchial glands without inflammatory cell infiltration [11].

Even from laryngectomized patients the amount of fluid is rather small and is probably not "normal." Flushing the bronchial tree with saline at bronchoscopy is another way to obtain bronchial secretion, but this is unsuitable for volumetric measurement and its relation to spontaneous secretion is not known. Since the amount of secretion from the normal bronchial tree is so small, chemical studies are usually performed on sputum from patients with one of the hypersecretory states—chronic bronchitis, asthma, bronchiectasis, or cystic fibrosis.

Saliva is present in sputum as a contaminant, but in many studies it seems of little significance. The regular sputum producer probably adds relatively little saliva, since, if sputum stands several hours there is little degradation, whereas if fresh saliva is added degradation occurs quickly [12]. Furthermore, saliva is of low viscosity, has a low total and macromolecular dry weight yield, and is low in "markers" of the various components of bronchial secretion—fucose, sulfate mannose, neuraminic acid, and albumin [10,13,14]. Saliva, by dilution, influences the results of chemical analysis of sputum if these are expressed by reference to original volume of sputum, but such effect seems to be small. Early in our studies we tried to use amylase, the enzyme in saliva, as a "marker" of sputum contamination by saliva, but its concentration varied so widely and so quickly in the normal, and its pattern was so inconsistent when related to features such as diurnal variation or smoking, that amylase was clearly of no use. Fortunately the presence of saliva seems of little practical importance except when sputum is stored.

B. Origin of Airway Secretion

In addition to the special mucus secretions, "airway secretion" includes contributions from serum, appropriately designated tissue transudate in the normal, exudate in disease. The mucosubstances or glycoproteins are the special feature of airway secretion and its most important single constituent, since they confer its viscoelastic properties [15]. Although serum yields five times the macromolecular dry weight of mucoid bronchial secretion the latter is 10 times more viscous [14].

The cells producing these substances are the mucous cells of the surface epithelium (to which the term "goblet cell" is applied), the serous cell (only recently identified at this site), and the Clara cell [1,16]. Tubules lined with mucous or serous cells are found in the submucosal glands. In the human airway this group of cell types shows a range of histochemical staining for acid glycoprotein and a similar range is found in the secretions in the lumen [1,17-20]. The final polymer, or polymers, of glycoprotein as recovered from sputum must therefore include several types of building blocks, whether or not these can be isolated separately. It may be that these striking histochemical differences are masked in the larger aggregates, being apparent only as a "polydispersity."

Sputum glycoproteins have features different from those of serum glycoproteins. From recent results it is apparent that epithelial surfaces may synthesize a small amount of glycoprotein of serum type in that, while biochemically similar, their antigenicity is different [21]. Albumin and immunoglobulin glycoproteins are constituents of sputum; some of the latter are derived from serum, but others are synthesized by cells within the airway wall [21-23].

Surface tension-reducing substance in bronchial secretion may be formed either in the alveoli and ascend in the periciliary layer, or in the peripheral airway by the Clara cell [24,25]. The balance of evidence is against the latter. Pattle [14,26] has demonstrated surfactant activity in sputum by his bubble stability technique and Warembourg and his colleagues [27] have identified in the fatty layer of sputum lipids and phospholipids identical to those isolated from pulmonary surfactant by Abrams [28] and Morgan et al. [29].

C. Collection and Preparation of Sputum

The main difficulty in comparing biochemical and rheological reports in the literature arises from differences in the methods of preparing the sputum for analysis. Since sputum degrades spontaneously, with change of macromolecules and reduction in viscosity, it is essential for reliable and repeatable results that special care be taken in its collection and storage. Mucoid sputum will change little over some hours on the bench, purulent sputum more quickly. Bacteria in the saliva seem to be the main source of degrading enzymes.

Since it is undesirable to leave sputum at room temperature for long periods, we collect it from each subject at intervals of 1-3 hr. The volume and weight of each sample are recorded and the period over which it is produced. Rheological testing is usually carried out immediately, on a fresh sample, although it may be satisfactory on a frozen sample, since Charman [30] has shown that, for testing with the Ferranti-Shirley viscometer, rapid freezing and thawing do not increase the variance within a sample (see p. 300), although less freezing as in a domestic refrigerator does.

Before biochemical analysis it is virtually always necessary to "treat" sputum in some way or another; for rheological studies it is not necessary or desirable. For biochemical testing the sputum should be frozen immediately. The subject is provided with a Dewar flask containing solid carbon dioxide (dry ice), a series of bottles labeled with name, date, and the hours over which sputum is to be put into the particular bottle. When not in use the bottle remains in the flask. The patient is provided with a pair of padded cotton gardening gloves so that he can remove the sputum pot from the dry ice, and manipulate its lid without discomfort. Sputum for biochemical analysis can be stored and transported.

II. Rheological Features

Rheology was defined by Bingham [31] as "the study of deformation and flow of matter." Biorheology now covers the study of many biological substances— sputum, blood, synovial fluid, cervical mucus, DNA, lymph—and its techniques are used to analyze the properties of a number of biological systems such as bone and cartilage and contracting muscle. The rheological or flow properties of sputum have increasingly been accepted as important in relation to its function of "flowing" along the ciliary tips.

In 1955 Blanshard [32] carried out some of the earliest studies of viscosity that can be regarded as within the modern era. He emphasized some technical difficulties associated with its measurements and recommended the use of cone and plate viscometer. Denton and Litt [33] measured viscosity to test the early suggestion that bronchial secretions were more viscous in cystic fibrosis than in other diseases. Palfrey and Davies [34] used the Weissenberg rheogoniometer to study synovial fluid in the late 1960s, while Davis and Dippy [35] and Sturgess et al. [36] applied this to sputum. Palmer and his colleagues [49], while making simple measurements with the Ferranti-Shirley viscometer, studied clinical respiratory problems.

Rheological studies of sputum are widely regarded as being unreliable; probably because of the methods of collection, storage, and pretreatment that have been used. Sputum was often left on the bench many hours, perhaps even days, before testing; specimens from different patients were pooled and often homogenized before testing. Since the act of shearing breaks down the "structure of sputum," any second test on a sputum sample was often different from the first so that homogenization as pretreatment or repeated shearing to achieve constant behavior was often used. To achieve repeatable results only fresh samples should be used. In our studies sputum is freshly produced, freshly collected, and freshly tested. Each sputum sample represents sputum from only 1 subject, collected over several hours. Where comparison is made between specimens from the same patient, these are collected at the same time each day. Mucoid samples should be compared with mucoid and purulent with purulent. "Homogenization of sputa by any of the various methods that can be used will, of course, give consistent results but these cannot be regarded as similar to the fresh state" [36].

It is important, when considering reports of rheological properties of sputum, to be clear whether it is concerned with whole, fresh sputum, or with freeze-dried, homogenized and reconstituted material, or with a solution of a macromolecular constituent. Such studies are a contribution to sputum behavior, but cannot replace the study of whole sputum.

Many physical features can be considered with respect to mucus or sputum
—its stickiness, fluidity, pourability, adhesiveness, wettability, rubberiness, consistency, elasticity, and gumminess. It seemed best to start with measurements
of viscosity and elasticity since these are the major determinants of rheological
behavior, and for their measurement satisfactory techniques are available, although other qualities may also prove to be of biological significance.

A. Viscosity and Elasticity

To analyze the rheological or flow behavior of sputum, it is necessary to consider
viscosity and elasticity; each of these is influenced by the time factor in testing.
Elasticity and viscosity are each concerned with deformation of a material by a
force. Although they can be analyzed separately, at any one time the behavior
of a substance, such as sputum, represents a balance between its viscous and
elastic properties. Certain physical concepts need to be considered before describing equipment available for testing. The purpose of this monograph is not
to give extended discussion of the theory of rheology and its mathematical aspects. detailed accounts can be found in several monographs and articles [37-40].

1. Viscosity

Viscosity is concerned with flow material, hence fluids. Newton was the first
to describe this property when he wrote, "Resistance which arises from the lack
of slipperiness of the parts of a liquid, other things being equal, is proportional
to the velocity with which the parts of the liquid are separated from one another"
(see Ref. 31).

Viscosity describes the ratio between the stress set up and the deforming
force that has been applied. Viscosity is then,

$$\frac{\text{stress}}{\text{force applied to produce stress}}$$

To assess the drag between the layers of a fluid, a rotational viscometer is frequently used. A common and acceptable form is the cone and plate. The coefficient of viscosity is then calculated as:

$$\text{Viscosity (poise)} = \frac{\text{shear stress (dyn cm}^{-2})}{\text{shear rate (sec}^{-1})}$$

The unit for shear rate, the reciprocal second, may be puzzling, since shear rate
represents the velocity gradient length/time between the platens (Length). When
the formula

$$\frac{\text{length}}{\text{(time) (length)}}$$

is resolved, length cancels out, but time (second) is left as a reciprocal.

In an ideally viscous substance or a Newtonian fluid, the viscosity is the same no matter what shearing force is applied; that is, with increasing shear rate, shear stress is proportional and a linear relationship exists between them. Newtonian fluids such as lubricant oils have the same viscosity no matter the shear rate applied; and so a test at one shear rate suffices to characterize the substance.

Newton found that the viscosity of certain liquids did not show a linear relationship to the force applied. Such are non-Newtonian fluids; in them, the viscosity varies according to the force applied. Sputum, like most biological fluids, is non-Newtonian, which means that testing must be performed over a range of shear rates to give a profile of viscosity related to shearing force. Various patterns of non-Newtonian behavior can be identified.

Pseudoplastic flow or shear thinning. Viscosity falls as the shear rate increases. The progressive fall may be due to realignment of molecules so that the relative resistance between adjacent layers diminishes.

Dilatancy is a rather rare phenomenon of increase in viscosity with increase in shear rate—whipping cream offers an example of this.

Plastic or Bingham flow does not start until a yield stress is exceeded. Force is needed to overcome the static friction which causes adhesion of particles within the substance. Until this is reached the substance behaves as a solid. This stress limit is called "yield value" and has been studied extensively by Bingham.

Thixotropy is demonstrated when a fall in viscosity, produced by shearing, is recovered on standing. Thixotropic paint is an example of this since stirring with a paintbrush reduces viscosity. On the wall, the paint "recovers."

2. Elasticity

Hooke described elasticity in the same century in which Newton described viscosity. The effect of an extending force on a wire or spring is to produce lengthening, the extension showing a linear relation to the force applied. If, with removal of the force, the material returns to its original length it is an "ideal elastic body." At the time of extension of an elastic substance, energy is stored; at the time of recovery, energy is expended.

To assess elasticity, it is necessary to follow recovery from deformation, that is, the input and decay of the applied force; this must be measured and the strain in the material under test be related to it. This can be estimated by the application of an oscillatory force (Fig. 2).

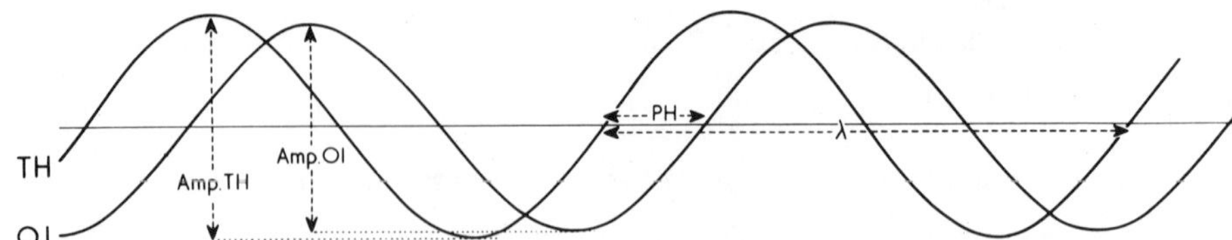

FIGURE 2   Recording obtained by oscillatory testing with the Weissenburg rheogoniometer. The applied oscillation is recorded as a sine wave (OI); the movement transmitted through the test fluid also appears as a sine wave (TH).

Another time-related property of sputum is stress relaxation:

1.   If sudden, but constant, force is applied to the material under test, the material is stressed. After release of the force time is necessary for the material to recover, "the relaxation time."

2.   If a small constant force is applied, stress within the material increases for some time producing the "creep" phenomenon. If this force is removed and recovery allowed, as above, "creep relaxation" can be measured.

B. Apparatus to Test Sputum

A number of instruments have been designed to measure viscosity, fewer to measure elasticity. Some of these are listed in Table 1. We have used two types of instrument, each of cone and plate variety, which together measure a wide range of shear rate. The Ferranti-Shirley viscometer tests only by rotation and at high shear rates. It was developed to test the viscosity of rayon and is widely used in industry. It does not measure elasticity of sputum. The other, the Weissenberg rheogoniometer, tests low shear rates by either oscillation or rotation. It was developed in the United Kingdom by Weissenberg during World War II to measure the viscosity of flame-throwing substances.

1. The Ferranti-Shirley Viscometer

The fluid to be tested is placed between two horizontal platens, one of which is cone-shaped, and the other flat. Since the force applied is inversely proportional to the thickness of liquid and directly proportional to the distance from the center of the platen, this geometrical shape permits application of a shear force constant throughout the liquid (Fig. 3). This instrument tests uniform acceleration over a period of time (sweep time). It has a wide range of shear rates at which viscosity can be tested. The stress setup within the liquid is measured through a torque spring and a rheogram is recorded on an X-Y recorder.

TABLE 1 Some Instruments Designed to Test Sputum

Apparatus	Parameters measured	Authors
Consistometer	Consistency	Scott-Blair and Glover [41]
Perforated syringe	Consistency, yield effect	Elmes and White [42]; Hirsh and Hamilton [43]
Inclined tube	Adhesiveness	Chodosh et al. [44]
No apparatus	Pourability	Keal [45]
Capillary flow viscometer	Relative viscosity, yield value	Volker [46]
Capillary viscometer	Relative viscosity, elastic recoil	Dulfano [47]
Magnetic rheometer (creep)	Elasticity, viscosity	Denton et al. [33]
Rheometer (torque relaxation)	Elastic recovery, yield and approximate	Puchelle et al. [48]
Viscometer cone and plate (Ferranti-Shirley)	Apparent, yield value	Reid [2]; Davis and Dippy [35]; Palmer et al. [49]; Sturgess et al. [36]; Keal [50]
Chemical balance rheometer Microforce rheometer (stress relaxation)	Viscosity, elasticity	Davis et al. [51]
Concentric cylinder rheometer	Viscosity, yield value, elastic recovery	Benis et al. [52]
Weissenberg rheogoniometer, (creep, oscillatory, rotatory)	Viscosity, elasticity, yield value	Sturgess et al. [36]; Davis [37]; Davis and Dippy [35]

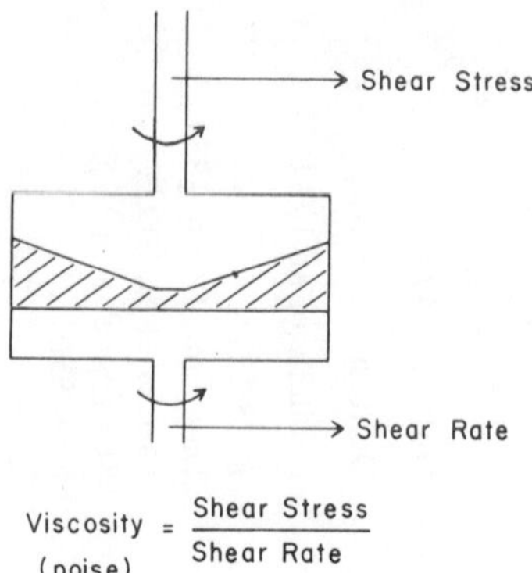

FIGURE 3 Diagrammatic representation of a cone and plate viscometer. The fluid under test (shaded) is placed between the two platens.

The apparent viscosity (η) can be calculated at any given shear rate.

$$\eta = c \times \frac{\text{shear rate}}{\text{shear stress}}$$

where c is the constant dependent on the cone.

The apparatus, if used for sputum, needs to be equipped with a 200 g cm torque spring [49]. A cone of 7 cm diam is suitable for less viscous specimens and one of 4 cm for the more viscous. Taking into account its limitations, the Ferranti-Shirley viscometer gives a good preliminary assessment of a sputum sample and enables large numbers of samples to be tested.

2. *The Weissenberg Rheogoniometer*

This apparatus tests either by oscillation or rotation. The specimen is placed between a lower conical and an upper flat platen. The upper platen is held by a torsion bar clamped to the head; the lower platen oscillates through a small angle and is driven through a 60-speed gearbox and clutch system. When oscillation is applied, an inductive transducer detects tangential shear stress by measuring the torsional displacement of the bar. The two signals, input and output amplitude, are fed to a suitable recorder. The applied oscillation is recorded as a sine wave (01 in Fig. 2); the movement transmitted through the test fluid also appears as a sine wave (TH). The amplitudes (Amp) of the two waves are compared, as well as the phase difference between them. The phase difference is the distance between the waves PH and the length of one cycle of oscillation (λ). In this way, differences in amplitude and phase are detected for each frequency and can be used to calculate viscosity (η) and elasticity (γ).

$$\eta = AB \, \frac{\Delta T}{I} \, \frac{I}{\sin \phi} \qquad \text{in dyn sec cm}^{-2}$$

$$\gamma = CD \, \frac{\Delta T}{I} \, \frac{I}{\cos \phi} \qquad \text{in dyn cm}^{-2}$$

where T is the amplitude of movement of the upper platen; I is the amplitude of movement of the lower platen; A and C are constants related to the torsion bar; and B and D are constants depending on the platen use.

Out-of-phase waves are characteristic of a highly elastic fluid, in-phase waves of a highly viscous fluid.

Several other ways of testing, including the creep test, are possible with the Weissenberg rheogonimeter. Such an apparatus is ideal for research and permits a precise screening over a wide range of shear rates. Perhaps the only inconvenience is the price, particularly when a transfer function analyzer is added.

C.   Rheological Features of Sputum

1.   Variation between Aliquots, Samples, and Disease

In our studies of the viscoelastic behavior of sputum [13,53], we first established the extent of variation between aliquots from the same specimen and between specimens from the same subject. We also determined the variation within single sputum samples of mucoid or purulent type and between samples obtained from four disease states: chronic bronchitis, bronchiectasis, cystic fibrosis, and asthma. Answers to these points are based on measurements made on the Ferranti-Shirley viscometer at a shear rate of 1350 sec^{-1} [13,53]:

1.   The coefficient of variation in apparent viscosity between two or three aliquots of the same mucoid specimen was never above 30%.

2.   The coefficient of variation of viscosity between samples from patients with chronic bronchitis, studied frequently over a 6-month period, was 40%, whether the sputum was mucoid or purulent.

3.   The coefficient of variation in apparent viscosity between patients with the same disease was approximately 60% for each disease, except for asthma where the coefficient was greater.

2. Effect of Temperature

A decrease in viscosity with increasing temperature is seen over the shear rate
range applied by the Ferranti-Shirley viscometer. Sturgess [36], studying mucoid
sputum that had been allowed to come to room temperature, found that on
heating it to 37°C, the viscosity was half that at 20°C. Charman and Reid [30]
further investigated the effect of freezing, storing, and thawing: (a) Freezing by
liquid nitrogen and storage at –13°C, or (b) freezing and storage both at –70°C
were satisfactory. Significant reduction in viscosity occurred from the other
combinations tested. Such alteration had not been reported by other authors
who studied sputum after freezing and storing [44].

3. Viscosity

As mentioned earlier sputum is a non-Newtonian fluid, showing pseudoplastic or
shear thinning. The act of testing destroys the nature of the secretion so that,
on retesting a sample, a different result is obtained (Fig. 4). These original properties are not recovered on standing [36]. Sputum often displays plastic behavior,
that is, a certain force is necessary before it flows at all—it has a "yield value."
Using the Ferranti-Shirley viscometer, yield values have been reported as a characteristic of sputum by Palmer [49]. Charman and Reid [53] found it in fewer

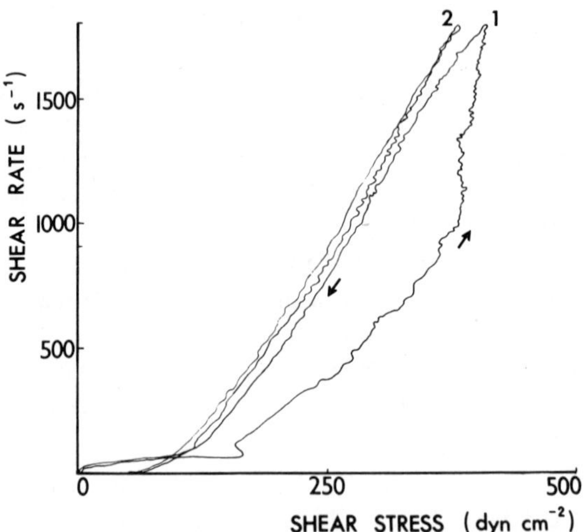

FIGURE 4 Rheograms of two tests on the same aliquot obtained with the
Ferranti-Shirley viscometer. The complex pattern shown by acceleration has
been destroyed by shearing and is not apparent on deceleration or second
testing.

than one-third of the specimens studied, regardless of the macroscopic type or disease. Sturgess [54] found that, when present, the yield value was proportional to the relative viscosity.

4. Shear Rates Applied Within the Bronchial Tree

There is as yet no agreement about the shear rates that are applied to mucus within the bronchial tree. The shear rate applied by cilia has been suggested as 1 sec^{-1} by Dulfano [55], whereas Pham et al. [56] suggested 3 sec^{-1}.

5. Viscosity Over a Range of Shear Rate

Relating viscosity to shear rate, the lower shear rates are found to be more discriminatory and several zones, each with characteristic features can be distinguished. It is now possible to suggest a chemical basis for some zones.

Figure 5 summarizes the viscosity levels found over a wide range of shear rate. The following four zones can be identified:

> Zone 1—up to 0.5 sec^{-1}
> Zone 2—0.5 sec^{-1} to 10 sec^{-1}
> Zone 3—10 sec^{-1} to 200 sec^{-1}
> Zone 4—above 200 sec^{-1}

In Fig. 6 sputum behavior in zones 1 and 2 is illustrated. Zone 1 (up to 0.5 sec^{-1}): includes a region of fluctuating viscosity that, in the plot, gives a "notched plateau." This occurs at about 0.1 sec^{-1} but is complete by 0.5 sec^{-1}. Up to 0.1 sec^{-1} the effect of shearing produced little change in viscosity [1,57, 58]. Zone 2 (from the end of the notched plateau to 10 sec^{-1}): after the "notched plateau," the sputum becomes more susceptible to shearing; the reduction in viscosity is faster. Zones 1 and 2 are both characterized by a time-related increase in viscosity (Fig. 7) [57,58]. If, after shearing at any rate within this range, the sputum is allowed to stand, a marked increase in viscosity occurs. This effect is not seen after shearing above a rate of 10 sec^{-1}. The consistency of the shear rate at which these phenomena occur suggests that a particular bond is being broken and that either this or another link is being reformed, but one that is susceptible to this shear rate range, since ultimately all are destroyed and viscosity then continues to fall.

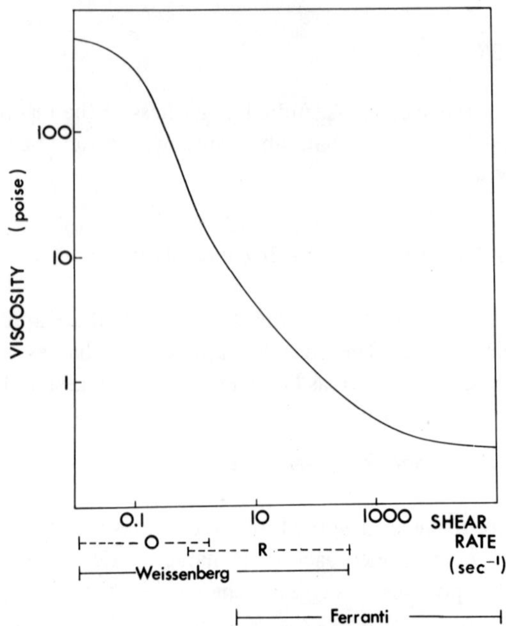

FIGURE 5 Sputum viscosity over a wide range of shear rates applied by either the Weissenberg rheogoniometer (O, oscillation; R, rotation), or the Ferranti-Shirley viscometer (log-log scale).

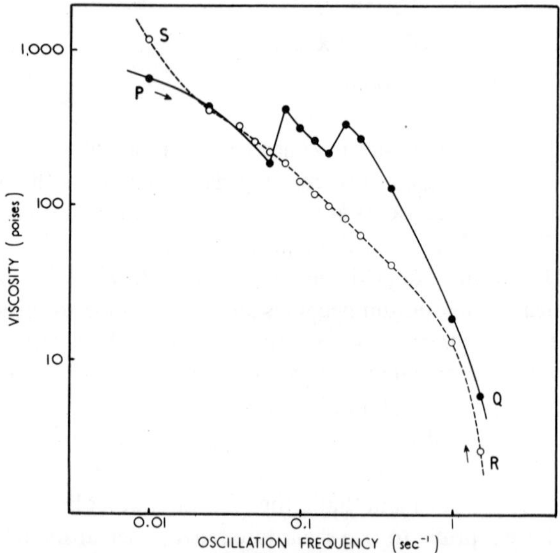

FIGURE 6 Zones 1 and 2. Sputum tested at ascending shear rates, PQ, show a notched plateau, a region of fluctuating viscosity. This is destroyed by shearing, as it is not apparent in another aliquot tested at descending shear rates, RS.

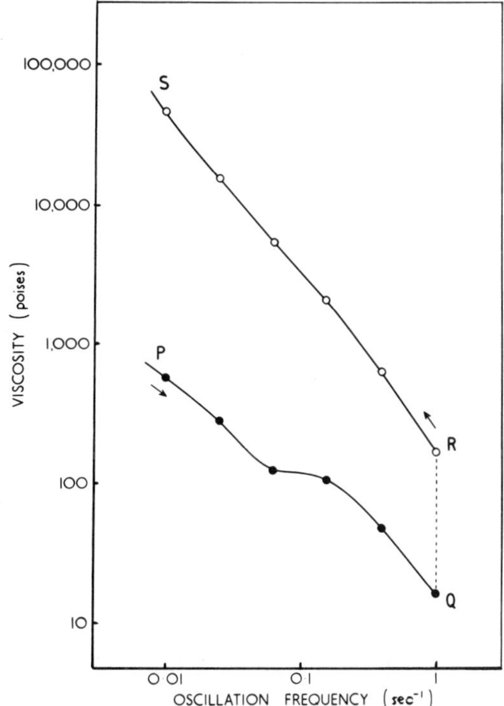

FIGURE 7   One aliquot of sputum tested PQ and then RS but left between the plateaus for 1 hr at QR. A time-related increase in viscosity occurs.

Mixing urea, a breaker of H bonds, with sputum (Mitchell-Heggs, personal communication) reduces viscosity to about the level achieved by shearing at 10 sec^{-1} (Fig. 9). The value of shear rate varies slightly for mucoid and purulent sputum, being rather higher for the latter. This suggests that up to this level the effect of shearing may be largely effected through breaking of hydrogen bonds. It is unlikely that this is all, since urea, while it reduces the plateau, does not completely destroy it.

Zone 3 (between 10 and 200 sec^{-1}): shearing between 10 and 200 sec^{-1} is associated with a steady fall in viscosity (Fig. 5). In the early part of this range sputum is at its most susceptible. The relative importance of realignment of molecules and of rupture of physicochemical bonds—or both—is not known, but standing after shearing does not lead to any significant increase in viscosity. It seems that crosslinks between molecules have been broken.

Zone 4 (above 200 sec^{-1}): in the fourth zone, the effect of shearing is much less at its beginning than in Zone 3, and, finally, the sputum can be regarded as nearly Newtonian—a suspension/solution of the various constituents, of the now degraded substance.

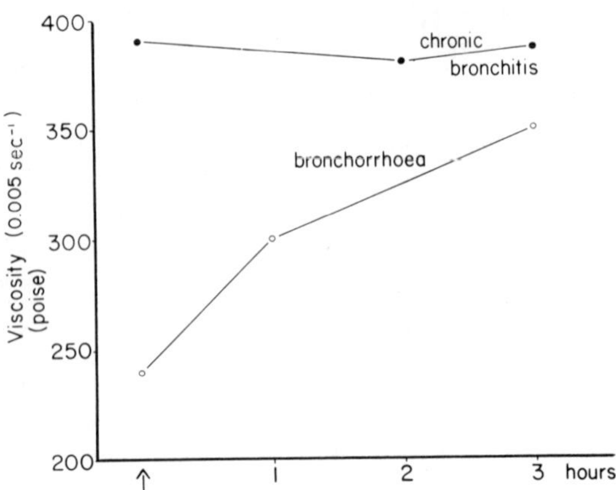

FIGURE 8   Sputum from patients with bronchorrhoea may show an increase in viscosity spontaneously or on standing.

6. Plateau

A plateau appeared over the same frequency range in all specimens regardless of viscosity. Shearing sputum through this range destroyed this behavior. On retesting the same aliquot at ascending shear rates, the plateau was not present, nor did it appear in another aliquot over this range at descending shear rate. This suggests that a certain type of bond or crosslinkage is responsible for the "notched plateau" and that they are disrupted by shearing (Fig. 6). This plateau was found in both mucoid and purulent sputum and in the diseases studied— asthma, cystic fibrosis, chronic bronchitis, and bronchiectasis [59]. It has also been found in saliva and gastric and nasal mucus, but not in human cervical mucus of pregnancy (this has about the same viscosity as sputum) or in synovial fluid [34].

To exclude an instrumental artifact [60], Mitchell-Heggs [59] showed that the plateau occurred in sputum samples measured with different platens, that its presence was not related to the angle of amplitude of oscillation and that it was apparent on testing with another Weissenberg rheogoniometer. Its presence in some biological fluids and not others of similar viscosity also suggests that it reflects sputum structure. It seems likely that over this frequency range sputum does not behave according to accepted classical rheological theory.

7. Time-Related Increase in Viscosity After Shearing

Shearing sputum at low shear rate and then allowing it to stand between the platens produces a striking increase in viscosity, which does not occur if sputum

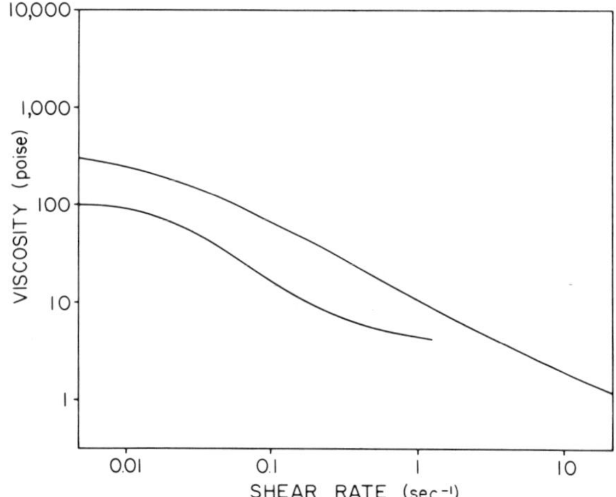

FIGURE 9 Comparison of the effect of urea (lower curve), a breaker of H bonds, with that of shearing at low rates. Urea lowers viscosity equivalent to a low shear rate.

is left on the bench without previous shearing (save in bronchorrhoea; see below) or if shearing is carried out at higher rates. The time-related increase is shown only in the range of zones 1 and 2 described above (Figs. 7 and 8). Drying is not the cause, since evaporation of a sputum sample to about 50% of its weight only produces a 2- to 3-fold increase in viscosity. Such an increase in viscosity could be produced by bronchial constriction with retention of secretion within the bronchus and may explain the wide range of viscosity seen in asthma.

8. Elasticity

Sputum elasticity has not been as widely studied as viscosity. Sputum is a highly elastic gel, even superelastic when compared with other elastic gels. Its superelasticity can be attributed to the random coil arrangement of the macromolecules of glycoprotein, elongating under stress, but storing energy which permits recovery [1].

Of the four zones in the viscosity/frequency plot of sputum, elasticity can be demonstrated at the frequencies corresponding to zones 1 and 2.† In zone 1, from 0.004 to 0.025 sec^{-1}, elasticity increases slowly. Little change occurs over the plateau region but then in zone 2 elasticity increases sharply (Fig. 10).

---

†Most work is based on techniques already described—particularly oscillatory testing with the Weissenberg rheogoniometer.

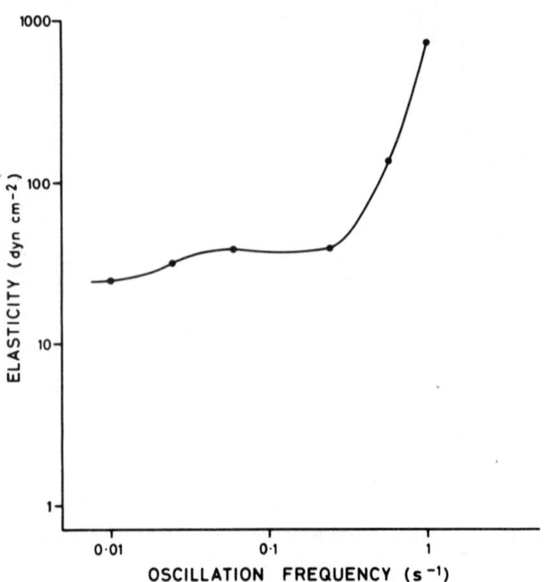

FIGURE 10   Over zones 1 and 2 sputum also shows elastic behavior.

The plateau is not as obvious in the elasticity plot as in that of viscosity. At low frequencies elasticity and viscosity of any sample are well correlated. For cystic fibrosis and chronic bronchitis sputum, as the frequency increases, this correlation decreases but for asthma and bronchiectasis it remains high. It has been suggested [51] that the correlation between viscosity and elasticity is a reflection of the decreased shear susceptibility of asthmatic and bronchiectatic sputum. The tertiary molecular structure of glycoproteins with abundant protein crosslinking may be similar in these diseases but different from that in chronic bronchitis and cystic fibrosis.

9.   "Pourability" or "Fluidity" Grade

Early in our studies Keal [45,62] found a significant correlation between pourability or fluidity and neuraminic acid levels, suggesting that pourability is a useful measure of viscosity, at least in mucoid sputum. Thus, a more precise measure of viscosity was desirable since fluidity seemed, in some way, to reflect the chemical composition of sputum.

Keal [45] recognized four grades of increasing pourability:

> Grade I: viscous, adhering closely to the container
>
> Grade II: viscous, moving slowly with gravity

Grade III: sliding steadily in one lump

Grade IV: flowing freely with gravity

It was useful to add three more grades, each between the above [59].

Subsequently Keal [45] was able to show a satisfactory inverse correlation between these grades and viscosity. It is of interest that this correlation holds only for mucoid sputum, perhaps because the "capsule" of clear liquid surrounding purulent sputum may be important in determining its behavior.

III. Biochemical Constituents of Sputum

A. Preparation of Sputum or Bronchial Fluid For Biochemical Analysis

Bronchial mucus is a viscoelastic semisolid substance of gelatinous consistency. Mucoid sputum is 95% water and consists of a gel of mucus or glycoprotein dispersed in a continuous sol phase. Microscopically, if stained appropriately, it is seen as a mesh of fibers, mainly glycoproteins in which embedded particles, debris, and granules are identified.

Sputum usually contains saliva, the source of many bacterial enzymes. Although oral bacteria are known to contain a neuraminidase which releases neuraminic acid from the salivary glycoprotein [12], the release of neuraminic acid from sputum has been little studied. The effect of various types of sputum pretreatment currently in use on neuraminic acid release and destruction has been followed by Woodcraft, Robert, Das, and Reid (unpublished work). Boiling at 100°C prevents neuraminic acid release and destruction, evidently by destroying enzymes. Urea (6M) prevents neuraminic acid release, probably by inhibiting sialidase, but neither dithiothreitol nor proteolytic enzymes have this effect. Homogenization or lyophilization appear not to affect neuraminic acid release or destruction, whereas toluene has some inhibitory effect. Purulent and mucoid sputum show little difference in the rate of release and breakdown. Since toluene seems to prevent release of neuraminic acid from sputum, it would seem that bacteria are the source of the enzymes. This is confirmed by the fact that enzymes are most prevalent in the saliva portion of sputum.

The major technical problem in analyzing sputum is its physical inhomogeneity and viscosity which prevent accurate pipetting, interfere with electrophoretic resolution, and clog the chromatographic column. According to the type of analysis, lengthy preparation of sputum may be necessary. Biochemical studies may analyze whole sputum or simply separate the sol from the gel. In either case, the main difficulty is to solubilize the gel glycoprotein.

The concentration of any constituent of sputum can be expressed by reference to the original volume or weight of sputum, to its total dry weight or to its macromolecular dry weight. The first and last of these are probably the most useful.

1. Homogenization of Total Sputum

Homogenization of sputum by ultrasonic disintegration or treatment with a Potter pestle [22,23] is widely used for simple biochemical and bacteriological studies. While these techniques decrease viscosity, they do not necessarily increase solubility enough for complete purification studies.

2. Solubilization of Mucus

If macromolecules are to be analyzed in total sputum, without separate characterization of the sol and gel phases, then solubilization of the gel glycoprotein can be achieved by treatments that are relatively mild in that they allow breakdown of mucus into smaller fragments and yet do not denature the protein molecules. Magnetic stirring with 6 M urea for over 3 hr solubilizes sputum gel probably by the destruction of hydrogen bonds. Treatment with proteolytic enzymes, such as papain or pronase, are also suitable for purification studies. Increasing the ionic concentration of certain salts such as calcium or sodium also makes it possible to solubilize mucins. $CaCl_2$ (50%) or sodium salt (1-2%) have been used to solubilize submaxillary mucins [64], cesium bromide to liquefy sputum [65] and also iodides and ammonium chloride. Iodine compounds solubilize sputum probably by promoting proteolysis of its proteins by proteases present in the sample [66], e.g., trypsin and chymotrypsin. The activation of proteases by iodides or the use of proteolytic enzymes (e.g., trypsin, chymotrypsin, or streptokinase activated plasmin) may not give suitable material for the analysis of proteins and glycoproteins since they degrade the protein moieties of the bronchial glycoproteins.

The most effective liquefying agents are the thiol compounds, the common ones being N-acetylcysteine, mercaptoethanol, mercaptoethane sulfonate, or dithiothreitol. These reduce disulfide bonds and lower the viscosity. They are capable of liquefying mucus and DNA, but do not affect fibrin, blood clots, or living tissues. The mucolytic effect is greatest at pH 7-9 and liquefaction is generally complete within 5-10 min. Dithiothreitol is used at a concentration 0.1 M [67]. Sometimes it is useful to block the reduced cysteine groups by alkylation with iodoacetamide, ethylenimine, or ethylmaleimide to prevent reoxidation by atmospheric oxygen. Sulfitolysis or performic acid oxidation of disulfide bridges may also be used to liquefy bronchial secretions [68].

3. Dialysis

Since it is the macromolecular constituents of sputum that are of chief interest, often the first procedure in preparation of sputum for biochemical study is dialysis to remove unbound small molecular components. The usual method is to dialyze sputum against distilled water changed twice daily for 3 days. The pore size used is 24 Å. The dialyzed material and the dialysate may each be freeze-dried. The amount of nondialyzable matter in mucoid sputum from patients with various kinds of lung disease has been reported as between 0.5 and 3 g% [69].

The small molecules that are removed by dialysis are mainly electrolytes such as sodium, chloride, and potassium (which are present in higher concentrations in whole pulmonary secretions than in serum) and small amounts of amino acids and sugars (Table 2). Other biologically active small molecules are the kinins, histamine, prostaglandins, and cyclic AMP.

The proportion of various nondialyzable macromolecules is [71]:

2-3%	(g/100 ml)	mucin
0.1-0.5%	(g/100 ml)	proteins and
0.3-0.5%	(g/100 ml)	fats

TABLE 2 Concentration of Small Molecules in Sputum from Laryngectomized Subjects[a]

Molecule	Sol phase of sputum[b]	Whole sputum	Percent dialyzable
Sodium[c]	211.1 ± 33.8	165 ± 42.0	85.1
Potassium[c]	16.6 ± 3.4	13.2 ± 5.4	87.9
Chloride[c]	156.7 ± 24.6	162 ± 60	95.9
Calcium[c]	2.5 ± 1.1	3.1 ± 1.0	71.0
Carbohydrate[c]	19.3 ± 9.4	—	33.6
Phosphorus[d]	—	27 ± 1.6	—
Histamine[e]	—	10 – 40	—
Osmolality[f]	359 ± 56	—	—

[a]From Ref. 70.
[b]Sol phase in the supernatant after centrifugation of sputum at 25,000 rpm for 4 hr at 4°C.
[c]Concentration: mM/1000 g.
[d]Concentration: meq/1000 g.
[e]Secretion from human subjects. Concentration: μg/g dry weight.
[f]Expressed in meq/liter.

In noninfected tracheobronchial secretion, only very small amounts of DNA are present and are not removed by dialysis. RNA has not been detected in any kind of sputum [72]. During the procedure of dialysis, partial solubilization of the gel phase is achieved.

4. Sol and Gel Phase

Separation of the sol and gel phases of sputum is achieved by centrifugation [73, 74], (generally to 40,000 × g), which produces clear separation into the two phases of chronic bronchitis sputum. With bronchorrhea sputum such separation is achieved at a lower speed. Each phase contains a complex mixture of proteins and glycoproteins (Table 3) and a distribution of macromolecules between the two phases varies as between diseases. By using speeds up to 120,000 × g and then washing the gel with saline, a much less "contaminated" gel phase is obtained—that is, more sol is squeezed out. The main difference in composition between sol and gel phase is the ratio of soluble protein to insoluble mucin; the gel contains more insoluble mucin. Repeated washing of the gel with water or saline progressively solubilizes the glycoprotein as does alternate freezing and thawing. The separation of sol and gel phase is facilitated by diluting whole sputum with 2-3 vol water before centrifugation.

5. Estimation of Sugar, DNA, Sulfate and Sulfhydryl Groups in Sputum

For biochemical studies, when constituent sugars are to be identified or a given marker substance [e.g., neuraminic acid and fucose mannose (see Sect. 8 below)] and when denaturation of the macromolecule does not affect the results, the following treatment is satisfactory and is the one used for these studies. An aliquot of sputum is boiled (10 min at 100°C) and then dialyzed against distilled water for 72 hr at 4°C with several changes of water. The dialysate can then be concentrated and used for the analysis of small molecules and inorganic components. Dialyzed macromolecules are dispersed by ultrasonic disintegration followed by treatment with pronase (2 hr at 37°C). Pronase-digested soluble material is used directly for the analysis of glycoprotein carbohydrate, such as neuraminic acid, fucose, etc.

Total protein in sputum is generally measured in sputum homogenized ultrasonically by the Lowry method [75]. Sputum albumin is conveniently estimated by the bromocresol green method [76].

6. Isolation of Glycoproteins

Drastic treatment must be avoided in preparing sputum glycoproteins for study if viability is to be preserved. On thawing, solubilization has been achieved

TABLE 3 Distribution of Large Molecular Components Between the Phases of Sputum from Chronic Bronchitic Patients[a]

Phase	Precipitable carbohydrate (mg/100 ml galactose)	Total protein (mg/100 ml albumin)	Albumin component (percent of total protein)	Glycoprotein component (percent of total protein)	Lysozyme component (percent of total protein)
Sol phase Limits of determined values	90–265	410–630	trace to 9	60–80	19–31
Average values	171	515	<5	72	25
Total sputum Limits of determined values	219–800	460–1145	trace to 6	75–92	8–21
Average values	422	815	<5	82	16

[a]From Ref. 73. Estimations were carried out on total homogenized sputum and on a sol phase prepared by ultracentrifugation.

TABLE 4 Methods Used in the Isolation and Fractionation of Glycoproteins from Sputum

Method	Authors
Paper electrophoresis	Warfvinge [79]; Brogan [80]; Bukantz and Berns III [81]; Gernez-Rieux, Havez, Voisin, and Cuvelier [82]; Nicholas [83]
Cellulose acetate	Ryley [84]; Atassi, Barker, Houghton, and Mullard [85]
Columns of glass beads	Atassi, Barker, and Stacey [86]
Electrophoresis in gels of acrylamide	Ryley [84]
Electrophoresis on agarose	Biserte, Havez, and Cuvelier [87]; Masson, Heremans, and Prignot [88]; Gotz and Fisher [89]
Precipitation by Cetyl trimethylammonium bromide followed by acid fractionation	Atassi, Barker, and Stacey [90]; Havez, Roussel, Degand, and Randoux [91]
Precipitation by various concentrations of ethanol	Roberts [63]
Gel filtration agarose (Sepharose)	Havez and Biserte [92]; Havez, Roussel, Degand, and Randoux [91]
Ion-exchange chromatography, e.g., DEAE-cellulose	Atassi, Barker, and Stacey [90]; Biserte, Havez, and Cuvelier [87]; Masson, Heremans, and Prignot [88]

by reduction with thiol compounds [77], by magnetic stirring in presence of 6 M urea [63], or by proteolytic enzymes (e.g., papain) [78]. Table 4 lists some of the methods used singly or in combination in the isolation and fractionation of sputum glycoproteins.

Recently Creeth et al. [93,65] isolated and purified glycoproteins from ovarian cyst fluid and sputum by cesium chloride and cesium bromide density-gradient ultracentrifugation. This method is based on the higher buoyant density of glycoproteins as compared with proteins. Sputum samples are dispersed in cesium bromide of density 1.5 g/ml, diluted to a density of 1.4 g/ml, and subjected to preparative centrifugation for 2 days at 42,000 rpm. This segregates the components, the free protein rising to the meniscus region, and the glycoprotein forming one or more zones in the lower half of the tube. These protein and glycoprotein fractions can be collected separately. Cesium chloride has a higher resolving power than cesium bromide and is therefore usually pre-

ferred for separation of protein from glycoprotein. Sputum cannot be dispersed
in cesium chloride, whereas cesium bromide is effective and so is our reagent of
choice.

7. Characterization of Sputum Glycoproteins

Sequential digestion of a carbohydrate unit in conjunction with methylation
and periodic oxidation at various stages of enzymatic degradation permits complete structural elucidation of the carbohydrate unit of the glycoprotein. Extensive proteolytic digestion of sputum glycoprotein, followed by fractionation,
yields the carbohydrate as glycopeptides. Their treatment with specific glyco-
sidases establishes the sequence and anomeric configuration of the saccharide
components. The enzymes exoglycosidases which act by removing only the
terminal, nonreducing sugar from the carbohydrate chain have been purified
and characterized [94]. Using N-acetylhexosaminidase in conjunction with the
long-known neuraminidase, an undegraded protein moiety can be prepared
enzymatically from glycoprotein. The gas-liquid chromatographic method can
estimate, in a single procedure, all the relevant sugars in sputum glycoproteins
[95,96] and combined with amino acid analysis, identifies the structural composition of sputum glycoproteins.

8. Biochemical Analysis as an Indication of Origin of Sputum

Certain constituents of sputum can be taken as a marker of either the epithelial
glycoprotein or the serum contribution (Tables 5 and 6). The polysaccharide
side chains of epithelial glycoproteins contain fucose and sulfate, which are
found only in low amounts in serum, while mannose concentration is relatively
high in serum glycoproteins and virtually absent from bronchial mucin [77,94].
Neuraminic acid is found in both and its serum and sputum concentration is
similar. Fucose is a marker of bronchial glycoprotein, mannose of serum glyco-

TABLE 5 Carbohydrate Composition of Purified Bronchial Glycoprotein
(Bronchial Cysts)[a]

	Dry weight (%)			
Glycoprotein	Mannose	Fucose	Sialic acid	Sulfate
Fucomucin	—	10.7	1.1	—
Sulfomucin	—	7.1	1.2	7.2
Sialomucin	—	6.2	21.1	3.3

[a]From Ref. 77.

TABLE 6 Carbohydrate Composition of Some Plasma Glycoproteins Present in Sputum[a]

Glycoprotein	Sugar components (mol/mol of protein)			Total carbohydrate (%)
	Mannose	Fucose	Sialic acid	
IgG[b]	5	2	1	2.5
IgA[b]	14	2	5	5.7
IgM[b]	35	2	9	9.2
α_1-Acid glycoprotein	12	3	15	35
Haptoglobin	31	2	30	14
Ceruloplasmin	22	2	9	7
Transferrin	8	–	4	5.2
α_2-Macroglobulin	94	13	48	8.6
Hemopexin	16	–	19	23
β_1-Glycoprotein	15	2	–	30

[a]From Ref. 94.
[b]Expressed as mol/mol of monomer.

protein and neuraminic acid of both. The neuraminic acid/fucose ratio indicates the relative proportion of the two and in a given patient is of value in tracing change in these two components.

The sputum produced after inhalation of prostaglandin $F_{2\alpha}$ is almost free of mannose, although mucoid chronic bronchitis sputum always contains more, suggesting that this pharmacologically stimulated bronchial fluid contains a negligible transudate component. Mannose levels are higher in purulent than in mucoid sputum.

Any albumin in sputum comes from serum and so is a marker of transudate or exudate. Since asthmatic sputum contains more albumin than bronchitic, Ryley and Brogan [73] suggested that transudate is more important in asthma than in chronic bronchitis. It has been observed that fluorescein-labeled albumin appears on the surface of the bronchial mucosa within a few minutes and Bonomo and D'Addabbo [97], using radioactive albumin, concluded that the passage of albumin into the sputum in chronic bronchitis explains the fall in radioactive albumin in plasma.

B. Serum and Bronchial Protein

Bronchial glycoprotein is usually studied in sputum since it is only the hypersecreting patient that produces enough material for analysis. Mucoid sputum

TABLE 7 Proteins and Glycoproteins in Sol Phase of Mucoid Sputum from Chronic Bronchitis

Specific protein or glycoprotein	Average concentration (mg/100 ml) in chronic bronchitis[a]
Albumin	29.7 ± 8.0
IgG	13.3 ± 8.2
IgA	80.0 ± 41.8
Secretory IgA	—
Lysozyme	43.0 ± 27.6[b]
Transferrin	2.9 ± 1.0
Haptoglobin	2.2 ± 1.6
Lactoferrin	27.0 ± 23.0[b]
α_1-Antitrypsin	2.7 ± 1.6
α_1-Antichymotrypsin	1.5 ± 1.1
α_1-Acid glycoprotein	1.2 ± 1.0
β_{1C}-Globulin	1.4 ± 1.3

[a]Sol phase of the sputum, mean ± SD [74].
[b]Small amounts of ceruloplasmin, hemopexin, kallikrein, fibrinogen, α_2-macroglobulin, β_{2A}-globulin, salivary α- and salivary β-globulin are also present in sputum [98].

from patients with chronic bronchitis are commonly used [74,87,90]; this includes all types of macromolecules shown in Table 7, salivary contamination and pus adds others [99]. Proteins and glycoproteins are here dealt with separately, although in descriptions of serum they are often dealt with together. Epithelial glycoproteins such as those secreted by the bronchial tree have a relatively higher sugar content than serum glycoproteins.

Albumin is virtually the only serum protein present in sputum. Lysozyme is the only protein known to be synthesized by the bronchial tree [106].

Other proteins have been identified in the protein moiety of purulent sputum. Lactic dehydrogenase and proteases are thought to be released during inflammation from bronchial epithelial cells, as well as from macrophages and leukocytes. The large amounts of DNA in purulent sputum inhibits protease activity. Increase in lactic dehydrogenase and in its isoenzyme pattern, and also in DNA concentration, have been reported in purulent sputum [100]. From bacteriological and clinical assessment it seems that lactic dehydrogenase and DNA concentration as well as the number of DNA fibers increase when inflammation is more severe.

C. Glycoproteins

1. *Nature*

Mucosubstances may be one of the two types of polypeptide/sugar compound,
the proteoglycans (also known as mucopolysaccharides or glycosaminoglycans)
or the glycoproteins. These polymers have in common a single polypeptide
chain to which are attached one or more polysaccharide side chains. The two
groups can be distinguished by other characteristics. The sugar side chains of
proteoglycans are linear, have alternate sequences of sugar residues and contain
uronic acid and sulfate. The sugar side chains of glycoproteins are branched
and have no simple sequence of sugar residues. Heparin, chondroitin sulfate,
dermatan sulfate, and hyaluronic acid are all examples of proteoglycans. In
serum and bronchial secretions it is glycoproteins that are found, not proteoglycans.

In addition to the features mentioned above, the peptide core of glycoproteins has a large amount of threonine and serine, and the alkali-labile polysaccharide side chains generally contain neuraminic acid, fucose, galactose,
N-acetylgalactosamine, and N-acetylglucosamine. Tracheobronchial glycoprotein has the same general composition as other glycoproteins from saliva, or
gastrointestinal or genital secretions. Epithelial glycoproteins have a high carbohydrate content, 40-70 g% of their total dry weight [99] compared with below
30% in serum glycoproteins such as immunoglobulins.

Glycoproteins may be neutral or acid. In epithelial glycoprotein the acid
radical is either neuraminic acid or sulfate (Fig. 11). In human bronchial mucin
the neuraminic acid is the N-acetyl form. Neutral glycoproteins contain large
amounts of fucose in the terminal positions of the oligosaccharide chains and
have fewer charged radicals. Neutral glycoproteins may also contain small
amounts of neuraminic acid. The predominant amino acids in the polypeptide
chains are serine and threonine. Neutral and acid glycoproteins are present in
both sol and gel phases of mucoid bronchitic sputum [99] and exhibit blood
group activity [9]. In sputum activity of the ABO (H) variety has been reported
by several authors [101,102].

It is possible to separate acidic from neutral mucins by free flow electrophoresis; the separation of sialomucin from sulfomucin can be effected in the
same way but only after treatment of the mixture with neuraminidase, which
reduces electrophoretic mobility of sialomucin [91].

The basic structure of acidic glycoproteins is very similar to that of neutral
with large amounts of threonine and serine with attached polysaccharide chains.
Amino acids like aspartic acid, glutamic acid, and cysteine are more often found

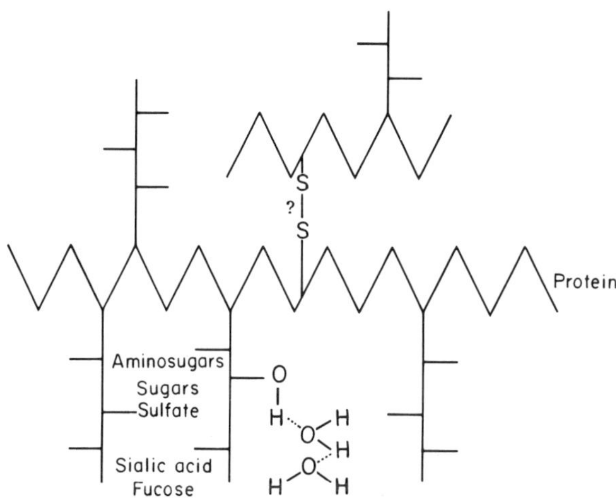

FIGURE 11 Diagrammatic representation of the acid glycoprotein molecule. Sialic acid and sulfate are found terminally on the oligosaccharide side chain. Hydrogen bonds and other linkages occur between side chains of the same and other molecules. Chemical behavior suggests that disulfide bonds may crosslink the molecules.

in the acidic glycoproteins. Sialic acid-containing mucins inhibit hemagglutination by certain viruses, especially influenza virus [103].

Analysis of typical oligosaccharide side chains of neutral, sialo- and sulfomucin of bronchogenic cyst fluid is shown in Table 8 [77]. Glycoproteins isolated by these authors from bronchogenic cysts consist of peptide chains rich in hydroxyamino acids [25-40%) but lack cysteine. Arginine was the only N-terminal group detected. Alkali-labile O-glycosidic linkages to serine and threonine were the main linkages in the purified glycoproteins. Predominantly sulfated glycoproteins had fewer carbohydrate-peptide linkages, longer carbohydrate chains and higher amounts of arginine when compared with the fucose and neuraminic acid-containing glycoproteins. The site of attachments of sulfate in bronchial mucins is not known. In ovine colonic mucus sulfate has been shown to be attached to glucosamine [104], whereas in hog gastric mucosa it is attached to C-6 of N-acetylglucosamine [105].

2. Serum Glycoproteins

Serum contains many glycoproteins, the immunoglobulins being the best-known type that find their way into bronchial secretion, α_1-acid glycoprotein, α_1-antitrypsin, α_1-antichymotrypsin, haptoglobin, transferrin and β_{1C}-globulin. There

TABLE 8 Typical Composition of Oligosaccharide Side Chain of Bronchogenic Cyst Fluid—Neutral, Sialo- and Sulfomucin[a]

Sugar	Neutral mucin	Sialomucin	Sulfomucin
Fucose	10.7	6.2	7.1
Sialic acid	1.1	21.1	1.2
Sulfate	0	3.3	7.2
Hexose	13.7	25.0	20.4
Hexosamines	18.7	25.9	17.3
GlcN/GalN ratio	1.09	0.98	9.0

[a]Concentrations are given as the percentage of dry weight of purified mucins. From Ref. 77.

is a selective concentration of α_1-antichymotrypsin in sputum as compared with serum. Ryley and Brogan [74] suggested α_1-antichymotrypsin in sputum may inhibit neutral protease originating from leukocyte disintegration.

Evidence has accumulated that several glycoproteins, e.g., lactoferrin, secretory IgA, salivary α_1- and β_1-globulins are synthesized by respiratory tract cells [99]. Organ culture studies have shown that lactoferrin and secretory IgA are produced in respiratory epithelial cell [106].

3. Bronchial Acid Glycoprotein

The acid glycoprotein in bronchial secretion is still relatively poorly characterized and reports are rather different. It can be said however, that it is this compound or group of compounds that is mainly responsible for the viscosity of sputum and many of its characteristics. The method of preparation of sputum will affect the form in which "purified" glycoprotein is recovered.

Histochemical studies are of help in the interpretation of biochemical results and contributes to our knowledge of the types of molecule polymerized in bronchial glycoprotein. In the secretory cell of the human submucosal gland it is unusual to find any cell that includes only neutral glycoprotein: neuraminic acid and sulfate can usually be identified (Figs. 12 and 13). NANA is in two forms, one susceptible to removal by the enzyme sialidase, the other resistant to its digestion.

Sulfate is identified in four different histochemical types. A cell may contain only one acid type, e.g., sialidase-susceptible sialomucin, when all acid glycoprotein staining is removed by the enzyme sialidase. Or more than one may be identified, as when the sensitive form is present with the siali-dase-resistant form: staining of the latter does not occur after acid hydrolysis

Type of Acid Glycoprotein		I	II	III	IV	V	VI
		Individual Cell Type					
(a) Sialomucin							
Sialidase susceptible		+	+	+			
(b) Sialidase resistant			+	+	+		
(c) Sulphomucin							
S35 uptake	+ ve			+	+	+	
stains	– ve						
(d) S35 uptake	+ ve				+	+	+
stains	– ve						

FIGURE 12   Types of acid glycoprotein identifiable histochemically and their intracellular combination.

Type of Sulfate	Staining Response		
	HID	$AB_{pH\ 2.6}$	$AB_{pH\ 1.0}$
I	+	+	+
II	+	–	+
III	–	+	+
IV	–	–	–

FIGURE 13   The four types of sulfate which can be identified histochemically within the human bronchial gland.

which removes all sialic acid. Sulfate may be present in the same cell as sialic acid. The various patterns of staining are shown by secretion both within the cell and within the lumen. The acid radicals may be found mixed within a cell but then only in certain combinations. In many cells only sialic acid can be identified indicating that the cell is making a pure sialomucin. The absence of sulfate can be demonstrated both histochemically and by organ culture.

Havez and his colleagues reported from their biochemical studies that epithelial glycoproteins of sputum include a fucomucin and two distinct types of acid glycoproteins, a sialomucin and a sulfomucin. Yet in each acid glycoprotein both acid groups are present, it is the predominant one that confers the name. The fact that some cells make only a sialomucin suggests that the material studied biochemically by Havez and his colleagues, although "purified" represented a polymerization of secretions from several types of cell. To solubilize the glycoproteins Havez and his colleagues pretreated the sputum to break S–S bonds. This may well have reduced the size of the polymer.

Creeth and his colleagues [65] have isolated glycoproteins from sputum by cesium chloride and cesium bromide density-gradient ultracentrifugation. After dispersal in cesium bromide and removal of most of the protein in the first preparative centrifugation, they used cesium chloride for further fractionation. Using this method they isolated one main epithelial glycoprotein component from each sputum. This method they claim is able to separate the glycoprotein from sputum without any structural alteration and free from protein. Certainly this method probably interferes less with glycoprotein structure that any other yet available. Purified glycoprotein obtained by this method showed a molecular weight in the region of 3-7 $\times$ 10^6 which is high compared to other epithelial glycoprotein, e.g., ovarian cyst fluid, suggesting that it does represent covalent aggregates of smaller units, as was found by Starkey et al. [107] in gastric mucoproteins. The amino acids of the bronchial glycoprotein included some cysteine, more than the negligible amount found by Degand et al. [77] which also supports the idea of covalent aggregation of sputum glycoprotein. The glycoprotein from mucoid chronic bronchitic sputum is probably a sulfo-sialoglycoprotein. A similar result was obtained earlier in our laboratory by Roberts [63].

In some cases, sputum glycoprotein isolated by the technique of Creeth and his colleagues showed two distinct components or conformers, one less dense than the other. During the separation process interconversion of the two conformers was observed, the more dense being favored under conditions of high-pressure or slow reaction with the solvent component.

D. Sol and Gel

The separation of sputum into its two phases of sol and gel is readily accomplished by centrifugation. Mostly the epithelial glycoprotein, being relatively insoluble, is found mainly in the gel, the serum glycoproteins being soluble, in the sol. Each phase is contaminated to some extent by the other. Schultze and Heremans [99] separated chronic bronchitic sputum into a sol and gel phase by dialysis after freezing and thawing several times and then low-speed centrifugation of the material. Fractionation by DEAE cellulose ion exchange chromatography led to the isolation of the proteins, albumin and lysozyme, and several glycoproteins, including the immunoglobulins (IgA, IgG), lactoferrin, transferrin, some salivary ones, and about six acid glycoproteins [99,87].

By high-speed centrifugation (118,000 $\times$ g for 4 hr at 4°C), Ryley and Brogan [73] isolated the sol from chronic bronchitis and asthma. In the sol they identified albumin, lysozyme, and glycoprotein "complex" as the principal high molecular weight components. Recently, using quantitative cross immunoelectrophoresis, they [74] were able to measure quantitatively albumin, α_1-acid glycoprotein, α_1-antitrypsin, α_1-antichymotrypsin, haptoglobin, transferrin, β_{1C}-globulin, and immunoglobulins, IgA and IgG. By the same technique they also measured the concentration in serum and saliva. The results showed that relative to other constituents, α_1-antichymotrypsin and IgA were higher in the sol phase of sputum than would be expected if they came from passive transudation.

The gel phase contains the bulk of the glycoprotein. It has proved intractable to most extraction procedures, and so, degradative methods such as enzyme digestion or thiol treatment are usually used. Gel phase has been less adequately studied than the sol.

After disruption of the gel phase by acid hydrolysis or reduction with 2-mercaptoethanol followed by fractionation various components have been found in the gel including some albumin, large amounts of sialomucin, immunoglobulins, and serum β_{2A}-globulin [63,99,108,109]. Roberts [63] studied the gel fraction from chronic bronchitic sputum after urea solubilization. After fractionation with different concentrations of ethanol and gel filtrations on Sephadex 4B he was able to isolate only one type of sulfo-sialo-glycoprotein which was similar in composition to those of other epithelial glycoproteins [110]. In the gel fraction of sputum he was also able to show the presence of α_1-acid glycoprotein, haptoglobulin, lactoferrin, and immunoglobulins A and G. Further studies in our laboratory on the sol and gel fraction after centrifugation at 160,000 X g showed that the dry macromolecular weight, NANA and fucose concentrations in the gel are at least four times higher than those of the sol fraction. Large amounts of albumin, IgA and IgG were found to be associated with the gel fraction of mucoid chronic bronchitic sputum which are not removed by washing with saline and recentrifuging at 160,000 X g. The amount bound to the gel was relatively much less in bronchorrhea sputum which supported the idea that binding of serum protein and glycoprotein with bronchial glycoprotein may be a major determinant of the viscosity of sputum since the viscosity of chronic bronchitic sputum was much higher than that of bronchorrhea sputum.

The possibility that lactoferrin or IgA could crosslink bronchial glycoprotein chains has been suggested by Gernez-Rieux and colleagues [111]. Presence of serum glycoprotein with the "purified" mucin glycoprotein was also observed during fractionation of mucin from different sources, such as porcine submaxillary mucin [112,113], bovine cervical mucin [114], and ovomucin [115].

Purified mucins are fairly resistant to proteases [92] and are poor in lysine, arginine, and cysteine (the last being the only disulfide bridge-forming amino acid). However, proteases that generally act on lysine-arginine bonds, as well as agents disrupting the disulfide bond both reduce viscosity which suggests that these are acting not on the polypeptide of the mucin glycoprotein but on some other protein crosslinked to the glycoprotein.

E. Physiochemical Properties of Mucus Glycoproteins

Epithelial or mucus glycoproteins are characterized by their large molecular size, their capacity to form viscoelastic gels at relatively low concentration, and their high degree of polydispersity. In mucins the glycoproteins have a fiber-like structure [116,117]. It is the great number of fine and coarse bundles of fibers, irregularly interwoven, that imparts cohesion and elasticity and is also responsible for viscosity, adhesiveness, and tear resistance [118]. DNA fibers originating mainly from the nucleic acids of polymorphonuclear leukocytes [72,119] with some contribution from necrotic inflammatory mucosal cells [117,120] are also found in purulent sputum.

Recent work has shown that the viscoelastic properties of bronchial secretions are essential for normal clearance of mucins from the airways of the normal subject as from the patient. The viscosity and cohesiveness of mucus arises from the rigidity of the fiber-like structure as well as from their crosslinking to form a lattice. Rigidity can be explained by the abundance of negative ionic charges arising from neuraminic acid and sulfate residues on the carbohydrate side chain, mutual repulsion of these charges causing "stiffening" of the molecule with an increase in viscosity. Lowering the level of ionization by raising ionic strength reduces viscosity and so does the removal of neuraminic acid [121]. But the reduction of disulfide bonds or proteolysis is even more effective than neuraminidase in reducing the viscosity of human bronchial secretions [108]. This presumably is either because the length of the glycoprotein polymer is reduced or because crosslinking between protein and glycoprotein chains is disrupted.

IV. Variation in Biochemical Constituents with Macroscopic Type and Disease

A. Variation with Macroscopic Type of Sputum

Classification of sputum by macroscopic appearance presents considerable difficulty since it is a subjective assessment and a high variance has been reported between observers and even by a single observer [122,123].

Miller and Jones' [123] classification is probably the most widely used:

M1   pure mucoid specimen with no suspicion of pus

M2   predominantly mucoid sputum with a suspicion of pus

P1   purulent Grade 1—pus amounting to less than one-third of specimen

P2   purulent Grade 2—pus amounting to one-third or two-thirds of the specimen

P3   purulent Grade 3—pus amounting to more than two-thirds of the specimen

Other classifications are used, but basically they correspond with Miller and Jones (Table 9). It seems generally agreed that five grades can, in practice, be distinguished.

1. Methods for Assessing Sputum Purulence

The first methods for assessing the presence of pus in sputum were those of Hippocrates, from the smell produced when sputum was burnt, and that purulent sputum sank when floated on water. Macroscopic purulence is due to the content of polymorphonuclear leukocytes and deoxyribonucleoprotein fibers arising from disintegrated cells, particularly from polymorphs. Total and differential cell counts and quantitative assessment of DNA fibers have each been used to assess the degree of purulence, [119,120,123] cell counts being particularly useful on mucopurulent sputum from asthmatic patients since the pus may consist largely of eosinophils [124-126]. The concentrations of deoxyribonucleic acid (DNA) and of lactic dehydrogenase (LDH), particularly isoenzymes 4 and 5, offer a criterion of inflammation [120]. LDH correlates with the number of fibers and with the concentration of DNA (Table 10). Mucoid sputum from patients with cystic fibrosis has been found to contain DNA [127] (Table 11).

2. Chemical Constituents of Sputum

Any attempt to compare the results reported in the literature is difficult and disappointing; the reasons are as follows:

1.   Lack of standardization of methods for collecting sputum, for its storage and variation in types of pretreatment

2.   Nature of the material studied (sputum is a complex and inhomogeneous fluid, therefore inadequate sampling may lead to high variance)

3.   Discrepancies in diagnosis of lung disease particularly as between chronic bronchitis and intrinsic asthma

Histochemical and biochemical studies [19,20,63,77,128-131] have shown that bronchial acid glycoproteins contain fucose, sulfate, and N-acetylneuraminic

TABLE 9 Classification of Sputum by Macroscopic Appearance[a]

Miller and Jones [123]	May [125]	Charman and Reid [53]
M1	M	M
M2	MP±	MMP
P1	MP+	MP
P2	MP++	MPP
P3	MP+++	P

[a]Method of grading used in our studies

TABLE 10 Correlation Between DNA Fibers in Fluorescence Microscopy and DNA Concentration and LDH in Sputum[a]

DNA fibers[b]	DNA (mg/ml)	LDH (IU)
0	1.1 ± 0.8[c] $(0-2.95)$[d]	168 ± 140 $(1-479)$
+	2.7 ± 0.65 $(0.8-3.5)$	265 ± 58 $(98-352)$
++	3.2 ± 2.2 $(0.6-10.7)$	462 ± 118 $(210-657)$
+++	9.9 ± 2.2 $(4.4-14.7)$	948 ± 304 $(441-1537)$

[a]From Ref. 120.
[b]+, Small number of DNA fibers present; ++, intermediate number of DNA fibers present; +++, large number of DNA fibers present.
[c]Mean, twice standard deviation.
[d]Range.

TABLE 11 DNA Concentrations in Sputum Related to Macroscopic Types and Disease

Macroscopic types	No. of specimens	DNA (mg/ml) Mean	DNA (mg/ml) Range
Mucoid (chronic bronchitis and asthma)	20	0.004	0 –0.05
Mucoid (cystic fibrosis)	20	0.12	0.05-0.27
Mucopurulent (cystic fibrosis and bronchiectasis)	20	0.56	0.06-2.38
Purulent (cystic fibrosis and bronchiectasis)	20	0.94	0.30-1.71

acid (NANA), but mannose has not been identified in purified glycoprotein from bronchial cysts [77]. Serum glycoproteins are rich in mannose and neuraminic acid but contain little or no fucose [94] (Tables 5 and 6). Fucose and sulfate can therefore be taken as "marker" substances of bronchial glycoproteins, neuraminic acid of both bronchial and serum glycoproteins, and mannose of serum. Absolute levels of marker substances and their ratios give an indication of the origin of the airway fluid in different diseases and of the response to drugs.

3. Macromolecular Dry Weight Yield

The macromolecular dry weight represents the total undialyzable material present in sputum. Mucoid chronic bronchitic sputum has been used for most of the biochemical studies on airway fluid since it is easy to obtain and the amount normally produced (5-25 ml/day) [132] allows several chemical estimations to be carried out on the same specimen.

Recently inhalation of prostaglandin $F_2\alpha$ ($PGF_2\alpha$) has been shown to be a useful method for obtaining "sputum" from healthy subjects [10]; $PGF_2\alpha$ sputum represents bronchial fluid from a normal bronchial tree and can act as control when studying sputum in disease (Table 12).

The dry macromolecular weight of mucoid chronic bronchitic sputum represents between 0.8 and 2.6% of the total weight of the sputum sample [69,133-135]. The macromolecular weight increases with the degree of purulence (Fig. 14) due to a higher content of DNA, proteins, and carbohydrates [45,69,98,133,134,136].

4. Osmolality

The water-soluble fraction (sol phase) of airway fluid is slightly hypertonic when compared with serum [137]. The osmolality of the water-soluble fraction of

TABLE 12 Macromolecular Dry Weight Yield (mg/ml) Related to Macroscopic Types

| | | Chronic bronchitic sputum | |
	$PGF_2\alpha$ sputum n = 4	Mucoid n = 24	Mucopurulent n = 26	Purulent n = 13
Mean	6.0	15.9	21.2	51.2
SE	0.8	0.9	1.5	4.5
Range	4.2-7.4	8.0-26.0	12.0-15.0	19.0-84.0

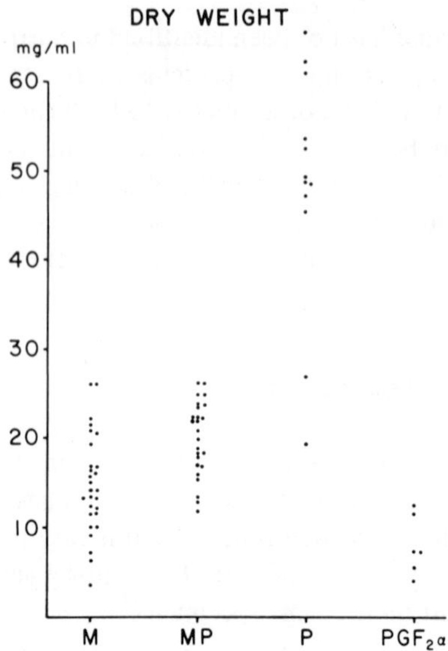

FIGURE 14   The yield of macromolecular material of mucoid, mucopurulent, and purulent sputum

nonpurulent sputum from laryngectomized patients and purulent secretions from patients with cystic fibrosis or bronchiectasis was estimated, no significant difference was found in osmolality between the two types of sputum (Table 13). The osmolality of whole sputum was lower than that of the water-soluble fraction; although purulent specimens had a higher osmolality than mucoid, the difference between them was not statistically significant [138].

5. pH

The pH of bronchial secretions has been measured in situ at bronchoscopy and in sputum. Steinman [139] claimed that several factors might influence the acid-base equilibrium of the tracheobronchial secretions, i.e., carbonic acid, rate of secretion and degree of inflammation. He found that the pH varied with the region of the tracheobronchial tree, being more alkaline (6-6.7) in trachea and more acid (5.7-6) in the main bronchi. When inflammation was present the secretions were more alkaline than either (7.4). In contrast, Guerrin et al. [140] reported that the pH, also measured in situ, became acid (5.2) when pus was present. The pH when measured in sputum is slightly more alkaline; values of 7.4-8.2 have been reported by various authors [141,142]. In sputum produced after inhalation of PGF$_2\alpha$ the pH ranges from 8 to 9 with a mean value of 8.28 [10]. The reason for the difference in the various reports is not clear.

TABLE 13 Osmolality (mOsm/liter) of Water-Soluble Fraction and Whole Sputum

	Macroscopic types	
Fraction	Mucoid	Purulent
Water-soluble[a]	359 ± 56	371 ± 70
Whole sputum[b]	235 ± 60	273 ± 36

[a]Potter et al. [137].
[b]Lopez-Vidriero [138].

TABLE 14 Electrolytes Concentrations in Mucoid and Purulent Sputum (mmol/liter)[a]

	Na	K	Ca	Cl	PO_4
Mucoid	165	13.2	3.1	162	18
Purulent	101	28	3.6	75	63

[a]From Boat and Mathews [70].

6. Electrolytes

Studies carried out on whole sputum and on its water-soluble fraction indicate that electrolytes are bound mainly to soluble macromolecules, i.e., to those in the sol, although small amounts, particularly of calcium and chloride, are bound to macromolecules of the gel phase [137]. Sodium and chloride concentrations are lower in purulent secretions than in mucoid, while potassium and calcium increase when pus is present [69,137]. The increased phosphorus content reported in purulent sputum [69,134] is partly due to the increase in DNA content (Table 14).

7. Marker Substances

a. N-Acetyl Neuraminic Acid (NANA)

NANA levels increase in chronic bronchitic sputum [45,143] with the degree of purulence (Tables 15 and 16). In chronic bronchitis similar findings have been reported by Atassi et al. [144] and Lopez-Vidriero [133] and also in asthma, bronchiectasis, and cystic fibrosis [138]. The increase in NANA levels in purulent sputum is doubtless due to a relatively high contribution from serum transudate or inflammatory exudate since levels of mannose, a marker of serum

TABLE 15 Mean Neuraminic Acid Levels for Sputum Samples from a Series of
Patients Related to the Percent of Sputum Specimens Containing Pus

Sputum specimens purulent (% of 20-30 spec.)	No. of Cases	NANA (μmol/ml)
20	6	1.7
21–30	13	2.0
31–40	11	2.6
40	18	2.8

glycoproteins, increase with the degree of purulence, while fucose, a marker
substance of bronchial glycoprotein, changes very little [138].

b. Fucose

The fucose content of sputum seems to bear no relationship to the degree of
purulence (Fig. 15), suggesting that bronchial mucus secretion is little altered by
infection [138].

c. Sulfate

Histochemically bronchi from patients with cystic fibrosis show an increase in
the number of sulfated mucous cells [130,145]. Sulfate, like fucose, is a marker

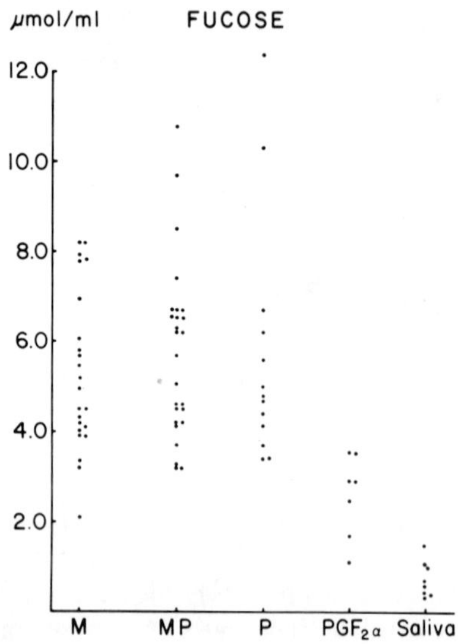

FIGURE 15 The fucose content of mucoid, mucopurulent, and purulent sputum.

TABLE 16 Chemical Constituents of Sputum: Comparison Between Diseases

Constituents	$PGF_2\alpha$	Chronic bronchitis[a]	Asthma	Bronchiectasis[b]	Cystic fibrosis[b]	Bronchorrhea[b]
Dry weight (mg/ml)	6.04	15.90	16.36	39.90	62.60	10.10
NANA (μmol/ml)	0.64	2.50	2.50	3.90	4.20	1.68
Fucose (μmol/ml)	2.77	5.30	4.70	6.90	5.90	2.37
Sulfate (μmol/ml)	0.86	1.84	1.31	2.32	2.02	1.10
Mannose (μmol/ml)	0.24	0.87	0.72	1.01	1.52	0.83
NANA/fucose ratio	0.25	0.40	0.60	0.60	0.80	0.71

[a]Mucoid.
[b]Purulent.

of bronchial glycoprotein. In purulent sputum sulfate levels are higher than in mucoid [138]: it seems that a more sulfated glycoprotein is produced when infection is present.

d. Mannose

Mannose levels after $PGF_2\alpha$ are significantly lower than in mucoid chronic bronchitic sputum, suggesting that the airway fluid from a "normal" bronchial tree contains negligible amounts of serum transudate. Purulent sputum from chronic bronchitis, asthma, bronchiectasis, and cystic fibrosis contains more mannose than do mucoid samples [138] (Fig. 16).

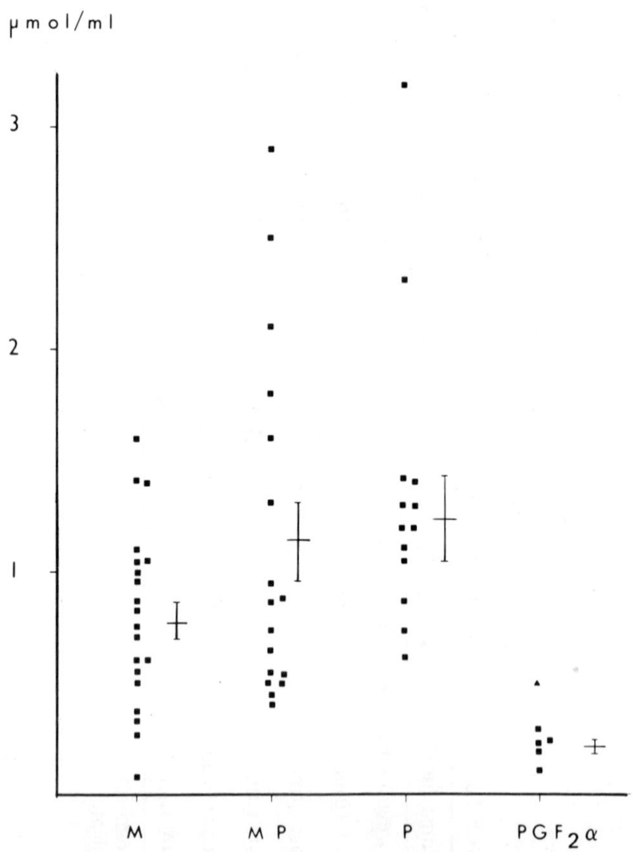

FIGURE 16   The mannose content of mucoid, mucopurulent, and purulent sputum.

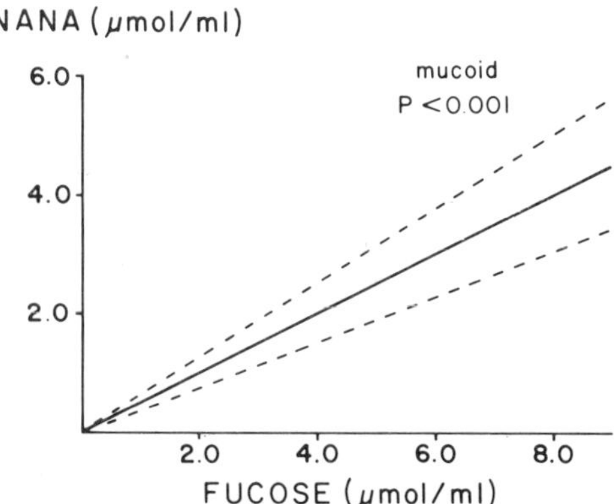

FIGURE 17   Regression line with 95% confidence limits for the relationship between fucose and N-acetylneuraminic acid (NANA).

e.   NANA/Fucose Ratio

Lopez-Vidriero et al. [133] reported a highly significant correlation between fucose and NANA in mucoid chronic bronchitis sputum; the regression line (with 95% confidence limits) can be used as an index for assessing the serum transudate component in bronchial fluid (Fig. 17). the NANA/fucose ratio in purulent secretions is significantly higher than in mucoid (Table 16). The rise, being due to an absolute increase in NANA content, is probably from serum since fucose levels vary little [138].

f.   Secretory IgA

Immunofluorescent methods applied to the normal bronchial wall and to tissue from cases with chronic pulmonary disease [21] have shown that cells producing IgA and IgG, particularly IgA, are increased in chronic pulmonary diseases. Medici and Bürgi [146] measured the secretory IgA concentration in sputum from chronic bronchitis and found a significant correlation between IgA levels and the degree of infection, assessed by sputum levels of DNA and LDH. Mean values of IgA ranged from 52 mg/100 ml when little inflammation was present to 225 mg/100 ml when inflammatory process was at its peak. Similar findings have been reported by Puchelle et al. [147].

B.   Sputum Differences in Various Diseases

Before comparing sputum in various diseases, the range of diurnal and seasonal variation is considered.

Diurnal variation. A diurnal variation of high early morning values with a gradual fall during the day has been described for sputum volume, dry weight,

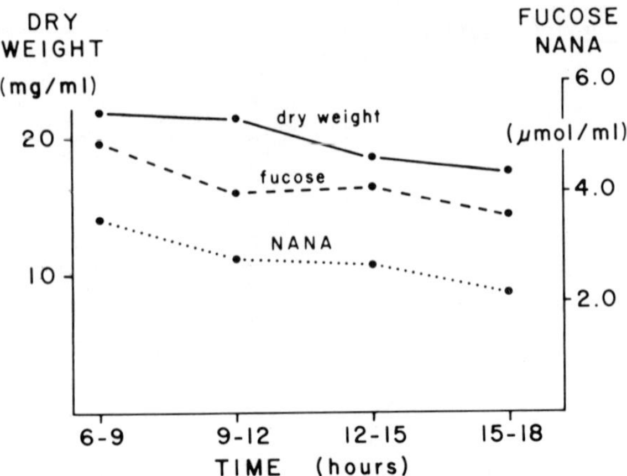

FIGURE 18   Pattern of diurnal variation for dry macromolecular weight, fucose, and NANA in chronic bronchitis.

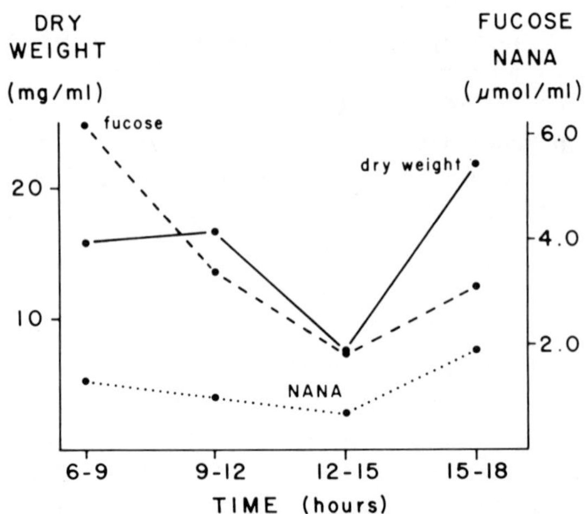

FIGURE 19   Pattern of diurnal variation for dry macromolecular weight, fucose, and NANA in asthma.

NANA, and fucose concentrations [148] (Figs. 18 and 19). These findings suggest that, while in chronic bronchitis there is a gradual decrease during the day in both the bronchial and serum component, in asthma both bronchial and serum glycoproteins, particularly the latter, increase later in the day. In chronic bronchitis Havez et al. [131] reported a similar pattern for secretory IgA, absolute values being higher in the morning and lowest between 4 and 8 p.m. A diurnal variation up to 100 mg/100 ml was reported by these authors.

Daily variation. The average percentage coefficient of day-to-day variation for the chemical constituents of sputum in chronic bronchitic patients was found to be lower than that for the diurnal variation, particularly for fucose, NANA/fucose ratio, and NANA [148]. Puchelle et al. [147] studied two chronic bronchitic patients over 3 consecutive days and found that the daily biochemical profile of sputum was fairly constant for each patient.

A daily variation of serum proteins and glycoproteins and bronchial glyco-proteins has been observed in patients with chronic bronchitis [131]. The day-to-day albumin levels were very stable, while secretory IgA levels varied. Three groups of patients could be identified according to the secretory IgA levels and the daily variation:

Group 1:   S IgA levels always below 100 mg/100 ml

Group 2:   S IgA levels varied between 100 and 220 mg/ml

Group 3:   S IgA levels always above 200 mg/100 ml

Annual variation. Seven chronic bronchitic patients were studied during two consecutive winters. No significant difference was observed in sputum volume, dry weight yield, NANA, or fucose levels [148]. It would seem that the rate of secretion and the type of glycoprotein produced by an individual patient are very constant if clinically the patient is in a steady state.

Seasonal variation. Keal [45,143] studied 16 chronic bronchitic patients at monthly intervals for 3 consecutive years. In some patients dry weight and neuraminic acid levels showed a winter rise (Figs. 20 and 21). Since all the specimens studied were mucoid, the increase in dry weight and NANA was not due to an infection but probably represents a higher concentration of mucus in the bronchial fluid.

1. Chronic Bronchitis

Chemical constituents have been identified and estimated in whole secretion, sol and gel phase as well as in purified bronchial glycoprotein. Electrolyte con-centrations and pH in the sol phase of chronic bronchitis sputum are similar to those found in other diseases [73,137] (Table 17). It has been suggested that albumin levels in the sol phase are higher in asthma than in chronic bronchitis [137], and the proportion of albumin has been used to distinguish chronic bronchitis from asthma. No significant difference has been found between chronic bronchitis and other diseases for serum glycoproteins, i.e., immuno-globulins [98].

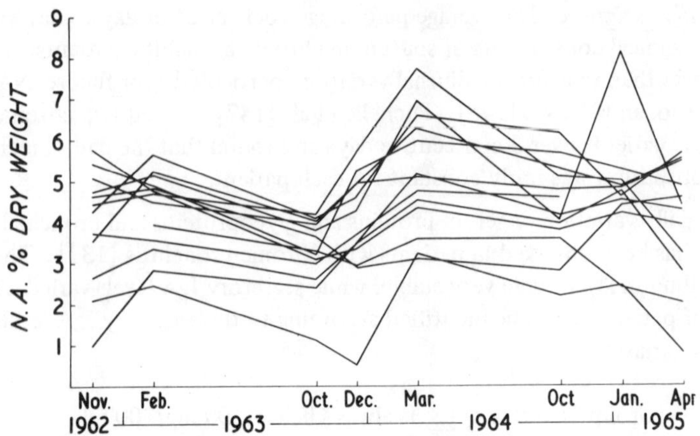

FIGURE 20   The seasonal variation in neuraminic acid as percent of the macro-molecular solids from the sputum in a group of chronic bronchitics.

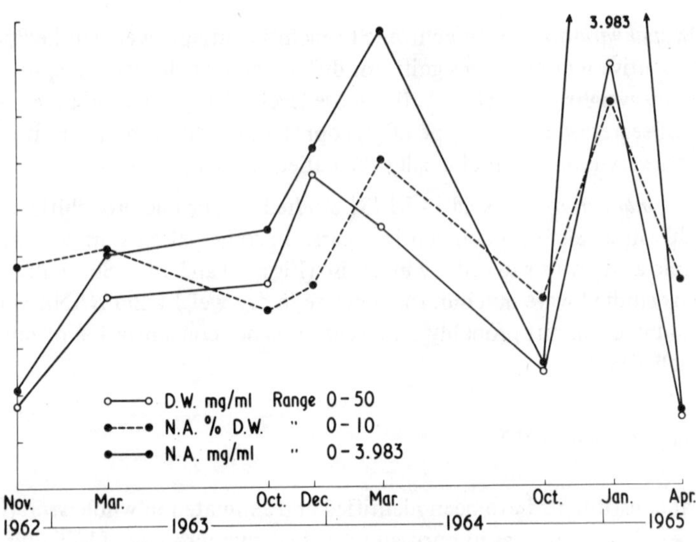

FIGURE 21   Seasonal variation: serial analyses of sputum from an individual showing concomitant rise in all three indices in winter months.

Dry macromolecular weight and absolute values of NANA and fucose show a wide range as between patients, and there is a great overlap between diseases [138] (Figs. 21-24). When mucoid chronic bronchitic sputum is compared with sputum produced after inhalation of PGF_{2a} three points emerge:

TABLE 17   pH and Electrolytes Concentrations (mmol/liter) of the Sol Phase of Sputum (Mean Values)

pH and electrolytes	Chronic bronchitis[a]	Asthma[a]	Bronchiectasis[b]	Cystic fibrosis[b]
pH	7.5	6.6	8.0[c]	
Na	100	100	141	131.1
K	27	23	23	35.4
Cl	88	100	97.8	77.6
Ca	2.5[d]	2.1[d]	3.08	2.73

[a]From Ryley and Brogan [73].
[b]From Potter et al. [137].
[c]From Basch, Holinger, and Poncher [141].
[d]From Hoffman and Elbert [149].

TABLE 18   Viscosity and Chemical Constituents of Serum, Saliva and Mucoid Sputum

Viscosity and chemical constituents	Serum	Saliva	PGF_{2a} sputum	Chronic bronchitis (mucoid)	Extrinsic asthma (mucoid)
Viscosity ($1350\ sec^{-1}$ poise)	0.01	0.02	0.34	0.41	0.40
Dry weight (mg/ml)	80.44	3.60	6.04	15.90	16.60
NANA (μmol/ml)	2.10	0.18	0.54	2.50	2.60
Fucose (μmol/ml)	0.36	0.79	2.77	5.30	4.90
Sulfate (μmol/ml)	0.91	0.33	0.86	1.85	1.31
Mannose (μmol/ml)	13.40	0.18	0.24	0.87	0.72
NANA/fucose ratio	5.27	0.25	0.25	0.40	0.60

                                          M. T. Lopez-Vidriero, I. Das, and L. M. Reid

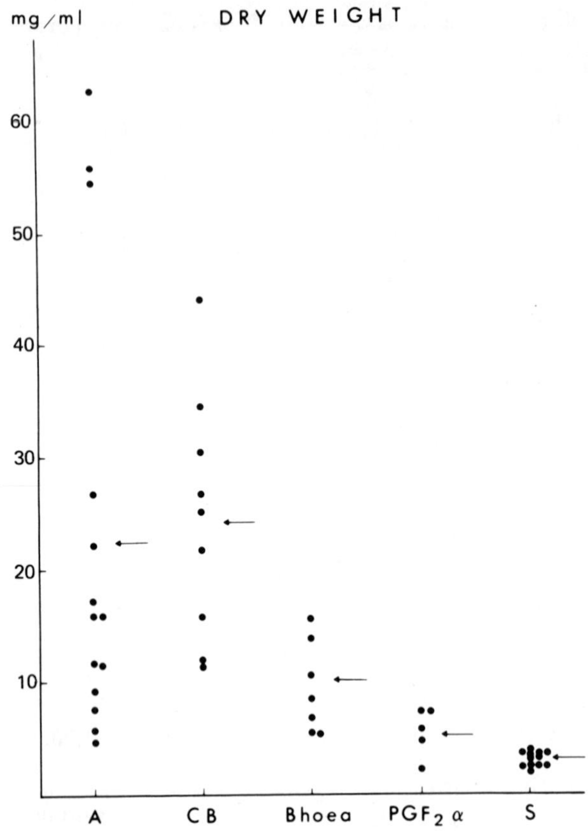

FIGURE 22   The yield of macromolecular material of saliva, PGF$_{2a}$ sputum and mucoid sputum from patients with chronic bronchitis, asthma, and bronchorrhea.

1.   The mannose content in chronic bronchitis is higher than in PGF$_{2a}$ sputum, suggesting relatively larger contribution from serum transudate.

2.   Absolute values of NANA in chronic bronchitis are about four times as great and fucose levels are double. With disease there is both increase in acid glycoprotein from airways and in serum transudate.

3.   Absolute levels of sulphate are significantly higher in mucoid chronic bronchitic sputum than in PGF$_{2a}$. Histochemical studies have shown that in patients with chronic bronchitis and in experimental animals exposed to sulfur dioxide or cigarette smoke there is a relative increase in the cells producing a sulfated glycoprotein [150,151].

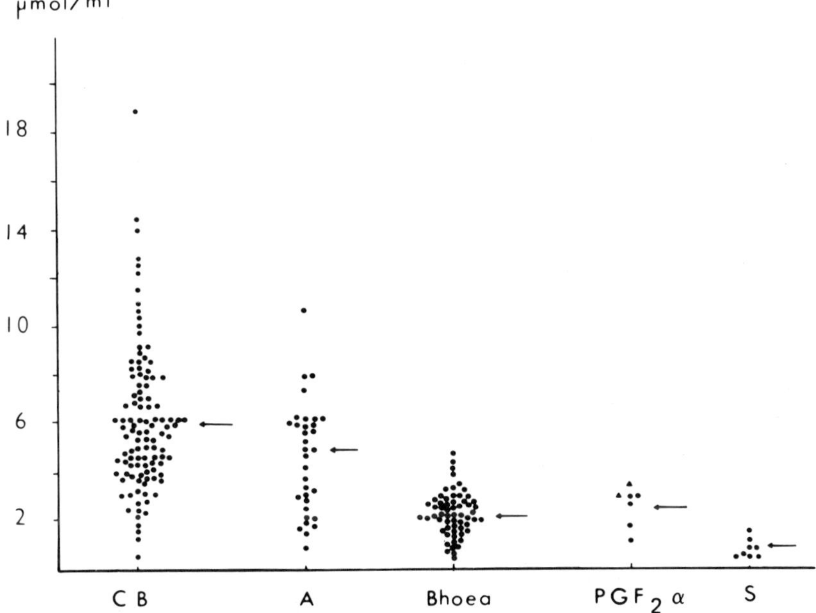

FIGURE 23 The fucose content of saliva, PGF_{2a} sputum, and mucoid sputum from patients with chronic bronchitis, asthma, and bronchorrhea.

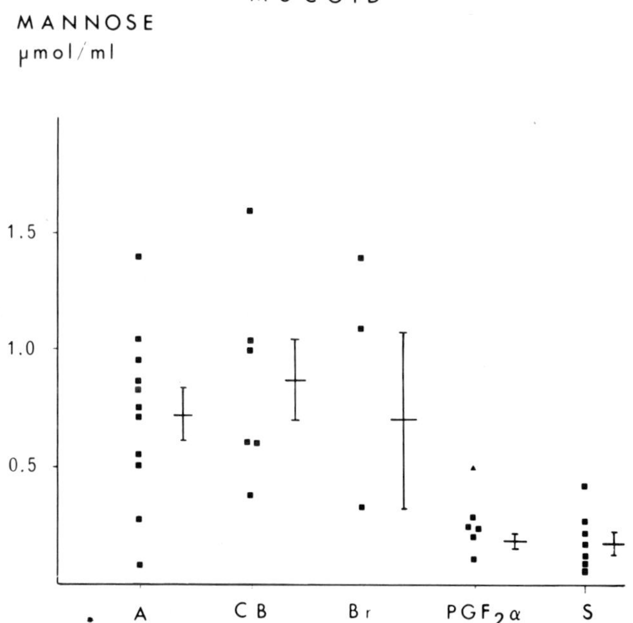

FIGURE 24 The mannose content of saliva, PGF_{2a} sputum, and mucoid sputum from patients with chronic bronchitis, asthma, and bronchorrhea.

2. Asthma

The first problem encountered when studying sputum from asthmatic patients
is the difficulty in differentiating clinically between intrinsic asthma and "wheezy
bronchitis." Speizer et al. [152] have reported conflicting results between
necropsy findings and clinical assessment. The macroscopic appearance of
asthmatic sputum can be misleading because of the presence of eosinophils—
mucoid samples may look purulent and be classified as such [124,125]. This
could explain some of the divergent results in the literature.

Inorganic ions concentrations and pH of sol phase from asthmatic patients
are similar to those reported for chronic bronchitis (Table 17) [73,149,153].
The number of proteins identified in the sol phase of asthma sputum is greater
than in chronic bronchitis and concentrations of albumin, α_1-acid glycoprotein,
transferrin, haptoglobin, and α_1-antitrypsin are significantly higher in asthma
suggesting that there is a greater contribution of serum transudate [153,74].
Secretory IgA levels in sputum from asthmatic subjects have been found to be
within the same range in atopic or non atopic subjects [154].

Keal [50] reported that the range of values for neuraminic acid is greater
in asthma than in chronic bronchitis. Further studies have demonstrated that
the intrinsic asthma group shows a wider range of values for dry weight, NANA,
and fucose than the extrinsic asthma group [138], but the difference between
mean values for the two groups is not significant. Absolute values of dry weight
and chemical constituents in asthma sputum fall within chronic bronchitis levels
except for the NANA/fucose ratio which is higher in asthma, another pointer to
the relative increase in serum transudate (Table 18).

The expectoration of bronchial casts by patients is not uncommon. Keal
[143] described two types of bronchial casts; one with dry weight and NANA
concentration within chronic bronchitis levels and another with a very high dry
weight and NANA content, although the NANA expressed as a percentage of
dry weight was within bronchitic range. The latter group probably arises in
part from dehydration.

3. Cystic Fibrosis

It used to be thought that the bronchial secretions from patients with cystic
fibrosis were more viscous than from other diseases and that the basic defect
was an abnormality of epithelial glycoprotein synthesis. Cystic fibrosis sputum
is not more viscous than from other diseases such as chronic bronchitis, asthma,
and bronchiectasis if the same type of macroscopic sputum is compared [58,
138] (Fig. 24). Electrolyte content of cystic fibrosis sputum is similar to that

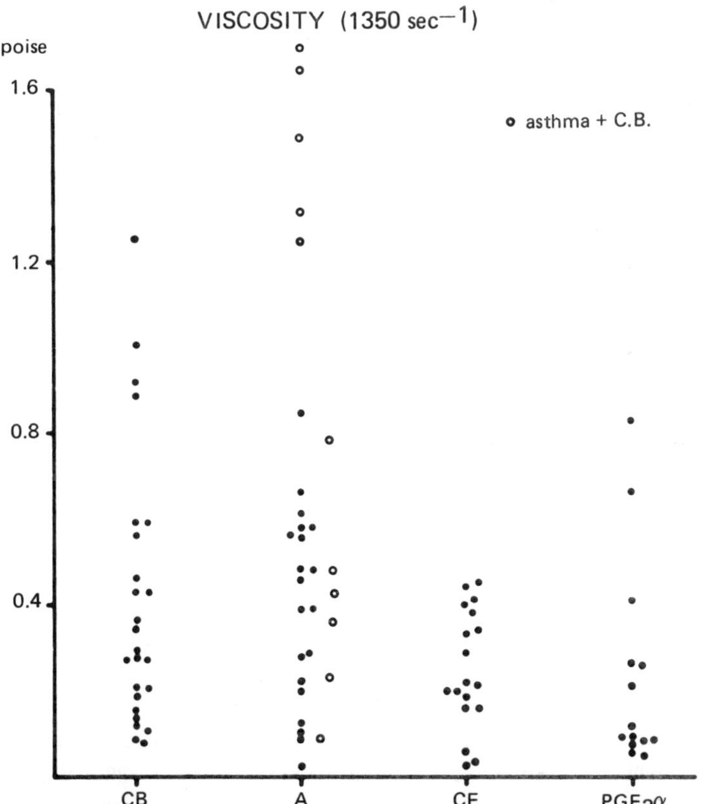

FIGURE 25 Viscosity values at 1.350 sec^{-1} for PGF_{2a} sputum and mucoid sputum from patients with chronic bronchitis, asthma, and cystic fibrosis.

found in other diseases where infection is present, such as bronchiectasis [137]; although an increase in calcium concentration was reported by Chernick and Barbero [134], this finding has not been confirmed by other authors [137]. Some differences are found between cystic fibrosis sputum and other diseases. The dry macromolecular weight of purulent cystic fibrosis sputum is higher than that of purulent bronchiectasis or chronic bronchitis [138]. It seems very likely that this increase in dry weight is mainly due to a higher DNA content as reported by Potter et al. [137] and Mathews et al. [69]. Semiquantitative analysis of DNA fibers in purulent sputum has shown that the sputum viscosity is related to the number of fibers rather than to the total DNA concentration [127].

Brogan et al. [98] identified two groups of cystic fibrosis patients from the levels of soluble proteins in sputum, one group with albumin concentration

of less than 50 mg/100 ml and another with more. The latter they thought due
to a pulmonary allergic reaction causing a severe inflammatory response. Histochemical and autoradiographic studies show that the proportion of mucus cells
staining for sulfated mucins is increased [130,145] and sialic acid sensitive to
sialidase is reduced or absent in cystic fibrosis [130,145,150] but no chemical
abnormalities have been identified in the types of acidic glycoproteins produced
in this disease. Levels of sialic acid, fucose, and sulfate in purulent cystic fibrosis
sputum fall within the ranges found for other purulent sputa from chronic
bronchitis and bronchiectasis [138] (Table 16).

4. Bronchiectasis

Levels of chemical constituents of bronchiectatic sputum are similar to those
found in purulent chronic bronchitis and cystic fibrosis [69,134,137,138]
(Table 16).

5. Bronchorrhea

Bronchorrhea has been defined as a condition in which more than 100 ml of
sputum is produced within 24 hr [50,155]. It may be associated with tuberculosis [156], alveolar cell carcinoma [155,157-160], chronic bronchitis [161,
162] or asthma [50]: or it may be idiopathic [163]. The large amount of
sputum produced is transparent, mucoid, resembles uncooked egg white, is often
topped by a frothy layer and separates spontaneously into two layers. Levels
of dry weight, NANA, fucose, sulfate, and mannose fall between saliva and
mucoid chronic bronchitis levels (Table 16). These, being always higher than
saliva, enable bronchorrhea to be distinguished from hypersalivation [14,155,
164]. The large sputum volume may be due to a change in the secretion from
mucus-secreting cells, or to an increase in serum transudate, or to a combination
of these two. In alveloar cell carcinoma the excess of secretion is probably autonomous and not under nervous control since the volume is not affected either by
atropine or fluid deprivation [155].

6. Pulmonary Edema

Keal [164] studied two groups of patients with mitral stenosis, one group pre-
senting with recurrent respiratory infection and the other with recurrent episodes of pulmonary edema. Chemical analysis showed that in the group of patients with pulmonary edema the concentration of NANA was closer to that
found in serum, fucose levels being lower than those found in mucoid chronic

bronchitis and $PGF_{2\alpha}$ sputum. The group with recurrent infections fell within
the mucoid chronic bronchitis range suggesting some gland hypertrophy. Brogan
et al. [153] found that the total protein concentration in the sol phase of
chronic bronchitics was higher if heart failure was present.

V. Biochemical Basis for Physical Properties

The arrangement of molecules of bronchial glycoprotein as a random coil and
the presence of crosslinkages between acid glycoprotein and other molecules,
such as proteins and immunoglobulins, are responsible for the rheological behavior of bronchial secretion. In Table 19 are summarized the rheological properties of sputum that are influenced by the concentration of chemical constituents and those that reflect the fundamental molecular arrangement and physical
properties.

A. Absolute Level of Viscosity

The viscosity of various epithelial secretions has been related to the neuraminic
acid content, e.g., bovine cervical mucus, pseudomyxomatous gels, and chronic
bronchitis sputum [114,142,165-167]. The apparent viscosity of mucoid
chronic bronchitis sputum, at shear rate of 1350 sec^{-1}, correlated directly with
the amount of neuraminic acid (NANA), dry weight, and fucose, particularly
with fucose (Table 20) [133]. It is unlikely that the physiochemical properties
of fucose a pentose sugar, explain the significant correlation with viscosity;
rather it is that fucose is a marker of the epithelial acid glycoprotein. This is
supported by the fact that in mucopurulent sputum viscosity is significantly
correlated with dry weight and fucose, while NANA shows a low degree of correlation since NANA represents both bronchial acid glycoprotein and tissue
fluid transudate/exudate [133]. Puchelle et al. [147] have reported a significant

TABLE 19 Rheological Properties of Sputum

Reflecting macromolecular concentration (e.g., fucose, dry weight, NANA)	Reflecting physiochemical arrangement
Absolute level of viscosity	Plateau viscosity
Absolute level of elasticity	Plateau elasticity
Yield value	Notching
	Slope of viscosity
	Time-dependent effect
	Urea effect

TABLE 20 Linear Correlation Coefficients Between Viscosity and Chemical Constituents (Chronic Bronchitis Sputum)

Macroscopic types of sputum	Dry weight (mg/ml)	NANA (μmol/ml)	Fucose (μmol/ml)
Mucoid	+ 0.68[a]	+ 0.66[a]	+ 0.73[a]
Mucopurulent	+ 0.62[a]	+ 0.38[b]	+ 0.79[a]

[a]$P < 0.001$.
[b]$P < 0.02$.

correlation, in chronic bronchitis sputum, (macroscopic type not specified) between viscosity and dry weight, galactose, certain hexosamines, and neuraminic acid all of which are also "marker" substances of the bronchial acid glycoprotein. They found a significant correlation between viscosity and concentration of secretory IgA and albumin in chronic bronchitis sputum perhaps because they contribute to crosslinkages.

B. Absolute Level of Elasticity

Puchelle et al. [147] reported in chronic bronchitic sputum a significant correlation between elasticity, expressed as "retarded recoverable strain," and absolute levels of secretory IgA, dry weight hexosamines, and neuraminic acid. In samples of sputum from different diseases asthma, chronic bronchitis, cystic fibrosis, and bronchiectasis Mitchell-Heggs et al. [61] found that absolute levels of elasticity measured at low shear rates with a Weissenberg rheogoniometer did not differ significantly with macroscopic appearance. These results suggest that elasticity reflects a basic physiochemical property of sputum rather than variation in the proportions of its constituents since these vary widely with macroscopic type [138]. Solutions of DNA have been reported to be very elastic [168], but no significant correlation has been demonstrated between elasticity and DNA concentration in sputum [147].

C. Pourability

In the case of mucoid sputum the clinical assessment of pourability has some predictive value for viscosity [45]. For mucoid chronic bronchitis sputum the pourability grades show a significant and inverse correlation with the viscosity measured with Ferranti-Shirley viscometer [4,5,143] and with the concentration of NANA and dry weight [143]. It is of some interest that for purulent sputum it does not hold: the capsule of clear fluid that surrounds purulent sputum may explain this.

VI. Correlation of Chemical Constituents and Rheological Properties with Clinical Features

The pathophysiological causes of airways obstruction in chronic bronchitis may be irreversible—fibrosis, distortion, narrowing and obliteration of bronchial tree— or reversible—bronchial fluid and inflammation. It has been generally accepted that the bronchial tree in healthy subjects is lined by a thin layer of mucus of about 5 μm [169,170], but the concept of a continuous mucus blanket is a false one. It seems that the cilia beat in a liquid periciliary layer and that any thick secretion is disposed as discrete globular masses along the bronchial tree [3] (Fig. 26).

In chronic bronchitis one of the basic changes is the increase in goblet cells in airway epithelium, particularly striking in the small peripheral airways, since, normally, they are virtually absent in airways of this size. An amount of mucus that may plug a small airway may encroach little on the lumen of a large one.

In chronic bronchitic patients [171] mucus is a significant cause of obstruction of the small airways. An increase in vital capacity after periods of expectoration (spontaneous or assisted) has been reported [56], but the changes

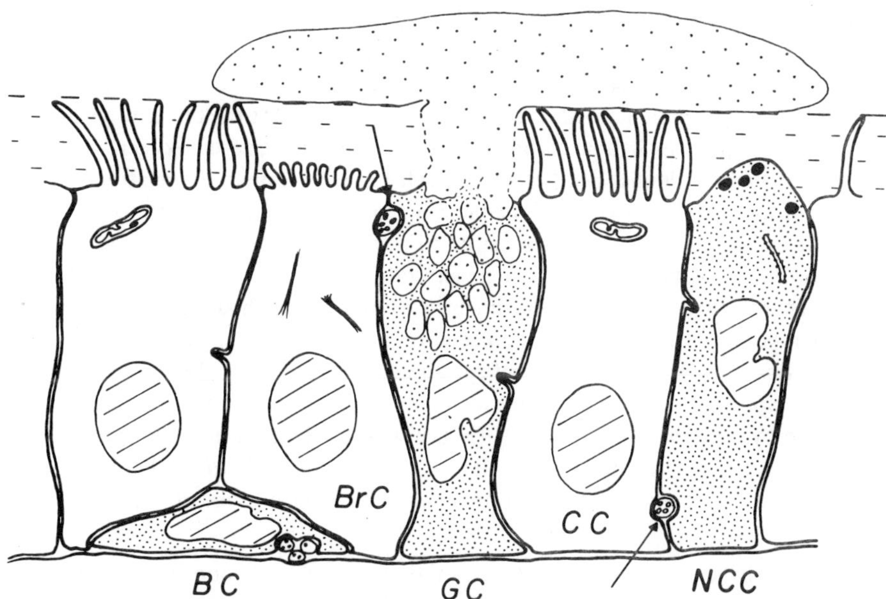

FIGURE 26 Diagrammatic representation of bronchial epithelium, periciliary layer and thick secretion. Ciliated cell (CC), goblet cell (GC), nonciliated cell (NCC), brush cell (Brc), basal cell (BC).

observed in airways obstruction were minimal. Brille, Hatzfeld, and Lehmann
[172] compared lung function tests and sputum volume in patients with bronchiectasis and found that the vital capacity increased after postural drainage,
the degree of improvement being inversely related to the volume of sputum expectorated. Similar findings have been reported in patients with cystic fibrosis
[173].

A. Two-Phase Gas Flow

When a mixture of gas and liquid flows through a tube, the gas-liquid mixture
may flow in a variety of different patterns, and it has been shown in experimental models that the type of two-phase flow most likely to occur in the bronchial
tree containing excessive bronchial secretion is annular [174]. The viscoelastic
properties of the liquid layer influence the resistance to airflow [175,176]. It
is tempting to extrapolate these findings in experimental models to human
studies.

B. Correlation Between Rheological Properties of Sputum and
Clinical Features

Rheological properties of sputum have been correlated with different aspects
of lung function, lung mechanics, distribution of ventilation, and gas exchange.
Polgar and Denton [173] suggested that the highly viscous mucus found in the
bronchial tree of patients with cystic fibrosis could account for the airways obstruction, increase in arterial carbon dioxide and reduced arterial oxygen reported in these patients. Feather and Russell [177] examined the relation
between sputum viscosity, and dynamic lung volumes in patients with cystic
fibrosis and found the sputum viscosity was negatively correlated with the
forced vital capacity (FVC) and forced expiratory volume in 1 sec (FEV_1). In
patients with chronic bronchitis or intrinsic asthma Chodosh and colleagues
[178] reported inverse significant correlation between vital capacity and airway
obstruction, and slowness of flow of sputum down an inclined glass. A relation
between apparent viscosity of sputum and airways obstruction has also been
reported by other workers [56,179] (Fig. 27). Gas exchange also seems to be
influenced by the viscoelastic properties of the sputum [56], the higher the
viscosity the more marked the abnormalities of gas exchange measured as
$PaCO_2$.

Rheological, chemical, and lung function studies carried out in a group
of chronic bronchitic patients [179] showed that apparent viscosity was closely
correlated with chemical constituents, particularly fucose, dry weight yield
and NANA, but only NANA content showed a significant inverse correlation
with airways obstruction measured as $FEV_1/FVC\%$ (Table 21; Fig. 28) suggest-

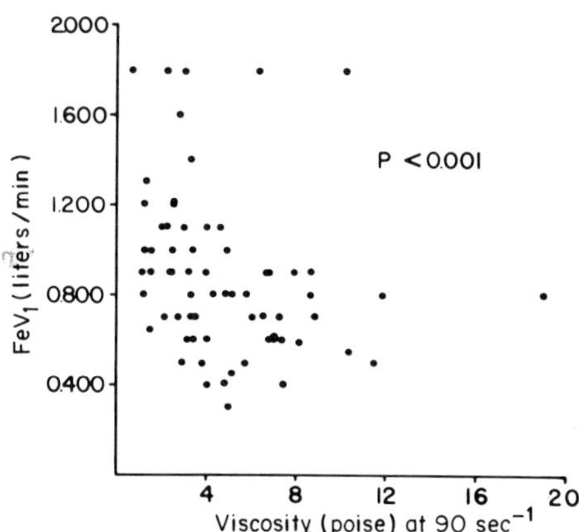

FIGURE 27 Relationship between forced expiratory volume in 1 sec and sputum viscosity.

TABLE 21 Linear Correlation Coefficients Between Airways Obstruction and Apparent Sputum Viscosity

	Viscosity (poise)
FVC	− 0.161
FEV$_1$	− 0.458[a]
FEV$_1$/FVC	− 0.298[b]

[a]$P < 0.001$.
[b]$P < 0.02$.

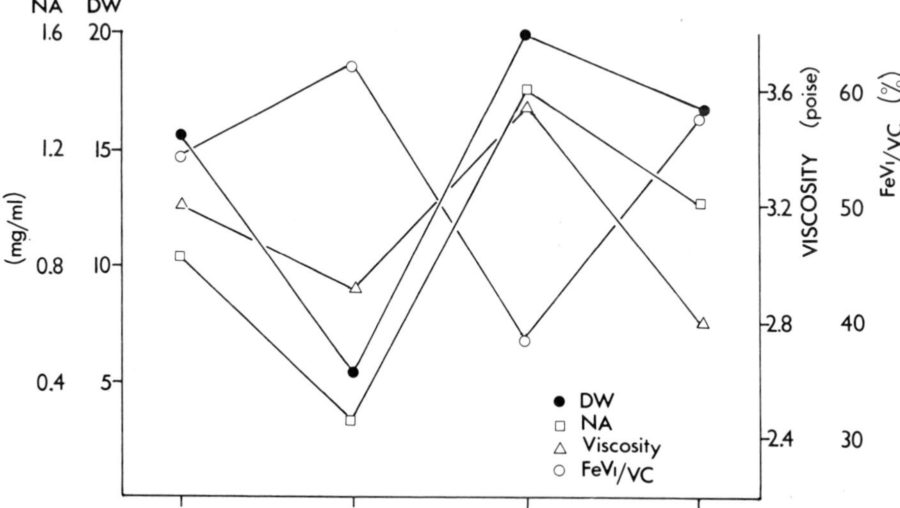

FIGURE 28 Variation in sputum viscosity with airways obstruction, dry weight yield, and neuraminic acid concentration of sputum.

ing "inflammation," probably from pollution rather than infection, since these patients clinically showed no evidence of an infective incident. Exacerbation of chronic bronchitis symptoms are rarely related to clear clinical evidence of infection [164].

References

1. L. Reid, Chronic Bronchitis—A report on mucus research, *Proc. R. Inst. Gt. Br.*, **43**:438-463 (1970).
2. L. Reid, Bronchial secretion. General comments, *Bull. Physiopathol. Respir. (Nancy)*, **9**:15-30 (1973).
3. L. Reid, Histopathological aspects of bronchial secretion, *Scand. J. Respir. Dis. [Suppl.]*, **90**:9-15 (1974).
4. L. Reid, Rheology—Relation to the composition of sputum, *Scand. J. Respir. Dis. [Suppl.]*, **190**:27-35 (1974).
5. L. Reid, An experimental study of hypersecretion of mucus in the bronchial tree, *Br. J. Exp. Pathol.*, **44**:437-445 (1973).
6. R. Jones, P. Bolduc, and L. Reid, Goblet cell glycoprotein and tracheal gland hypertrophy in rat airways: The effect of tobacco smoke with or without the anti-inflammatory agent phenylmethyl-oxadiazole, *Br. J. Exp. Pathol.*, **54**:229-239 (1973).
7. J. Sturgess and L. Reid, The effect of isoprenaline and pilocarpine on (a) bronchial mucus-secreting tissue and (b) pancreas, salivary glands, heart, thymus, liver and spleen, *Br. J. Exp. Pathol.*, **54**:388-403 (1973).
8. R. Jones, A. Baskerville, and L. Reid, Histochemical identification in pig bronchial epithelium (a) normal and (b) hypertrophied from enzootic pneumonia, *J. Pathol.*, **116**:1-11 (1975).
9. Thomas-Mawdesley-Thomas and P. Healey, Experimental bronchitis in lambs exposed to cigarette smoke, *Arch. Environ. Health*, **27**:248-250 (1973).
10. M. T. Lopez-Vidriero, I. Das, P. Smith, and L. Reid, Bronchial secretion from normal human airways after inhalation of prostaglandin F_{2a}, (Submitted for publication).
11. L. Reid, Bronchial mucus production in health and disease. In *The Lung*. International Academy of Pathology Monograph No. 8. Baltimore, Williams & Wilkins, 1967, Chap. 8.
12. S. A. Leach, Release and breakdown of sialic acid from human salivary mucin and its role in the formation of dental plaque, *Nature*, **199**:486-487 (1967).
13. J. Charman, M. T. Lopez-Vidriero, E. Keal, and L. Reid, The physical and chemical properties of bronchial secretion, *Br. J. Dis. Chest*, **68**:215-227 (1974).
14. S. G. Spiro, M. T. Lopez-Vidriero, J. Charman, I. Das, and L. Reid, Bronchorrhoea in a case of alveolar cell carcinoma, *J. Clin. Pathol.*, **28**:60-65 (1975).

15. R. A. Gibbons and F. A. Glover, The physicochemical properties of two mucoids from bovine cervical mucin, *Biochem. J.,* **73**:217-225 (1959).

16. P. Jeffery and L. Reid, The ultrastructure of normal large bronchi, *Bronches,* **23**:369-380 (1973).

17. D. Lamb and L. Reid, Histochemical types of acidic glycoprotein produced by mucous cells of the tracheobronchial glands in man, *J. Pathol.,* **98**:213-229 (1969).

18. D. Lamb and L. Reid, Quantitative distribution of various types of acid glycoprotein in mucous cells of human bronchi, *Histochem. J.,* **4**: 91-102 (1972).

19. R. Jones and L. Reid, The effect of pH on Alcian blue staining of epithelial acid glycoproteins. I. Sialomucins and sulphomucins (singly or in simple combination), *Histochem. J.,* **5**:9-18 (1973).

20. R. Jones and L. Reid, The effect of pH on Alcian blue staining of epithelial acid glycoproteins. II. Human submucosal gland, *Histochem. J.,* **5**: 19-27 (1973).

21. F. J. Martinez-Tello, D. G. Braun, and W. A. Blane, Immunoglobulin production in bronchial mucosa and bronchial lymph nodes particularly in cystic fibrosis of the pancreas, *J. Immunol.,* **101**:989-1003 (1968).

22. L. Bonomo, V. Liso, F. Tedesco, F. P. Schena, and V. Gillardi, Secretions of proteins into the sputum in chronic bronchitis, *Protides Biol. Fluids,* **16**:371-377 (1968).

23. P. Turgeon, J. Robert, and F. Turgeon, Les immunoglobulines secretories tracheo-bronchiques, *Union Med. Can.,* **98**:934-939 (1969).

24. B. Meyrick and L. Reid, Electron microscopic aspects of surfactant secretion, *Proc. R. Soc. Med.,* **66**:386-387 (1973).

25. A. H. Niden, Bronchiolar and large alveolar cell in pulmonary phospholipid metabolism, *Science,* **158**:1323-1324 (1967).

26. R. E. Pattle, Surface lining of lung alveoli, *Physiol. Rev.,* **45**:48-79 (1965).

27. H. Warembourg, R. Havez, S. Sezille, P. Scherpereel, P. Roussel, and P. Degand, Les lipides de l'expectoration, isolement et caracterisation du surfactant pulmonaire dans l'expectoration. Hypersecretion Bronchique Colloque International de Pathologie Thoracique. Lille, 1968, pp. 181-192.

28. M. E. Abrams, Isolation and quantitative estimation of pulmonary surface active lipoprotein, *J. Appl. Physiol.,* **21**:718-720 (1966).

29. T. E. Morgan, T. N. Finely, and H. Fialkow, Comparison of the composition and surface activity of "alveolar" and whole lung lipids in the dog, *Biochem. Biophys. Acta,* **106**:403-413 (1965).

30. J. Charman and L. Reid, The effect of freezing, storing and thawing on the viscosity of sputum, *Biorheology,* **10**:295-301 (1973).

31. G. Scott Blair, *Elementary Rheology,* London and New York, Academic Press, 1969, p. 1.

32. G. Blanshard, The viscometry of sputum, *Arch. Middlesex Hosp.,* **5**:222-241 (1955).

33. R. Denton, W. Forsman, S. H. Hwang, M. Litt, and C. E. Muller, Viscoelasticity of mucus, *Rev. Respir. Dis.,* **98**:380-391 (1968).

34.   A. J. Palfrey and J. B. White, The viscosity of synovial fluid during oscilla-
      tory movement, *Biorheology,* **5**:189-198 (1968).
35.   S. S. Davis and J. E. Dippy, The rheological properties of sputum, *Bio-
      rheology,* **6**:11-21 (1969).
36.   J. M. Sturgess, A. J. Palfrey, and L. Reid, Rheological properties of sputum,
      Rheol. Acta, **10**:36-43 (1971).
37.   J. R. van Wazer, J. W. Lyons, K. Y. Kim, and R. E. Colwell, *Viscosity and
      Flow Measurement, a Laboratory Handbook of Rheology,* New York,
      Interscience, 1966.
38.   M. Litt, Basic concepts of mucus rheology, *Bull. Physiopathol. Respir.
      (Nancy),* **9**:33-46 (1973).
39.   H. L. Gabelnick and M. Litt, *Rheology of Biological Systems.* Edited by
      H. L. Gabelnick and M. Litt, Springfield, Ill., Charles C Thomas, 1973.
40.   G. W. Scott-Blair, *An Introduction to Biorheology.* Amsterdam, Elsevier,
      1974.
41.   G. W. Scott-Blair and F. Glover, More early pregnancy tests from studies
      of bovine cervical mucus, *Br. Vet. J.,* **113**:417-423 (1957).
42.   P. C. Elmes and J. C. White, Deoxyribonuclease in the treatment of puru-
      lent bronchitis, *Thorax,* **8**:295-300 (1953).
43.   S. R. Hirsh and L. H. Hamilton, Evaluation of changes in sputum consis-
      tency with a new instrument, *Am. Rev. Respir. Dis.,* **94**:784-789 (1966).
44.   S. Chodosh, T. C. Medici, and K. Enslein, Comparison of five methods for
      measuring sputum physical characteristics, *Bull. Physiopathol. Respir.,* **9**:
      127-139 (1973).
45.   E. Keal, Methodes d'etude des modifications de la secretion bronchique
      et de la viscosite, *Poumon Coeur,* **26**:52-58 (1970).
46.   D. Volker, A new method for measuring sputum viscosity with a capillary
      flow viscometer, *Bull. Physiopathol. Respir. (Nancy),* **9**:103-111 (1973).
47.   M. J. Dulfano, Viscoelastic studies using capillary methods, *Bull. Physio-
      pathol. Respir. (Nancy),* **9**:91-101 (1973).
48.   E. Puchelle, A. M. Benis, and J. M. Zahm, Measurement of the viscoelastic
      properties of sputum with a concentric-cylinder rheometer, *Bull. Physio-
      pathol. Respir. (Nancy),* **9**:113-126 (1973).
49.   K. N. V. Palmer, D. Ballantyne, M. L. Diament, and W. F. D. Hamilton,
      The rheology of bronchitic sputum, *Br. J. Dis. Chest,* **64**:185-191 (1970).
50.   E. E. Keal, Biochemistry and rheology of sputum in asthma, *Postgrad.
      Med. J.,* **47**:171-177 (1971).
51.   S. S. Davis, B. Warburton, and J. E. Dippy, The chemical balance as a
      rheometer for biological fluids, *J. Pharm. Sci.,* **59**:1764-1769 (1970).
52.   A. M. Benis, E. Puchelle, and P. Sadoul, Adaption of a concentric cylinder
      viscometer to the routine measurement of the viscoelastic properties of
      sputum. In *Rheology of Biological Systems.* Edited by H. L. Gabelnick
      and M. Litt. Springfield, Ill., Charles C Thomas, 1973, pp. 218-259.
53.   J. Charman and L. Reid, Sputum viscosity in chronic bronchitis, bronchi-
      ectasis, asthma and cystic fibrosis, *Biorheology,* **9**:185-199 (1972).

54. J. M. Sturgess, The control of the bronchial mucous glands and their secretion. Ph.D. Thesis, London University, 1970.

55. M. J. Dulfano, *Sputum.* Edited by M. J. Dulfano. Springfield, Ill., Charles C Thomas, 1973, pp. 201-242.

56. Q. T. Pham, R. Peslin, E. Puchelle, D. Salmon, G. Caraux, and A. M. Benis, Respiratory function and the rheological status of bronchial secretions collected by spontaneous expectoration and after physiotherapy, *Bull. Physiopathol. Respir. (Nancy),* **9**:293-314 (1973).

57. P. Mitchell-Heggs, A. J. Palfrey, and J. Charman, Viscosity of normal synovial fluid. *Abst VII.* European Congress of Rheumathology, 1971.

58. J. M. Sturgess, A. J. Palfrey, and L. Reid, The viscosity of bronchial secretion, *Clin. Sci.,* **38**:145-156 (1970).

59. P. F. Mitchell-Heggs, The rheological properties of sputum at low shear rates, Ph.D. Thesis, London University, 1974.

60. S. S. Davis, Techniques for the measurement of rheological properties of sputum, *Bull. Physiopathol. Respir. (Nancy),* **9**:127-139 (1973).

61. P. Mitchell-Heggs, A. J. Palfrey, and L. Reid, The elasticity of sputum at low shear rates, *Biorheology,* **11**:417-426 (1974).

62. A. G. Ogilvie, C. Strong, P. O. Leggat, and D. J. Newell, *A Ten-Year Prospective Study of Chronic Bronchitis in the North-East of England,* London, Churchill, 1973.

63. G. P. Roberts, Isolation and characterization of glycoprotein from sputum, *Eur. J. Biochem.,* **50**:265-280 (1974).

64. W. Pigman and A. Gottschalk, Submaxillary gland glycoproteins. In *Glycoproteins. Their Composition, Structure and Function.* Edited by A. Gottschalk, Amsterdam, Elsevier, 1966, pp. 434-445.

65. K. R. Bhaskar, J. M. Creeth, I. Das, M. T. Lopez-Vidriero, and L. Reid, Isolation and characterisation of glycoproteins from sputum. *International Symposium of Glycoconjugates: Function in Animals.* Abstracts, p. 37, 1975.

66. J. Lieberman and N. B. Kurnick, The induction of proteolysis in purulent sputum by iodides, *J. Clin. Invest.,* **43**:1892-1904 (1964).

67. S. R. Hirsch, J. E. Zastrow, and R. C. Kory, Sputum liquefying agents: A comparative in vitro evaluation, *J. Lab. Clin. Med.,* **74**:346-353 (1969).

68. C. H. W. Hirs, Performic acid oxidation, *Methods Enzymol.,* **11**:197-199 (1967).

69. L. W. Mathews, S. Spector, J. Lemm, and J. Potter, Studies on pulmonary secretions. The overall chemical composition of pulmonary secretions from patients with cystic fibrosis, bronchiectasis and laryngectomy, *Am. Rev. Respir. Dis.,* **88**:199-205 (1963).

70. T. F. Boat and L. W. Mathews, Chemical composition of human tracheobronchial secretions. In *Sputum, Fundamentals and Clinical Pathology.* Edited by M. J. Dulfano. Springfield, Ill., Charles C Thomas, 1973, Chap. 7.

71. P. L. Masson and J. F. Hermans, Sputum proteins. In *Sputum, Fundamentals and Clinical Pathology.* Edited by M. J. Dulfano. Springfield, Ill., Charles C Thomas, 1973, pp. 413-475.

72. J. L. Potter, S. Spector, L. Mathews, and J. Lemm, Studies on pulmonary secretions. III. The nucleic acid in whole pulmonary secretions from patients with cystic fibrosis, bronchiectasis and laryngectomy, *Am. Rev. Respir. Dis.,* **99**:909-916 (1969).

73. H. C. Ryley and T. D. Brogan, Variation in the composition of sputum in chronic chest diseases, *J. Exp. Pathol.,* **49**:625-633 (1968).

74. H. C. Ryley and T. D. Brogan, Quantitative immunoelectrophoretic analysis of the plasma proteins in the sol phase of sputum from patients with chronic bronchitis, *J. Clin. Pathol.,* **26**:852-856 (1973).

75. O. H. Lowry, N. J. Rosebrough, A. L. Farr, and R. J. Randall, Protein measurement with the Folin phenol reagent, *J. Biol. Chem.,* **193**:265-275 (1951).

76. B. T. Dammas, W. A. Watson, and H. G. Biggs. Albumin standards and the measurement of serum albumin with bromocresol green, *Clin. Chem. Acta,* **31**:87-97 (1971).

77. P. Degand, P. Roussel, A. Lamblin, and R. Havez, Purification et etude des mucines de cystes bronchogeniques, *Biochim. Biophys. Acta,* **320**:318-330 (1973).

78. J. T. Gallagher and P. W. Kent, Structure and metabolism of glycoproteins and glycosaminoglycans secreted by organ cultures of rabbit trachea, *Biochem. J.,* **148**:187-196 (1975).

79. L. E. Warfvinge, Paper electrophoresis of sputum in bronchiolar carcinoma (pulmonary adenomatosis), *Acta Med. Scand.,* **153**:49-52 (1955).

80. R. D. Brogan, The high molecular weight components of sputum, *Br. J. Exp. Pathol.,* **41**:279-288 (1960).

81. S. C. Bukantz and A. W. Berns, Studies with sputum. III. Paper electrophoresis and gel diffusion of sputum in the diagnosis of pulmonary carcinomatosis. Identification of the fast moving component as albumin, *Am. Rev. Respir. Dis.,* **85**:351-363 (1962).

82. C. Gernez-Rieux, G. Biserte, R. Havez, . Voisin, and R. Cuvelier, Etude biochémique de l'expectoration bronchique, *Pathol. Biol.,* **11**:729-741 (1963).

83. R. Nicholas, Kliniseh-experimentelle Sputumuntersuchungin bei chronischer Bronchitis, *Med. Thorac.,* **21**:223-252 (1964).

84. H. C. Ryley, An electrophoretic study of the sol phase of sputum, *J. Lab. Clin. Med.,* **75**:382-390 (1970).

85. M. Z. Atassi, S. A. Barker, L. E. Houghton, and K. G. Mullard, Mucoproteins of bronchial mucins, *Nature,* **192**:1269-1270 (1961).

86. M. Z. Atassi, S. A. Barker, and M. Stacey, Neuraminic acid and its relation to chronic bronchitis. V. Glass column electrophoresis of sputum, *Clin. Chem. Acta,* **7**:706-709 (1962).

87. G. Biserte, R. Havez, and R. Cuvelier, Glycoproteids des secretions bronchiques, *Exposes Annu. Biochim. Med.,* **24**:85-120 (1963).

88. P. L. Masson, J. F. Heremans, and J. Prignot, Studies on the protein of human bronchial secretions, *Biochem. Biophys. Acta,* **111**:466-478 (1965).

89. H. Gotz and M. Fischer, Verhalten der elektrophoretisch, biochemisch und immunologisch definierbaren Proteine und Sputums unter Sekretolyse, *Clin. Chim. Acta,* **30**:53-64 (1970).

90. M. Z. Atassi, S. A. Barker, and M. Stacey, Neuraminic acid and its relation to chronic bronchitis. IV. Isolation of homogeneous mucoproteins, *Clin. Chim. Acta,* **7**:588-591 (1962).

91. R. Havez, P. Roussel, P. Degand, A. Randoux, and G. Biserte, Biochemical exploration of bronchial hypersecretion, *Protides Biol. Fluids,* **16**:343-360 (1968).

92. R. Havez and G. Biserte, Etude biochimique des secretions bronchiques. In *Hypersecretion Bronchique.* Colloque International de Pathologie Thoracique, Lille, 1968, pp. 43-67.

93. J. M. Creeth and M. A. Denborough, The use of equilibrium-density-gradient methods for the preparation and characterization of blood group specific glycoprotein, *Biochem. J.,* **117**:879-891 (1970).

94. R. G. Spiro, Glycoproteins, *Adv. Protein Chem.,* **00**:419-424 (1973).

95. J. R. Clamp, Analysis of glycoproteins. In *The Metabolism and Function of Glycoproteins.* Edited by R. M. S. Smellie and J. G. Beeley. London, The Biochemical Society, 1974, pp. 3-16.

96. I. Das, G. Laballeur, R. Picot, and L. Reid, Determination of carbohydrates in bronchial secretion by gas-liquid chromatography. (In preparation).

97. L. Bonomo and A. D'Addabbo, ^{131}I-Albumin turnover and loss of protein into the sputum in chronic bronchitis, *Clin. Chim. Acta,* **10**:214-222 (1964).

98. T. D. Brogan, H. C. Ryley, L. Neale, and J. Yassa, Soluble proteins of bronchopulmonary secretions from patients with cystic fibrosis, asthma and bronchitis, *Thorax,* **36**:72-79 (1975).

99. H. E. Schultze and J. F. Heremans, *Molecular Biology of Human Proteins,* Vol. 1, Amsterdam, Elsevier, 1966, pp. 816-831.

100. H. V. Burgi, U. Wiesmann, R. Richterich, J. Regli, and T. Medici, New objective criteria for inflammation of bronchial secretion, *Br. Med. J.,* **2**:654-656 (1968).

101. S. C. Bukantz and A. W. Berns, Studies with sputum: I. Initial observations on the chemical nature and blood group substances of asthmatic sputum, *J. Allergy,* **29**:29-43 (1958).

102. R. Havez, P. Roussel, P. Degand, Y. Delmas-Marsalit, and G. Biserte, Etude de substances de groupe sanguin A isolees du mucus bronchique, *Bull. Soc. Chim. Biol. (Paris),* **51**:245-259 (1969).

103. Z. Dische, Reciprocal relationship between fucose and sialic acid in mammalian glycoproteins, *Ann. N.Y. Acad. Sci.,* **106**:259-270 (1963).

104. P. W. Kent, J. P. Ackers, and J. C. Marsden, Structural studies on the principal glycoprotein of the sheep colonic epithelium, *Biochem. J.,* **105**:24 (1967).

105. B. L. Slomiany and K. Meyer, Isolation and structural studies of sulphated glycoproteins of hog gastric mucosa, *J. Biol. Chem.,* **247**:5062-5070 (1972).

106.   T. F. Boat, R. N. Iyer, M. N. Macintyre, D. M. Carlson, and L. W. Matthews, Evaluation of a culture method for study of human tracheobronchial secretions, *Am. Rev. Respir. Dis.*, **103**:915 (1971).

107.   B. J. Starkey, D. Snary, and A. Allen, Characterization of gastric mucoproteins isolated by equilibrium density-gradient centrifugation in caesium chloride, *Biochem. J.*, **141**:633-639 (1974).

108.   R. Havez, P. Roussel, P. Degand, and G. Biserte, Etude des structures fibrillaires de la secretion bronchique humane, *Clin. Chim. Acta*, **17**:281-295 (1967).

109.   T. Anzae, J. Ibayashi, C. M. Carpenter, and L. Hyde, Beta A-globulin as a molecular constituent of insoluble bronchial mucus gel, *Am. Rev. Respir. Dis.*, **88**:503-508 (1963).

110.   W. Pigman and Y. Ashimoto, Composition of bovine submaxillary mucins, *Biochem. Biophys. Acta*, **69**:579-580 (1963).

111.   C. Gernez-Rieux, G. Biserte, R. Havez, C. Voisin, P. Roussel, and P. Degand, Etude de l'activite in vitro de differents agents reduisant la viscosité de l'expectoration, *Acta Tuberc. Pneumol. Belg.*, **55**:138-164 (1964).

112.   A. Tettamanti and W. Pigman, Purification and characterization of bovine and ovine submaxillary mucins, *Arch. Biochem. Biophys.*, **134**:41-50 (1968).

113.   M. De Salequi and H. Plonska, Preparation and properties of porcine submaxillary mucins, *Arch. Biochem. Biophys.*, **129**:49-56 (1969).

114.   R. A. Gibbons, The composition of mucus with special reference to its rheological properties, *Protides Biol. Fluids*, **16**:299-305 (1966).

115.   D. S. Robinson and J. B. Monsey, Studies on the composition of egg-white ovomucin, *Biochem. J.*, **121**:537-547 (1971).

116.   R. A. Gibbons, Physio-chemical methods for the determination of the purity, molecular size and shape of glycoproteins. In *Glycoproteins, Their Composition, Structure and Function.* Edited by A. Gottschalk. Amsterdam Elsevier, 1966, pp. 29-95.

117.   H. Bürgi, Fibre systems in sputum, *Bull. Physiopathol. Respir. (Nancy)*, **9**:191-196 (1973).

118.   H. Bürgi, Die Viskositat des purulenten und sterilen Sputums bei chronischer Asthmabronchitis, *Med. Thorac.*, **21**:156-167 (1964).

119.   K. Adler, O. Wooten, W. Philippoff, E. Lerner, and M. J. Dulfano, Physical properties of sputum III. Rheologic variability and intrinsic relationships, *Am. Rev. Respir. Dis.*, **106**:86-96 (1972).

120.   H. Bürgi, U. Wiesmann, R. Richterich, J. Regli, and T. Medici, New objective criteria for inflammation in bronchial secretion, *Br. Med. J.*, **2**:654-656 (1968).

121.   A. Gottschalk, Correlation between composition, structure, shape and function of a salivary mucoprotein, *Nature*, **186**:949-951 (1960).

122.   P. C. Elmes, A. A. C. Dutton, and C. M. Fletcher, Sputum examination and the investigation of chronic bronchitis, *Lancet*, **1**:1241-1244 (1959).

123. D. L. Miller and R. Jones, A study of techniques for the examination of sputum in a field survey of chronic bronchitis, *Rev. Respir. Dis.,* **88**: 473-483 (1963).

124. R. J. May, Pathogenic bacteria in chronic bronchitis, *Lancet,* 2:839-842 (1954).

125. R. J. May and D. S. May, Bacteriology of sputum in chronic bronchitis, *Tubercle,* **44**:162-173 (1963).

126. R. J. May, Chemotherapy of chronic bronchitis and allied disorders. Edited by D. Taverner and J. Trounce. London, English University Press, 1968.

127. R. Picot, I. Das,and L. Reid, Rheological properties of purulent sputum: Correlation with total DNA content and DNP fibres. (In Preparation).

128. D. Lamb and L. Reid, Goblet cell increase in rat bronchial epithelium after exposure to cigarette and cigar tobacco smoke, *Br. Med. J.,* 1: 33-35 (1969).

129. D. Lamb and L. Reid, Histochemical and autoradiographic investigation of the serous cell of human tracheo-bronchial tree, *J. Pathol.,* **100**:127- 138 (1970).

130. D. Lamb and L. Reid, The tracheobronchial submucosal glands in cystic fibrosis: A qualitative and quantitative histochemical study, *Br. J. Dis. Chest,* **66**:239-247 (1972).

131. R. Havez, A. Laine-Bassez, A. Hyem-Levy, and J. Lebas, Definition bio- chimique des proteines dans l'expectoration, *Bull. Physiopathol. Respir. (Nancy),* **9**:219-235 (1973).

132. T. Ashcroft, Daily variation in sputum volume in chronic bronchitis, *Br. Med. J.,* 1:288-290 (1965).

133. M. T. Lopez-Vidriero, J. Charman, E. Keal, D. J. de Silva, and L. Reid, Sputum viscosity: Correlation with chemical and clinical features in chronic bronchitis, *Thorax,* **28**:401-408 (1973).

134. W. S. Chernick and G. J. Barbero, Composition of tracheobronchial secretions in cystic fibrosis of the pancreas and bronchiectasis, *Pediatrics,* **24**:739-745 (1959).

135. T. D. Brogan, The high molecular weight components of sputum, *Br. J. Exp. Pathol.,* **41**:288-297 (1960).

136. J. L. Potter, L. Matthews, J. Lemm, and S. Spector, Human pulmonary secretions in health and disease, *Ann. N.Y. Acad. Sci.,* **106**:692-697 (1963).

137. J. L. Potter, L. W. Matthews, S. Spector, and J. Lemm, Studies on pul- monary secretions: II. Osmolality and the ionic environment or pul- monary secretions from patients with cystic fibrosis, bronchiectasis and laryngectomy, *Am. Rev. Respir. Dis.,* **96**:83-87 (1967).

138. M. T. Lopez-Vidriero and L. Reid, Sputum viscosity: Correlation with chemical constituents in chronic bronchitis, asthma, bronchiectasis and cystic fibrosis (In preparation).

139. E. Steinmann, La secretion bronchique et le pH, *Bronches,* **6**:126-129 (1956).

140. F. Guerrin, C. Voisin, V. Macquet, H. Hobin, F. Wattel, and P. Boulanger, Resultats de la pH metrie in situ Hypersecretion bronchique. Methods nouvelles d'exploration. Incidences therapeutiques. Colloque International de Pathologie Thoracique. Lille, 1968, pp. 249-256.

141. F. P. Basch, P. Holinger, H. Poncher, Physical properties of sputum, *J. Dis. Child.*, **62**:981-990 (1941).

142. H. Kwart, W. W. Moseley, and M. Katz, The chemical characterization of human tracheobronchial secretions: A possible clue to the origin of fibrocystic mucus, *Ann. N.Y. Acad. Sci.*, **106**:709-721 (1963).

143. E. E. Keal and L. Reid, Neuraminic acid content of sputum in chronic bronchitis, *Thorax*, **27**:643-653 (1972).

144. M. Z. Atassi, S. A. Barker, L. E. Houghton, and M. Stacey, Neuraminic acid and its relationship to chronic bronchitis, *Clin. Chim. Acta*, **4**:741-747 (1959).

145. R. Lev and S. S. Spicer, A histochemical composition of human epithelial mucins in normal and hypersecretory states including pancreatic cystic fibrosis, *Am. J. Pathol.*, **46**:23-47 (1965).

146. T. Medici and H. Bürgi, The role of immunoglobulin in endogenous bronchial defense mechanisms in chronic bronchitis, *Am. Rev. Respir. Dis.*, **103**:784-791 (1971).

147. E. Puchelle, J. M. Zahm, and R. Havez, Relation des proteines et mucines bronchiques avec les proprietes rheologiques, *Bull. Physiopathol. Respir. (Nancy)*, **9**:237-256 (1973).

148. M. T. Lopez-Vidriero, Variation in neuraminic acid and fucose content of sputum in disease and during treatment. Ph.D. Thesis, London University (1976).

149. H. Hoffman and H. Ebelt, Der elektrlytgehalt des Sputums und des Serums bei Patienten mit Asthma bronchiale und bei Patienten mit chronischer Bronchitis, *Allerg. Asthma forsch.*, **14**:227-239 (1968).

150. L. Reid and R. de Haller, The bronchial mucous glands—their hypertrophy and change in intracellular mucus. 4th International Converence on Cystic Fibrosis of the Pancreas (Mucoviscidosis), Berne/Grindelwald 1966, Part I. *Mod. Probl. Pediat.*, **10**:195-199 (1967).

151. D. Lamb and L. Reid, Mitotic rates, goblet cell increase and histochemical changes in rat bronchial epithelium during exposure to sulphur dioxide, *J. Pathol.*, **96**:87-111 (1968).

152. F. E. Speizer, R. Doll, and P. Heaf, Observation on the recent increase in mortality from asthma. *Br. Med. J.*, **1**:335-339 (1968).

153. T. D. Brogan, H. C. Ryley, L. Allen, and H. Hutt, Relation between sputum sol phase and diagnosis in chronic diseases, *Tubercle*, **26**:418-423 (1971).

154. J. Salvaggio, M. Lopez, P. Arquembourg, R. W. Waldman, M. Sly, Salivary nasal wash and sputum IgA concentrations in atopic and non atopic individuals, *J. Allergy Clin. Immunol.*, **51**:335-343 (1973).

155. M. T. Lopez-Vidriero, J. Charman, E. Keal, and L. Reid, Bronchorrhoea, *Thorax*, **30**:624-630 (1975).

156. S. Gee, On the causes and forms of bronchitis. In *Medical Lectures and Aphorisms*, London, Smith Elder, 1902.

157. E. H. Wood, Jr., Unusual case of carcinoma of both lungs, *Radiology*, **40**:193-195 (1943).

158. R. Kennamer, Pulmonary adenomatosis, *JAMA*, **145**:815-818 (1951).

159. C. F. Storey, K. P. Knudson, and B. J. Lawrence, Bronchiolar carcinoma of the lung, *J. Thorax Surg.*, **26**:331-406 (1953).

160. G. S. Schools and E. Ray, A case report of bronchoscopic observation of severe bronchorrhoea, *Chest*, **39**:643-645 (1965).

161. R. Kourilsky, Recherches recentes sur les bronchorrhees chroniques, *Presse Thermale Climat.*, **97**:218-226 (1960).

162. A. Calim, Bronchorrhoea, *Br. Med. J.*, **4**:274-275 (1972).

163. P. H. S. Hartley and I. J. Davis, A case of pituitous catarrh, *Br. Med. J.*, **1**:1052 (1923).

164. E. E. Keal, The neuraminic acid content of sputum, its variation in disease and contribution to the physical properties. M D Thesis, London University (1971).

165. R. A. Gibbons, Chemical properties of two mucoids from bovine cervical mucin, *Biochem. J.*, **73**:209-217 (1959).

166. L. Odin, Mucopolysaccharides of epithelial mucus. Ciba Foundation Symposium on the Chemistry and Biology of Mucopolysaccharides. London, Churchill, 1958, p. 234.

167. R. Munies, T. C. Grubb, and R. E. Caliari, Relationship between sputum viscosity and total sialic acid content, *J. Pharm. Sci.*, **57**:825-827 (1968).

168. E. L. Uhlenhopp, Rheopexy of denatured bacterial DNA solutions, *Biorheology*, **12**:137-142 (1972).

169. T. Dalhamn, Mucous flow and ciliary activity in the respiratory tract of healthy rats and rats exposed to respiratory irritant gases, *Acta Physiol. Scand. [Suppl. 123]*, **36**:1-161 (1956).

170. K. H. Kilburn, A hypothesis for pulmonary clearance and implications, *Am. Rev. Respir. Dis.*, **98**:449-463 (1968).

171. J. C. Hogg, P. T. Macklem, and W. M. Thurlbeck, Site and nature of airway obstruction in obstructive lung disease, *N. Engl. J. Med.*, **278**:1355 (1968).

172. D. Brille, C. Hatzfeld, and M. G. Lehman, Fonction respiratoire et suppurations bronchiques, *J. Fr. Med. Chir. Thorac.*, **6**:556-570 (1952).

173. G. Polgar and R. Denton, Cystic fibrosis in adults. Studies of pulmonary function and some physical properties of bronchial mucus, *Am. Rev. Respir. Dis.*, **85**:319-327 (1972).

174. J. G. Jones, S. W. Clarke, and D. R. Oliver, Two-phase gas-liquid flow in airways, *Br. J. Anaesth.*, **41**:192-193 (1969).

175. S. W. Clarke, J. G. Jones, and D. R. Oliver, Resistance to two-phase gas-liquid flow in airways, *J. Appl. Physiol.*, **29**:464-471 (1970).

176. S. W. Clarke, The role of two-phase flow in bronchial clearance, *Bull. Physiopathol. Respir. (Nancy)*, **9**:359-372 (1973).

177. E. A. Feather and G. Russel, Sputum viscosity and pulmonary function in cystic fibrosis, *Arch. Dis. Child.*, **45**:807-808 (1970).

178.	S. Chodosh, T. C. Medici, S. Ishikawa, and K. Enslein, Relationship of physical sputum characteristics and ventilatory capacity in chronic bronchial disease, *Chest,* **63**, Suppl. (1973).
179.	M. T. Lopez-Vidriero, Individual and group correlations of sputum viscosity and airways obstruction, *Bull. Physiopathol. Respir. (Nancy),* **9**:339-346 (1973).

10

Physiological and Pharmacological Control of Airways Secretions

E. E. KEAL

Brompton Hospital
Cardiothoracic Institute
London, England

I. Introduction

In choosing the term "airways secretions" the editors will presumably have in mind the products of those specialized secretory structures of the respiratory mucous membrane—the submucosal glands and goblet cells. While this is a logical progression from the anatomical studies of these structures it is a regrettable fact that, in the intact animal, their separate products are rarely available for study in pure form to assess physiological or drug-induced changes in function. It is hoped that the recently introduced techniques of fiberoptic bronchoscopy will permit entry to one more orifice—the duct of single mucous glands.

Many authors have recognized the complexity of airways secretion, bronchial fluid or sputum (Fig. 1) and have given way to frustrated comments such as those of John Forbes [1] in 1834: "It is, indeed, surprising to observe how little attention is paid by most of our practical authors to this most important symptom, while every trifling and insignificant variety of pulse, tongue, stools, etc. is dwelt on with tiresome minuteness."; and Gunn [2] in 1927 "If it were as easy to determine the quality of bronchial secretion as it is to measure the amount of urine, there would be by this time no such vagueness as still exists."

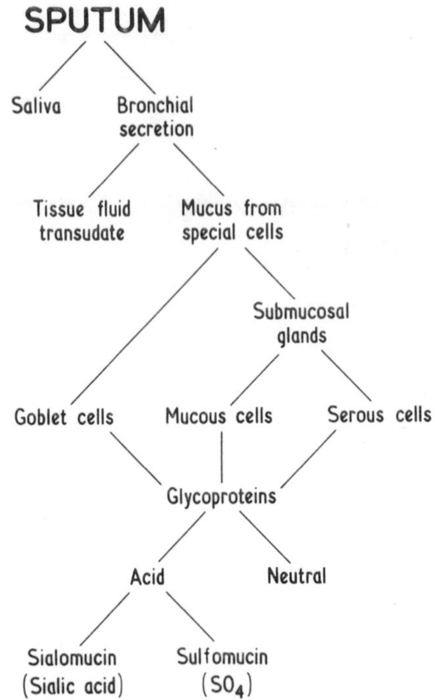

FIGURE 1 Simplified schema of some sputum constituents.

At best, the material available for study from the intact animal is respiratory tract fluid or, more precisely, bronchial fluid since it is usually collected from the major airways. In these circumstances it is well to remember that the bronchial mucosa may have a resorptive as well as a secretory function. Boyd and Shepherd [3] showed that a tracheal pocket formed in front of a small cannula ligated into the trachea could absorb half or more of the volume of respiratory tract fluid reaching it.

Bronchial fluid is a mixture of goblet cell and mucous gland secretions but also contains the secretions of nonspecialized cells and tissue fluid transudate or exudate. In most cases, however, this fluid must be obtained under nonphysiological conditions. Patients undergoing bronchoscopy are usually premedicated with atropine or scopolamine and the volume of normal mucus obtained is usually so small that it has to be flushed through the sucker with the loss of a volumetric baseline. Larger volumes may be obtained via tracheostomy, but in these cases the protective function of the upper respiratory tract will have been bypassed and the response of the bronchial mucosa will be largely dependent on the control of the inspired air. In animal studies, bronchial fluid may be studied through tracheal windows or collected from tracheal pouches; the same problems arise as with tracheostomy collection of fluid and allowance must be made for species differences in the relative numbers of goblet cells and mucous glands and in the types of glycoprotein they contain. Current studies take advantage of these species differences in that the nervous and pharmacological control of mucous glands can be studied in animals having no goblet cells. Conversely, goblet cell control may be studied in birds (such as the goose) which appear to have no submucosal glands.

It has been estimated that in humans the volume of the submucosal glands is some 40 times that of the goblet cells, but, while the former may contribute the greater part of the bronchial fluid or sputum available for study, the lesser amounts of goblet cell secretion in the peripheral airways may be of greater physiological importance. This is particularly so in disease states where, for example, the hypersecretion of chronic bronchitis is associated with a tremendous goblet cell metaplasia in smaller airways. Assessment of drug activity based on the volume of fluid produced will tell us nothing of its effect on the target organ —gland or goblet cell—or whether its effects are due to alteration of tissue fluid exudate or transudate. Characterization of bronchial fluid by chemical analysis must also take into account possible differences in source before any change is attributed to inherent abnormality of mucus synthesis in any specific disease, or to its change by drug therapy.

Material available for study in man is termed "sputum" and its definition will never be exact. The normal volume of respiratory tract fluid is estimated to be 0.1-0.3 ml/kg body weight, or about 10 ml/day in laryngectomized patients [4]. Boyd and Ronan [5] collected 600 ml/24 hr in animals. The more commonly quoted figure of 100 ml [6] was derived from studies of lightly anesthetized animals held in a head-down position to allow secretions to drain from the mouth. Such amounts are presumably swallowed and do not give rise to "normal" expectoration. Sputum consists of bronchial fluid contaminated by cellular debris, the products of infection and by upper respiratory tract secretions. The latter includes saliva (indeterminate amounts) which contains enzymes able to degrade the glycoproteins of the bronchial fluid. A neuraminidase from the bacterial flora of the mouth may split off neuraminic acid from the glycoprotein molecule, and this may be lost in dialysis prior to estimation [7,8]. Different diseases will make different contributions to sputum: mucous gland and goblet cell secretion in simple chronic bronchitis; tissue fluid transudate in allergic and inflammatory states; DNA in the presence of infection. Like must be compared with like, mucoid with mucoid, and purulent with purulent before any biochemical or rheological changes can be attributed to specific pathological states or to the effect of drugs.

The effect of pharmacological agents may be to increase or decrease the volume of sputum or to change its biochemical constituents or its rheological

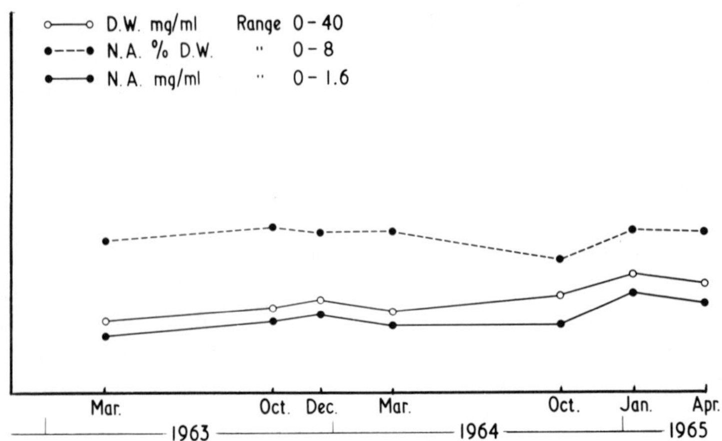

FIGURE 2 The dry weight and neuraminic acid content of sputum from a patient with chronic bronchitis to illustrate the consistent results found in many cases over a period of three winters.

behavior. These factors may be studied in short- or long-term experiments, but in either case naturally occurring variations must be considered. In some chronic bronchitics, the dry weight and neuraminic acid content of the sputum remains remarkably constant when estimated at intervals over a period of 3 years (Fig. 2). In others there is a seasonal variation (Fig. 3), which perhaps is induced by change in climatic conditions or atmospheric pollutants (Fig. 4) but is independent of acute bacterial infection [9]. Similar spontaneous biochemical variation has been shown and correlated with measurements of sputum viscosity and of airways obstruction in stable bronchitics studied at monthly intervals throughout one winter [10].

A. Physiology vs. Pharmacology

The physiological control of airway secretion is inseparable from the pharmacological effect of drugs to the extent that naturally occurring chemical mediators are sometimes used. These same substances are also involved in the maintenance of bronchial muscle tone, so that in many functional studies of the intact airway the effects are inseparable. For example, we cannot tell whether the mucus plugging and cast formation in the smaller bronchi in status asthmaticus is due to an excess of normal secretion, an abnormal secretion, or simply due to the entrapment of normal volumes of normal secretion by intense bronchoconstriction. Experimental studies have shown that mild shearing of mucus followed by stasis causes an 80-fold increase in apparent viscosity at low shear rates—a situation which may be present in asthma [11].

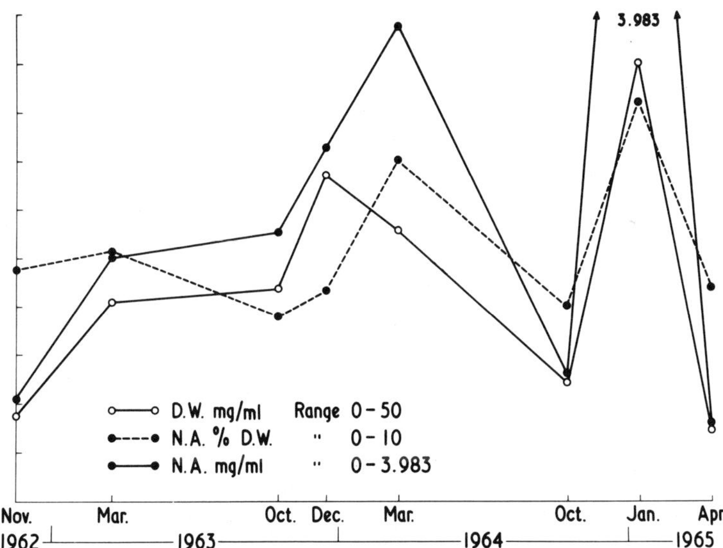

FIGURE 3 The dry weight and neuraminic acid content of sputum from a patient with chronic bronchitis to illustrate the seasonal variation found in 16 of 48 patients studied over a period of three winters. The winter peaks were not associated with acute bacterial infections and not necessarily with exacerbations of bronchitis.

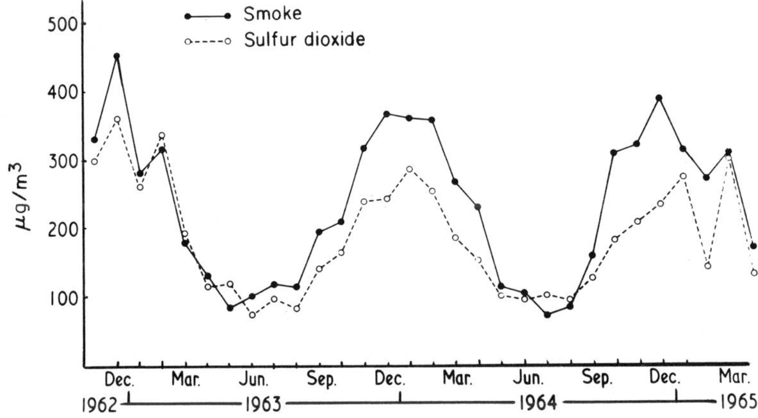

FIGURE 4 The smoke and sulfur dioxide concentrations in the atmosphere during the studies referred to in Figs. 2 and 3.

B.   Methods of Study of Airways Secretion

During the past 15 years, techniques have been developed to help overcome many of the difficulties in understanding secretory control. These will include:

1.   Morphological and histochemical studies of bronchial mucus glands and goblet cells, the changes in disease and the response to therapy

2.   Organ culture of bronchial mucosa to determine the secretory activity of mucosal glands and goblet cells and its alteration by the addition of pharmacological agents to the culture medium

3.   Electron microscopy of individual secretory cells with quantitation of the uptake and throughput of metabolites, particularly glucose (oligosaccharide synthesis) and threonine (incorporation into the peptide chain of the glycoprotein [12,13]).

4.   Direct observation of mucus formation on the exposed mucosa and measurement of the volume of airways secretion (or sputum) both in the experimental animal and in human disease states together with its modification by drugs

5.   Quantitation of the mucociliary escalator, its response to hypersecretion and the effect of drugs.

6.   Estimation of chemical constituents of sputum, their variation in disease and response to therapy

7.   Measurement of the rheological properties of sputum, their relation to chemical constituents and the response to therapy

8.   Measurement of the ventilatory capacity of the lung as it may reflect changes in the biochemical and rheological properties of bronchial secretion and their alteration by pharmacological agents

II. Historical

The earliest reported study of bronchial secretion was in 1882 by Rossbach [14] who cut the trachea lengthwise to expose the mucous membrane in unanesthetized cats. After blotting the surface dry, he measured the time lapse before the surface was again covered with fluid. Calvert [15] in 1896 used a similar technique to establish an "interval of secretion" of 3 to 5 minutes between drying the mucous membrane and the reappearance of surface secretion in the normal cat trachea. He used this as a baseline to compare the effect of drugs injected into the circulation and showed that alkalis, including sodium carbonate, ammonium carbonate and chloride, and potassium iodide as well as ipecacuanha stimulated secretion. Neither Rossbach [14] nor Calvert [15] were able to say whether the fluid layer might be the result of bronchial secretion or represented a transudate from the bronchial wall. The influence of vasodilatation and vasoconstriction was studied by Calvert [15] by placing ice bags or hot poultices on the abdomen of the cat. After cooling, new blood vessels appeared in the tracheal mucosa and secretion increased. With hot poultices, the blood vessels diminished in size with a reduction in secretion. The influence of vasodilatation on gland secretion was emphasized by Dos Ghali [16], who suggested that the glands were controlled by the effect of the nervous system on blood vessels. Policard and Galy [6] showed that vasoconstriction induced by ergotamine resulted in drying of the mucous membrane.

In 1898, Fuchs-Wolfring [17] first observed that pilocarpine, a stimulant of cholinergic receptors [18], exhausted the serous cells of the rabbit tracheal gland leaving them completely devoid of granules. Kountz and Koenig [19] in 1929 investigated the effect of direct nerve stimulation on the decerebrate dog and concluded that vagal stimulation increased bronchial secretion, while sympathetic stimulation either diminished secretion or had no effect. Again, it was not determined whether the effect was directly on the glands or mediated by change in the blood vessels.

Florey, Carleton, and Wells [20] first suggested a functional distinction between mucous glands and goblet cells; their work was based on an observation by Larsell [21] that the glands, in contrast to the goblet cells, were innervated. They found that stimulation of either the central or distal end of the cut vagus activated the glands of the trachea and that pilocarpine caused similar activity. The secretion was inhibited by the administration of atropine. They were unable to demonstrate any effect of nerve stimulation on the goblet cells, but found that pilocarpine increased goblet cell content without promoting discharge. Tappan and Zalar [22] questioned the alleged differences in innervation of mucous glands and goblet cells when studied at cellular level. They quote Hollander's work utilizing acetylcholine to obtain greater amounts of mucus from isolated gastric surface mucus cells and Werner [23], who used acetylcholine to collect larger quantities of colonic goblet cell mucus. Tappan and Zalar suggest acetylcholine is a common mediator, although the site of the biochemical reaction may vary.

The first report on the control of bronchial mucous glands in man was by Klassen et al. [24], who, by bronchoscopy, showed that vagal stimulation caused bronchoconstriction and bronchorrhea. Brantigan et al. [25,26] reported that total denervation of the lung reduced the viscidity of sputum. Four to five years after denervation, atrophy of the mucous glands was observed, but no change was seen in the goblet cells.

It must be stressed that all physiological and pharmacological investigations of bronchial gland secretion so far mentioned involve other physiological events. Observations on the total secretion will be affected by alterations in the volume of transudate which would be affected by changes in blood vessels or bronchomotor tone. It has been shown [27] that blood pressure changes in the salivary glands influence secretory rates.

III. The Morphology and Histochemistry of Airways Secretion Control

Simple chronic bronchitis is defined in clinical terms as hypersecretion of mucus sufficient to cause cough and expectoration in the absence of specific lung disease [28]. The pathological changes are well documented [29,30]; there is hypertrophy of mucosal glands and an increase in the number of goblet cells with extension to the smaller peripheral airways. No abnormal glycoproteins have been demonstrated, but there is an increase in sulfomucins and sialidase-resistant sialomucins relative to the sialidase-sensitive and neutral glycoproteins. The etiology of chronic bronchitis is complex and involves genetic, environmental, and infective factors. In the experimental animal, some of these factors can be isolated as causal factors of the hypersecretory state.

A. Work Hypertrophy

An overall increase in gland size may result from hypertrophy or hyperplasia, or a combination of both. One of the most widely accepted theories of gland increase is that of a functional response to increased work. Harris [31] states, "It is tempting to suppose that a general increase in turnover of substrate might induce synthesis of a large number of enzymes and thus form the basis of hypertrophy."

B. Tobacco Smoke and Sulfur Dioxide

Tobacco smoke and SO_2 will produce changes indistinguishable from those of chronic bronchitis when inhaled by specific germfree rats even though the lungs remain sterile. Jones et al. [32] suggested that an alteration in glycoprotein pattern is the earliest change by which the effect of tobacco smoke may be assessed and that this change may be reversible. Dogs exposed to cigarette smoke via tracheostomies showed an increase in goblet cells at 44 days, although

these were subsequently replaced by layers of cuboidal cells [33]. Section of the
superior and recurrent laryngeal nerves or heavy atropinization does not prevent stimulation of the tracheal mucous glands by application of mustard oil
[20]. Apparently, the activity of the tracheal glands is due to direct action of
the irritant upon the gland cells. Falk et al. [34], using isolated rabbit trachea
and ciliated esophageal mucosa of the frog, showed similar histological changes
in response to a variety of irritants and correlated these with mucus flow rate
as measured by soot particle transport. There was stimulation in the first minute
after exposure followed by depression for about 16 min and a recovery period of
30 min. Pretreatment of the excised frog esophagus by the application of acetylcholine or eserine salicylate had little effect on the initial increment of mucus
flow but completely abolished the subsequent depression of flow. Lightowler
and Williams [35] correlated changes in particle transport with the histological
changes in rat respiratory epithelium after exposure to sulfur dioxide. Particle
transport was diminished before ciliary activity, suggesting that the main effect
was on the mucus-secreting structures.

C. Phenylmethyloxadiazole

Phenylmethyloxadiazole (PMO), a substance first used as an antitussive agent,
protects the rat against the ciliostatic effect of tobacco smoke when administered
with the tobacco in the cigarette [36]. Its parent compound, oxalamine citrate,
has a similar effect when given orally [37] suggesting that these compounds
modify the tissue response to irritation. Addition of PMO to tobacco also prevents the increase in goblet cells found in tracheal epithelium of rats exposed to
tobacco smoke alone [38]. Paradoxically, there was a greater increase in the
cell size and acinar diameter of the submucosal glands than that produced by
tobacco smoke alone. PMO gave no protection against the shift from neutral to
acid glycoprotein in the mucous glands and goblet cells usually associated with
the hypersecretory state induced by smoke, sulfur dioxide, or injected isoprenaline.

D. Infection

The association of mucus gland hypertrophy, hypersecretion of mucus, and
bacterial infection of the lower respiratory tract is well recognized in chronic
bronchitis. The most common infecting organism is the *Haemophilus influenzae*
[39,40]. As recently as 1948 bacterial infection was thought to be the initiating
factor in chronic bronchitis, but all subsequent work suggests that tobacco smoke
and atmospheric pollutants initiate mucous gland hypertrophy. Thus bacterial
infection is a secondary phenomenon which may be related to the virulence-

enhancing effect of mucus [41,42]. The role of viral infection is less certain. It is a major factor in causing exacerbations of chronic bronchitis [43] and Gregg [44] has suggested that recurrent viral infection in children may be a precursor of adult chronic bronchitis.

Hers et al. [45] described the histopathology of the respiratory tract in human influenza. Desquamation of the mucosa is followed by regeneration and hypertrophy of mucus-secreting structures. A similar sequence of events is reported by Bang and Bang [46] in the nasal mucosa of chicks infected with laryngotracheitis virus. During the recovery phase (at 4 weeks) the mucosa is still thickened with unevenly distributed mucous glands and a heavy population of goblet cells. In the larynx and trachea, no sloughing is apparent at any stage, but at 3 weeks there is intense staining of the mucous gland with PAS and ropes of mucus are present in the lumen.

Jones, Baskerville, and Reid [47] studied the respiratory epithelium of pigs infected with *Mycoplasma pneumoniae* and found bronchial gland hypertrophy, which may occur in response to a local stimulus mediated through nervous pathways. *Mycoplasma* is known to produce peroxide [48], which acts directly on epithelial cells and may act as an irritant to receptors within the airways.

E. S-Carboxymethylcysteine

Huyen [49] showed that in the rat exposed to sulfur dioxide the shift in glycoprotein toward a sialidase-resistant sialomucin and a sulfomucin which is usually found was inhibited by the concomitant administration of S-carboxymethylcysteine by mouth.

F. The Effect of Isoproterenol on Mucus-Secreting Structures

Of great interest is the recently described effect of parenteral injection of isoproterenol into the intact animal in producing changes in the bronchial mucosa similar to those seen in human chronic bronchitis and in the experimental animal exposed to tobacco smoke or sulfur dioxide.

Selye et al. [50] showed that the injection of isoproterenol into rats caused profuse salivation and a fivefold increase in the weight of the submaxillary salivary glands. The effect on gland weight was confirmed by Brown-Grant [51]. Wells [52] confirmed Selye's findings and compared the effect to that of repeated incisor teeth amputation thought to be mediated by sympathetic stimulation. Cessation of a series of incisor amputations [53] or of isoproterenol administration [54] both resulted in regression of the enlarged glands. The chronic administration of parasympatholytic drugs—atropine and scopolamine—had no striking effect on gland weight and did not interfere significantly with the response to repeated incisor amputation [53].

Surgical extirpation of the sympathetic nerve supply to the salivary glands or the administration of adrenergic blocking agents, such as diphenyl-chlorethylamine (Dibenamine) [54], reserpine, and bretylium tosylate [53], inhibited the response of the salivary glands to amputation of the incisors.

Using pathogen-free rats, Sturgess and Reid [55] followed the effect of 6 or 12 daily injections of isoproterenol (10 or 25 mg) and of pilocarpine (10 mg). Both drugs caused an increase in bronchial submucosal gland size (Figs. 5 and 6) and an increase in the number of goblet cells (Figs. 7 and 8), previously thought not be under nervous control [20]. Isoproterenol increased goblet cells containing acid glycoprotein; pilocarpine increased all types. Isoproterenol increased small acini in the glands; pilocarpine increased large ones. After 12 injections of pilocarpine, the secretory cells were "exhausted" and relatively empty of secretions.

Similar changes were seen in the pancreas and salivary glands with both drugs. These results were compared to the changes seen in cystic fibrosis: hypertrophy of mucous cells in the salivary glands [56] and mucous gland hypertrophy with an increase in acid glycoproteins [57]. These findings lend support to the concept that an imbalance of the autonomic nervous system may be implicated in cystic fibrosis [58].

While these results suggest a parallel response of the submaxillary salivary glands and the bronchial mucosal glands in the rat, there is no evidence that a similar relationship exists in man. The life of the "sputologist" would be much easier if it were so. The effect of injected isoproterenol on the rat bronchial mucosal glands does, however, raise questions regarding the effect of inhaled isoproterenol on the human or animal bronchial mucosa and further study is required. Jones (personal communication) demonstrated goblet cell metaplasia in the bronchial mucosa of a child whose death from asthma was thought to be associated with the excessive use of an isoproterenol inhaler.

IV. The Volume of Respiratory Tract Secretions

No accurate measurements have been made of the volume of normal respiratory tract fluid produced in man. The most widely quoted estimate of 100 ml in 24 hr was obtained by extrapolation from animal studies by Policard and Galy [6]. Toremalm [4] collected 10 ml/day from tracheotomized accident victims. The volume of fluid obtained in such circumstances will reflect the balance between secretion and reabsorption [59]. Boyd and Sheppard [3] showed that a

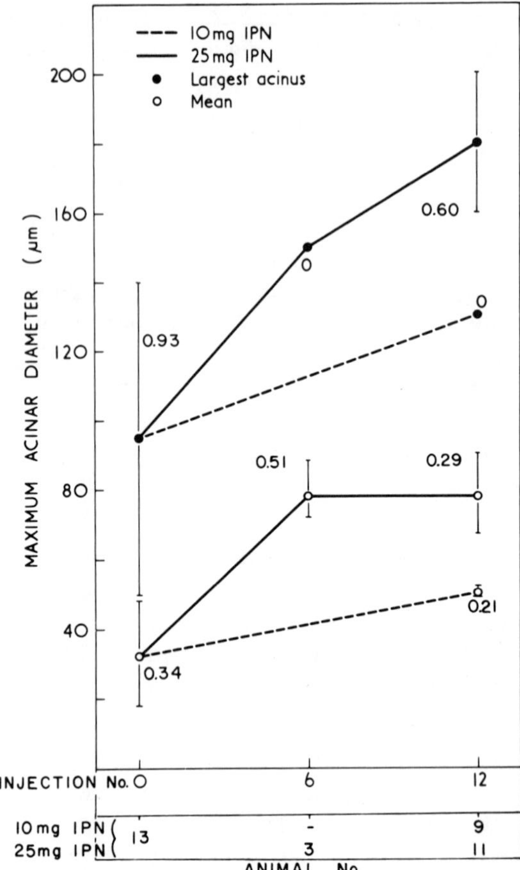

FIGURE 5   Rat bronchial submucosal glands. Increase in maximum acinar diameter after injections of 10 or 25 mg of isoproterenol. The mean diameter of the largest acinus in each animal increases with increasing number of injections whereas the mean for each section increases after 6 injections with no further change after 12 injections. Each point, mean acinar diameter; vertical line, the range; each figure, the standard error for each group.

tracheal pocket could absorb half the volume of respiratory fluid reaching it, while Kilburn [60] postulated that, with a mucus layer of 5 μm depth, the enormous decrease in the surface area of mucus membrane from the periphery to the trachea must require considerable reabsorption of fluid.

Boyd [61], summarizing several studies, reported a positive correlation between the volume of respiratory tract fluid and various factors, including body temperature, cannula size, cardiac output, inhaled air humidity, and

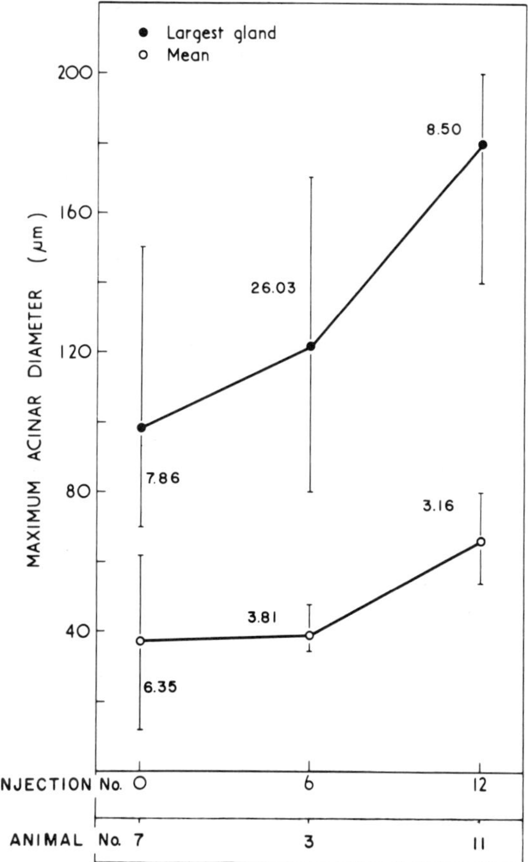

FIGURE 6 Rat bronchial submucosal glands. Increase in maximum acinar diameter after 6 or 12 injections of 10 mg pilocarpine. The value for the largest gland from each rat has increased after 6 injections, the mean value for each section increasing only after 12 injections.

inhaled air temperature. Surprisingly, there was in inverse correlation with body weight and no correlation with rate and depth of respiration.

Dehydration of a patient with severe chronic bronchitis is associated with difficulty in expectorating small amounts of viscid sputum and rehydration usually eases expectoration. It is not clear whether this effect is on secretory gland activity or on tissue fluid transudate. Hyperhydration in animals does not increase the volume of secretions [62], but, in man, the ingestion of 1 liter or more of water in 10 min by normally hydrated subjects leads to an increase in parotid flow [63]. It is not known what the effect would be on sputum volume,

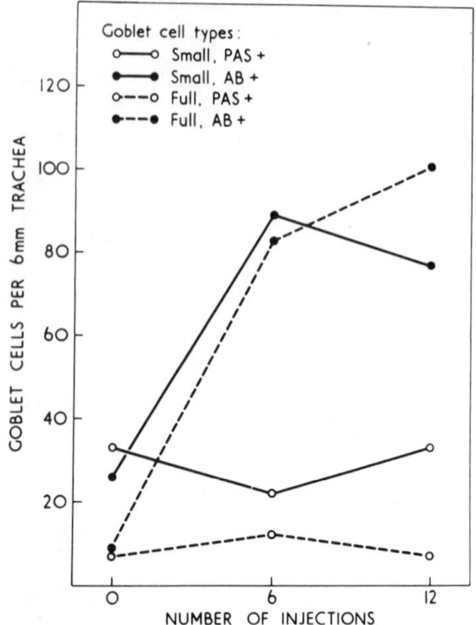

FIGURE 7　Goblet cell increase in rat trachea after 6 and 12 injections of 25 mg of isoproterenol. Small and full AB + cells (containing acid glycoprotein) increase, with little change in PAS + cells (containing neutral glycoprotein). Each point, mean value of 6 rats.

although Blanshard [64] showed a reduction in sputum viscosity following an increased fluid intake.

A. Seasonal Variation in the Volume of Respiratory Tract Fluid

Reference has already been made to the "spontaneous" seasonal variation in certain chemical constituents and rheological properties of sputum [9]. Boyd [65] first observed a seasonal variation in the effect of drugs on the volume of respiratory tract fluid in various animals. Pilocarpine, given subcutaneously to guinea pigs, cats, dogs, and rabbits, is less effective in the winter months of the year. Boyd et al. [66] found that glyceryl guaiacolate augmented volume only in the autumn months of the year. Other experiments show that a majority of inhaled volatile oils have the greatest effect in the months of September and October. With eucalyptus oil and Friar's balsam, there is no seasonal variation in response [67].

Boyd and Sheppard [68] found that the intragastric administration of Bisolvon (bromhexine) (5 mg/kg) to fasting rabbits augments the volume output

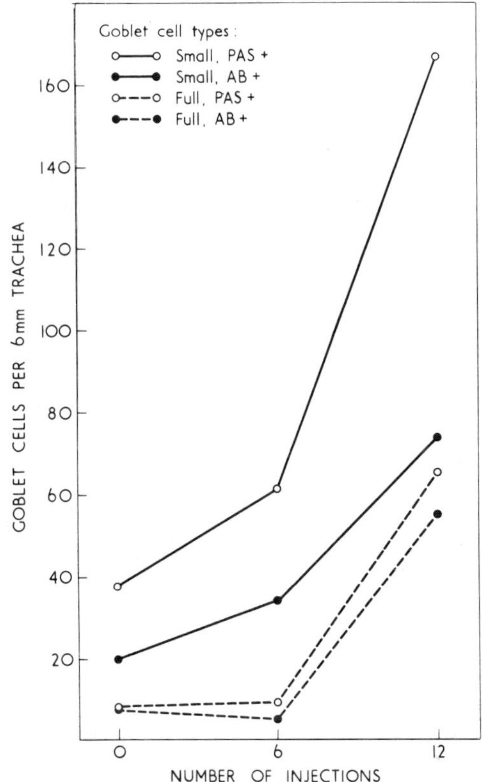

FIGURE 8 Goblet cell increase in rat trachea after 6 and 12 injections of 10 mg of pilocarpine. Both AB + cells (containing acid glycoprotein) and PAS + goblet cells (containing neutral glycoprotein) are increased, particularly the small PAS +. Each point, mean value of 6 rats.

of respiratory tract fluid in 4 hr, the response in November being twice as great as in March, with intermediate values in June. This finding is difficult to correlate with that of Gent et al. [69]: that Bisolvon, given orally to patients with chronic bronchitis between February and May, has a greater effect on respiratory function than when given between May and July.

V. Organ Culture Studies of Human Bronchial Mucous Gland Activity

A true understanding of the control of airways secretion must begin at the cellular level. The technique of organ culture allows study of different cell types under controlled conditions [70,71]. Sturgess and Reid [71] obtained 69 specimens of normal and diseased bronchi either at surgical resection or at necropsy within 4 hours of death. Strips of mucosa and submucosa, including the submucosal glands, were cultured by the technique of Trowell [72]. Tritium-labeled glucose and threonine were incorporated into the culture medium and their passage into and through the cells was followed by autoradiography. An index of the mean activity of the mucous cells of the glands, the secretory index, was calculated as the percentage of cells which were discharging labeled glucose into the lumen after 4 hours incubation. The secretory activity of serous cells was quantitated by the progress of radioactive tracer through the cell estimated from the number of silver grains overlying each of four zones from base to apex of the serous acini. Thus we define "secretory activity" as the rate of incorporation of metabolites within the cell as well as the rate of discharge of the final product. The degree of mucous gland hypertrophy was assessed by the gland/wall ratio [30].

The secretory index is a measure of overall gland behavior and is found to vary considerably from subject to subject but to be remarkably constant at different levels of the bronchial tree in any one subject (Table 1). There is a close positive correlation between the secretory index and the degree of gland hypertrophy as measured by the gland/wall ratio (Fig. 9). The gland/wall ratio is found to be closely related to acinar diameter and this, in turn, is related to cell height in both mucous and serous cells. Lamb [73] has shown that the concentration of radioactivity incorporated into unit volume of cytoplasm is greater in hypertrophied glands. Sturgess and Reid [71] therefore conclude that increase in cell size, increase in the number of cells secreting at a fast speed, and increase in the amount of secretion per unit of cytoplasm each contributes to the hypersecretion associated with mucous gland hypertrophy.

TABLE 1 Comparison Between the Secretory Index in Glands Situated at Different Levels in the Bronchial Tree[a]

Treatment	Trachea	Main bronchus	Lobar bronchus
Control	24.3	25.1	22.8
Acetylcholine	58.9	58.0	56.4
Atropine	10.0	8.8	12.0
Guanethidine	28.0	26.2	27.1
Isoprenaline	25.2	24.0	23.6

[a]Each value represents the mean secretory index after 4 hours of incubation with [^{3}H]glucose, calculated from six explants of a gland. Each drug was tested on a different bronchus. (Reproduced from Ref. 74 by permission of the publishers of *Clinical Science.*)

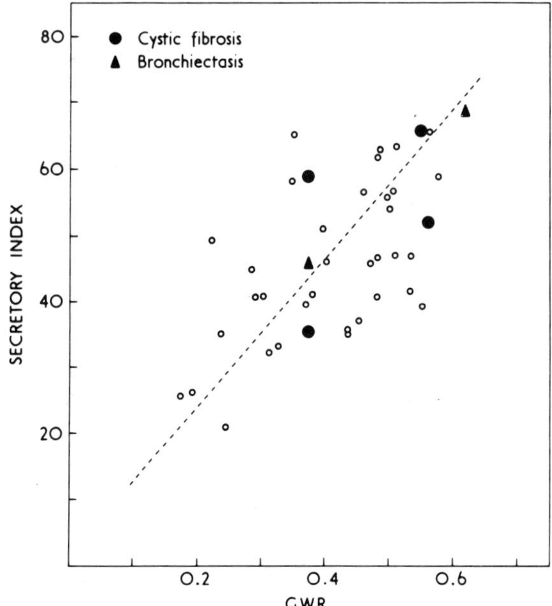

FIGURE 9 The degree of gland hypertrophy (GWR) as related to secretory index in the normal gland and in chronic bronchitis (P = 0.001). Each point represents a bronchus from a different patient. The cases of cystic fibrosis and bronchiectasis fall within the same range.

A. The Effect of Drugs on the Secretory Activity of the Human Bronchial Submucosal Gland in Organ Culture

Using the methods described above, Sturgess and Reid [74] were able to study the effect of drugs incorporated into the culture medium on the secretory activity of human bronchial mucosal glands.

1. Parasympathomimetic Drugs

Acetylcholine, pilocarpine, and carbachol all increase the secretory index in all glands studied (Table 2) and, where investigated, a dose-related response is seen. The same concentration of acetylcholine (1 μg/ml) causes increases between 18 and 140% in different bronchi, but a similar response is found in glands from three different levels—trachea, main bronchus, and lobar bronchus—of the same bronchial tree (Table 1).

TABLE 2 Effect of Parasympathomimetic and Parasympatholytic Drugs on the Secretory Index of Mucous Cells[a]

Parasympathomimetic drugs	Concentration (μg/ml)		
	1	10	100
Acetylcholine	+18	+38	
	+19	+41	
	+23		
	+25		
	+38		
	+40		
	+41		
	+43		
	+48	+72	
	+50		
	+84	+152	+180
	+92	+133	+156
		+143	
	+140		
Pilocarpine	+46	+54	
		+28	
Carbachol	+36	+48	

Parasympatholytic drugs	Concentration (μg/ml)		
	0.5	5	50
Atropine	−18	−29	−35
	−29		
	−58		
	−59		
	−73		
		−20	−31
		−33	
		−43	
		−55	
		−60	
		−71	
Scopolamine	−36	−48	

[a]The values represent percent increase (+) or decrease (−) in the secretory index as compared with the corresponding control tissue. Each row of values represents different specimens from the same bronchus. (Reproduced from Ref. 74 by permission of the publishers of *Clinical Science.)*

2. Parasympatholytic Drugs

The effect of atropine and scopolamine was to reduce the secretory index in the
mucous cell population, but in no case was the secretion inhibited completely.
A dose-related response to atropine was demonstrated up to a concentration of
50 μg/ml, but increments above this level cause no additional inhibition. Again
there is a wide variation in the degree of inhibition in bronchi from different
subjects (Table 2), but a consistent response in glands from different levels from
the same bronchial tree (Table 1). The serous cells are relatively little affected by
the parasympatholytic drugs.

3. Sympathomimetic and Sympatholytic Drugs

No sympathomimetic or sympatholytic drug tested, including alpha- and beta-
blocking agents (Table 3), has any effect on the secretory index of either mucous
or serous cells.

4. Nonsteroid Antiinflammatory Agents

Sodium salicylate shows a dose-related reduction in the secretory index of the
mucous cells and of serous cells. Subsequent work suggests that this effect may
be due to the antagonistic action of nonsteroid antiinflammatory agents (sali-
cylic acid, indomethacin, and phenylbutazone) on prostaglandin activity in the
lung.

5. Corticosteroids

In only some of the specimens of bronchus treated with hydrocortisone is the
secretory index significantly reduced but, for these bronchi, all explants demonstrated a dose-related response. The cortisone derivatives betamethasone, dexa-
methasone, and triamcinolone, although more potent than hydrocortisone, give
neither a more consistent nor a greater effect than hydrocortisone. This group
of drugs differs from those affecting the autonomic nervous system in that not
all subjects are affected.

6. Phenylmethyloxadiazole

The protection afforded by phenylmethyloxadiazole against some of the effects
of tobacco smoke on the bronchial mucosa [36-38] led to a further study com-

TABLE 3 Drugs Tested in Organ Culture[a]

Drug	Concentration range (μg/ml of culture medium)
Parasympathomimetic	
Acetylcholine	0.1–100
Pilocarpine nitrate	1–100
Carbachol	1–10
Parasympatholytic	
Atropine sulphate	0.5–50
Scopolamine hydrochloride	1–10
Sympathomimetic	
Epinephrine	1–100
Norepinephrine	1–100
Isoprenaline sulfate	1–100
Sympatholytic	
Phenoxybenzamine	1–100
Phentolamine	1–100
Propranolol	1–100
Guanethidine sulfate	1–100
Humoral agents	
Bradykinin	1–100
Mefenamic acid	1–100
Histamine	1–100
Antihistamine	1–100
Corticosteroid hormones	
Hydrocortisone hemisuccinate	1–100
Betamethasone	1–10
Dexamethasone	1–10
Triamcinolone	1–100
Prednisolone	1–10
Mucolytic agents	
N-Acetylcysteine	1–1000
Vasicine	1–1000
Miscellaneous	
Potassium iodide	1–100
Disodium cromoglycate	1–100
Nicotine	1–100
Sodium salicylate	1–1000

[a]Reproduced from Ref. 74 by permission of the publishers of *Clinical Science.*

TABLE 4 The Effect of Tobacco Smoke and Phenylmethyloxadiazole (PMO) on Mean Acinar Diameter and Secretory Index

	Mean acinar diameter (μm)	Secretory index
Control	52.3 ± 11.5	37.8 ± 5.6
Tobacco	76.3 ± 13.6[a]	63.7 ± 6.8[a]
Tobacco + PMO	66.8 ± 8.3[a]	42.0 ± 9.7
PMO control	48.6 ± 10.3	24.9 ± 6.8[a]

[a]Values significantly different from control (p = < 0.01).

bining organ culture with histology. Reid et al. [75] showed that the increase in acinar diameter is associated with a reduction in the secretory index (Table 4). Brown and Tong [76] showed that PMO conferred stability on the erythrocyte membrane. Jones et al. [38] postulate a similar action of PMO in preventing a nonsecretory cell from developing into a secretory one (goblet cell metaplasia), while simultaneously impairing discharge from cells producing an acid glyco-protein, which leads to retention of secretion.

So far studies using phenylmethyloxadiazole have had no clinical application, but the techniques employed to assess the control of airways secretion by irritants may be applied to the testing of alternative smoking substances.

7. Dibutyryl Cyclic Adenosine Monophosphate

Whimster and Reid [77] cultured explants of bronchial mucosa for 2 hours in medium containing tritiated glucose before washing and transfer to medium containing dibutyryl cyclic adenosine monophosphate (DBcAMP) or other test substances for 1 hour. The results are expressed as discharge index (DI) rather than secretory index as used by Sturgess and Reid [71]. The criterion for discharge is the presence of silver grains on the luminal surface of the cell. The progress of radioactive label through the cell was also assessed after a 2 hour incubation with tritiated glucose followed by 1 hour in plain culture medium.

The results are shown in Table 5. The mean discharge indices of the control tissue varies considerably from case to case, but no relationship to the degree of gland hypertrophy was found in this series. Treatment with DBcAMP produces a response which is related to gland size; a significant increment in DI is seen only in explants from subjects with normal size glands (GWR < 0.35), with no increase in severely hypertrophied glands (GWR > 0.60).

Theophylline, a phosphodiesterase inhibitor, increases the concentration of cAMP and has produced a significant and consistent increase in DI [78]. This

TABLE 5 Effect of DBcAMP and Other Drugs on the Discharge Index (DI) of Mucous Cells (Mean Discharge Indices)

Case	Gland/wall ratio	Control[a]	DBcAMP[a,b]	Other drugs [a,b]
1	0.46	1.8 (0.6)	40.5 (5.6)s	
2	0.43	38.4 (5.4)	40.5 (3.2)	
3	0.34	14.3 (3.3)	23.1 (3.7)	
4	0.51	6.1 (1.7)	17.4 (2.7)s	
				Theophylline
5	0.27	26.0 (6.0)	42.9 (11.5)s	55.0 (9.7)s
6	0.66	19.7 (5.0)	12.2 (6.8)	46.4 (2.5)s
7	0.60	22.0 (2.7)	24.5 (4.3)	45.7 (8.9)s
8	0.46	22.9 (4.3)	31.9 (6.4)	33.3 (4.2)
9	0.41	10.8 (4.1)	9.6 (3.4)	21.6 (8.4)
10	0.34	15.6 (4.5)	25.9 (5.5)	28.9 (4.4)
				Opipramol
11	0.64	20.4 (4.9)	17.5 (4.1)	12.8 (4.3)
12	0.51	12.9 (1.4)	14.0 (3.1)	31.0 (5.7)s
13	0.40	11.6 (2.6)	25.8 (5.3)	18.2 (4.3)
				Chlorpromazine
14	0.61	25.1 (3.2)	14.9 (3.1)s	8.4 (3.1)s
15	0.69	7.4 (2.0)	4.8 (2.4)	7.0 (1.1)
16	0.58	22.1 (4.7)	24.6 (11.0)	24.1 (3.3)
				Lithium chloride
17	0.42	17.5 (5.4)	6.1 (0.4)	9.2 (3.8)
18	0.55	56.7 (9.2)	54.9 (5.6)	78.9 (7.0)
19	0.56	12.5 (5.2)	47.1 (11.7)s	50.8 (10.5)s

[a]Numbers in parentheses indicate standard error
[b]s = significant difference from control: $p < 0.05$.

effect is not related to gland/wall ratio, and is thought to reflect a potentiation of the cAMP system [79], not previously investigated in human organ culture systems.

An acinar analysis of the numbers of distended, intermediate, and flat cells, before and after incubation with DBcAMP, together with an assessment of the progress of radioactive label through the cells suggests that the effect of DBcAMP and of theophylline is to bring about discharge of distended cells.

8. Ouabain

The secretory index as defined and quantitated by Reid and Sturgess [71] represents the total activity of the cell including uptake, synthesis, transport, and discharge. Recent studies to analyze synthesis separately from discharge utilized

TABLE 6 The Percentage Reduction in Secretory Index (SI) for [^{3}H] Glucose and [^{3}H] Threonine in the Mucous Cells with Ouabain Added to Tissue Culture Medium

	Percentage reduction	
	[^{3}H] Glu	[^{3}H] Thr
Normal gland (gland/wall ratio <0.30) (Mean SI = 0.26)		
Pulse + ouabain	50	90
Pulse before ouabain	15	45
Hypertrophied gland (gland/wall ratio >0.45) (Mean SI = 0.60)		
Pulse + ouabain	50	55
Pulse before ouabain	5	35

ouabain, an inhibitor of the ATPase responsible for Na$^+$/K$^+$ exchange, in tissue culture medium [75]. Gallagher et al. [80] report that ouabain added to rat submaxillary gland cultures causes an accumulation of secretory material in the cells, thus inhibiting secretion.

Reid et al. [75] used pulse labeling of mucosal explants with radioactive precursors [^{3}H] glucose and [^{3}H] threonine added to the culture medium for 60 min. Ouabain was added either with the pulse or following it. In all cases, there is a reduction of the secretory index for both precursors when ouabain was given with the pulse (Table 6). When the pulse was given before ouabain, there is little effect on the secretory index suggesting that cell discharge is not dependent on a ouabain-sensitive mechanism. A concomitant reduction in grain density over the mucous cells (Table 7) suggests that uptake and synthesis are, to some extent, dependent on a ouabain-sensitive mechanism. The greater effect on [^{3}H] threonine incorporation than on [^{3}H] glucose may reflect the slower uptake of threonine by the cell which is seen in electron microscopic studies [82] (see below). It is also apparent that the effect of ouabain is greater in the normal than in the hypertrophied gland.

9. Other Drugs

Other drugs, such as mucolytic agents (see Table 3) were chosen because of their claimed physiological or therapeutic effect on the bronchial tree. None of these produced any effect in organ culture. The mucolytic agents N-acetylcysteine and a vasicine derivative tested for periods up to 15 hours showed no effect on mucous or serous cell secretion.

TABLE 7 The Percentage Reduction in Grain Density Counts of [^{3}H] Glucose and [^{3}H] Threonine Over Mucous Cells with Ouabain Added to Tissue Culture Medium

	Percentage reduction	
	[^{3}H] Glu	[^{3}H] Thr
Normal gland (gland/wall ratio <0.30)		
Pulse + ouabain	50	85
Pulse before ouabain	20	40
Hypertrophied gland (gland/wall ratio >0.45)		
Pulse + ouabain	40	40
Pulse before ouabain	0	25

B. Relationship Between Drug Activity and Gland Hypertrophy

The widely varying response of the glands in explants from different bronchi and the similar response in explants from different levels of the same bronchial tree was explained by a correlation between drug activity and the degree of hypertrophy of the glands as measured by the gland/wall ratio [30]. The results are illustrated for acetylcholine (Fig. 10) and for atropine (Fig. 11). A similarity of gland size throughout the individual bronchial tree had previously been reported by Lamb and Reid [81].

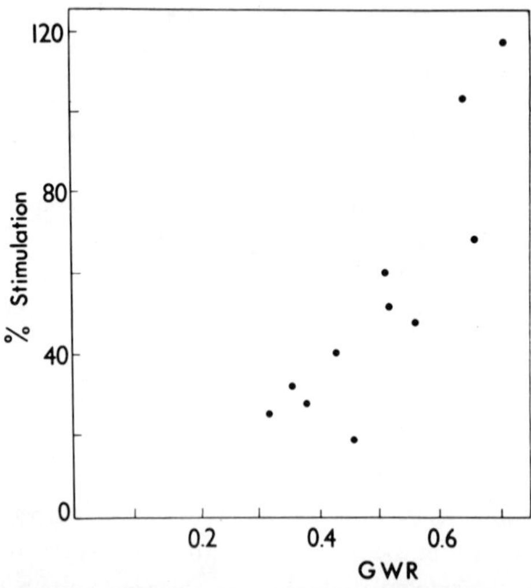

FIGURE 10 Relationship between the degree of gland hypertrophy (gland/wall ratio) and the percentage stimulation (increase in secretory index) caused by acetylcholine (1 μg/ml). Each point represents specimens from a different bronchus. The effect of acetylcholine is greater on the hypertrophied gland (P = 0.001).

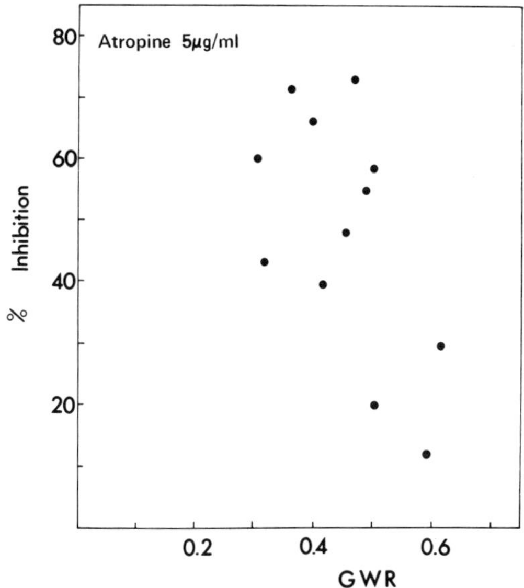

FIGURE 11   Relationship between the degree of gland hypertrophy (gland/wall ratio) and the percentage inhibition (decrease in secretory index) caused by atropine (5 µg/ml). Each point represents specimens from a different bronchus. The effect of atropine is less in the hypertrophied gland (P = 0.001).

VI. Electron Microscopic Studies of Airway Secretion Control

The techniques of organ culture already described, together with the addition of pharmacological agents to the culture medium, may be used with electron microscopic analysis to trace the intracellular pathways of uptake, synthesis, transport, and discharge of radioactive metabolites within a single cell.

A. Incorporation of Radioactive Metabolites into Organ Culture

In a recent study Meyrick and Reid [82] used pulse labeling and electron microscope autoradiographs to follow the in vitro incorporation of [3H] threonine and [3H] glucose by mucous and serous cells of human bronchial submucosal glands. By the selection of explants from the culture medium at intervals up to 8 hours, they showed that tritiated threonine was first localized in the endoplasmic reticulum and later migrated to the Golgi apparatus. The passage of tritiated threonine through the cell is slower than that of tritiated glucose, which is first localized in the Golgi apparatus. Both radioactive precursors were next identified in vacuoles and, finally, in secretory granules before discharge. Thus the oligosaccharide of both mucous and serous cell glycoproteins appears to be synthesized mainly in the Golgi apparatus, there added to the polypeptide core which is synthe-

sized in the endoplasmic reticulum. The mucous cell incorporated strikingly
more of both precursors than the serous cell.

The incorporation of physiological and pharmacological agents into a
system such as this can clearly be expected to throw further light on the control
of intracellular secretory mechanisms than is possible with light microscopy
alone.

B. Electron-Microscopic Changes with Bromhexine (Bisolvon)

One of the most widely studied drugs used in the modification of bronchial
secretions has been the vasicine derivative bromhexine [83,84]. Merker [85]
reported electron microscopic studies of rat bronchial epithelium after intra-
peritoneal injection of Bisolvon in large dosage (4 or 20 mg/kg body weight).
He noted enlargement, loosening, and reduced electron density of the secretory
granules of the tracheal goblet cells which he attributed to an increased water-
binding capacity of the mucopolysaccharides. He also noted an increase in
multivesicular bodies, a specific type of primary lysosome which may fuse with
the secretory granule either within the cell or in the lumen following discharge.

Gieseking and Baldamus [86] gave Bisolvon (12 mg daily) to nine patients
with either pulmonary tuberculosis or bronchial carcinoma for periods of 7 to 21
days between preliminary bronchoscopy and subsequent lung resection. Comparing bronchial biopsy specimens before and after treatment with Bisolvon, they
found no increase in the number or secretory activity of goblet cells (as reported
by Merker [85] in the rat), but there was increased secretory activity of the mucous and serous cells of the bronchial glands. In the serous cells, the number of
secretory granules was increased and showed the appearance of lysosomes. These
granules disintegrated in the glandular lumen when in contact with the mucous
substances, producing a surrounding zone of diminished density of the mucus.
The authors conclude that the secretory effect of Bisolvon is due to the increased
formation of lysosomal secretory granules in the serous cells of the bronchial
gland.

VII. Clinical Studies on the Modification of Bronchial Secretion

A. Introduction

Clinical studies on the modification of bronchial secretions by pharmacological
agents give only very indirect evidence about the underlying secretory mechanisms
and their control. Most clinical studies have been concerned with the modification of bronchial secretion in disease states. The aim of therapy has usually been

to alter the rheological properties of bronchial mucus, to reduce the degree of airways obstruction, to enhance the function of the mucociliary escalator, and to promote expectoration. In other patients, particularly asthmatics and persons with profuse mucoid sputum, the reduction of sputum volume has been the primary objective.

The methods of measuring such changes are frequently imprecise and indirect. They include in vitro measurement of the rheological properties of sputum, chemical analysis, quantitation of the mucociliary escalator, and the measurement of airways obstruction.

B.   The Rheology of Sputum

The rheological properties of mucus and their measurement have been described in Chap. 9. Viscosity and elasticity are the properties most often measured, but they may not fully characterize the behavior of mucus within the bronchial tree [87]. Bürgi et al. [88,89] stressed the importance of the fiber system in mucoid and purulent sputum, and Chodosh and colleagues [90], using an inclined plane method of measurement, described a factor believed to represent sputum stickiness or "adhesiveness." The effects of tube caliber, the thickness of the lining fluid layer on airways resistance and the entrainment of fluid by the airstream (cough) have been investigated by Clarke et al. [91,92] in a model system of rigid tubes. The additional complication of everchanging caliber has so far resisted analysis.

C.   Rheology and Biochemistry

The relationship of viscosity to the chemical composition of mucus has been extensively investigated in recent years [93-100], but only a few studies show that pharmacological agents have a specific effect on individual components, altering the rheological properties. Only circumstantial evidence can elucidate the underlying secretory control mechanisms.

D.   The Mucous Layer

Lucas and Douglas [101] postulated a double layer of mucus in the bronchi periciliary sol phase in which the cilia beat and a superficial blanket of gel which is moved forward by the tips of the cilia. Asmundsson and Kilburn [102] found no studies to confirm the histological evidence for this double layer and there is no general agreement on the origin of a sol layer. While Lucas and Douglas [101] thought that it may originate from the serous cells of the mucous glands or arise as a transudate from the nonspecific epithelial cells, Kilburn [60] suggested its production by alveolar type II and by Clara cells. Negus [103] suggested a common source of the two layers, the sol layer being differentiated because of thixotrophy by the beating of the cilia. While this seems unlikely, its source is not yet certain.

E.   The Sol and Gel Phases of Sputum

Whatever the conditions existing in bronchial lumen sputum in situ, it undoubtedly separates into sol and gel fractions in vitro. These fractions can be further separated by ultracentrifugation, while preserving the concentration of their components without denaturation [106]. The gel phase contains the bronchial glycoprotein, some serum proteins and protein bound to the bronchial glycoprotein [105]; the sol phase contains any soluble components of bronchial secretion along with serum proteins. Separate analysis of these two fractions then gives information of their source. Ryley and Brogan [104] have shown a higher proportion of sol phase in the sputum of asthma than in chronic bronchitis. This suggests that a transudate type of secretion predominates in asthma, whereas a glandular type of secretion is the main element in chronic bronchitis.

Using such methods of analysis comparative studies before and after the use of drugs may further explain the pharmacological control of individual components of bronchial secretion.

F.   The Effect of Drugs on Different Components of the Mucociliary Escalator

Asmundsson and Kilburn [102], in their excellent review of the mechanisms of respiratory tract clearance stress that drugs influencing ciliary function may also modify the mucus secretion. Thus, their precise mode of action cannot be determined simply by measurement of mucociliary transport.

It is against this background of biological complexity and controversial methodology that the claimed effects of therapeutic agents must be judged (see also Dulfano and Philipoff [107]).

G.   Sputum Rheology, Biochemistry, and Respiratory Function: Their Modification by Pharmacological Agents

1. Inhalants

While agents given by inhalation undoubtedly alter the characteristics of mucus present in the bronchial tree, they tell us nothing of the control mechanism for mucus secretion. Many drugs given parenterally have been shown to alter the rheological and chemical properties of sputum. Short-term experiments may produce a beneficial effect on mucus clearance and airways obstruction, but any long-term advantage to the patient with hypersecretion of mucus and severe airways obstruction is less apparent.

2. Fluid Balance

Severe dehydration has been shown to cause thickening of the mucous blanket in the nasal mucosa of birds [46]. Blanshard [64], in one of the earlier studies of factors affecting sputum viscosity, showed that in a normally hydrated patient an increase in fluid intake from 53 to 132 oz/day resulted in a reduction of mean daily viscosity of the sputum from 460° to 155° of deflection on a rotational viscometer. These findings are in keeping with the clinical observation that the dehydrated patient in respiratory failure has difficulty in clearing retained secretions, which can be alleviated by intravenous fluids and presumably reflects the dependence of the secretory structures on adequate tissue fluid.

In contrast, the profuse bronchorrhea associated with some cases of alveolar cell carcinoma appears to be autonomous and independent of fluid restriction. Spiro et al. [108] described a case in which steroid therapy failed to reduce the daily output of 1 liter of sputum; the fluid intake was reduced in an attempt to lessen the patient's distress. Figure 12 shows that the daily urine volume fell

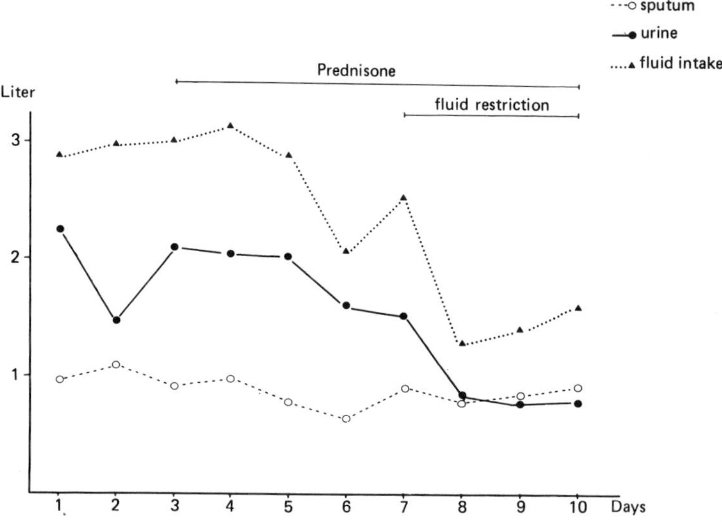

FIGURE 12 Bronchorrhea; alveolar cell carcinoma. Sputum volume unaffected by steroids or by fluid restriction sufficient to cause reduction of urine volume below that of sputum.

below that of the sputum. During fluid restriction, neither sputum nor plasma osmolality changed, but that of urine was doubled. Surfactant studies suggested an alveolar origin for most of the fluid.

3. Pharmacological Agents in the Treatment of Bronchorrhea

Bronchorrhea, or pituitous catarrh [109,110], has been defined by Keal [111] as the production of more than 100 ml of mucoid sputum in 24 hr. The figure is an arbitrary one but is well above the average sputum volume of 25 ml/day obtained in patients with chronic bronchitis [112,113]. It may arise without preceding lung disease or, more commonly, is seen in association with both intrinsic and extrinsic asthma, with chronic bronchitis or emphysema [111] and, in a special form, in some patients with alveolar cell carcinoma [114-116]. Such patients are a convenient group for the study of the effect of drugs on sputum volume, chemical constituents and rheological properties.

These studies paid particular attention to the macromolecular dry weight of the sputum and to the neuraminic acid content because of their previously described relationship to viscosity [93-100]. Fucose is present in the mucoid secretions of simple chronic bronchitis in similar concentration to that of neuraminic acid. It is virtually absent from serum where neuraminic acid is found in similar concentration to that in bronchial mucus. Thus lower levels of fucose with a higher neuraminic acid/fucose ratio are found in the inflammatory conditions of late chronic bronchitis, bronchiectasis and cystic fibrosis than in the mucoid secretions of simple chronic bronchitis [95] (Table 8). Changes in the neuraminic acid/fucose ratio as a result of therapy may reflect change in the relative proportions of bronchial gland secretion and tissue fluid transudate.

Dische et al. [117] found that the neuraminic acid/fucose ratio in pure submaxillary saliva of dogs varied with the nature of the stimulus. While the in-

TABLE 8 The Neuraminic Acid (NA) and Fucose (F) Content and the Neuraminic Acid/Fucose Ratio in the Sputum of Patients with Early and Late Bronchitis, Bronchiectasis, and Cystic Fibrosis

Disease	Sputum (μmol/ml)		
	NA	F	NA/F ratio
Early chronic bronchitis (31)	2.499	5.727	0.44
Late chronic bronchitis (22)	2.616	3.543	0.74
Bronchiectasis (8)	1.766	3.21	0.55
Cystic fibrosis (15)	2.447	3.986	0.61

Number of patients in parentheses.

tensity of nerve stimulation did not alter the ratio, a great variation occurred depending on the dose of pilocarpine administered. After injection of 0.3 mg pilocarpine, the ratio was greater than 3; after 4 mg pilocarpine, producing maximal flow rate, the ratio fell to 0.2. They suggest this is due to differences in the mechanism of regulation of the mucus secretion by direct nerve action or by drugs.

a.   Corticosteroids

Differing patterns of response to corticosteroids are in part related to associated lung pathology and in part to the level of neuraminic acid in the sputum. In general, higher levels of neuraminic acid are found in association with chronic bronchitis and continuous or intrinsic asthma. The response to steroids is less than that in patients with extrinsic asthma and with lower levels of neuraminic acid. Figure 13 illustrates the changes in a female (age 38) with extrinsic asthma.

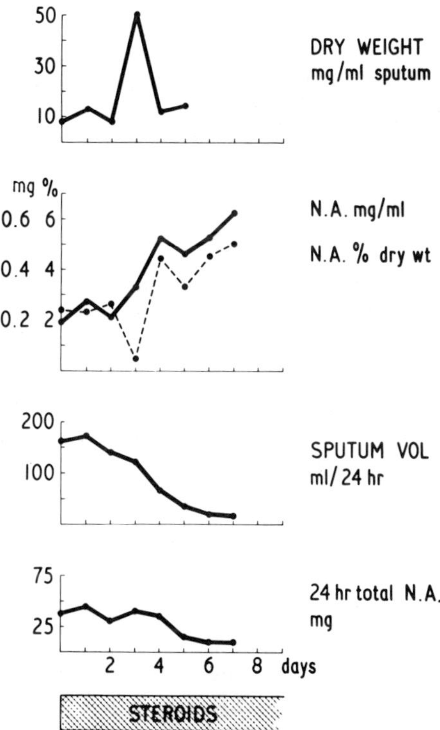

FIGURE 13 Bronchorrhea; female age 38; extrinsic asthma. Initially low neuraminic acid content of sputum associated with good response to steroids in reduction of sputum volume. The high dry weight and low percentage of neuraminic acid on day 3 was due to a purulent specimen on that day.

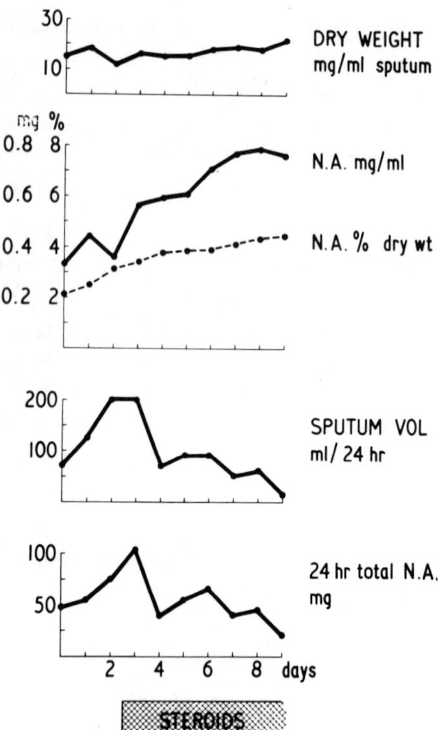

FIGURE 14   Bronchorrhea; male age 30; intrinsic asthma. Good response to steroids with reduction in sputum volume and increase in neuraminic acid content of sputum. The dry weight is unchanged.

The reduction in sputum volume is associated with an increase in the neuraminic acid content but no change in the yield of macromolecular solids. The high dry weight and low neuraminic acid percentage on day 3 are due to the presence of pus in that one specimen. Figure 14 shows a similar response in a male (age 30) with intrinsic asthma and Figure 15 correlates the same changes with an increase in the peak expiratory flow rate.

In contrast to these findings in patients with a low neuraminic acid content in the sputum, Fig. 16 typifies the lack of response in a 50-year-old man with intrinsic asthma, where the neuraminic acid forms 4% of the macromolecular material. The point and circle (BR) are the neuraminic acid levels found in mucus obtained at bronchoscopy on the same day, showing that sputum analysis accurately reflects that of intraluminal mucus. Figure 17 illustrates a similar failure to respond to steroids in a 64-year-old male with chronic bronchitis and emphysema associated with bronchorrhea and higher levels of neur-

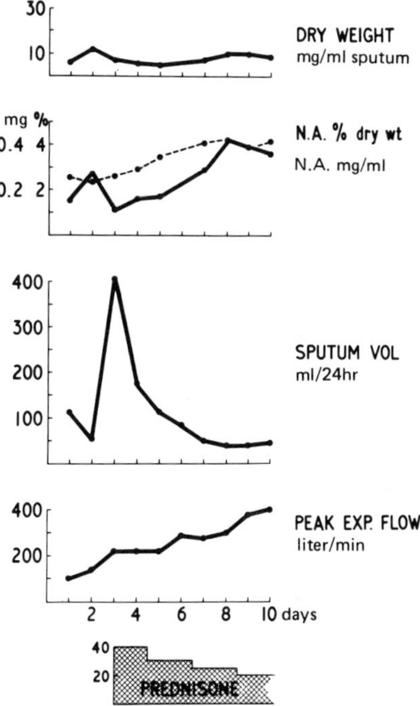

FIGURE 15 Bronchorrhea; female age 34; intrinsic asthma. The reduction in sputum volume and increase in neuraminic acid content is associated with relief of airways obstruction.

aminic acid. When the neuraminic acid/fucose ratio is considered (Fig. 18), there is a concomitant rise in both sugars suggesting that in this case of intrinsic asthma the effect of steroids is a reduction in mucus gland secretion without any alteration in the source of the sputum.

These differences in clinical response to steroids in patients with bronchorrhea may reflect the same dual response found by Sturgess and Reid [74] in organ culture of human bronchial submucosal glands and attributed to differences in tissue sensitivity. In those cases responding to steroids, membrane stabilization may reduce the secretory rate of the mucous cells with more complete synthesis of sialomucin and fucomucin. The constancy of the dry weight appears to rule out any process of dehydration of the mucus which would lead to increased concentration of macromolecular material.

Alternatively, the different responses to steroids may be dependent on the source of the sputum. A nonantigenic or antigenic-induced transudate is suppressed by prednisone in extrinsic asthma and in some cases of intrinsic asthma, but prednisone has no effect in other cases of intrinsic asthma or chronic

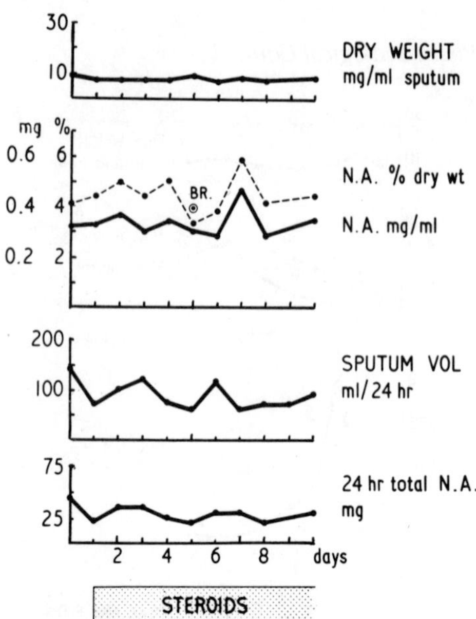

FIGURE 16 Bronchorrhea; male age 50; intrinsic asthma. Lack of response to steroids in patient with initial high level of neuraminic acid in the sputum. The point and circle BR represent the percentage and concentration of neuraminic acid in bronchial fluid obtained at bronchoscopy for comparison with those in the sputum.

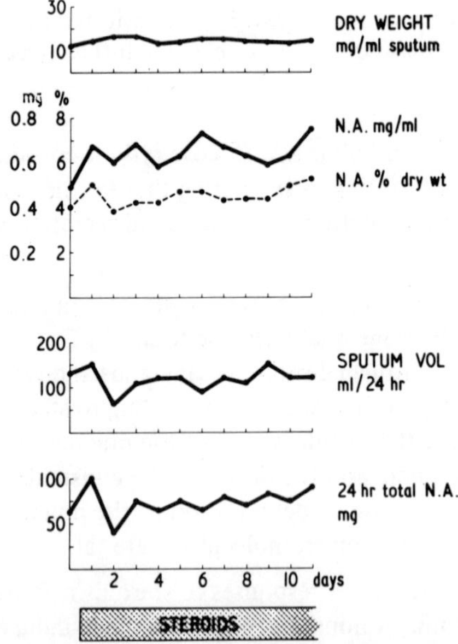

FIGURE 17 Bronchorrhea; male age 64; chronic bronchitis and emphysema. No response to steroids in patient with high neuraminic acid level in the sputum.

390

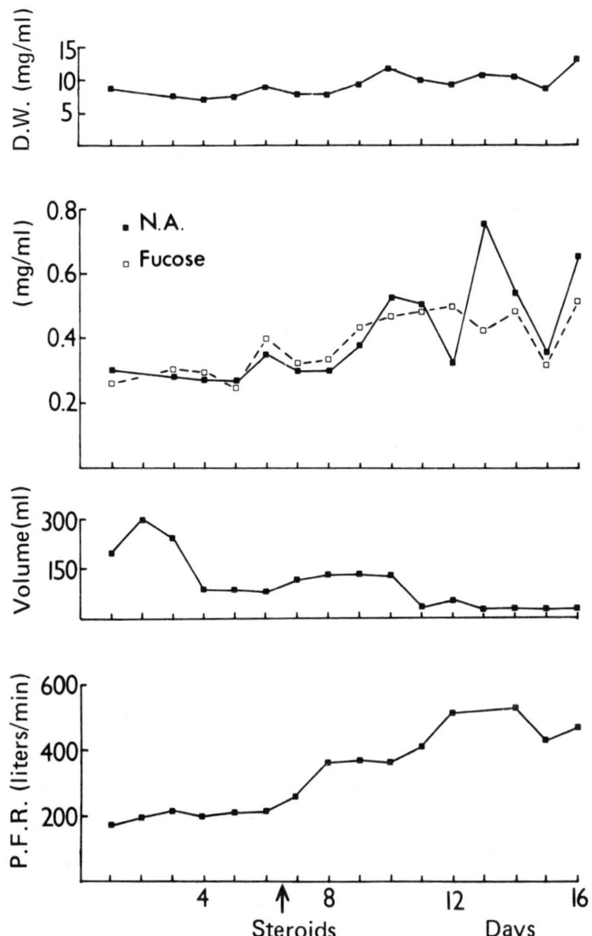

FIGURE 18   Bronchorrhea; intrinsic asthma. Concomitant rise in neuraminic acid and fucose content of sputum in response to steroids.

bronchitis, where mucous gland hypertrophy may contribute to the bulk of the sputum.

b.   The Relationship of Bronchorrhea to Airways Obstruction

In the cases described above, the reduction in sputum volume is paralleled by a reduction in airways obstruction. Although it might be assumed that the two are related, Fig. 19 illustrates that this might not be so in all cases. A sputum volume of over 200 ml/day is associated with minimal airways obstruction (peak flow rate 425 liters/min) and a reduction in sputum volume by steroid adminis-

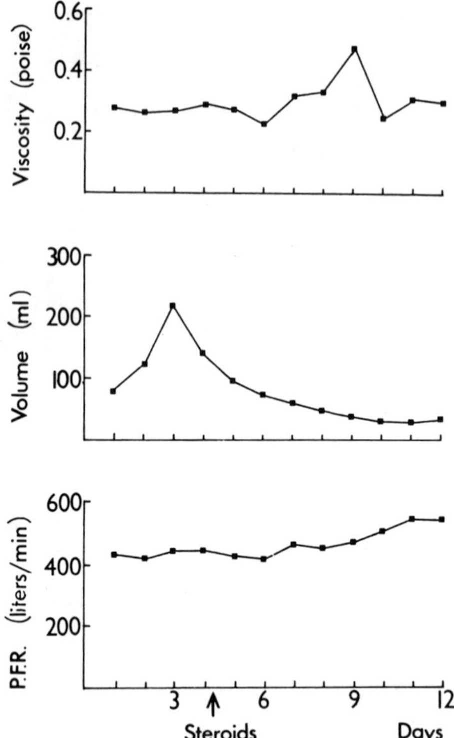

FIGURE 19 Bronchorrhea; intrinsic asthma. A large sputum volume reduced by steroids but associated with minimal airways obstruction. There is no change in the viscosity of the sputum.

tration causes only slight reduction in airways obstruction and no increase in sputum viscosity.

It is evident that these indirect methods have so far given only circumstantial evidence of the effect of drugs on secretory control mechanisms. Current studies are concerned with the identification of more specific marker substances of serum and of mucous gland secretion and their changing concentrations, in sputum of mixed origin, in response to drug therapy.

c. The Effect of Atropine on Sputum Production

In recent years there has been a revival of interest in the role of the parasympathetic nervous system in nonantigenic and antigenic-induced bronchoconstriction and in the role of atropine in blocking the mechanism [118]. Although studied in the early 19th century [119], more recent reports that atropine may

alter the physical properties of the mucous layer [120] or slow ciliary beating
and mucus transport rate [121] have led to caution in its use.

Blanshard [122] reports no change in sputum viscosity in patients with
chronic bronchitis after atropine. Lopez-Vidriero et al. [118] studied the effect
of atropine on sputum viscosity and biochemistry in 9 patients—5 with asthma,
2 with bronchiectasis, and 2 with chronic bronchitis. Atropine was administered
by inhalation in 3 patients, by intramuscular injection in 5 patients and orally in
1 patient. The dose varied from 0.3 to 2.4 mg daily, but in every case was sufficient to cause a dry mouth or significant rise in pulse rate. Measurements were
made on single control and treatment days. There was a reduction in sputum
volume in only 1 patient, during the 3-hr period following intramuscular administration of atropine. In all patients, the 24-hr sputum volume was unchanged
and in no case was there any change in the sputum viscosity, dry weight, and
neuraminic acid or fucose content for corresponding periods on control and
treatment days. In one patient with intrinsic asthma and bronchorrhea (Fig. 20),
who had been treated with corticosteroids for 5 years, an increase in prednisone
from 10 to 60 mg daily had no effect on sputum volume. When oral atropine
was increased from 0.6 to 2.4 mg daily over a period of 3 weeks, there was
marked reduction in sputum volume but again no change was found in sputum

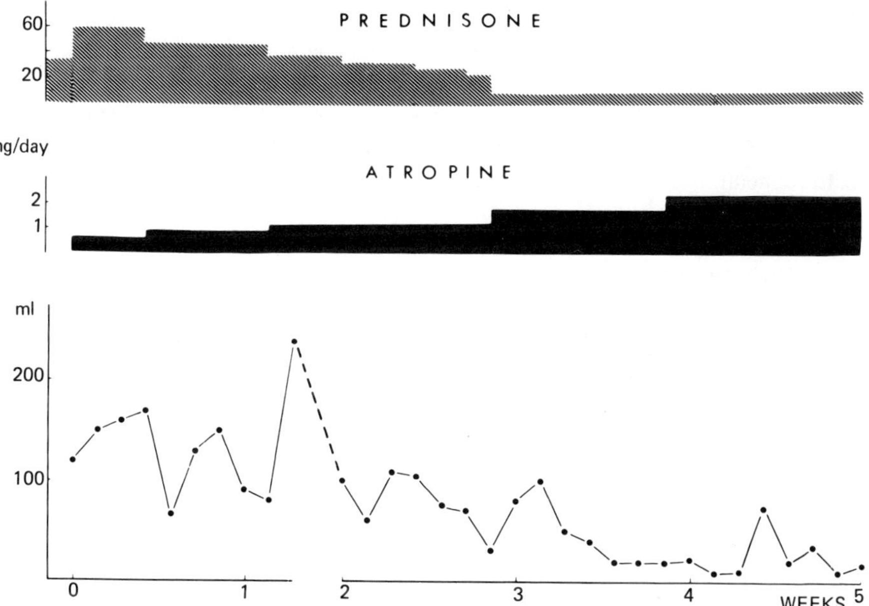

FIGURE 20 Bronchorrhea; intrinsic asthma. Reduction in sputum volume by
atropine after failed response to high dosage steroids.

viscosity or its chemical constituents. The decrease in volume without change in the neuraminic acid/fucose ratio suggests an inhibitory effect of atropine on bronchial gland secretion.

The authors comment that any difficulty in expectoration experienced by the patients after atropinization seemed to be due to drying up of salivary secretions and loss of the "lubricating" action of saliva described by Hillis [123] rather than to any change in the physical properties of the bronchial mucus.

VIII. Future Investigation of Secretory Control Mechanisms

The study of secretory control mechanisms continues to demand a multidisciplinary approach. The study of total secretion in the form of sputum, in as near normal a state as possible without preliminary physical or chemical degradation, is essential to consideration of its role within the bronchial lumen. Sputum is, however, a crude mixture and more detailed analysis of its components and of protein and glycoprotein content is required to establish their relationship to physical properties. It may then be possible to relate changes in these fractions induced by physiological and pharmacological agents to the role of these agents in secretory control.

The incorporation of drugs and radioactive precursors into organ culture of human bronchial mucosa offers the most immediate prospect of elucidating secretory control. When combined with electron microscopy there will be opportunity to evaluate the effect of drugs on the secretory activity of single mucous cells. The chemical analysis of secretory products extracted from organ culture medium is likely to be more rewarding than the study of sputum.

References

1.   J. Forbes, Notes in the Translation of R T H Laennec's *A Treatise on the Diseases of the Chest and on Mediate Auscultation,* 4th ed., London, Longmans, 1834, p. 203.
2.   J. A. Gunn, Action of expectorants, *Br. Med. J.,* 2:972-975 (1927).
3.   E. M. Boyd and E. P. Shephard, The effect of steam inhalation of volatile oils on the output and composition of respiratory tract fluid, *J. Pharmacol. Exp. Ther.,* 163:250-256 (1968).
4.   N. G. Toremalm, The daily amount of tracheobronchial secretions in man. A method for continuous tracheal aspiration in laryngectomised and tracheotomised patients, *Acta Otolaryngol. [Suppl.] (Stockh.),* 158:43-53 (1960).
5.   E. M. Boyd and A. Ronan, Excretion of respiratory tract fluid, *Am. J. Physiol.,* 135:383-386 (1942).

6.   A. Policard and P. Galy, Glandes et sécrétions bronchiques, *Les Bronches: Structure et Mecanismes a l'Etat normal et pathologique,* Paris, Masson, (1945).

7.   S. A. Leach, Release and breakdown of sialic acid from human salivary mucin and its role in the formation of dental plaque, *Nature,* **199**:486-487 (1963).

8.   S. Woodcraft, G. P. Roberts, I. Das, and L. Reid, The origin of neuraminidase in sputum and its effects on the isolation of glycoprotein. (In preparation).

9.   E. E. Keal and L. Reid, Neuraminic acid content of sputum in chronic bronchitis, *Thorax,* **27**:643-653 (1972).

10.   M. T. Lopez-Vidriero, J. Charman, E. E. Keal, D. J. de Silva, and L. Reid, Sputum viscosity: Correlation with chemical and clinical features in chronic bronchitis, *Thorax,* **28**:401-408 (1973).

11.   J. Sturgess, A. J. Palfrey, and L. Reid, Rheological properties of sputum, *Rheol. Acta,* **10**:36-43 (1971).

12.   P. W. Kent and A. Allen, Biosynthesis of intestinal mucins, *Biochem. J.,* **106**:301, 645-658 (1968).

13.   R. Havez and G. Biserte, Étude biochimique des secretions bronchique. In *Hypersecretion Bronchique. Colloques International de Pathologie Thoracique.* Lille, 1969, p. 43.

14.   C. Rossbach, Cited by Calvert. *Berl. Klin. Wochschr.,* (1882).

15.   J. Calvert, Effect of drugs on the secretion from the tracheal mucous membrane, *J. Physiol.,* **20**:158-164 (1896).

16.   M. Dos Ghali, Thèse Med. Paris, cited by Policard and Galy (1945).

17.   S. Fuchs-Wolfring, Über den feineren Bau der Drüsen des Kehlkopfes und der Luftröhre, *Arch. Mikrosk. Anat.,* **52**:735-761 (1898).

18.   H. H. Dale, The action of certain esters and ethers of choline and their relation to muscarine, *J. Pharmacol. Exp. Ther.,* **6**:147-190 (1914-1915).

19.   W. B. Kountz and K. Koenig, Studies of bronchial secretion, *J. Allergy,* **1**:429-433 (1929-1930).

20.   H. W. Florey, H. M. Carleton, and A. O. Wells, Mucus secretion in the trachea, *Br. J. Exp. Pathol.,* **13**:269-284 (1932).

21.   O. Larsell and R. S. Dow, Innervation of the human lung, *Am. J.Anat.,* **52**:125-146 (1933).

22.   V. Tappan and V. Zalar, The pathophysiology of bronchial mucus, *Ann. N.Y. Acad. Sci.,* **106**:722-745 (1963).

23.   I. Werner, Studies on glycoproteins from mucous epithelium and epithelial secretions, *Acta Soc. Med. Upsal.,* **58**:1-55 (1953).

24.   K. P. Klassen, D. R. Morton, and G. M. Curtis, The clinical physiology of the human bronchi. III. The effect of vagus section on the cough reflex, bronchial caliber and clearance of bronchial secretions, *Surgery,* **29**:483-490 (1951).

25.   O. C. Brantigan, M. D. Kress, and E. A. Mueller, A surgical approach to pulmonary emphysema, *Chest,* **39**:485-499 (1961).

26.   R. V. Goco, M. B. Kress, and O. C. Brantigan, Comparison of mucous

glands in the tracheobronchial tree of man and animals, *Ann. N.Y. Acad. Sci.,* **106**:555-571 (1963).

27.   V. E. Henderson, The action of drugs on salivary secretion, *J. Pharmacol. Ex. Ther.,* **2**:1 (1910).

28.   Medical Research Council. Definition—Classification of chronic bronchitis for clinical and epidemiological purposes, *Lancet,* **1**:775-779 (1965).

29.   L. Reid, Pathology of chronic bronchitis, *Lancet,* **1**:275-279 (1954).

30.   L. Reid, Measurement of the bronchial mucous glands: A diagnostic yardstick in chronic bronchitis, *Thorax,* **15**:132-141 (1960);

31.   H. Harris, Cell growth and multiplication. In *General Pathology,* 3rd ed. Edited by H. W. Florey, London, Lloyd-Luke, 1964, p. 514.

32.   R. Jones, P. Bolduc, and L. Reid, Goblet cell glycoprotein and tracheal gland hypertrophy in rat airways: The effect of tobacco smoke with or without the anti-inflammatory agent phenylmethyloxadiazole, *Br. J. Exp. Pathol.,* **54**:229-239 (1973).

33.   J. M. Frasca, O. Auerbach, V. R. Parks, and J. D. Jamieson, Electron microscopic observations of the bronchial epithelium of dogs. II. Smoking dogs, *Exp. Mol. Pathol.,* **9**:380-399 (1968).

34.   H. L. Falk, P. Kotin, and W. Rowlette, The response of mucus-secreting epithelium and mucus to irritants, *Ann. N.Y. Acad. Sci.,* **106**:583-608 (1963).

35.   N. M. Lightowler and J. R. B. Williams, Tracheal mucus flow rates in experimental bronchitis in rats, *Br. J. Exp. Pathol.,* **50**:139-149 (1969).

36.   T. Dalhamn and R. Rylander, Reduction of cigarette smoke ciliatoxicity by certain tobacco additives, *Am. Rev. Respir. Dis.,* **103**:855-857 (1971).

37.   T. Dalhamn, Inhibition of the ciliostatic effect of cigarette smoke by oxolamine citrate, *Am. Rev. Respir. Dis.,* **94**:799-800 (1966).

38.   S. Jones, P. Bolduc, and L. Reid, Protection of rat bronchial epithelium against tobacco smoke, *Br. Med. J.,* **2**:142-144 (1972).

39.   J. R. May, The bacteriology of chronic bronchitis, *Lancet,* **2**:534-537 (1953).

40.   J. R. May, *Chemotherapy of Chronic Bronchitis and Allied Disorders,* London, English Universities Press, 1968.

41.   W. J. Nungester and L. F. Jourdonais, Mucin as an aid in the experimental production of lobar pneumonia, *J. Infect. Dis.,* **59**:258-265 (1936).

42.   L. Olitzki, Mucin as a resistance lowering substance, *Bacteriol. Rev.,* **12**:149-172 (1948).

43.   N. R. Grist, Viruses and chronic bronchitis, *Scot. Med. J.,* **12**:408-410 (1967).

44.   I. Gregg, A study of recurrent bronchitis in childhood, *Respiration [Suppl.],* **27**:133-138 (1970).

45.   J. F. P. Hers, N. Masurel, and J. Mulder, Bacteriology and histopathology of the respiratory tract and lungs in fatal Asian influenza, *Lancet,* **2**:1141-1143 (1958).

46.   B. G. Bang and F. B. Bang, Responses of the upper respiratory mucosae to dehydration and infection, *Ann. N.Y. Acad. Sci.,* **106**:625-630 (1963).

47. R. Jones, A. Baskerville, and L. Reid, Histochemical identification of glyco-proteins in pig bronchial epithelium: (a) normal and (b) hypertrophied from enzootic pneumonia, *J. Pathol.*, **116**:1-11 (1975).

48. J. D. Cherry and D. Taylor-Robinson, Large quantity production of chicken embryo tracheal organ cultures and use in virus and mycoplasma studies, *Appl. Microbiol.*, **19**:658-662 (1970).

49. V. N. Huyen, Etude histochimique chez l'animal de l'activite de la S-carboxy-methyl-cysteine, *Bull. Physiopathol. Respir. (Nancy)*, **9**:463-464 (1973).

50. H. Selye, R. Vieilleux, and M. Cantin, Excessive stimulation of salivary gland growth by isoproterenol., *Science*, **133**:44-45 (1961).

51. K. Brown-Grant, Enlargement of salivary glands in mice treated with isopropyl noradrenaline, *Nature*, **191**:1076-1078 (1961).

52. H. Wells, The neural regulation of salivary gland growth, *Ann. N.Y. Acad. Sci.*, **106**:654-667 (1963).

53. H. Wells, C. Handelman, and E. Milgram, Regulation by sympathetic nervous system of accelerated growth of salivary glands in rats, *Am. J. Physiol.*, **201**:707-710 (1961).

54. H. Wells, Submandibular salivary gland weight increase by administration of isoproterenol to rats, *Am. J. Physiol.*, **202**:425-428 (1962).

55. J. Sturgess and L. Reid, The effect of isoprenaline and pilocarpine on (a) bronchial mucus secreting tissue and (b) pancreas, salivary glands, heart, thymus, liver and spleen, *Br. J. Exp. Pathol.*, **54**:388-403 (1973).

56. J. M. Shackleford and H. P. Bentley, Changes in the carbohydrate histo-chemistry of the salivary glands and pancreas in cystic fibrosis. In *Research on Pathogenesis of Cystic Fibrosis*. Edited by P. A. di Sant Agnese. Bethesda, National Institute of Arthritis and Metabolic Research, 1964.

57. D. Lamb and L. Reid, The tracheobronchial submucosal glands in cystic fibrosis: A qualitative and quantitative histochemical study, *Br. J. Dis. Chest*, **66**:239-247 (1972).

58. G. J. Barbero, Pathogenesis of cystic fibrosis: Some current views. Cystic Fibrosis. *Ciba Found. Study Group*, **32**:2 (1968).

59. W. Messerklinger, Uber die funktionellen Vorgange bei der Sekretion und Resorption der Schleimhaut der oberen Luftwege und ihre akute Storungen, *Z. Laryngol. Rhinol. Otol.*, **30**:247-268 (1951).

60. K. H. Kilburn, A hypothesis for pulmonary clearance and its implications, *Am. Rev. Respir. Dis.*, **98**:449-463 (1968).

61. E. M. Boyd, Pharmacological agents and respiratory tract fluid. In *Sputum, Fundamentals and Clinical Pathology*. Edited by M. Dulfano. Springfield, Ill., Charles C Thomas, 1973.

62. E. M. Boyd and C. E. Boyd, Expectorant activity of water in acute asthma attacks, *Am. J. Dis. Child.*, **116**:397-399 (1968).

63. I. L. Shannon, Hyperhydration and Parotid flow in Man. SAM-TR-66,97. US Air Force Sch, Aerospace Med 1-4.

64. G. Blanshard, The viscosity of sputum, *Arch. Middlesex Hosp.*, **5**:222-241 (1955).

65.   E. M. Boyd, Expectorants and respiratory tract fluid, *Pharmacol. Rev.*, **6**: 521-542 (1954).

66.   E. M. Boyd, E. P. Sheppard, and C. E. Boyd, The pharmacological basis of the expectorant action of glyceryl guaiacolate, *Appl. Ther.*, **9**:55-59 (1967).

67.   E. M. Boyd, *Respiratory Tract Fluid*, Springfield, Ill., C. C Thomas, 1972.

68.   E. M. Boyd and E. P. Sheppard, On the expectorant activity of Bisolvon, *Arch. Int. Pharmacodyn. Ther.*, **163**:284-295 (1966).

69.   M. Gent, P. A. Knowlson, and F. J. Prime, Effect of bromhexine on ventilatory capacity in patients with a variety of chest diseases, *Lancet*, **2**:1094-1096 (1969).

70.   J. Sturgess, The control of bronchial glands and their secretion, Ph.D. Thesis, University of London, 1970.

71.   J. Sturgess and L. Reid, Secretory activity of the bronchial mucous glands in vitro, *Exp. Mol. Pathol.*, **16**:362-381 (1972).

72.   O. A. Trowell, The culture of mature organs in a synthetic medium, *Exp. Cell. Res.*, **16**:118-147 (1959).

73.   D. Lamb, Intracellular development and secretion of mucus in the normal and morbid bronchial tract, Ph.D. Thesis, University of London, 1969.

74.   J. Sturgess and L. Reid, An organ culture study of the effects of drugs on the secretory activity of the human bronchial submucosal glands, *Clin. Sci.*, **43**:533-543 (1972).

75.   L. Reid, B. Meyrick, and S. Coles, Glycoprotein Synthetic pathways in and drug effects on the human bronchial mucosa in vitro. In *Organ Culture in Biochemical Research*. Edited by M. Balls and M. A. Monnickendam. London, Cambridge University Press, 1975.

76.   H. Brown and H. S. Tong, Erythrocyte membrane stabilisation by Compound III. Lorillard Research Center, 1971.

77.   W. F. Whimster and L. Reid, The influence of dibutyryl cyclic adenosine monophosphate on human bronchial mucous gland discharge, *Exp. Mol. Pathol.*, **18**:234-240 (1973).

78.   R. W. Butcher and E. W. Sutherland, Adenosine $3',5'$-phosphate in biological materials: I. Purification and properties of cyclic $3',5'$-phosphate in human urine, *J. Biol. Chem.*, **237**:1244-1250 (1962).

79.   T. Posternak, E. W. Sutherland, and W. F. Henion, Derivatives of cyclic $3',5'$-adenosine monophosphate, *Biochim. Biophys. Acta*, **65**:558-560 (1962).

80.   J. T. Gallagher, J. C. Marsden, and A. W. Robards, The effect of an inhibitor of Na^+-K^+-Active transport on the secretory activity of rat submaxillary gland cell cultures, *Z. Zellforsch. Mikrosk. Anat.*, **117**:314-321 (1971).

81.   D. Lamb and L. Reid, Quantitative distribution of various types of acid glycoproteins in mucous cells in human bronchi, *Histochem. J.*, **4**:91-102 (1972).

82.   B. Meyrick and L. Reid, In vitro incorporation of [^{3}H] threonine and [^{3}H] glucose by the mucous and serous cells of the human bronchial

submucosal gland. A quantitative electron microscope study, *J. Cell Biol.*, **67**:320-344 (1975).

83. H. Bürgi, Erste klinisch-experimentelle Erfahrungen mit dem Mucolytikum Bisolvon, *Schweiz. Med. Woshenschr.*, **95**:274-278 (1965).

84. R. A. Bruce and V. Kumar, The effect of a derivative of vasicine on bronchial mucus, *Br. J. Clin. Pract.*, **22**:289-292 (1968).

85. H. J. Merker, Elecktronenmikroskopische Untersuchungen uber die Wirkung von N-cyclohexyl-N-methyl-(2-amino-3,5-dibrombenzyl)-ammoniumchlorid auf das Bronchialepithel der Ratte, *Arzneim. Forsch.*, **16**: 509-516 (1966).

86. R. Gieseking and U. Baldamus, Elektronenmikroskopische Befunde an der menschlichen Bronchialschleimhaut nach Behandlung mit Bisolvon, *Pneumonologie,* **137**:1-18 (1968).

87. A. M. Benis, Q. T. Pham, R. Peslin, and E. Puchelle, Systematisation des variables impliques dans les maladies pulmonaires chroniques obstructives: Role de la rheologie des secretions, *Bull. Physiopathol. Respir. (Nancy)*, **9**:277-291 (1973).

88. H. Burgi, Die Viskositat des purulenten und sterilen Sputums bei chronischer Astmnabronchitis, *Med. Thorac.*, **21**:156-167 (1964).

89. R. A. Bruce and V. Kumar, The fibre systems in sputum: Significance and simple techniques of demonstration, *Lab. Pract.*, **16**:316-317 (1967).

90. S. Chodosh, T. C. Medici, and K. Enslein, Comparison of five methods for measuring sputum physical characteristics, *Bull. Physiopathol. Respir. (Nancy)*, **9**:127-139 (1973).

91. S. W. Clarke, J. G. Jones, and D. R. Oliver, Resistance to two-phase gasliquid flow in airways, *J. Appl. Physiol.*, **29**:464-471 (1970).

92. S. W. Clarke, Role de l'ecoulement en deux phases dans la clearance bronchique, *Bull. Physiopathol. Respir. (Nancy)*, **9**:359-372 (1973).

93. R. A. Gibbons and F. A. Glover, The physicochemical properties of two mucoids from bovine cervical mucin, *Biochem. J.*, **73**:217-225 (1959).

94. R. Munies, T. C. Grubb, and R. E. Caliari, Relationship between sputum viscosity and total sialic acid content, *J. Pharm. Sci.*, **57**:824-827 (1968).

95. E. E. Keal, The neuraminic acid content of sputum: Its variation in disease and contribution to the physical properties. M.D. Thesis, University of London, 1971.

96. T. C. Medici, S. Chodosh, S. Ishikawa, and K. Enslein, Proprietes physique du crachat et fonction respiratoire dans la bronchite chronique, *Bull. Physiopathol. Respir. (Nancy)*, **9**:315-323 (1973).

97. M. T. Lopez-Vidriero, J. Charman, E. E. Keal, D. J. de Silva, and L. Reid, Sputum viscosity: Correlation with chemical and clinical features in chronic bronchitis, *Thorax*, **28**:401-408 (1973).

98. P. Degand, P. Roussel, G. Lamblin, G. Durand, and R. Havez, Donnees biochimiques et rheologiques dans l'expectoration. I. Definition biochimique des mucines dans l'expectoration, *Bull. Physiopathol. Respir. (Nancy)*, **9**:199-216 (1973).

99. R. Havez, A. Laine-Bassez, A. Hayem-Levy, and J. Lebas, Donnees

Biochimiques et rheologiques dans l'expectoration. II. Definition bio-chimique des proteines dans l'expectoration, *Bull. Physiopathol. Respir. (Nancy),* **9**:219-235 (1973).

100.   E. Puchelle, J. M. Zahm, and R. Havez, Donnes biochimiques et rheolo-giques dans l'expectoration. III. Relation des proteines et mucines bronchiques avec les proprietes rheologiques, *Bull. Physiopathol. Respir. (Nancy),* **9**:237-256 (1973).

101.   A. M. Lucas and L. C. Douglas, Principles underlying ciliary activity in the respiratory tract. II. A comparison of nasal clearance in man, monkey and other animals, *Arch. Otolaryngol.,* **20**:518-541 (1934).

102.   T. Asmundsson and K. H. Kilburn, Mechanisms of respiratory tract clear-ance. In *Sputum: Fundamentals and Clinical Pathology.* Edited by M. J. Dulfano. Springfield, Ill., C. C Thomas, 1973, pp. 107-180.

103.   V. E. Negus, The function of mucus, *Acta Otolaryngol. (Stockh.),* **56**: 204-214 (1963).

104.   H. C. Ryley and T. D. Brogan, Variation in the composition of sputum in chronic chest disease, *Br. J. Exp. Pathol.,* **49**:625-633 (1968).

105.   G. Biserte, R. Havez, and R. Cuvelier, Les glycoproteides des secretions bronchiques, *Exp. Ann. Biochim. Med.,* **24**:85-120 (1963).

106.   H. C. Ryley and T. D. Brogan, Quantitative immunoelectrophoretic analysis of the plasma proteins in the sol phase of sputum from patients with chronic bronchitis, *J. Clin. Pathol.,* **26**:852-856 (1973).

107.   M. J. Dulfano and W. Philipoff, Physical properties. In *Sputum: Funda-mentals and Clinical Pathology.* Edited by M. Dulfano. Springfield, Ill., C. C Thomas, 1973, pp. 201-242.

108.   S. G. Spiro, M. T. Lopez-Vidriero, J. Charman, I. Das, and L. Reid, Bronchorrhoea in a case of alveolar cell carcinoma, *J. Clin. Pathol.,* **28**: 60-65 (1975).

109.   R. T. H. Laennec, *Traite de l'Auscultation Mediate et des Maladies des Poumons et du Coeur,* 3rd ed., Paris, Chaude, 1831, p. 151.

110.   P. H.-S. Hartley and I. J. Davies, A case of pituitous catarrh, *Br. Med. J.,* **1**:1052-1053 (1923).

111.   E. E. Keal, Biochemistry and rheology of sputum in asthma, *Postgrad. Med. J.,* **47**:171-177 (1971).

112.   T. Ashcroft, Daily variations in sputum volume in chronic bronchitis, *Br. Med. J.,* **1**:288-290 (1965).

113.   D. L. Miller, C. M. Tinker, and C. M. Fletcher, Measurement of sputum volume in factory and office workers, *Br. Med. J.,* **1**:291-293 (1965).

114.   L. E. Warfringe, Paper electrophoresis of sputum in bronchiolar carcinoma (pulmonary adenomatosis), *Acta Med. Scand.,* **153**:49-52 (1959).

115.   C. Gernez-Rieux, C. Voisin, V. Macquet, C. Clayes, G. Piat, and P. Fournier, Cancer alveolaire avec bronchorree incoercible (etude clinique et biologique), *Lille Med.,* **6**:3-9 (1961).

116.   R. Asselain, D. Uzzan, P. Roussel, and P. Degand, Etude clinique, ana-tomique et biochimique d'une observation de cancer bronchiolo-alveo-laire hypersecretant. Colloque Internationale de Pathologie Thoracique, Lille, 1968, pp. 101-114.

117. Z. Dische, C. Pallavicini, J. J. Cizek, and S. Chien, Changes in the control of the secretion of mucus glycoproteins as possible pathogenic factors in cystic fibrosis of the pancreas, *Ann. N.Y. Acad. Sci.*, **93**:526-540 (1962).

118. M. T. Lopez-Vidriero, J. Costello, T. J. H., Clark, I. Das, E. E. Keal, and L. Reid, The effect of atropine on sputum production. *Thorax*, **30**: 543-547 (1975).

119. G. G. Sigmond, Lectures on materia medica and therapeutics now in course of delivery at the Windmill School of Medicine. Lecture 17, *Lancet*, **2**:393 (1836).

120. F. P. Basch, P. Holinger, and H. G. Poncher, Physical and chemical properties of sputum; influence of drugs, steam, carbon dioxide and oxygen, *Am. J. Dis. Child.*, **62**:1149-1171 (1941).

121. A. M. Blair and A. Woods, The effects of isoprenaline, atropine and disodium cromoglycate on ciliary mobility and mucous flow measured in vivo in cat, *Br. J. Pharmacol.*, **35**:379-380 (1969).

122. G. Blanshard, Sputum viscosity and postoperative pulmonary atelectasis, *Chest*, **37**:75-81 (1960).

123. B. R. Hillis, The assessment of cough-suppressing drugs, *Lancet*, **1**:1230-1235 (1952).

MUCOCILIARY CLEARANCE MECHANISMS

We have now seen how the ambient air can be modified prior to access to the lungs and have examined some of the characteristics of respiratory cilia, mucous membranes, and respiratory tract secretions. In Chapters 11, 12, and 13 we direct our attention to the mechanism by which the mucociliary apparatus clears respiratory surfaces and to the mucosal capacity for repair after injury.

Some clarification of terms is needed. The word "mucus" is widely used to designate the fluid lining the airways. This usage is justifiable if we remember that the fluid lining airways surfaces is heterogeneous and its exact nature varies in different locations. Another word which can lead to confusion is "clearance." It is generally used to denote the removal of materials from all parts of the respiratory tract. Unfortunately the term "clearance" is also employed to refer to the decrease in the viability of bacteria in the lung. We suggest that it be limited to the actual physical removal of substances from the respiratory tract.

Not long ago it was generally accepted that the airways are lined with a continuous "carpet" of mucus. This concept may be somewhat oversimplified. A continuous "carpet" of mucus might become folded upon itself while passing over the complex contours of the nose and through areas where large surfaces channel into smaller, as at the junction of bronchi. In fact, since the bronchial "carpet" ascends from a multitude of conducting airways into the trachea, one might expect its thickness to become much greater. Other questions are apparent. Did the "carpet" originate in peripheral airways? What was the nature of its communication with alveolar surfactant lining? How do cilia beat in the viscid material thought of as mucus?

In 1934 Lucas and Douglas recognized that respiratory "mucus" was at least a two-layer system. Although we do not yet have a complete understanding of the mechanism, there is growing agreement on certain probable characteristics. Cilia appear to beat in a thin serous fluid which normally reaches to the fully extended cilia tips. A more viscid material is secreted by goblet cells and passes through the periciliary fluid to lie in drops and plaques on the surface. The fluid secreted from mucous glands may contribute to both the serous periciliary layer or the surface layer. In any event, the periciliary fluid appears to be a continuous layer of relatively uniform depth. The mucus itself may be a discontinuous layer, sparse in peripheral airways and covering more and more of the surface as the total cross section of the pulmonary conducting airways is reduced.

From the alveoli to the point at which cilia first appear, surface tension forces may lead to a flow of surfactant toward the bronchioles. Occasional cilia in respiratory bronchioles may assist in this outward movement. Still, the region of very small peripheral airways remains one of mystery. It is probably the seat of very early change in many pulmonary diseases but, because of the difficulty in studying it either in vivo or in vitro, we still have little certain knowledge of either its normal physiology or the early changes associated with disease.

Maximal velocity of the fluid lining the airways probably occurs just at the tips of the extended cilia and decreases closer to the cell surface. Islands of mucus move by adhering to the moving periciliary fluid or by actual contact with cilia tips. The joining of ciliary streams (as at bronchial spurs and, perhaps, where paranasal sinus fluid enters the nose) may produce whirls, eddies, and resultant relative stasis. Although all clearance streams are not parallel, they all normally move in the general direction of the pharynx; but in some conditions (as in chronic bronchitis) they may completely stop or reverse direction in certain areas.

Rates of clearance vary widely in apparently normal experimental animals and man. Differences exist between and within individuals but the rate of transport is a characteristic of an individual and changes relatively little over time. What rate, either average or local, can be so slow as to impair respiratory defenses is an important question, but correlation between rates of mucociliary clearance and susceptibility to respiratory disease is lacking. Now that we have a better understanding of the clearance mechanism and improved methods for measuring it in humans, the extent to which faulty clearance systems are involved in the pathogenesis of respiratory disease may be discovered.

Donald F. Proctor

11

Nasal Mucociliary Systems

BETSY G. BANG and FREDERIK B. BANG

The Johns Hopkins University
School of Hygiene and Public Health
Baltimore, Maryland

I. Aspects of Mucus in Mucociliary Systems

There are four major characteristics of the mucociliary system which are common to vertebrates and to many invertebrates: (a) secretion of mucus to form a surface cover; (b) associated ciliary beat to transport the secretion; (c) coordinated directional movement of surface secretions by the cilia; and (d) the presence of leukocytes in these secretions [13]. All four of these obtain in the vertebrate nose. All have been derived from the mucociliary filter-feeding mechanisms of the invertebrates; the patterns in the vertebrate mucociliated respiratory systems probably began to evolve in the ciliated pharyngeal funnel of neotenous larval forms of tunicates [15].

A. Mucus In Vivo and In Vitro

Mucus is not a standardized viscoelastic substance, but a protean mixture of molecular species. Throughout evolution it has adapted to a variety of functions thanks to its flexible physiochemical properties. Its carbohydrate molecules may also provide information to the underlying epithelia [47]. In some animals, pheromonal information is delivered via epithelial mucus [45]. The regulation of mucus is a mystery. Even its localized normal composition fluctuates in response to environmental stimuli both internal and external.

Nasal mucus evidently has its own peculiarities: 16 amino acids are found present in consistently differential amounts [35]. Taylor reports that secretions from human nasal serous and mucous acinar cells and goblet cells all consist of acid mucopolysaccharides and mucoprotein, but only goblet cells are sulfated [63]. Lamb [56] had earlier found that sulfates were taken up by both acinar gland cells and goblet cells. Taylor made the important observation that capillary loops penetrate the basement membrane of the respiratory epithelium and lie between the epithelial cells (Fig. 1); in this way fluids pass directly onto the surface and surface mucus may be maintained at optimal dilutions.

Presumably systemic dehydration would deplete these fluids. In experimental dehydration of chickens there was progressive retardation of the mucociliary transport rate, then retention of mucus within the nasal acini. Mucus was anchored in the acini by heavy strands, indicating profound alteration of its physiochemical properties. When such severely dehydrated chickens were given water or were injected with pilocarpine, mucociliary transport was restored within minutes and acinar mucus was reasonably normal histologically by 60 minutes [8] (Fig. 2).

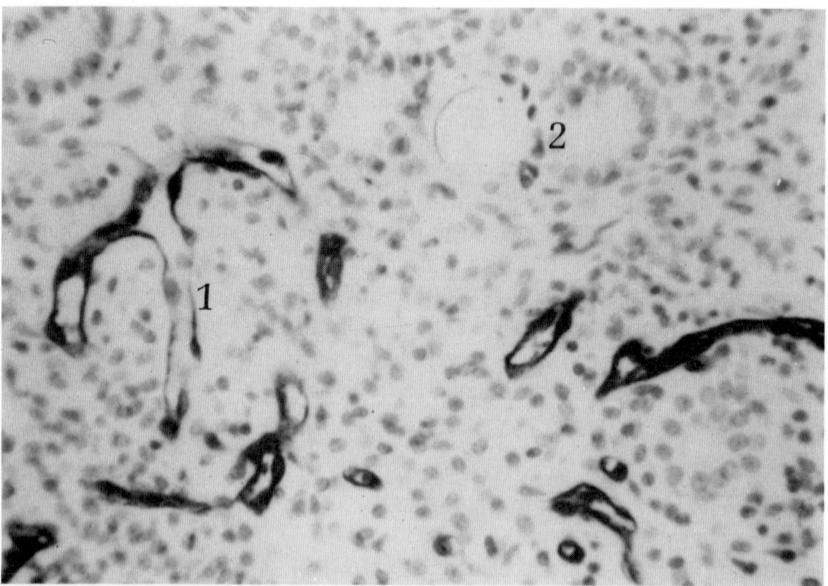

FIGURE 1 Capillaries lying among the basal cells of the epithelium. 1, Several ducts are present. 2, Section cut parallel to the basement membrane. Gomori's method counterstained with hematoxylin. ×450. (Reprinted courtesy of the author [63] and the Editors of *Laryngoscope*.)

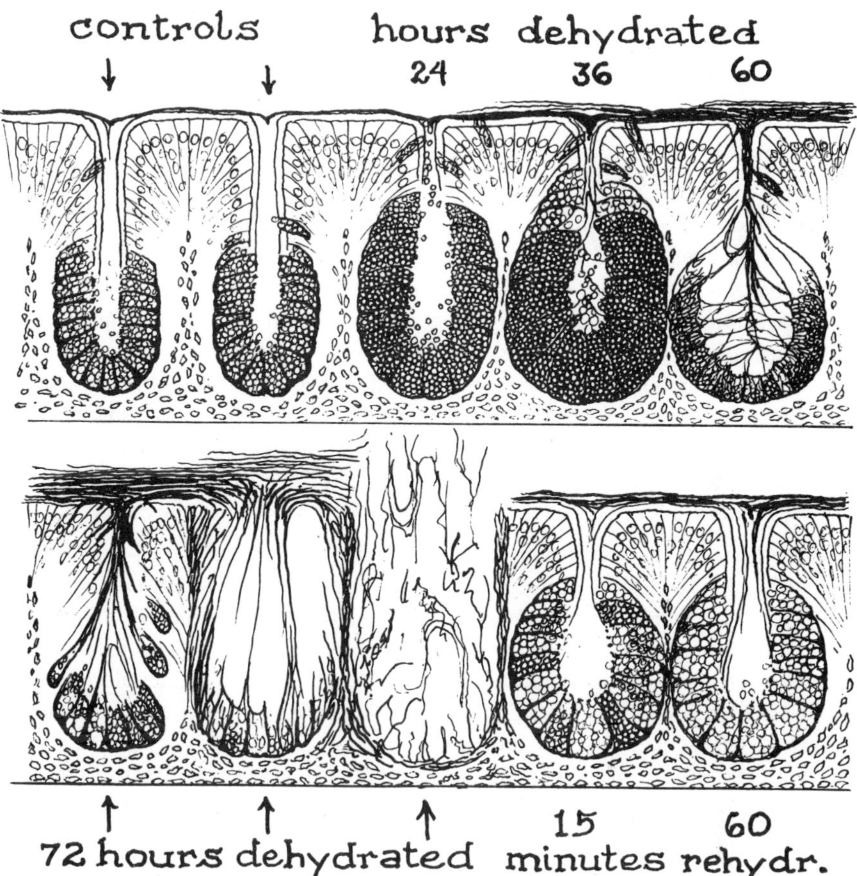

FIGURE 2 Sequence of effects of dehydration and rehydration on the concentration and retention of mucus in glands along the main (chicken nasal respiratory) airway. Small droplets do not represent mucus granules, but suggest relative intensity of periodic acid-Schiff staining. (Reprinted courtesy of the Editors of *Annals of the New York Academy of Science*.)

Human mucociliary flow rates, on the other hand, were consistently doubled over control rates in subjects who drank two cups of hot sweet tea in succession [12]. These increased rates may have resulted from increased volumes of fluid passing onto the surface by way of (distended?) epithelial capillaries.

Organ cultures of nasal and lower respiratory tissues have been very useful in studies of nasal acinar and goblet cell secretions, and in responses of mucosal tissues to drugs and infectious agents. Bang and Niven [14] first used organ cultures of human embryonic nasal and tracheal explants to study effects of myxoviruses on these tissues. Lamb found that acetylcholine caused secretory discharge from acinar but not from goblet cells in cultures of human bronchus, and that atropine delayed the acinar effect [58]. Boat and Kleinerman [18], among others, have also found that cholinergic mechanisms control nasal and tracheal acinar, but probably not goblet cell, secretions. They found that cholinergic agents stimulated equal release of mucous glycoproteins and nonmucin proteins from the mixed acinar glands.

The concept of a continuous sheet of mucus in the vertebrate lower respiratory tract has been revised by the studies of Iravani [37] who has demonstrated that it comprises aggregates of identifiable droplets. The report of Kim [45] on the ultrastructure of mucous and secretory cells in the rat sublingual gland also describes differences in mucous types. Keiser-Nielsen [42] proposed some time ago that the mucus in mucociliary systems is a gel-like sheet or blanket, the lower surface of which is shaggy, with strands which extend into the periciliary spaces. This concept is supported by current electron microscope studies in our laboratory which show that unhomogenous secretions of chick nasal turbinates have a smooth lumenal surface and an under surface in direct contact with the tips of the cilia [58a].

Methods for high resolution study of the surface configuration [43] of the mucous blanket, and of mucus secretions in invertebrates and vertebrates, are now technically available by scanning and transmission electron microscopy. The question of the biological activity of the numerous secretory products other than mucus which feed into the invertebrate blanket is, to a large extent, unexplored. The vertebrate situation is even more complex, for depending on the species, products of lacrimal, Harderian, exophthalmic, septal, lateral, and maxillary compound glands; and secretions from ciliated, goblet, olfactory gland cells, olfactory nerve cells, and olfactory supporting cells, are all incorporated in the nasal secretions. Leukocytes, immunoglobulins, and other molecules synthesized locally in nasal membranes are also carried in the blanket. If each of these secretory products has evolved one or more defensive properties against invading organisms, the mucous blanket is indeed a formidable arsenal. Some of the secretory products and their cellular constituents may be highly species-specific in function.

In regard to the histochemical staining properties of the intranasal and paranasal mucous glands, we would emphasize that among the dozen or so small mammals and over 100 species of birds whose nasal secretory systems we have examined histochemically, homologous glands often show significantly different, even reverse, species-specific, staining properties in regard to acid, neutral, sulfated, or mixed moieties. These staining characteristics often change during and after infection and during physiological alterations in the host, including gnotobiotic rearing and specific nutritional deficiency [9].

Yarington has pointed out that the staining characteristics in human salivary glands and their ducts also change in chronic infection. He suggests that the mucous and serous glands and their ducts are multipotential in regard to their secreted products [77].

B. Invertebrate Responses

There have been several excellent reviews of filter-feeding mechanisms in invertebrates, which are in turn included in a systematic quantitative review by Jörgensen [39]. These invertebrate filter-feeding mechanisms will therefore not be described here. In relation to mucociliary "loading," however, it seems relevant to recall that *Ostrea* has the capacity to handle large-heavy and small-light particles separately in two different ciliary currents, one of which goes to the mouth and the other (which rests until stimulated by heavy particles) to a disposal site [3]. Molluscan gills have long been a favorite model in which to study the coordinated directional movement of mucus by cilia. The various functions of the leukocytes in the mucus and body fluids in several invertebrates, and their relationship to immunologic responses in vertebrates, are reviewed in a recent volume on invertebrate immunology [27].

1. A Model for Study of Mucus Regulation

Study of the adaptive capability of mucociliary mechanisms in invertebrates has been somewhat neglected, with the outstanding exception of the studies of the Roumanian physician Cantacuzene, who studied responses of an invertebrate mucociliated cell system to bacterial infection and to foreign cells [23]. The urn cell is a ciliated mucus-secretory cell which originates in the lining epithelium of the celom of the marine invertebrate *Sipunculus nudus.* Thousands of these cells detach from the epithelial mucosa and become a system of free-swimming scavenger cells which remove foreign and cellular debris from the celomic fluid. This they do by secreting a small tail of sticky mucus (Fig. 3) in which such particles become stuck as the cells swim at random in the fluid. This is a selective stickiness, for the animal's own celomic cells do not adhere to the mucus tail, clearly distinguishing between "self" and "not self."

Cantacuzène found that urn cells would respond to injection of bacteria into the celom by hypersecreting long tails of mucus in which the bacteria were trapped and were eventually cleared from the celomic fluid.

Urn cells can be cultured quite easily in sterile celomic fluid [10]. A drop of cultured "urn fluid" contains 50 to 100 urns. Their activity can be directly observed in an open depression slide by light microscopy. They can be stimulated to hypersecrete mucus in this in vitro system, not only by adding bacteria to the drop of fluid, but by adding macromolecules from several specific secretory sources. One is derived from a factor which is released into the serum of a bacterially infected sipunculus in vivo; the factor is not active in unheated filtered serum, but if the acute bacteremic filtered serum is heated to 80-90°C for 5 min, a drop of this serum causes intensive hypersecretion in all urn cells in a drop

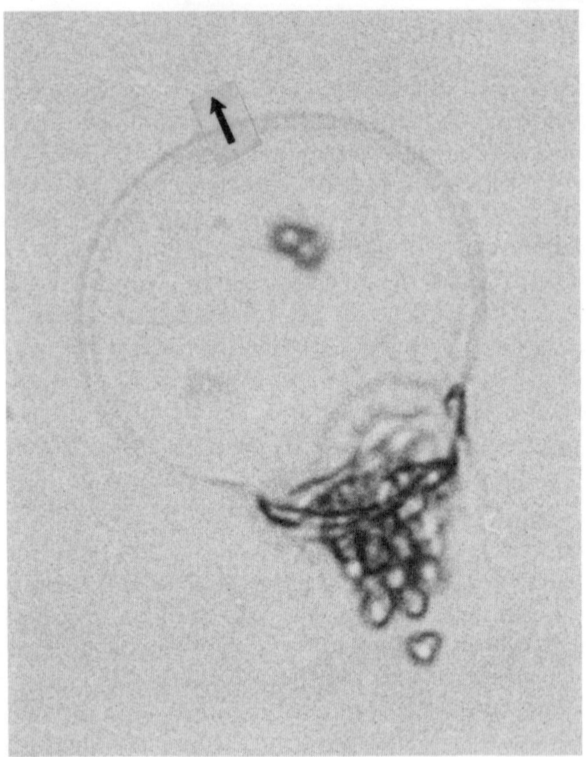

FIGURE 3 Living cultured urn cell swimming in its own host serum. It comprises a clear vesicle cell attached to a mucus secretory ciliated base cell. The normal small tail of sticky mucus has cell debris stuck in the mucus. Ciliary beat is too rapid to show in the photograph. Cell diameter about 30 μm. Unstained.

of culture fluid [10] (Fig. 4). This factor is not present in heated serum of uninfected sipunculus.

The active molecule in the heated serum has been found to be a large molecule. In work still under way, large molecules which induce morphologically different types of hypersecretion in urn cells have also been found in heated human serum and unheated human urine. The molecules in all three cases (heated sipunculus acute serum, heated human serum, unheated urine) have been characterized as large molecules by Spinco ultracentrifugation and by Amicon Centriflo cone ultrafiltration. All factors withstand freeze-thawing [11].

These studies seem directly relevant to the question of *macromolecular regulation of mucus secretion in animal systems.*

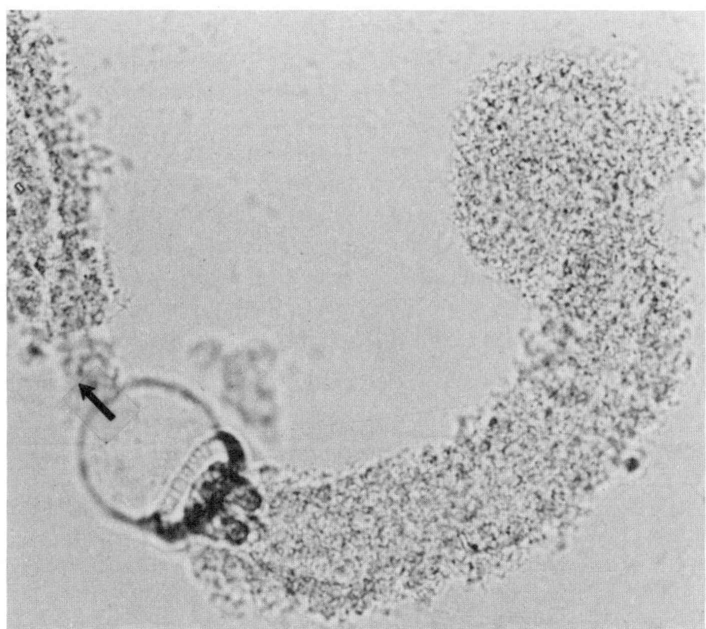

FIGURE 4 Living cultured urn stimulated by heated acute sipunculus serum to hypersecrete mucus. The urn is swimming in the direction of the arrow, secreting as it goes. Tails from 10 to 50 times the length of the vesicle cell can be secreted within 5–10 min. Cell diameter about 30 μm. Unstained.

Studies of the control of secretion in organ cultures could be profitably coordinated with studies of urn cell responses to secretions from normal and abnormal human subjects.

II. Nasal Mucociliary Activity

A. Aspects of Comparative Structure and Function

In small laboratory mammals (mouse, rat, guinea pig, ferret, hamster, rabbit), evolutionary pressures concerned chiefly with olfaction and dentition have defined the shape of the nasal cavity and the distribution of nasal tissue systems. The olfactory conchae, and their protective mucociliary conchae, are highly complex. Presumably the complex maxillary (inferior) conchae of all small laboratory animals provide far better protection of the lower tract epithelia (better filtration, absorption, and disposal of airborne particles and gases) than do the reduced defenses of the human nose.

The structure of the nostril and nasal vestibule, the relative mucous membrane surface area, and the relative complexity of the underlying conchae in the nasal fossae of mammals and birds are geared to the importance of olfactory discrimination in a given species. The olfactory conchae of highly olfactory-dependent species show varieties of simple or multiple branching or scrolling which extend the receptor surfaces according to the need for precise discrimination between odorants in an animal's milieu [2]. The size and complexity of the mucociliated conchae are in turn roughly proportional to those of the olfactory set. Adams [1] has summarized the relative olfactory-respiratory mucosa as reported for a few species of small mammals; the ratio is roughly 55% olfactory to 45% respiratory, with a shrew having the greatest olfactory rate at 69%.

In nature, the prototypes of the commonly used laboratory mammals all have moderately or extremely complex conchal structures. The conchae of humans and chickens are relatively simple. These observations are illustrated in Figs. 5 and 6. The fact that conchal complexity may vary greatly even in related species of mammals and birds [7], emphasizes that the need for refined discrimination exerts evolutionary pressure on intranasal gross structure, as shown in Fig. 7.

The three chief functional areas of the internal nose are the squamous cell-lined vestibule, the mucociliated "respiratory" membrane area which includes the sinuses and their ostia, and the olfactory membrane area. Each serves or is served by the mucociliary epithelial system.

1. The Nasal Vestibule

The nasal vestibule functions chiefly to baffle, narrow, direct, warm, and humidify the incoming airstream. In the latter case the vestibule acts as a reservoir

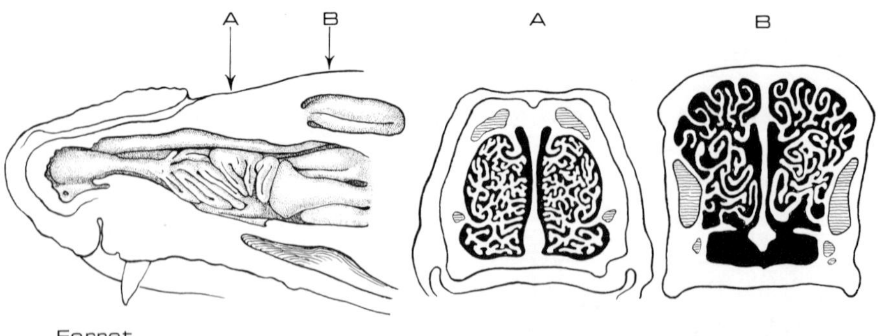

FIGURE 5 Complexity of the maxillary conchae of ferret (level A in lateral wall is level of vertical section A) in relation to development of olfactory conchae, Level B.

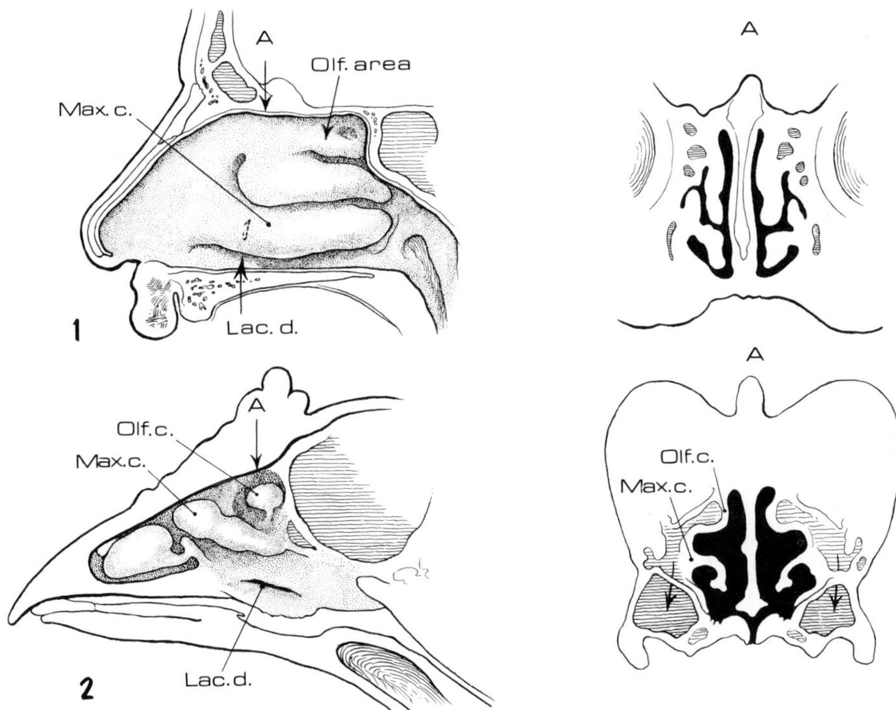

FIGURE 6 Relatively simple conchal structures of human and chicken. The avian anterior concha has no mammalian homolog. Olfactory epithelia are chiefly in the arch of the nasal fossa, the rest are mucociliated.

area for droplets of serous or mucoserous secretions. In nearly all mammals and birds these secretions are produced by serous or mucoserous lateral nasal glands, all of which discharge into the nasal vestibule. Broman [20] first suggested their possible role in humidifying the incurrent airstream, and Bøjsen-Møller [19] has proposed that the droplets of physiologic fluid which they produce are atomized with each inspired breath, thus providing the saline vapor required for efficient olfactory reception and respiration. Fish lack these glands; survival of land forms depends on their capacity to manufacture their own olfactory-respiratory saline vapor.

Although the lower primates have lateral nasal glands homologous with those of other mammals, the higher primates and man usually lack them after their fleeting appearance and subsequent resorption in very early embryonic life. The functional analog in humans is probably the serous moiety of the mixed mucous-serous nasal glands [19]. The ducts of these glands begin to develop at the anterior mucocutaneous junction area of the fetal nose by the 10th gestational week, when goblet cells are already actively secreting in the

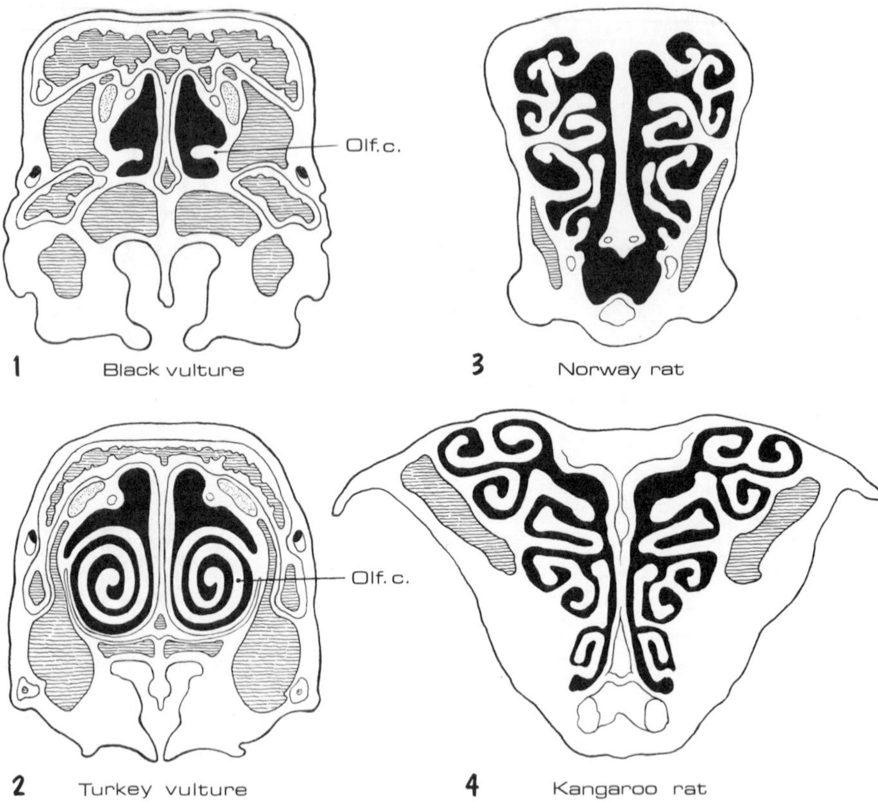

FIGURE 7   Relative evolution of olfactory conchae in the American black vulture (1) which hunts visually and the turkey vulture which hunts by olfaction (2); and in the common Norway rat (3) and the desert kangaroo rat (4) which have different habits and habitats.

same area. They increase numerically at this site, and then successively posterior-ward. The developing acini and their ducts begin secretion by the 16th week, and by the 21st gestational week the total number of prospective adult ducts and glands is established [6]. Meanwhile, larger thin-walled ducts and glands have developed in the maxillary sinus antrum and proximal sinus wall. The patterned distribution and the patterned direction in which the nasal gland ducts are directed (anteriorly, or fanwise, or choanally) are completely obscured in the adult, but show clearly in stained, cleared whole mounts of developing embryonic tissues [6]. The timetable of development of the nasal glands parallels that of the tracheal glands, which also develop in anteroposterior succession [66].

2. The Mucociliated Epithelium

The mucociliated epithelium of the small laboratory mammals is a sheet of ciliated cuboidal cells along most of its extent, with concentrations of almost pure goblet cells in the ventral portion of the membrane. There are occasional shallow mucous acini on the lower septum and nasal floor.

A mucociliated glandular maxillary sinus is present in nearly all nose-breathing birds and mammals; the ostia invariably adjoin the orally directed portion of the olfactory epithelium. In birds, the ostia are encircled by deep, thin-walled mucous glands qualitatively different from those in the nasal fossa proper. These sinuses and their ostia may constitute additional defense against drying and damage to the olfactory membrane, especially during prolonged or rapid oral breathing. Mice also have septal glands of a type neither mucous nor serous, which are heavily innervated and apparently have an intracellular ion transport system which discharges the products into the nasal vestibule via quite long ducts [24].

Chickens and humans have a type of mucociliated membrane quite different from that of the small mammals. In both, a ciliated pseudostratified cuboidal epithelium studded with goblet cells overlies a tunica propria liberally supplied with secretory acini which excrete onto the surface. In the chicken these acini secrete an acid mucus in the basal area and neutral mucus in the duct mouths [9], relationships which are often altered during abnormal environmental or pathologic conditions. It should be emphasized that in certain physiologic states, dehydration for example, mucus becomes anchored within the glands and there is no ciliary movement [8]. The human mixed acini in young adults normally have more mucous than serous units, in middle age serous units predominate, and in old age nearly all of the superficially positioned units are serous, the mucous ones deeper [66]. The trigeminal sensory receptor cells of the human nasal respiratory mucosa are all nonmyelinated and end in simple terminal arborizations in the lamina propria; they do not reach the surface [24].

3. The Olfactory Membrane

In terms of its cellular components—not the secretory and enzymatic products of these components—the olfactory membrane is basically similar in all vertebrates. Ontogenically, the proximal nerves develop free of any connection with the central nervous system. Increasingly precise methods of electron and scanning microscopy, of electrophysiology, and of enzyme chemistry continue to add new information to this field which was long neglected because of technologic restraints. The most inclusive reviews on the rapid developments in olfactory anatomy and physiology up to 1968 are a series of symposia on olfaction and taste sponsored by the Wenner-Gren Center [71], and the initial volume of a projected series on chemoreception [38].

As presently understood, the olfactory membrane consists of the ducts, olfactory glands, ciliated terminal receptor rods of the olfactory nerve cell axons, and surrounding groups of supporting cells which are covered at the lumen surface by long uncoordinated motile cilia; these overlie a deeper layer of basal cells which rest on a basement membrane; the lamina propria contains olfactory glands (Bowman's), connective tissues, stromal cells, and olfactory nerve fasciculi partly or completely ensheathed in perineural squamous epithelial cells. Blood vessels do not reach the epithelial cell layer in adults, but in fetal and infant olfactory membranes there are capillaries in the olfactory mucosa which are intimately associated with the perikarya of the sensory cells [55]. These later regress and the entire avascular mucosa rests on a basement membrane. The olfactory nerve cells, the supporting cells and the glands of Bowman are secretory, the basal cells are not. A report on the olfactory epithelium in rhesus monkeys indicates that olfactory basal cells may be functionally associated with the perineural epithelial cells [63]. This report states that unlike the basal cells of the mucociliary membrane, the olfactory basal cells do not give rise to surface cells, nor are they immature supporting cells. However, an ultrastructural and autoradiographic study of frog olfactory epithelium clearly showed that after resection of olfactory nerves there was extremely active mitosis of basal cells which differentiated to form a new population of neurons [32]. This author suggests that in this model the basal cells are capable of giving rise to both supporting and sensory cells, and at times even to cells which resemble respiratory membrane cells.

The locations and concentrations of numerous hydrolytic and oxidative enzymes in the mucous membrane tissues and in the mucoserous surface sheet have been identified in both adult [63,73] and embryonic [62] subjects. Current studies are beginning to elucidate the separate and sequential stages of enzymatic responses to olfactory stimuli; Shantha and Nakajima [63] suggest that the surface secretions themselves are metabolically active. It is known that the rate of secretion is stimulated to increase during odorant reception [31].

Since the olfactory membrane lacks an intrinsic vascular system, it is clear that nutrients must reach the cells in some other way. Electron microscope studies of rhesus olfactory tissues indicate that access is by way of substantial gaps in the perineural sheaths of certain groups of olfactory nerve fasciculi just distal to the basal cells, so that nutrients can reach the membrane through the basal cells [63]. These same "nude" areas may contribute to the considerable loss of olfactory capacity in advancing age, since the increase in connective tissues could cover them and choke and destroy the delicate nerve fibers.. Mulvaney and

Heist [52], in electron microscope studies of rabbit mucosae, have found that olfactory cell development may persist well into adult life, and Graziadei [32] found that there was normal turnover of both sensory and supporting cells (10 times more rapid in sensory than in supporting cells) in frogs. It is now clear that there are great differences in the normal dynamics of olfactory membrane cells of different animal species. Some of the implications of these differences will be discussed in reference to regeneration of olfactory tissues in Chapter 13.

B. Olfaction

Chemical communication systems have now been found in most of the principal animal phyla. In addition to reception of odors intrinsic to the environment, animals respond to specific chemical signals (pheromones) produced by their own species. Pheromones are secreted outside of the body from exocrine glands and they function to transmit specific information to another member of the same species by means of smell or taste. Both the signaling and the response are genetically programmed [74]. Among most mammals, pheromones give signals for recognition, alarm, trailing, state of estrus, dominance, and territoriality, and chief among the signals are those related to sexual attraction and sexual stimulation, including the remarkable feedback to population growth first recognized in mice [21,72].

To a great extent, this "pheromonal language" explains the dominance of the olfactory sense in the small mammals which have been adopted for laboratory use. There is sexual dimorphism in the skin glands of many species, and, not surprisingly, seasonal variation [34,53]. In humans, the remarkably low threshold for odorants [26] has, up to now, rarely been exploited except by perfumers and trackers, and the physiological importance of human olfaction is just beginning to be systematically evaluated. About one-half of human sensory perceptions are olfactory, eliciting not only alarm and recognition, but memory and emotional responses. The role of olfaction in reproductive success is currently being investigated in several laboratories. It is known that some differences in human odors are sex-linked and are pheromonal releasers of endocrine responses; one study concluded that use of the "pill" by women depresses responses of human males [51]. By all means the best discussion of human olfactory perception in the light of olfactory information theories is that of Engen [29], who cuts through decades of guesswork and presents a framework within which information theories can be fruitfully tested in human subjects.

III. Dynamics of Nasal Mucociliary Flow

Forty years ago Lucas and Douglas [49,50] demonstrated four important comparative points about mucociliary flow: (a) the direction of air flow is deter-

mined chiefly by the shape of the nose and location and angle of the external
nares; (b) the direction of air currents is a factor in the development of the pattern of mucociliary flow, and the two often move in countercurrent directions;
(c) the most vigorous regions of ciliary activity in nonprimate laboratory animals are in the upper parts of the nose while in primates rates are most rapid
along the floor; (d) the "mucus sheet" is composed of two layers, an outer
stratum of mucus which rests on a moving fluid layer of low viscosity in which
the cilia beat freely. They found that forward motion of the mucous blanket
along the nasal roof toward the nares was characteristic of the laboratory mammals, while all motion was choanally directed in humans and monkeys.

We would add two observations: First, there is a very rapid mucociliary
current emerging from the dorsal rim of the sinus ostium in mice, rats, and ferrets;
one important biological function of this rapid maxillary current is suggested by
the fact that after rats are exposed to heavy concentrations of cigarette smoke
there is an immediate inflammatory response by the sinus epithelia, followed by
hypersecretion of the maxillary gland [69]. Second, in humans, while the major
portion of the mucous blanket is carried choanally, in some individuals a significant amount is carried forward to accumulate at the rim of the nasal vestibule
[57,12].

Not surprisingly, in humans and chickens there is a direct correlation between the alignment of nasal cilia and the alignment of the underlying acini and
their ducts. This was first determined in chickens by combined experimental and
whole-mount staining techniques [4]. Both the flat septal surface and the conchal surface of the nasal membranes were exposed in sagittal sections of heads
of freshly killed birds. Brisk ciliary activity was evident, seen through the dissecting microscope. Dots of india ink were placed in various distal sites on the
membrane, and were carried anteriorly along consistent pathways which were
easily mapped [5]. Having mapped these pathways, the entire septum and the
maxillary concha were dissected out, fixed, and stained with periodic acid-Schiff
stain to identify the mucoproteins. These preparations when cleared in anise oil
showed a remarkable lineation of the acini which conformed completely with
the pathways of ciliary flow. The presence of lines of cilia in the clear spaces
between the rows of glands was confirmed in histologic sections.

Having determined these interrelations, movement of the entire sheet
could then be observed by carefully flooding india ink along the cribriform plate
of a freshly sectioned septum and observing the movement of the entire sheet,
which went forward in an arc. Most of the stained portion eventually moved into
the choana. Figure 8 shows this forward arc in a preparation which was fixed as
the sheet moved forward, and was then stained and cleared. These interrelationships between the positioning of acinar glands and lines of cilia are presumably
genetically programmed.

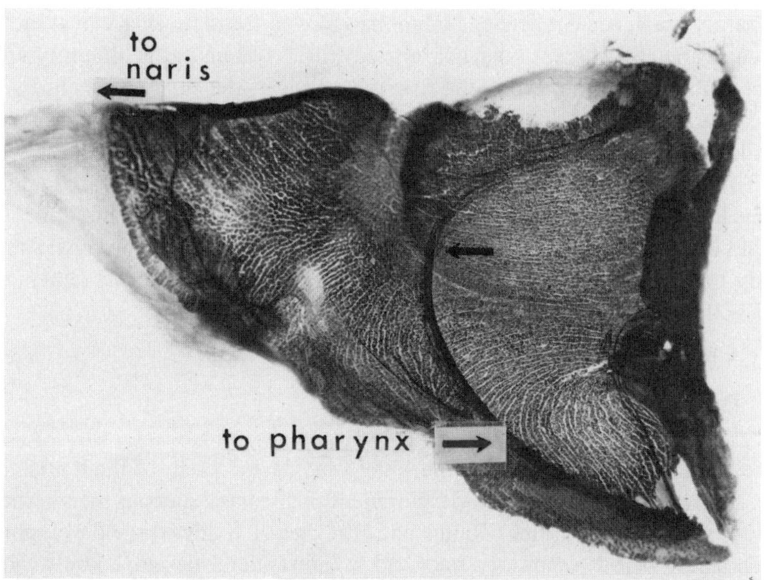

FIGURE 8 Whole mount of chicken nasal septum, acini stained with PAS. The clear lines between rows of acini are rows of ciliated cells. India ink was spread along the cribriform plate of the freshly dissected septum, cilia carried this forward in an arc (arrow), and at this point formol alcohol was flooded onto the surface, arresting ciliary transport.

In humans, the correlation between the patterned rows of cilia and the position of the underlying acini and their ducts is demonstrable in developing embryos. In a series of embryos aged from the 10th gestational week to full term, whole mounts of the developing nasal acini and their ducts were stained with periodic acid-Schiff stain, and the cleared preparations were observed as in the chicken preparations. The patterned distribution of the glands could then be nicely correlated with visible rows of patterned surface corrugations on the mucosae of two full-term infants. Again, these interrelationships were demonstrable in histologic sections [6]. Naessen has shown that these surface and histologic correlations also obtained in whole mounts and histologic sections of a 10-month-old infant [54].

The small laboratory mammals lack the submucosal nasal acini present in humans and birds. A similar alignment of cilia, however, is visible as minute corrugated rows seen at high power with indirect light in a dissecting microscope, and the lineation is consistent with the directional flow of the blanket. In vertebrates the olfactory glands presumably produce mucus of a highly favorable consistency for mucociliary flow. Sadé and his associates [60] have shown that mucus, specifically, is essential for the transport capability of cilia in the respiratory tract, since it acts as a coupling agent to translate ciliary beat into mucus clearance. There is no synchronized ciliary system in the olfactory area. There is no mucus produced in the first two-thirds of the ciliary epithelium adjacent to the olfactory area in most laboratory mammals. Mucus transport by ciliary traction therefore depends completely on coordination of these two independent components. The proximal third of the nasal mucociliated epithelium is liberally supplied with goblet cells, and the composition of the mucous blanket changes thenceforward. Simple histochemical staining demonstrates clearly that AB-positive, PAS-positive, and mixed varieties of mucus remain discrete within the nasal mucous blanket.

IV. Ontogeny of Nasal Epithelia

There is a wealth of literature on the embryonic development of the human nose, but recent information on development of the nasal mucous membranes per se is surprisingly wanting. Kanda and Hilding [4] followed the development of cilia in the rabbit respiratory tract and found that cilia developed between the 22nd and 28th days, appearing first in the nose, then in larynx, trachea, and bronchi in that order. Histologic sections of developing chick embryo noses show only squamous epithelia at 9 days; the cell layers in the olfactory area are about four times deeper than those in the respiratory area, and contain the anlagen of Bowman's glands. Cilia first show on day 16 at the olfactory-respiratory junction, and by day 18 a full ciliated-goblet cell cover is established in the respiratory area. The youngest of the human embryos used in the study of developing nasal glands [6] was approximately 10 weeks' fetal age and histologic sections of the nasal fossa showed that both the olfactory and respiratory epithelia were well differentiated, with fully formed cilia and with goblet cells (often ciliated) which stained for mucins. Since cilia do not appear in human bronchi until 11-13 weeks, it may be that the schedule of development of human respiratory membranes is similar to the rabbit's. Evidently the nasal membranes, which are of ectodermal origin, consistently develop in advance of the lower tract membranes which derive from gut-associated entoderm.

Over forty years ago Anderson Hilding asked, and answered, the question whether the several types of epithelia found in the nose were stable evolutionary types or whether they were interchangeable, depending on environmental factors. He surgically closed one nostril in a series of rabbits, then sequentially followed the effects by histologic comparison of the closed side and the open side where airflow had been doubled. There was great alteration in the epithelia in both sides: a marked increase in goblet cells and proportional lack of ciliated cells in the closed side, and loss of cilia with stratification of epithelia on the open side. He concluded that nasal epithelial types are not fixed but are changeable in direct relation to exposure to air: stratified and nonciliated in exposed sites, ciliated in moderately exposed areas, and predominantly goblet in sheltered areas [35]. In this context it is of interest that during fetal development of the human nasopharyngeal epithelium there is a gradual transition of the ciliated epithelium into stratified squamous epithelium in the caudal portion of the nasopharynx [40].

Other ways in which epithelial cells may be altered in vivo have been defined in organ cultures of embryonic skin, especially by Fell and her colleagues. In 1953, with Mellanby, she discovered a remarkable effect resulting when excess vitamin A was added to the medium: keratinization was arrested and the basal cells differentiated into mucus-secreting epithelium with ciliated areas [30].

There have been several subsequent studies directed toward understanding the role of vitamin A in epithelial responses. Vitamin A deficiency was produced in organ cultures of prostate glands of young mice by growing them in a defined medium which lacks protein [47]; a similar result was achieved in organ cultures of chick trachea by adding the vitamin A antagonist, citral, to the cultures [4]. In both systems the normal epithelia showed progressive squamous, then keratinizing, metaplasia. The sequence of events in metaplasia of the tracheal epithelia of vitamin A deficient rats has been followed in electron microscope studies by Wong and Buck [75].

Recent studies by Dingle et al. [28] have revealed some of the limitations in the sequence by which retinol-binding protein delivers retinol to peripheral sites, including the plasma membrane at the cell surface. Lack of vitamin A has profoundly different effects at different ages during growth in different animals.

V. Nasal Secretory Antibody

Among the determinants of resistance to infection, antigen-antibody response is one of the most crucial. The nose is the main portal of entry for antigens potentially harmful to the respiratory tract. It is provided with both circulating and locally secreted antibodies in the form of immunoglobulins, a first line of defense which is reinforced by the sentinel types of lymphoid tissue found at all main portals of entry into the body. Cellular immune responses follow activation of local antibody responses. Immune cells and their products are in the surface secretions, the epithelial cell layer, the submucosa, and the vessels of the lamina propria. Upon antigenic stimulation there is rapid local synthesis of secretory immunoglobulins even before clinical symptoms are manifest [22], and at much higher titers than in serum. The principal secretory immunoglobulin is IgA.

IgA antibodies in maternal serum are not available to the fetus [65,76]. The development of humoral immunity after birth largely depends on maturation plus experience with antigens. Most studies show that serum IgA levels reach adult values very slowly, varying from 4 to 16 years [65], but concentrations of local secretory IgA apparently increase much more rapidly: for example, Selner et al. showed 92% of normal infants had adult levels of IgA in saliva by 28 days of age [61]. Presumably this reflects local antigenic stimulation at the mucous membrane level; there are no data on the timetable of increases in nasal secretory IgA.

The concept of local immunity is nearly 50 years old [16], yet its relevance to local immunization has been fully recognized and exploited only in the last decade. It is now clear that, in distinction to serum IgA, secretory IgA is locally synthesized. Recent information about different aspects of secretory IgA is available in reviews by Bienenstock [17] on its biological significance, by Tomasi and Gray [65] on its structure and function, by Waldman [70] and Heremans [33] on its localized functions, by Chanock [25] on its clinical significance, and by Rossen et al. [59] on its relation to viral infections.

VI. Summary

In summary, this chapter has reviewed selected points about mucociliary systems, with emphasis on past and current research relevant to functions of the nasal mucous membranes:

1. Mucus secretion and ciliary beat are independent, but directional transport of mucus by cilia is an interdependent event.

2. Cilia will transport only mucus, and will effectively transport only mucus which has particular physical qualities.

3. The quality of mucus is altered by altered physiologic states and by infection.

4. Small laboratory mammals have nasal mucosae which have evolved to protect the olfactory receptor system; their intranasal conchae as well as their mucous membranes thus differ structurally from those of humans; the chicken nasal fossa and its membranes are more analogous to the human.

5. The epithelial stem cells in the respiratory portion of the nasal epithelium develop into mucous, ciliated, squamous, or keratinizing cells depending on environmental and nutritional factors. This does not obtain in the olfactory membrane.

6. The first line of defense against upper respiratory disease is rapid local synthesis of intranasal antibodies, reinforced by cellular immune responses.

7. The alignment of cilia and the alignment of mucous acini are structurally and functionally integrated in humans and chickens.

8. The anterior mucotaneous junction area of the nose is evidently of maximum importance in meeting incoming antigens, and as a resource area for the saline vapor required for mucociliary function, olfaction, and respiration.

9. Individual mucus-secretory cells in an invertebrate model system secrete mucus which recognizes cells as "self" and "not self." Factors which may regulate mucus secretion in humans have begun to be investigated in the same system.

References

1. D. R. Adams, Olfactory and non-olfactory epithelia in the nasal cavity of the mouse, Peromyscus, *Am. J. Anat.,* **133**:37-45 (1972).
2. E. D. Adrian, Olfactory discrimination, *Ann. Psychol.,* **50**:107-113 (1951).
3. D. Atkins, On the ciliary mechanisms and interrelationships of lamellibranchs, *Q. J. Microsc. Sci.,* **79**:339-373 (1937).
4. M. B. Aydelotte, The effects of vitamin A and citral on epithelial differentiation *in vitro, J. Embryol. Exp. Morphol.,* **11**:279-291 (1963).
5. B. G. Bang, The surface pattern of the nasal mucosa and its relation to mucus flow, *J. Morphol.,* **109**:57-72 (1961).
6. B. G. Bang, The mucous glands of the developing human nose, *Acta Anat.,* **59**:297-314 (1964).
7. B. G. Bang, Functional anatomy of the olfactory system in 23 orders of birds, *Acta Anat. [Suppl.],* **58**:1-76 (19).
8. B. G. Bang and F. B. Bang, Responses of upper respiratory mucosae to dehydration and infection. *Ann. N.Y. Acad. Sci.,* **106**:625-630 (1963).
9. B. G. Bang and F. B. Bang, Experimentally induced changes in nasal mucous secretory systems and their effect on virus infection in chickens, *J. Exp. Med.,* **130**:105-119 (1969).
10. B. G. Bang and F. B. Bang, Mucous hypersecretion induced in isolated mucociliated epithelial cells by a factor in heated serum, *Am. J. Pathol.,* **68**:407-422 (1972).
11. B. G. Bang and F. B. Bang, Invertebrate model for study of macromolecular regulators of mucous secretion, *Lancet,* ii:1292-1297 (1974).
12. B. G. Bang, A. L. Mukherjee, and F. B. Bang, Human nasal mucus flow rates, *Johns Hopkins Med. J.,* **121**:38-40 (1967).
13. F. B. Bang, Immune reactions among marine and other invertebrates, *Bioscience,* **23**:584-589 (1973).
14. F. B. Bang and J. S. Niven, A study of infection in organized tissue cultures, *Br. J. Exp. Pathol.,* **39**:317-322 (1958).
15. N. J. Berrill, *The Origin of Vertebrates,* Oxford, Clarendon Press, 1955.

16. A. Besredka, *Local Immunization*, Williams & Wilkins, Baltimore, Md., 1927.

17. J. Bienenstock, The significance of secretory immunoglobulins, *Can. Med. Assoc. J.*, **103**:39-43 (1970).

18. T. F. Boat and J. I. Kleinerman, Human respiratory tract secretions 2. Effect of cholinergic and adrenergic agents on in vitro release of protein and mucous glycoprotein, *Chest [Suppl.]*, **67**:325-345 (1975).

19. F. Bøjsen-Møller, The anterior nasal glands, *Int. Rhinol.*, **3**:117-127 (1965).

20. I. Broman, Über die Entwickelung der konstanten grösseren Nasenholendrüsen der Nagetiere, *Z. Anat.*, **60**:439-586 (1921).

21. H. M. Bruce, Smell as an exteroceptive factor, *J. Animal Sci. [Suppl.]*, **25**:83-89 (1966).

22. W. T. Butler, Changes in IgA and IgG concentrations in nasal secretions prior to the appearance of antibody during viral respiratory infections in man, *J. Immunol.*, **105**:584-591 (1970).

23. J. Cantacuzène, Recherches sur les réactions d'immunité chez les invertébrés, *Arch. Roumaines Pathol. Exp. Microbiol.*, **1**:7-80 (1928).

24. N. Cauna, K. Hinderer, and T. Wentges, Sensory receptor organs of the human nasal respiratory mucosa, *Am. J. Anat.*, **124**:187-210 (1969).

25. R. M. Chanock, The secretory immunologic system. Proceedings of a Conference on the Secretory Immunologic System. Dec. 10–13, 1969, Vero Beach, Florida, Washington, D.D., H.E.W., N.I.H., U.S. Govt. Printing Office, 1969.

26. A. Comfort, Communication may be odorous, *New Scientist Sci. J.*, **49**: 412-414 (1971).

27. E. L. Cooper (ed.), Contemporary topics in immunobiology. In *Invertebrate Immunology*, Vol. 4, New York, Plenum Press, 1974.

28. J. T. Dingle, H. B. Fell, and D. S. Goodman, The effect of retinol and of retinol-binding protein on embryonic skeletal tissue in organ culture, *J. Cell. Sci.*, **11**:393-402 (1972).

29. T. Engen, *Advances in Chemotherapy*. Edited by J. W. Johnston, Jr., D. G. Moulton, and A. Turk. New York, Appleton-Century-Crofts, 1970, pp. 361-383.

30. H. B. Fell and E. Mellanby, The effect of hypervitaminosis A on embryonic limb-bones cultivated in vitro, *J. Physiol. (Lond.)*, **116**:320-349 (1952).

31. P. P. C. Graziadei, Nasal mucous membranes, *Ann. Otol.*, **79**:433-422 (1970).

32. P. P. Graziadei, Cell dynamics in the olfactory mucosa, *Tissue Cell*, **5**: 113-131 (1973).

33. J. F. Heremans, Immunoglobulin formation and function in different tissues, *Curr. Top. Microbiol. Immunol.*, **45**:131-203 (1968).

34. E. R. Hesterman and R. Mykytowycz, Some observations on the intensity of odors of anal gland secretions from the rabbit *Oryctolagus cuniculus*, *CSIRO Wildlife Res.*, **13**:71-81 (1968).

35. A. Hilding, Experimental surgery of the nose and sinuses, *Arch. Otolaryngol.*, **16**:9-18 (1932).

36. A. C. Hilding, E. J. Cowles, and J. H. Stuart, Amino acid composition of normal human nasal mucus, *Ann. Otol. Rhinol. Laryngol.*, **82**:75-79 (1973).

37. J. Iravani and A van As, Mucus transport in the tracheobronchial tree of normal and bronchitic rats, *J. Pathol.*, **106**:81-93 (1972).

38. J. W. Johnston, Jr., D. G. Moulton, and A. Turk (eds.), Communication by chemical signals. In *Advances in Chemoreception*, Vol. 1, New York, Appleton-Century-Crofts, 1970.

39. C. B. Jörgenson, Quantitative aspects of filter feeding in invertebrates, *Biol. Rev.*, **30**:391-454 (1955).

40. R. Kanagasuntheram and M. Ramsbotham, Development of the human nasopharyngeal epithelium, *Acta Anat.*, **70**:1-13 (1968).

41. T. Kanda and D. Hilding, Development of respiratory tract cilia in fetal rabbits. Electron microscopic investigation, *Acta Otolaryngol. (Stockh.)*, **65**:611-624 (1968).

42. H. Keiser-Nielsen, *Mucin*, Kφbenhavn, Dansk Videnskobs Forlag A/S, 1953.

43. R. D. Kelley, R. A. F. Dekker, and J. G. Bluemink, Ligand-mediated osmium binding, *J. Ultrastruct. Res.*, **45**:254-258 (1973).

44. D. Kerjaschki, The anterior medial gland in the mouse septum: an uncommon type of epithelium with abundant innervation, *J. Ultrastruct. Res.*, **46**:466-482 (1974).

45. S. K. Kim, C. E. Nasjleti, and S. S. Han, The secretion process in mucus and serous secretory cells of the rat sublingual gland, *J. Ultrastruct. Res.*, **38**:371-389 (1972).

46. H. Kleerekoper, *Olfaction in Fishes*, Bloomington, Ind., Indiana University Press, 1969.

47. I. Lasnitzki, Hypovitaminosis A in the mouse prostate gland cultured in a defined medium, *Exp. Cell Res.*, **28**:40-51 (1962).

48. M. Lippman, Interdisciplinary investigation of mucus production and transport, *Ann. N.Y. Acad. Sci.*, **130**:963-964 (1966).

49. A. M. Lucas and L. C. Douglas, Direction of flow of nasal mucus, *Proc. Soc. Exp. Med. Biol.*, **31**:320-321 (1933).

50. A. M. Lucas and L. C. Douglas, Principles underlying ciliary activity in the respiratory tract, *Arch. Otolaryngol.*, **20**:518-541 (1934).

51. R. P. Michael and E. B. Keverne, Pheromones in the communication of sexual status in primates, *Nature*, **218**:746-748 (1968).

52. B. D. Mulvaney and H. Heist, Regeneration of rabbit olfactory epithelium, *Am. J. Anat.*, **131**:241-252 (1971).

53. R. Mykytowycz, Territoriality in rabbit populations, *Austral. Nat. Hist.*, **14**:326-329 (1964).

54. R. Naessen, The anatomy of drainage of nasal secretions, *Acta Otolaryngol.*, **84**:1231-1234 (1970).

55. R. Naessen, An enquiry on the morphological characteristics and possible changes with age of the olfactory region in man, *Acta Otolaryngol.*, **71**:49-62 (1971).

56. V. Negus, The function of mucus: A hypothesis, *Proc. R. Soc. Med.,* **60**: 75-77 (1967).

57. D. F. Proctor, I. Andersen, and G. Lundkvist, Clearance of inhaled particles from the human nose, *Arch. Intern. Med.,* **131**:132-139 (1973).

58. L. Reid, Mucus in respiratory disease, *Proc. R. Soc. Med.,* **60**:78-80 (1967).

58a. M. Reissig, B. G. Bang, and F. B. Bang, Electron and light microscopy of chick nasal mucous membranes, *Fed. Proc.,* (In press).

59. R. D. Rossen, J. A. Kasel, and R. B. Couch, The secretory immune system: Its relation to respiratory-viral infection, *Prog. Med. Virol.,* **13**:194-238 (1971).

60. J. Sadé, N. Eliezer, A. Silberberg, and A. C. Nevo, The role of mucus in transport by cilia, *Am. Rev. Respir. Dis.,* **102**:48-52 (1970).

61. J. C. Selner, D. A. Merrill, and H. N. Claman, Salivary immunoglobulin and albumin, *J. Pediat.,* **72**:685-689 (1968).

62. B. L. Shapiro, Enzyme histochemistry of embryonic nasal mucosa, *Anat. Rec.,* **166**:87-97 (1970).

63. T. R. Shantha and Y. Nakajima, Histological and histochemical studies of the rhesus monkey (*Macaca mulatta*) olfactory mucosa, *Z. Zellforsch.,* **103**: 291-319 (1969).

64. M. Taylor, The origin and function of nasal mucus, *Laryngoscope,* **84**: 612-636 (1974).

65. T. B. Tomasi and H. M. Grey, Structure and function of immunoglobulin A, *Prog. Allergy,* **16**:81-213 (1972).

66. H. H. Toppozada and H. A. Gaafar, Electron microscopy of human mixed acini, *J. Laryngol. Otol.,* **87**:639-645 (1973).

67. M. Tos, Development of tracheal glands in man, *Acta Pathol. Microbiol. Scand. [Suppl. 185],* **68** (1966).

68. A. van As and J. Webster, The organization of ciliary activity, *S. Afr. Med. J.,* **46**:347-350 (1972).

69. B. Vidic, S. S. Taylor, M. W. Rana, and B. D. Bhagat, The respiratory glandular system in the rat's lateral nasal wall in normal and polluted environments, *Verh. Anat. Ges.,* **66**:83-85 (1971).

70. R. H. Waldman, Local mucosal immunity: An editorial, *Am. J. Med. Sci.,* **260**:255-260 (1970).

71. *Wenner-Gren Center Symposia on Olfaction and Taste.* Vol. 1, 1963: Edited by Y. Zotterman. Vol. 2, 1965: Edited by T. Hayashi. Vol. 3, 1968: Edited by C. Pfaffman. New York, Pergamon Press.

72. W. K. Whitten, Pheromones and mammalian reproduction, *Adv. Reprod. Physiol.,* **1**:155-177 (1966).

73. J. Willemot, La muquese olfactive chez l'homme vivant, *Acta Otolaryngol.,* **71**:197-205 (1971).

74. E. O. Wilson, In *Chemical Ecology.* Edited by F. Sondheim and J. B. Simeone. New York, Academic Press, 1970, Chap. 7.

75. Y. C. Wong and R. C. Buck, An electron microscopic study of metaplasia in the rat tracheal epithelium, *Lab. Invest.,* **24**:55-66 (1971).

76. C. B. S. Wood, The development of immunity in fetal life and childhood, *J. R. Coll. Phys. (Lond.),* **6**:246-258 (1972).

77. C. T. Yarington, Jr., Diversity of mucus staining characteristics of human salivary glands, *Laryngoscope,* **82**:2103-2121 (1972).

12

Nasal Mucociliary Function in Humans

DONALD F. PROCTOR

The Johns Hopkins University
School of Hygiene and Public Health
Baltimore, Maryland

IB ANDERSEN and GUNNAR LUNDQVIST

University of Aarhus
Aarhus, Denmark

I.   Introduction

Since a large proportion of all inhaled particulates are deposited in the nose and many inhaled foreign gases and vapors are absorbed in the surface mucus of the nasal passage, the nose would be an intolerably contaminated dumping ground were it not provided with an effective mechanism for continual clearance of its surfaces and replenishment of its surface fluids. The mainstay of this clearance system is mucociliary function. Materials deposited upon or absorbed in the surface fluid may make their way through it and reach or penetrate surface cells prior to clearance. The more slowly the surface fluid is cleared into the naso-pharynx and thence into the stomach, the greater the opportunity for such penetration. Because the nose is ordinarily the first target of inhaled materials as well as the site which conditions inspired air, the task of maintaining clean, moist surfaces is greater there than in any other part of the respiratory tract and some special attention to nasal mucociliary function seems justified.

The nature of ciliary activity and the biochemistry and rheology of airway secretions have been the subject of intensive investigation, especially during the past few decades. Although such basic studies are of fundamental importance, it is wise to bear in mind that, in the final analysis, the major physiological role

of the ciliary beat and airway secretion is the effective clearance of surface fluid
and the concomitant removal of materials deposited upon respiratory surfaces.
It is the study of this function in the human nose with which this chapter is
concerned. The clearance mechanism is of obvious importance in the lungs (see
Chap. 14), but it is an inescapable fact that normally the nose is the site of first
contact with the inspired air and, therefore, the site most exposed to injurious
agents in the air.

To serve these functions effectively (air conditioning and cleaning, and
clearance of mucosal surfaces) and to provide for adjustments to rapidly changing conditions in the ambient air, the nose is equipped with a number of special
mechanisms not present in the remainder of the respiratory tract [36]. First,
the vasculature, especially the submucosal erectile tissue over much of the septum and turbinates, is capable of rapid and profound alterations in blood supply
(see Chaps. 3 and 4). Changes in the state of this vasculature can adjust the
nature of the airway itself, surface temperature, and, in part at least, the nature
of surface fluids. These adjustments are accomplished through changes in vasomotor tone and blood flow in submucosal vascular compartments and in surface
mucosal vessels, and, perhaps, alterations in capillary permeability.

Second, the paranasal sinuses, nearly surrounding the nasal passages (Fig.
1), are lined with mucosa producing a surface secretion which is continually
swept into the nose by ciliary activity all along the main path of the inspiratory
airstream. The paths of mucociliary streaming within a sinus are circuitous, but
all eventually lead to the orifices communicating with the nose. The frontal
sinuses drain into the most anterior portion of the middle meatus, the maxillary
antra just posterior to that, the ethmoids along the rest of the middle meatus
(to a lesser extent above the middle turbinate), and the sphenoids above the
posterior end of the middle turbinate [34]. Tears also pass into the nose through
the nasolachrynal duct but into the inferior meatus, not the main path of inspiratory airflow.

Third, it is possible that the mucociliary system in the nose is less prone to
injury than that of the tracheobronchial tree. In the donkey it has been shown
that cigarette smoke which bypasses the nose may impair bronchial clearance,
but when taken directly into the nose it fails to affect nasal clearance [18]. Such
a difference could be attributable to less permeable nasal secretions, the continual addition of paranasal secretions along the line of inspiratory airflow or
other as yet unidentified factors. The permeability of nasal secretions may be
affected by disease [47].

Fourth, the mucociliary system in the main nasal passage (essential for
removal of deposited materials, protection of olfactory surfaces, and maintenance of the fluid surface requisite for humidification) provides in the nose a two-

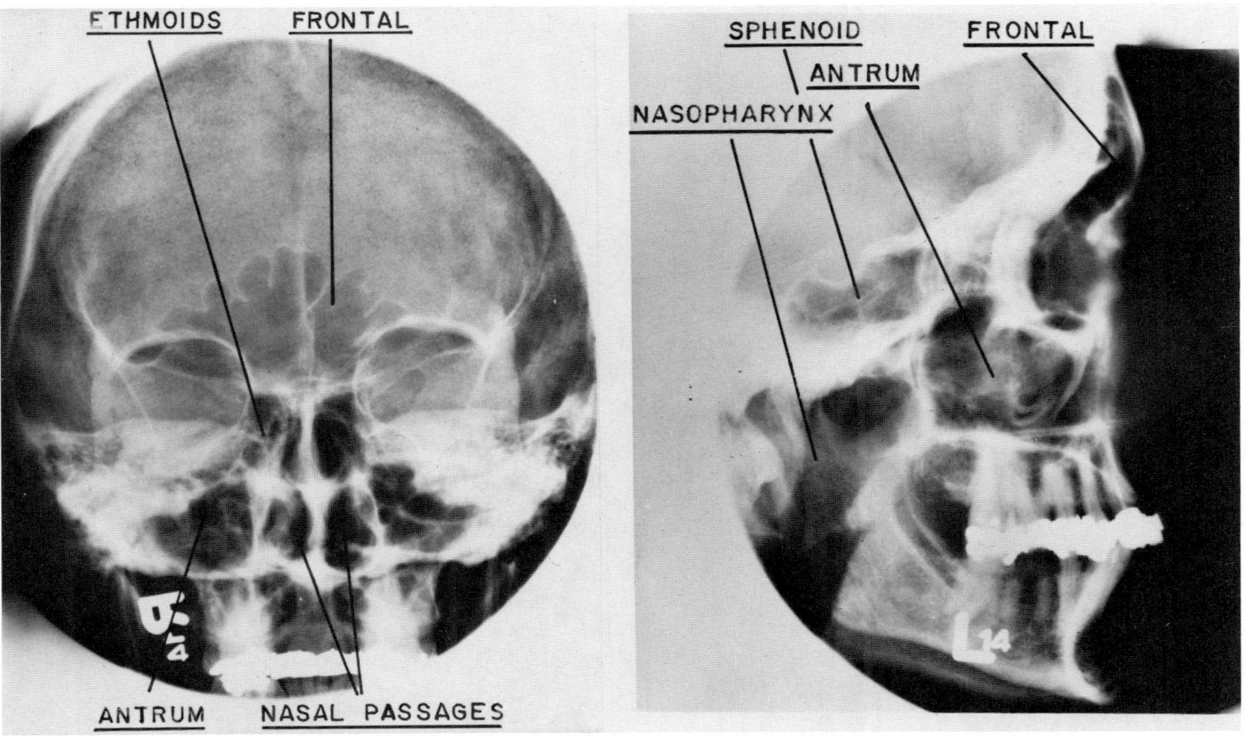

FIGURE 1 Radiographs of the head showing the clustering of the paranasal sinuses about the nasal airway.

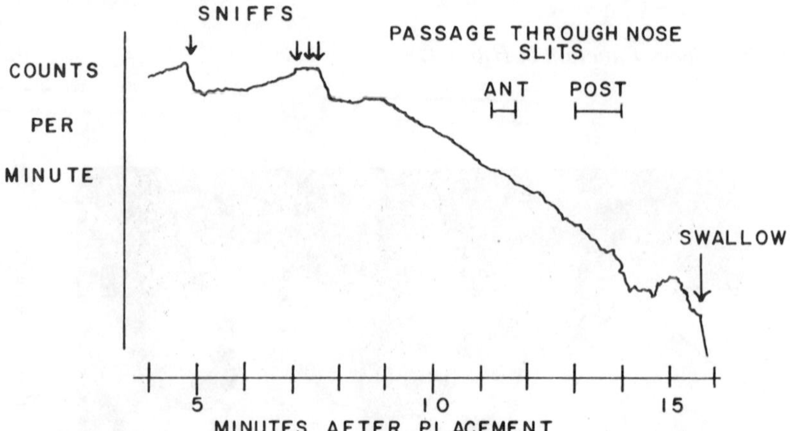

FIGURE 2   Record from a detector placed in front of the nose. Fall in radio-activity (counts per minute) as the particle moves backward is indicated by a falling tracing while a rise in the tracing indicated forward motion. As shown by the arrows repeated sniffs were required to move the particle backward onto the posterior moving stream. At two points distinct anterior motion is indicated. To the right, after passing slits of the collimated detectors, back and forth motion indicates arrival at the nasopharynx, and, on swallowing radioactivity disappears.

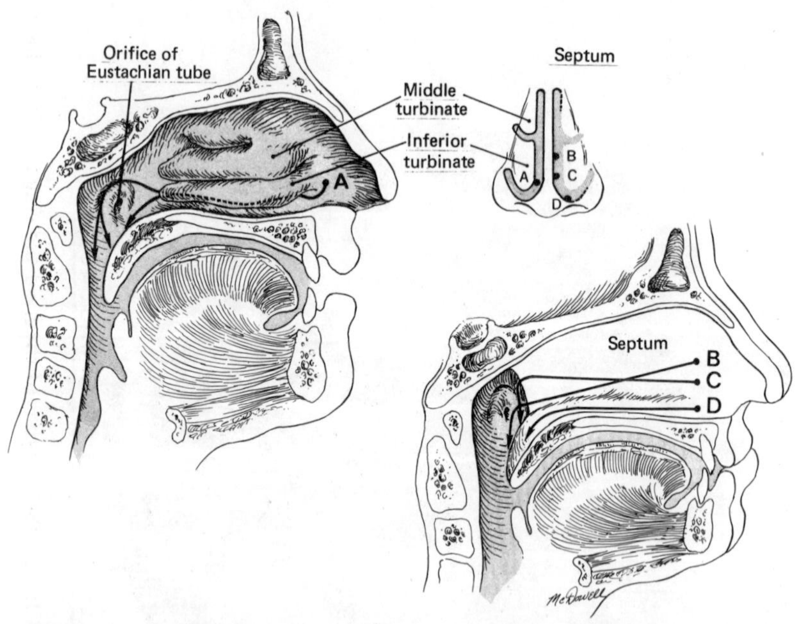

FIGURE 3   Diagram of varying paths of mucus flow. Note that at the naso-pharynx the stream may end behind the soft palate or course around the Eusta-chian orifice. Posteromedially to this orifice are the adenoids. (Used with the courtesy of the Editor, *American Review of Respiratory Disease* [43].)

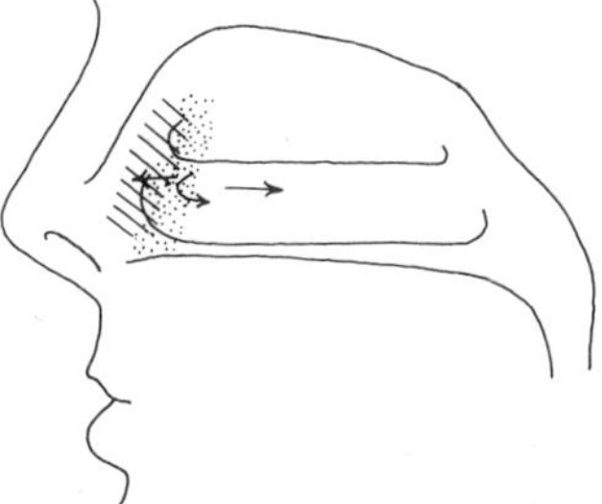

FIGURE 4 Diagram of lateral wall of nasal passage indicating the region without cilia (cross-hatched) and the anterior area in which mucus flow seems to move forward (stippled). The arrows show directions of mucus flow.

way clearance. Whereas the mainstream moves backward to the nasopharynx whence material is swallowed, in the anterior area (where most particle deposition occurs) clearance is forward to a region from which material can be removed by nose blowing, wiping, etc. (Figs. 2 and 4). Finally, in the path of the main backward moving stream, is lymphoid tissue (adenoids) which accumulates samples of inhaled materials deposited from the bending airstream or carried there by the mucociliary stream [3] (Fig. 3). There is evidence that development of immune processes may be initiated in both the adenoids and tonsils. The tonsils are also exposed to all materials which are swallowed [35,48,51].

II. Fate of Inhaled Materials

A. Nasal Deposition and Absorption

In the individual chronically exposed to a particle-laden atmosphere, we might expect those particles which deposit prior to access to the lungs to be distributed over the upper respiratory surfaces in the following fashion: Dependent upon particle size and breathing rate, most of the material will be found in the anterior unciliated region, either initially impacting there or transported there from the anterior ciliated mucosa. That this is the major site of impaction is attributable to the combination of a very high linear velocity with a bend in the airstream (see Chap. 3).

Some particles will be on the anterior ciliated area in transit anteriorly and some, just posterior to this region, will be making a slow transit forward and then backward onto the main ciliated passage (Fig. 4). Most of this material will be near the anterior end of the middle meatus quite close to the anterior ethmoid sinuses. A smaller portion of particles will lie along the main nasal passage being carried backward to the nasopharynx. Some of these will have deposited there from the inspired air, but most will be carried there from a more anterior deposition site. Most of these particles will lie along the middle meatus and on the septum opposite to this area. Finally, there will be two small accumulations of particles in the nasopharynx. One of these will be at the terminal points of the mucociliary clearance paths (Fig. 3) awaiting dispatch to the stomach at the next swallow. The other will be in the crypts of the adenoids, either impacted there from the inspiratory stream or carried there in the mucociliary stream (for anatomical orientation see Fig. 5 in Chap. 3).

During a recent study in Aarhus, Denmark, in which 16 normal subjects were exposed to a black dust-laden atmosphere, we had the opportunity to observe deposition patterns. Although they differed somewhat from subject to subject, we could see quite clearly the black coating of the anterior ends of both inferior and middle turbinates and, to a lesser degree, of the septum just opposite this region. It was of interest that there was no visible coating of nostril hairs, although there was some obvious deposition on the skin of the nostrils. The aerosol was heterogeneous with particle size varying between less than 2 to about 10 μm.

Of the total load of inspired particulates encountered, what proportion might be found at any moment in these various upper respiratory areas and what is the likely residence time in any area for any single particle? Some of the problems posed by that question are discussed in detail in Chaps. 5 and 14 and some crude answers are provided in the preceding paragraph. The simple fact is that we do not have completely satisfactory answers as yet.

There is evidence that the majority of naturally occurring particles, if breathed through the nose, will either pass in and out without respiratory deposition or be deposited in the nose. The exact proportion of particles in the 1-5 μm range which will penetrate to be deposited in the lung has yet to be determined, but scans after nasal breathing of a radioactively tagged water aerosol (4-6 μm) show a heavy concentration in the anterior nares (Fig. 7 in Chap. 4). That the great majority of those deposited find their way to this spot is demonstrated by the ease with which after such a study, radioactivity is cleared on nose blowing and cleaning. Residence time in any one spot is dependent upon multiple variables including mucociliary clearance rate.

One of us deliberately placed a radioactive substance in the anterior nasal region (roughly that indicated by the stippled and hatched areas in Fig. 4) and submitted to recurring scans for 95 minutes. During that time there was not the slightest evidence of movement of the material. On blowing the nose and cleaning the nostril with a handkerchief the radioactivity was completely cleared. Thus it appears that retention time for the majority of the particulates in the nose could be indefinitely prolonged were it not for our tendency to sneeze, blow our noses, and otherwise clean it periodically.

Inhaled gases and vapors which are water-soluble are absorbed in the surface fluid lining the main nasal passage. Just as in air conditioning (see Chap. 4), the narrow airstream, the large surface area, and the slow linear velocity all combine to afford an optimum opportunity for such absorption during the transit of inspired air through the main nasal passage. During another study in Aarhus [6] we found that essentially all SO_2 was removed from inspired air during the nasal passage even after 6-hr exposure to a level of 25 ppm.

B. Significance of the Respiratory Clearance Mechanism

From nearly all parts of the respiratory tract the general direction of mucociliary clearance is toward the hypopharynx. Mucus flow in the bronchi and trachea may be circuitous, and certainly at each spur where two bronchi join the path it must be devious, but the overall movement is away from the lung periphery and toward the larynx. At the larynx the stream converges posteriorly and passes over the posterior commissure into the hypopharynx and is then swallowed.

In the nose and paranasal sinuses, although the path may also be circuitous, overall mucus flow is backward to the nasopharynx. At each swallow, the soft palate moves upward and backward to wipe the nasopharyngeal wall, dispatching material from there to the stomach. As has already been mentioned, one exception to this direction of movement is the region near the anterior ends of the turbinates and meati where mucus flow is forward (Figs. 2 and 4).

It is possible that noxious materials offer a lesser hazard to the anterior squamous epithelium than to respiratory mucosa. The nostril is a relatively infrequent site of disease. Some material in the main passage may be cleared by sniffing or nose blowing, but, ordinarily, the nasal fluid is too sparse to be moved by these actions and mucociliary clearance must be relied upon.

The significance of retention or clearance of materials, whether in the upper or lower respiratory tract, depends upon the material. Whereas a single tubercle bacillus in an alveolus or a few droplet nuclei containing respiratory viruses may establish infection if body cells are reached and penetrated, relatively large quantities of such substances as silica are needed to produce significant lung disease. Carcinogens probably require prolonged or repeated contact with tissues to result in cancer; some inhaled materials, such as tin, appear innocuous to the lung even when retained in large quantities for long periods of time.

As for the nose, even with gases or vapors which dissolve in the fluid lining the main passage with short or periodic exposure, the concentration to which underlying cells are exposed will be influenced by the speed of passage of the main nasal mucociliary stream. How long it takes one of the common cold viruses to reach and penetrate nasal epithelial cells is unknown, and the same is the case for pollen.

In any event the more thoroughly respiratory surfaces are cleaned and the shorter the time of residence for any particle the safer we are. In view of all these considerations it would appear that the combination of removal of the majority of particulates in the anterior nose and the mucociliary clearance of the main passage combine to make the nose normally one of the mainstays of defense against airborne disease.

III. Previous Studies of Mucociliary Function

A great deal of work has been carried out in vitro and in experimental animals to determine the nature of ciliary activity and mucociliary function. That work will not be reviewed here (see Chaps. 11 and 13). Brief reference will be made to some of the attempts of others to evaluate nasal mucociliary clearance in man.

Many workers have addressed themselves to the problem of mucociliary clearance, and much sound and useful information has been obtained [10,16,26, 29,31,44]. Hilding mapped human nasal mucus flow paths as early as 1931, and most of his findings [24,25] have been substantiated by later investigations. But, an exact and detailed evaluation awaited Quinlan's development of a technique (described below) enabling the study of all parts of the nasal cavity without intranasal instrumentation [37,38,43] (Fig. 3).

Ewert [16,17] placed inert dye particles with a powder blower on the anterior nasal septal mucosa and observed their motion backward through a calibrated diploscope (providing opportunity for magnification and photography). He was able to follow the particles over a distance of 2 mm. A nasal speculum was used to provide the field of view, a factor which could alter nasal air flow patterns. This technique is ingenious and useful, but it limited his studies to a small area of anterior septal mucosa, perhaps especially prone to effects of ambient air changes not likely to alter function in the main passage.

Ree [44] and Bang [10] have placed colored materials (Edicol orange and Sky blue) on the anterior nasal mucosa and then looked for them in the nasopharynx. Such methods are useful, especially in field studies, but they fail to yield information on the nature of clearance between the point of placement and the nasopharynx. In addition they require frequent examination with tongue depressor or mirror to detect the posterior appearance, tedious for both examiner and subject.

Quinlan was the first to develop a technique which could be used in man to follow the entire path of nasal mucociliary flow. He placed a radioactively tagged particle on the anterior nasal mucosa and followed its subsequent motion with an Anger camera [43] (Fig. 5). Movement of the particle was displayed on and photographed from an oscilloscope screen. Our modification of his method is described below, and more fully in our publications [4-9].

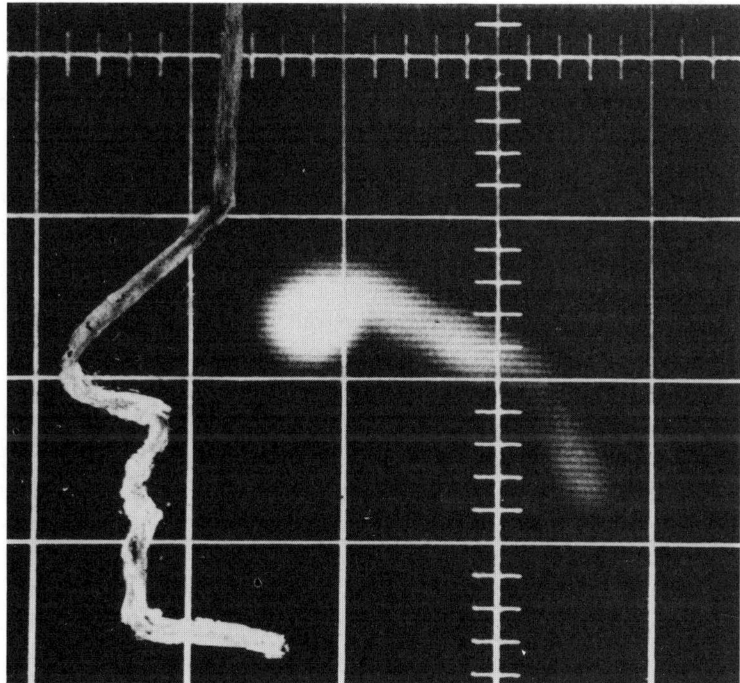

FIGURE 5 Photograph from the oscilloscope screen. Film exposed for 9 min after particle placement. Profile drawn on film for orientation.

IV. Studies by the Authors

Our own work in this field was begun on the premise that an exact knowledge of "normal" function, studied under carefully controlled environmental conditions, must underlie any understanding of the significance and causation of abnormalities. Beginning in 1970 we carried out a series of experiments with normal subjects conducted in the climate control chamber in Aarhus, Denmark [4-9,39,40]. In this chamber we are able to maintain constant environmental conditions, altering only one factor, the possible effect of which is under study.

We find two methods for measuring nasal mucociliary function to be most satisfactory. The first of these is the modification of the Quinlan technique and offers the most detailed information. The second is a method developed by us and, although providing less exact data, is admirably applicable to studies where radioactive techniques are not readily available. This method is far simpler and correlates well with the radioactive method but yields only a measure of the fastest time for total transit of the nose.

A. Methodology

In our modification of the Quinlan technique an inert resin particle (0.5 mm diam) tagged with 3-6 μCi of technetium-99m is placed, under direct vision, on the superior surface of the inferior turbinate some 3-5 cm from the nasal tip, [4,20,22,43]. This site was chosen because it is relatively accessible and it is in the main path of the inspiratory airstream. The placement must be just posterior to the area of anterior ciliary motion. If anterior motion is detected, sometimes a simple sniff will move the particle backward to the proper area. In other instances, on reinspection of the nose the particle had moved 1-2 cm anterior to the point of its original placement; it was removed with forceps and replaced.

Immediately after placement of the particle the subject was seated in a dental chair and the detection apparatus fixed to the head by two solid plugs fitting into the external auditory canals. The detection apparatus consists of three probes, one (uncollimated) immediately in front of the nose, and the other two (collimated) pointed directly across the nasal passage (Fig. 6). These latter two have lead collimators between their faces and the subject's head, the collimation allowing us to record the passage of the particle at six points along the nose. The distance between the first and last observation slits is 5.6 cm.

Thus, we obtain a record against time which yields information as to the motion of the particle backward (away from the anterior probe) and peaks of detected radioactivity as each of the collimated slits is passed. It is difficult to quantify the record from the anterior probe but simple to state exact passage times between each pair of slits. Depending upon the subject's anatomical dimensions, the sixth slit may face across the nasopharynx and thus fail to detect the particle prior to its being swallowed, and the first slit may straddle the anterior forward motion area.

In interpreting our measurements it should be borne in mind that they only show distance traveled in the anterior posterior axis. Ciliary paths which move at an angle to this axis (as in curling around a turbinate) will show up as apparent slowing of the detected motion.

In our second method a particle of saccharine, similar in size to the resin particle, is placed in the same manner. A stop watch is started and the subjects instructed to swallow once every 30 seconds and report the first taste of sweetness. They describe this sensation as sharp and unmistakable, and there is no sense of irritation in the nose. The disadvantages of this method are that it depends upon a subjective response, that it yields no information on variations in clearance speed along the nasal passage, and that it gives only the fastest flow rate. The saccharine undoubtedly dissolves in the airway surface fluid and probably covers an area of several square millimeters. It is the arrival at the front of this area which is the endpoint of the test, and how slowly the remainder

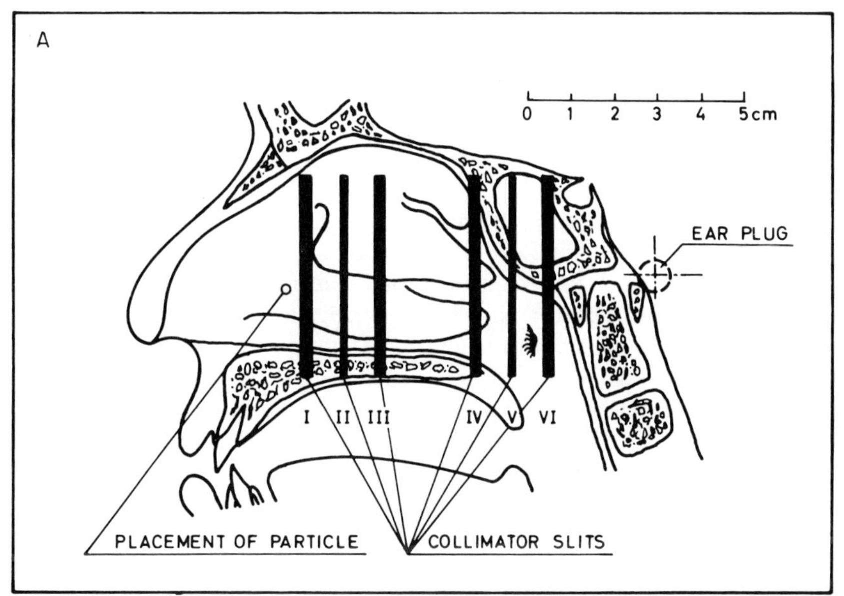

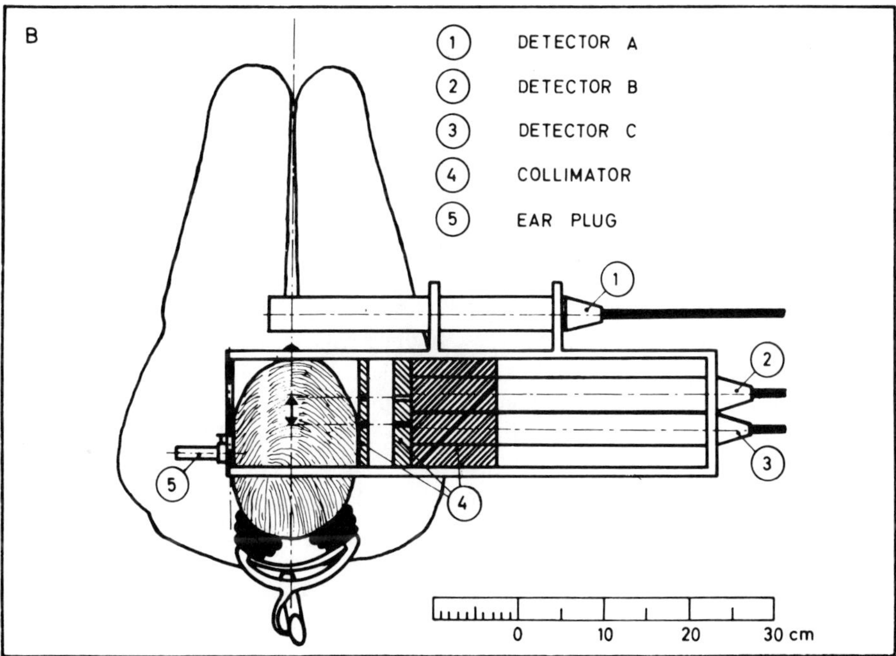

FIGURE 6 (A) Diagram of the nasal passage as seen by the collimated detectors. Initially we employed only the two slits III and IV; but, in all later studies we have used the "comb" collimator for more detailed records. (b) Diagram of the subject seated in chair with detector in place. (Used with the permission of the Editor, *American Review of Resipiratory Disease* [4].)

is cleared cannot be determined. The subjective element of the test can be reduced by using substances with different flavors, or by asking the subject to
describe the detected taste without giving him a clue as to what it will be.

The saccharine test does have major advantages. It is simple to perform,
and can be carried out in the field with only the aid of a light, nasal speculum,
applicator, and timer. It involves no unpleasantness for the subject and no radioactivity and thus could be ideal for clinical investigations, as in children with
cystic fibrosis, or on-site studies in industry. There is a close correlation between
the saccharine test times and those obtained in the same subject with the radioactively tagged particle, but the times for the former are ordinarily faster. Receptors for the taste of sweetness are said to be located in the anterior tongue.
The sharpness with which the subject detects the swallowed saccharine suggests
the possibility that taste buds sensitive to sweet may also lie at the base of the
tongue, on the palate, or on the epiglottis. Physiology texts describe taste buds
in these and other areas besides the tongue, but we have been unable to uncover
any other evidence that they may be sensitive to sweet.

We also tested particle placement on the nasal septum, the floor of the
nose, and the middle turbinate. Similar clearance rates are observed from each
point. The middle turbinate is quite sensitive and touching it may evoke sneezing. Normal subjects studied at different hours of the day show no significant
diurnal variation in clearance.

B. Subjects

Up to December 1974 we had studied 181 normal subjects in the Aarhus collabor-
ative studies (Table 1). Combining these findings with those of other workers it is
now possible to define "normal" nasal mucociliary function if we accept that
term to mean function found in young adult subjects without respiratory symptoms and without gross evidence of nasal disease. We have deliberately included

TABLE 1   Data on Normal Subjects Studied to Date (September 1974)

Studies in:	Males	Females	Age range (years)
Aarhus climate chamber	103	10	20–28
Stockholm	36	–	20–35
England	14	18	18–46
Totals	153	28	18–46
	181		24 (avg)

Average nasal mucus flow rate: 5.3 mm/min; range 0.5–23.6 mm/min.

subjects with minimal to moderate nasal septum deviations and find no correlation between this very common abnormality and the rate of nasal clearance. In an occasional subject we believe we have observed a particle "hanging up" where a septal deflection may touch the turbinate mucosa. It is clear that, even confining ourselves to "normal" subjects, individual variations in function are such as to make it desirable to use each subject as his own control in measuring the effects of some variable. In using one group as control and comparing it with a different group of test subjects very large numbers would be required to obtain statistically significant results.

Basic data on the subjects studied thus far are found in Table 1. Because of the possible effect of the menstrual cycle on mucosal function, we have confined most of our studies to male subjects [23]. Those females who have been included show no striking differences from the males. We have found no relation between smoking habits and nasal clearance, but only a very few of our subjects smoked as much as a pack a day. No smoking was allowed on the day of a study.

C. "Normal" Function

The average rate of clearance of the single particle is 6 mm/min, but this figure is probably of little significance in view of the wide range—from less than 1 mm/min to over 20 mm/min. In repeated measurements on the same subject, either in 1 day or on different days, consistence in flow rates is common enough so that one quickly learns to characterize slow or fast movers. Only rarely do we find both a very fast and a very slow measurement in the same subject, but occasionally one of several measurements may be three to four times faster or slower than other measurements in the same subject.

The total transit time from particle placement to the point at which it is swallowed varies from 4 or 5 minutes to over 30. We usually discontinue the measurement if total transit is not complete at 30 min. On examining such a subject an hour later the particle is invariably absent from the nose but can be detected in the stomach. Some records show relatively fast or slow movement throughout, while others have apparently slow movement in only part of the record. Such a finding can be attributed to the particle reaching a portion of the nose where mucus flow is retarded, or it may be attributed to a deviation of the mucociliary stream at an angle to the anterior posterior direction. In most records such an apparent showing is found near midpassage and seems likely to be due to a movement around the inferior turbinate. Frequently there is apparent stasis of the particle for a minute or two immediately after placement. This might be attributed to local trauma as the mucosal surface is touched with the applicator. It does not influence our results since we begin measurement only when the particle has moved to the first observation slit.

The most surprising finding has been that in all of our groups studied to date some 20% have exhibited slow flow (under 3 mm/min). In the study by Quinlan et al. [43] of 23 "normal" subjects none showed consistent clearance rates under 3 mm/min. Among an additional 7 subjects who had undergone laryngectomy, of 20 measurements the slowest rate was 4 mm/min and the fastest 13 mm/min (average 8.1 mm/min). We are tempted to attribute these findings in laryngectomees to the fact that, in contrast to the normal subjects, their noses have had no exposure to the ambient air for some months [14]. We have recently used the saccharine technique in studying 13 more laryngectomees. Their average time for clearance was 8 min with a range between 5 and 12.75 min. This reflects an average rate of about 12 mm/min (8.6–20 mm/min). Further studies in laryngectomees should offer an opportunity to pinpoint ambient air influences which, when introduced into their noses, impair function.

Our first Aarhus study was directed at the variability in nasal clearance when environmental air was clean and kept at constant temperature (23°C) and constant relative humidity (70%). In the second study changes after variation of relative humidity (RH) were examined [4,5] (Fig. 7): 58 subjects were exposed for 8-hr days with the lower humidities applied during the middle 4 hr. No changes in clearance rates were associated with lowering RH even to as low as 10%. The decrease and rise in humidity did not produce any altered perception

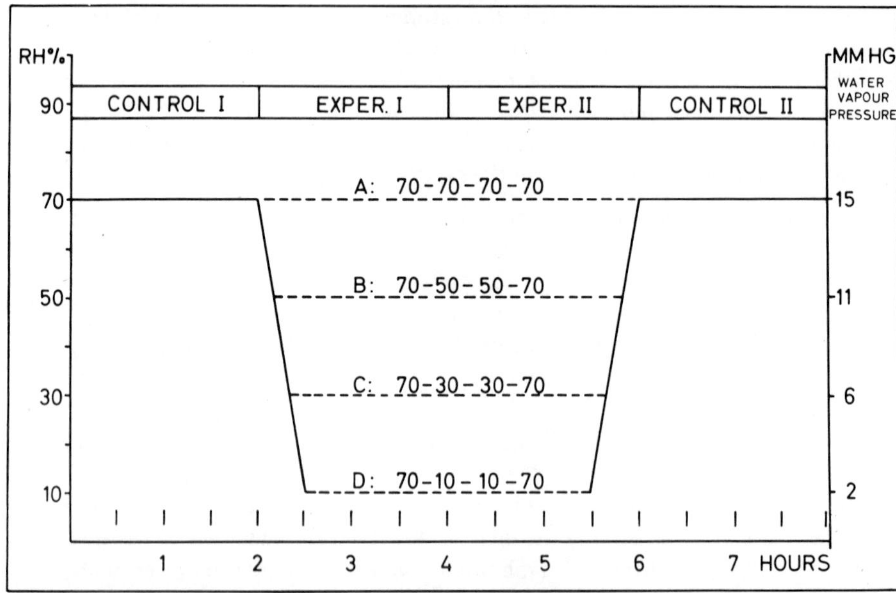

FIGURE 7   Design of the humidity study. (Used with the permission of the Editor, *American Review of Respiratory Disease* [4].)

of humidity, but very low humidity did cause a slight change in perception of temperature, usually a very slight sense of chilliness.

To confirm the significance of these findings another group of 8 subjects were studied in the chamber 24 hr/day for 6 days. During the middle period of 78 hr RH was lowered to 9% (temp. 23°C) and no change in clearance was observed. These subjects had no significant or consistent subjective complaints while in such extremely dry air [9].

During this study of prolonged exposure to dry air, we decided to determine whether an increase in nasal airflow (as with exercise) might overcome the nose's ability to humidify inspired air, dry the surface, and slow or stop mucociliary function. To test this, at the end of the 4th day in dry air, subjects were exercised for 20 min on a bicycle ergometer until they reached the maximum load possible with nasal breathing. Mucus flow rate measured after this exercise was no slower than before exercise. Thus, we feel that the humidifying capacity of the normal nose is greater than its breathing capacity, and that the moist surface essential to mucociliary function can be maintained even under these extreme demands.

Another point of interest was that 8 hr of high RH (70%) failed to produce a significant increase in rate even among those with initially slow movement. We conclude, in contradistinction to some other studies and some theoretical assumptions, that in normal subjects relative humidity in clean air at 23°C has no measurable influence on nasal mucociliary function.

We next undertook an investigation of the effect of a change in ambient temperature. The design of the study is shown in Fig. 8. We elected to keep the water content of the air constant and chose the amount producing saturation at our lowest temperature 5°C, 1.5 g/kg dry air. This produced an RH of 9% at 23°C.

Fifteen subjects were studied in groups of 3 or 4, each group rotating through the 4 days of the experiment in a different order. For each individual at least 4 days were allowed to pass between any two study days. Because of the difficulty in rapidly changing the chamber air to extremes of hot or cold, and because we were interested in the relative effects of inspired air only as opposed to whole body exposure, during the 2 days which ended with the temperature extremes only air conducted to a hood worn over the head was warmed or cooled. The chamber air remained at control conditions, 23°C and 9% RH.

It seems clear that either cooling or warming the air above or below 23°C (considered the norm) produced some change in nasal mucociliary clearance, usually a slowing. The change produced differed from subject to subject, but was never very pronounced. The majority of the subjects exhibited a fall in

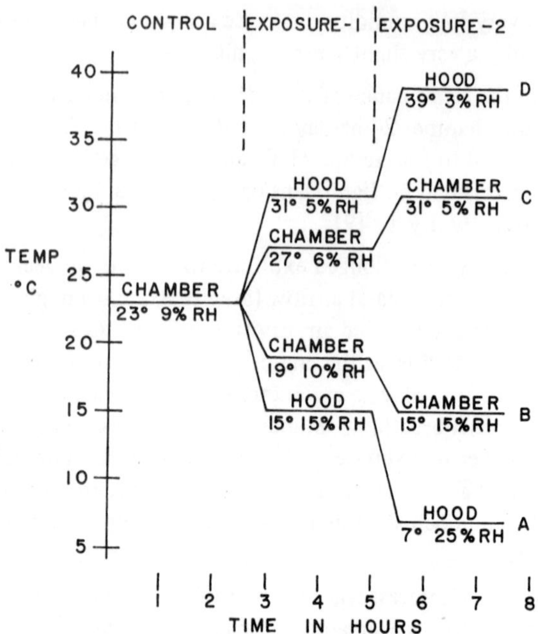

FIGURE 8   Design of the temperature study.

clearance on the first temperature change and a return toward normal on change
to the still lower or higher temperature later in the day, perhaps evidence of
adaptation.

We are left with important questions still to be resolved. How does the individual respond to sudden change? Is there adaptation? Does the smaller change
produce the more profound effect? Will longer exposures result in different findings? How do persons native to naturally occurring extremes of temperature
react?

D. Some Influences Upon "Normal" Clearance

Camner [13] has demonstrated in Stockholm that bronchial clearance rates of
monozygotic twins are more similar than those of other pairs of nontwin individuals. With his group we studied both nasal and bronchial clearance in pairs
of monozygotic twins and in nontwins [7,8] . No significant correlation between
nasal clearance rates in twins could be demonstrated using the particle technique,
but a similarity was found between the fastest clearance as measured by the
saccharine test (Fig. 9). This suggests that nasal clearance is less related to genetic
factors than is bronchial clearance, and, perhaps, more related to different reactions to ambient air influences.

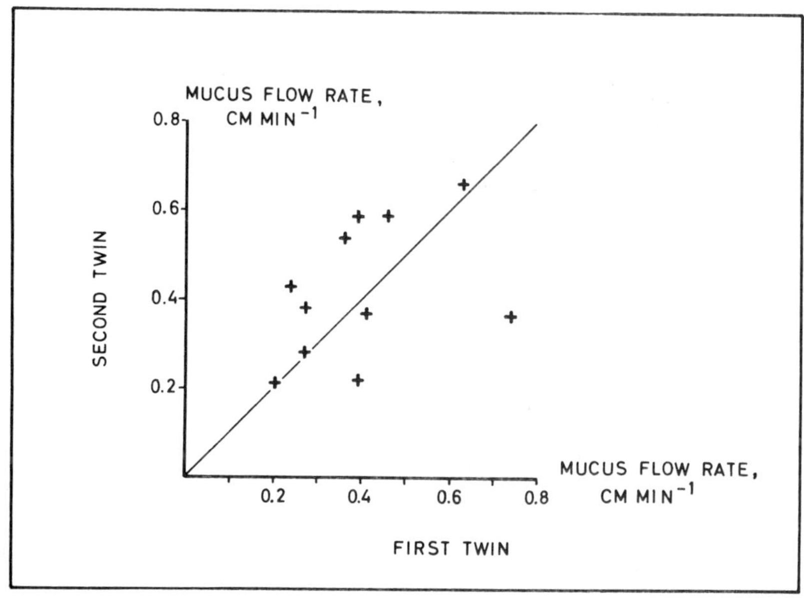

FIGURE 9 Nasal mucus flow rate in monozygotic twins. Each symbol represents a pair of twins, the rate on one indicated on the horizontal coordinate and the other on the vertical. This chart refers to the overall rate measured by the saccharine test. (Used with the permission of the Editor, *American Review of Respiratory Disease* [7].)

We also examined the possibility that nasal mucociliary function might be a reflection of clearance rates from the lung in the same subject, using both monozygotic twins and unrelated subjects. There was no significant relationship between lung and nasal clearance rates. Bronchial clearance was measured by the disappearance of an inhaled tagged aerosol, and the possibility exists that the differences found between nasal and bronchial clearance may be related to the methods used in the two measurements. It seems more likely to us that nasal clearance is more subject to repeated injury from upper respiratory infections and ambient air influences, and that, although rates might be similar at birth, by the time one reaches adult life nasal clearance has been more or less altered by environmental factors. There is other evidence that nasal and bronchial mucous glands do not respond in the same manner to environmental influences [12].

With this background on normal function we have begun to examine the effects of polluted air upon nasal function and the possible relationship between slow mucociliary clearance and susceptibility to airborne disease. In respect of the latter we are attempting long-term followups as to incidence of respiratory disease in several of our normal subjects, but many years must pass before these data will be significant.

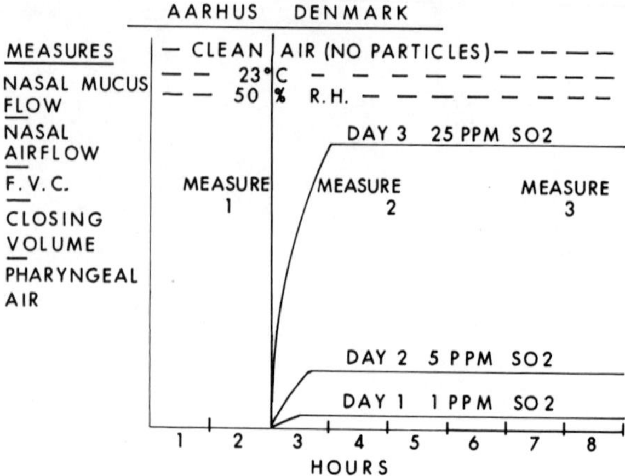

FIGURE 10 Design of the SO_2 study.

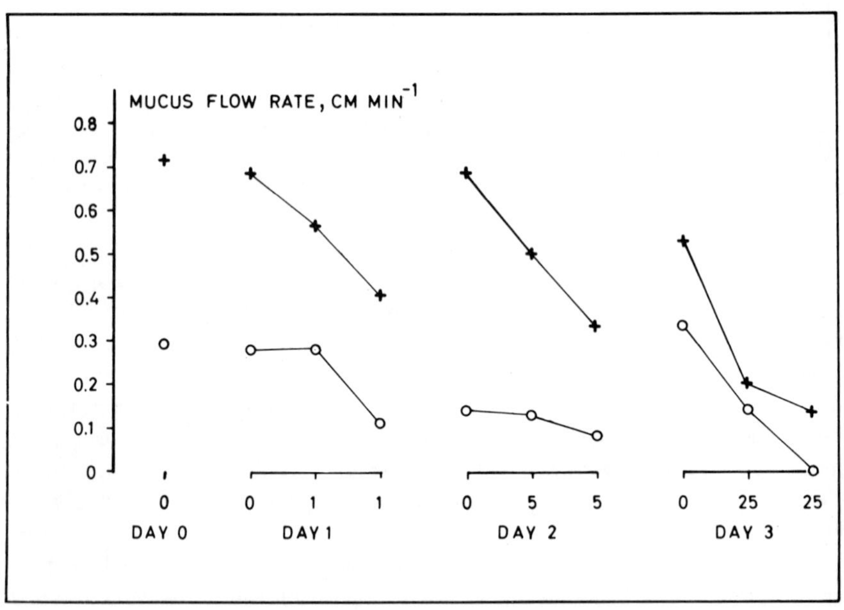

FIGURE 11 Degeneration in mucociliary flow with SO_2 exposure. The crosses refer to those with initially fast rates and the circles to the slow movers. Note that in the fast movers the initial control value on day 3 is low indicating a carry over of injury from 6 hr at 5 ppm SO_2 on the preceding day. (Used with the permission of the Editor, *Archives of Environmental Health* [6], copyright 1974, American Medical Association.)

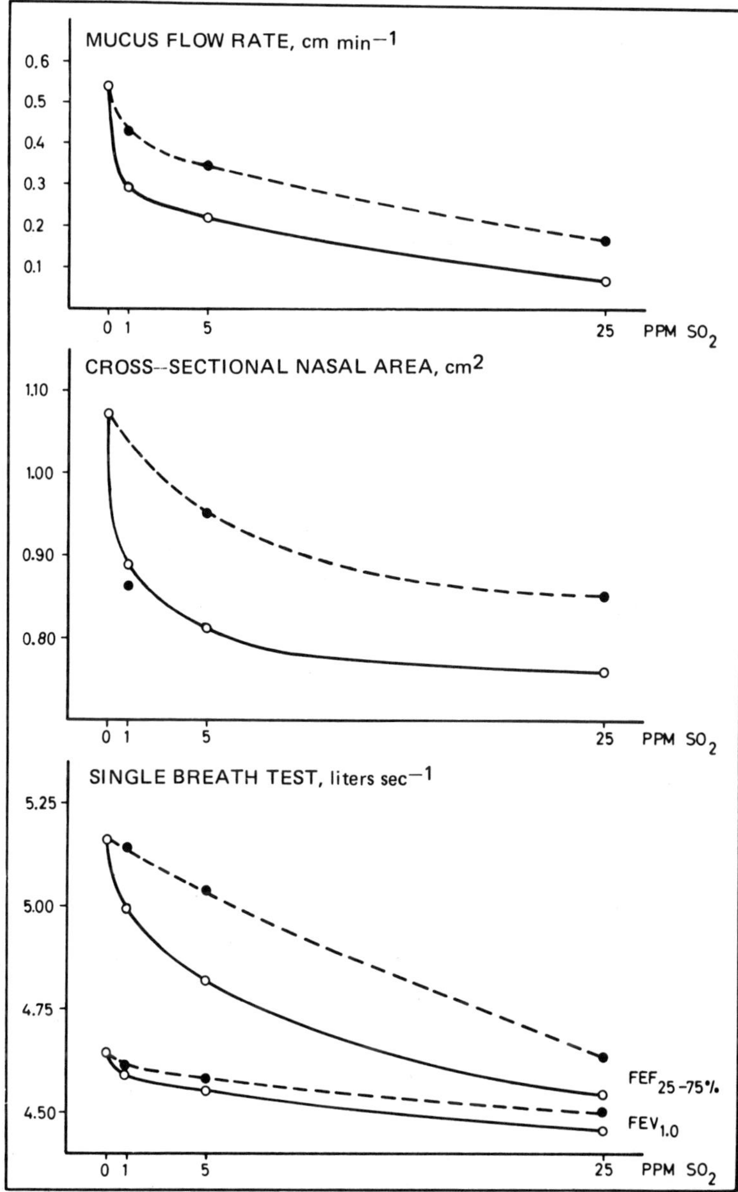

FIGURE 12 Results in the SO_2 study. (○) Mean of control values for all 3 days. The low mean value in the first 3 hr for nasal cross section at 1 ppm is attributable to two subjects who suffered a marked increase in nasal resistance, partially recovering during the last 3 hr.

Sulfur dioxide appears to impair defense mechanisms in two ways, by slowing nasal mucociliary clearance and by increasing nasal resistance to airflow [19, 30]. In another study carried out at the University of Aarhus, Denmark 15 young men were exposed for 6 hr on 3 consecutive days to 1, 5, and 25 ppm SO_2 [6]. Each day's exposure was preceded by 2 hr in the climate chamber with clean air at 23°C and 50% RH. Measurements of nasal mucus flow rate and nasal resistance to airflow were carried out during the control 2 hr and twice during the 6-hr exposure periods. SO_2 was presented as a gas in particle free air with temperature and humidity held at 23°C and 50% RH (Fig. 10).

A slowing of nasal clearance was noted at all levels of exposure, more marked in the last 3 hr, at the higher levels of exposure, in the anterior nasal passage where most of the absorption must occur, and most marked in those subjects whose control clearance was already slow in both relative and absolute terms (Fig. 11). The increase in nasal resistance to airflow was also most notable in the last 3 hr and at the higher levels of exposure. Nasal clearance was reduced and airflow resistance increased during the control period of the third day indicating a failure to recover 14 hr later from the earlier 6-hr exposure to 5 ppm (Fig. 12).

Sampling of air from the pharynx showed that the nose removed essentially all SO_2 from the inspired air even after 6 hr at 25 ppm. How long this could continue in the face of severely inpaired mucociliary function with drying and crusting of the surface which might occur is questionable.

The threshold limit value (TLV), that level thought to be safe for industrial workers' exposure during a 40-hr week, is in most places about 5 ppm. Our findings strongly suggest that this level may be injurious and that even 1 ppm may be hazardous.

We have now conducted a study of exposure to an inert dust (carried out in the same manner as the SO_2 study). Analysis of those data is as yet incomplete. Our plan is to follow with a study in which the exposure combines dust and SO_2. We also have planned for the coming year an on-site study of nasal function in such regions of extreme climate conditions as Israel, Malaysia, and Thule.

V. Discussion

In an attempt to elucidate the reasons for the very slow clearance found in some of our normals we have turned to animal studies [41] relating the biochemical and rheological properties of airway secretions with clearance rates [32,45]. We hope this work may lead to the point where similar experiments can be done in man. At present we are working on the hypothesis (one which seems to have some substantiation) that the cilia beat in a serous periciliary fluid and the resulting motion of this fluid clears surface mucus and particles [2,29]. The observations of Lucas and Douglas [29] suggest this possibility. When we described to Dr. Lynne Reid our finding in the experiment animal of slowed clearance associated with increased respiratory tract fluid she offered this hypothesis as a reasonable explanation.

To the extent to which such a double or multilayered system is involved in mucociliary function, transport will vary depending upon where the material is located. Particles resting on the surface may move at a different rate than materials at one or another depth in the surface fluid, and if the level next to the cell walls is reached it is likely that no transport at all will occur. Sufficient increase in the depth of the periciliary fluid may impair clearance of surface particles (Fig. 13). The use of two methods of measuring mucus flow rates yields information as to surface transport (the particle technique) and of transport of materials dissolved in the periciliary fluid (the saccharine technique).

An important question, as yet unresolved, is what determines whether or not small surface particles may penetrate into and through periciliary fluid and at what rate do soluble materials such as SO_2 penetrate to underlying cells, the speed of their motion along the surface and factors influencing it may be of major significance [47].

A practical example of the importance of these considerations to human health is the investigation of wood workers in the furniture industry in England [1,11,21]. Drs. Acheson and Hadfield uncovered the fact that there is a high

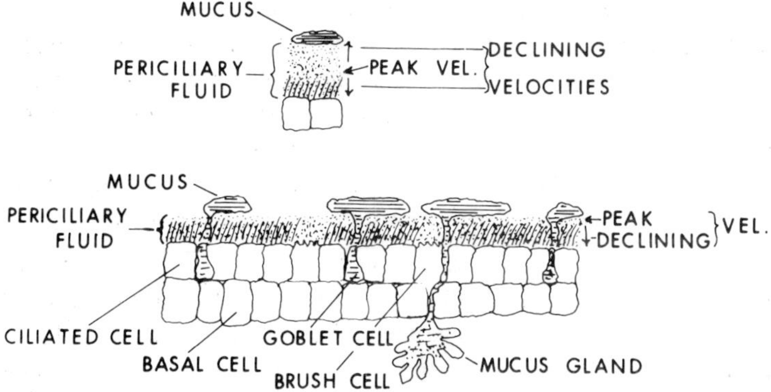

FIGURE 13 Diagram of a hypothetical suggestion as to slowing of clearance in relation to the depth of periciliary fluid. Below: the possible normal mechanism with surface materials being swept along with the top of the periciliary fluid. Above: if this fluid is increased in depth, a portion of it continues to move effectively; but the relatively remote surface materials move slowly if at all.

incidence of ethmoid sinus cancer in this industry. Dr. Black's group then demonstrated that there was a disruption in mucociliary function in the anterior nasal passage of such workers.

Thus, here is a well-documented instance strongly indicating that a breakdown in nasal defenses attributable to an environmental injury has led to serious disease. Chronic exposure to wood dust has led to changes in anterior nasal epithelium in the region where most inhaled particles are impacted. The normal clearance mechanism has been disturbed leading to chronic deposition. If carcinogens are among the materials so deposited it is not surprising if cancer develops; and similar disruptions in bronchial mucociliary clearance similarly accompanied by inhalation and chronic deposition may lead to lung cancer [15].

Iravani has studied changes in bronchial epithelium in rats developing chronic bronchitis [28]. He has found areas of disruption in the coordination of ciliary activity and areas of squamous metaplasia. Whether or not similar changes in the nose account for the nasal cancers in woodworkers (mentioned above) as well as other nasal disease, and whether or not injury to normal mucociliary function may play a part in the pathogenesis of airborne disease in general are important questions still to be resolved.

It would be useful for us to know whether clearance abnormalities develop on some constitutional basis and then, in turn, lead to susceptibility to airborne disease, or whether airborne injuries produce the clearance abnormality. There are a number of factors known to influence respiratory mucous membranes and especially those in the nose, but which of those influences result in impairment of a function such as mucociliary clearance is in large part still to be investigated.

In clinical medicine a wide variety of pharmacological agents are used systemically or topically to treat nasal symptoms. Whether or not such agents as vasoconstrictors, steroids, antihistamines, etc. impair mucociliary function in man we do not know. A number of topical anesthetics have been tested on mucous membranes [17], usually in vitro, but results have sometimes been conflicting. Prostaglandins play a role in nasal blood flow [46] and might affect mucociliary function, but the definitive studies remain to be done.

Hormonal [49] and emotional factors [27,50] have been demonstrated to influence nasal mucosa. It is well known that nasal congestion can accompany hypothyroidism [42]; and in some women pregnancy is accompanied by a vasomotor rhinitis [33]. We have seen patients in whom the development of swollen, pale nasal mucosa is one of the earliest signs of pregnancy, a condition which reverts to normal immediately after delivery. Emotional stress seems to affect all mucous membranes of the body, but most profoundly those of the nose. Wolff et al. [50] have noted that the degree and direction of this effect varies in different individuals.

Although we now have a sound background of knowledge regarding "normal" function there remains a great deal to be done. Many more studies will be required to discover the effects upon normal mucosa of various pollutants and other ambient noxious influences and their combinations. Still other studies are needed of patients suffering from respiratory disease or conditions, such as cystic fibrosis, predisposing to such disease. We urge that, whenever possible, such studies be carried out under carefully controlled conditions so that the effects of single factors can be learned and then the influences of their combinations investigated.

Acknowledgments

The authors wish to acknowledge the assistance of Preben Jensen without whose help much of the work herein reported would have been difficult if not impossible.

References

1. E. D. Acheson, Nasal cancer in woodworkers in the furniture industry, *Br. Med. J.*, 2:587-596 (1968).
2. K. B. Adler, O. Wooten, and M. J. Dulfano, Mammalian respiratory mucociliary clearance, *Arch. Environ. Health*, 27:364-367 (1973).
3. M. Y. Ali, Histology of the human nasopharyngeal mucosa, *J. Anat.*, 99: 657-672 (1965).
4. I. Andersen, G. Lundqvist, and D. F. Proctor, Human nasal mucosal function in a controlled climate, *Arch. Environ. Health*, 23:408-420 (1971).
5. I. Andersen, G. Lundqvist, and D. F. Proctor, Human nasal mucosal function under four controlled humidities, *Am. Rev. Respir. Dis.*, 106:439-449 (1972).
6. I. Andersen, G. Lundqvist, P. L. Jensen, and D. F. Proctor, Human response to controlled levels of sulfur dioxide, *Arch. Environ. Health*, 28:31-39 (1974).
7. I. Andersen, P. Camner, P. L. Jensen, K. Philipson, and D. F. Proctor, Nasal clearance in monozygotic twins, *Am. Rev. Respir. Dis.*, 110:301-305 (1974).
8. I. Andersen, P. Camner, P. L. Jensen, K. Philipson, and D. F. Proctor, A comparison of nasal and tracheobronchial clearance, *Arch. Environ. Health*, 29:290-293 (1974).
9. I. Andersen, G. R. Lundqvist, P. L. Jensen, and D. F. Proctor, Human response to 78 hours exposure to dry air, *Arch. Environ. Health*, 29: 319-324 (1974).
10. B. G. Bang, A. L. Mukherjee, and F. B. Bang, Human nasal mucus flow rates, *John Hopkins Med. J.*, 121:38-48 (1967).

11. A. Black, J. C. Evans, E. H. Hadfield, R. G. Macbeth, A. Morgan, and M. Walsh, Impairment of nasal mucociliary clearance in wood workers in the furniture industry, *Br. J. Ind. Med.,* **31**:10-17 (1974).

12. P. A. Burton and M. F. Dixon, A comparison of changes in mucous glands and goblet cells of nasal, sinus and bronchial mucosa, *Thorax,* **24**:180-185 (1969).

13. P. Camner, K. Philipson, and L. Friberg, Tracheobronchial clearance in twins, *Arch. Environ. Health,* **24**:82-87 (1972).

14. F. W. Dixon, N. L. Hoerr, and J. W. McCall, Nasal mucosa in laryngectomized patients, *Ann. Otol.,* **58**:535-547 (1949).

15. R. Doll, L. G. Morgan, and F. E. Speizer, Cancers of the lung and nasal sinuses in nickel workers, *Br. J. Cancer,* **24**:623-632 (1970).

16. G. Ewert, On the mucus flow rate in the human nose, *Acta Otolaryngol. [Suppl.] (Stockh.),* **200**:1-60 (1965).

17. G. Ewert, The effect of two topical anesthetic drugs on the mucus flow in the respiratory tract, *Ann. Otol.,* **76**:359-368 (1967).

18. R. Frances, D. Allessandro, M. Lippman, D. F. Proctor, and R. E. Albert, Effect of cigarette smoke on particle transport on nasociliary mucosa of donkeys, *Arch. Environ. Health,* **21**:25-31 (1970).

19. W. E. Giddens, Jr. and G. A. Fairchild, Effects of sulfur dioxide on the nasal mucosa of mice, *Arch. Environ. Health,* **25**:166-173 (1972).

20. R. Guillerm, J. L. Morcellet, R. Riu, P. Renon, R. Badré, and J. Hée, Étude du drainage muco-ciliaire nasal chez l'homme par scintigraphies séquentielles d'une particule marquée au Technetium-99m, *Ann. Otolaryngol. (Paris),* **88**:303-310 (1971).

21. E. H. Hadfield, Damage to the human nasal mucosa by wood dust. *Inhaled Particles III,* Vol. 2. Edited by W. H. Walton. Surrey, Unwin, 1971, pp. 855-861.

22. J. Hée and R. Guillerm, Influence des facteurs de l'environment sur l'activité ciliaire et le transport du mucus, *Bull. Physiopathol. Respir. (Nancy),* **9**:377-393 (1973).

23. I. D. Henderson, Cyclical changes in female nasal mucus, *J. Clin. Endocrinol.,* **16**:905-909 (1956).

24. A. Hilding, Ciliary activity and course of secretion currents of the nose, *Proc. Mayo Clin.,* **6**:285-287 (1931).

25. A. Hilding, The physiology of drainage of nasal mucus, *Arch. Otolaryngol.,* **15**:92-100 (1932).

26. J. R. Hill, The influence of drugs on ciliary activity, *J. Physiol. (Lond.),* **139**:157-166 (1957).

27. T. H. Holmes, H. Goodell, S. Wolf, and H. G. Wolff, *The Nose,* Springfield, Ill., Thomas, 1950.

28. J. Iravani and A. van As, Mucus transport in the tracheobronchial tree of normal and bronchitic rats, *J. Pathol.,* **106**:81-93 (1972).

29. A. M. Lucas and L. C. Douglas, Principles underlying ciliary activity in the respiratory tract. II. A comparison of nasal clearance in man, monkey and other mammals, *Arch. Otolaryngol.,* **20**:518-541 (1934).

30. S. W. Martin and R. A. Willoughby, Effect of sulfur dioxide on the respiratory tract of swine, *J. Am. Vet. Med. Assoc.*, **159**:1518-1522 (1971).

31. W. Messerklinger, Die normalen Sekretwege in der Nase des Menschen, *Arch. Klin. Exp. Ohren, Nasen Kehlkopfheilkd.*, **195**:138-151 (1969).

32. W. Messerklinger, Die Funktion der Nasenchleimhaut des Menschen, *Z. Erkr. Atmungsorgane*, **135**:311-315 (1971).

33. M. Mohun, Incidence of vasomotor rhinitis during pregnancy, *Arch. Otolaryngol.*, **37**:699-709 (1943).

34. R. Naessen, The anatomy of drainage of nasal secretions, *J. Laryngol.*, **84**:1231-1234 (1970).

35. H. F. Oettgen, R. Silber, P. A. Miescher, and K. Hirschhorn, Stimulation of human tonsillar lymphocytes in vitro, *Clin. Exp. Immunol.*, **1**:77-84 (1966).

36. D. F. Proctor, Physiology of the upper airway. *Handbook of Physiology, Respiration*, Vol. 1. Edited by W. Fenn and H. Rahn. Washington, D.C., American Physiological Society, 1964, Chap. 8.

37. D. F. Proctor and H. N. Wagner, Jr., Clearance of particles from the human nose, *Arch. Environ. Health*, **11**:377-371 (1965).

38. D. F. Proctor and D. L. Swift, The nose—A defense against the atmospheric environment. *Inhaled Particles III*. Edited by W. H. Walton, Surrey, Unwin, 1971, pp. 59-69.

39. D. F. Proctor, I. Andersen, and G. Lundqvist, Clearance of inhaled particles from the human nose, *Arch. Intern. Med.*, **131**:132-139 (1973).

40. D. F. Proctor, I. Andersen, G. Lundqvist, and D. L. Swift, Nasal mucociliary function and the indoor climate, *J. Occup. Med.*, **15**:169-174 (1973).

41. D. F. Proctor, E. F. Aharonson, M. J. Reasor, and K. Bucklen, A method for collecting normal respiratory mucus, *Bull. Physiopathol. Respir. (Nancy)*, **28**:351-358 (1973).

42. A. W. Proetz, Further observations of the effects of thyroid insufficiency on the nasal mucosa, *Laryngoscope*, **60**:627-633 (1950).

43. M. Quinlan, S. Salman, D. L. Swift, H. N. Wagner, Jr., and D. F. Proctor, Measurement of mucociliary function in man, *Am. Rev. Respir. Dis.*, **99**: 3-23 (1969).

44. J. H. L. van Rée and H. A. E. van Dishoeck, Some investigations on nasal ciliary activity, *Pract. Otorhinolaryngol. (Basel)*, **24**:383-390 (1962).

45. J. Sadé, N. Eliezar, A. Sliberberg, and A. C. Nero, The role of mucus in transport by cilia, *Am. Rev. Respir. Dis.*, **102**:48-52 (1970).

46. R. Stovall and R. T. Jackson, Prostaglandins and nasal blood flow, *Ann. Otol.*, **76**:1051-1060 (1967).

47. O. Strömme, On the absorption of particulate allergens as pollen through the nasal mucosa, and some remarks on bacterial nasal allergy, *Acta Allergol.*, **8**:282-288 (1955).

48. L. Surján and M. Surján, Immunological role of human tonsils, *Acta Otolaryngol.*, **71**:190-193 (1971).

49. M. Taylor, An experimental study of the influence of the endocrine system on the nasal respiratory mucosa, *J. Laryngol.*, **75**:972-977 (1961).

50. H. G. Wolff, S. Wolf, W. J. Grace, T. H. Holmes, I. Stevenson, L. Straub, H. Goodell, and P. Seton, Changes in form and function of mucous membrane occurring as part of protective reaction patterns in man during periods of life stress and emotional conflict, *Trans. Assoc. Am. Physicians,* **61**:313-334 (1948).
51. C. B. S. Wood, The development of immunity in fetal life and childhood, *J. R. Coll. Physicians (Lond.),* **6**:246-258 (1972).

13

Mucous Membrane Injury and Repair

FREDERIK B. BANG and BETSY G. BANG

The Johns Hopkins University
School of Hygiene and Public Health
Baltimore, Maryland

I. Destruction and Regeneration of Mucous Membranes

A. Normal Mucosal Turnover

The mucosa of the nasal chamber and of the rest of the respiratory tract is constantly being destroyed and renewed. The rate of renewal depends on the rate of destruction. While nasal mucosa may be destroyed by physical trauma, toxic chemicals, bacteria, or viruses, it is difficult to determine the "normal" rate of mucosal turnover in the nose. It seems strange that so little has been done to determine rates of turnover in the nasal mucosa under normal and altered conditions. Fabrikant and Cherry [32] developed some figures on human nasal mucosa from biopsy and surgical material and found that the labeling index (percent of cells which take up tritiated thymidine into the nucleus) was 6.1% for malignant neoplasia. These data were derived by immersing biopsy specimens into thymidine solution during exposure to oxygen under high pressure. He believed that only those cells which were in the process of DNA synthesis in vivo actually took up thymidine in vitro. The uptake data combined with other considerations led to a suggested doubling time of the normal cells of 184-213 hr and of the inflamed mucosal cells of 133-155 hr in respiratory mucosae of experimental animals.

In mice, labeling indices in nasal turbinates were 1.1%, on septa 0.66%, and much higher in the nasopharynx (8.9%) [64]. In chicks, cumulative indices over time, following a single pulse of tritiated thymidine, were 1.5% on the inner surface of the maxillary turbinate, 3% on the outer surface. Migration of replacement cells to the surface was also quite slow, since only 7% of labeled cells reached the surface of the pseudostratified epithelium by 48 hr, and only 40% by 7-18 days [78]. These data were not converted to turnover times. These data indicate that labeling indices in the mouse and chick nasal tissues are low in comparison with human indices, even though human and chick nasal epithelia are both pseudostratified. Chick and mouse mucociliated epithelia are better protected by vestibular baffle systems.

Studies of the turnover of cells in murine bronchi and trachea have yielded wide variations: 8-280 days for mice [17] and 33-200 days for rats [17]. Mean labeling indices in the rats differed somewhat between two strains (1.54 in one, 1.02 in the other, 6 tracheas per strain) at 24 hr. It is not clear how much variation there is from tissue to tissue, but Wells [107] has shown that there is a large difference between conventionally derived rats which usually carry a chronic respiratory infection, and those selected by cesarean section and reared especially for the absence of disease. The labeling indices of tracheal cells in conventionally reared rats varied from 1.3 to 2.6, a calculated turnover time of 11 to 22 days. The clean animals had lower (0.7 to 1.2) labeling indices with turnover times varying from 24 to 41 days. The variation from individual to individual within the different batches was several times the average for the batch.

B. Regeneration Following Experimental Trauma

The mucous epithelium of the nose is the first area to be affected by inhaled infectious agents, noxious gases, allergens, and other particulate materials. Following some of these insults there is cell destruction. Areas of differentiated cells may be sloughed, and repair processes ensue. Studies on responses to experimental trauma are valuable in analyzing responses to desquamation produced by infection, which will be discussed in the section on pathogenesis.

Most of the studies on mucous membrane regeneration after experimental trauma have been limited to sequential histological examination. Only one has examined function [15]. Values for recovery time range from a few days, after scraping the tracheal epithelium of chicks [15], calves [53] or rats [108], to several months in humans after surgical removal of the lining of the maxillary sinus [44]. The wide range of values relates not only to the degree of damage to the mucosa, but also to the criteria used in judging degree of recovery. It may therefore be of value to first summarize the different stages that are apparent in regeneration after trauma and then to discuss the factors which influence or foreclose regeneration. Stages of regeneration will be predicated on combined data from a number of studies on nasal, sinus, and tracheal mucous membranes in rat, chick, calf, ferret, and human subjects. Only one of these papers has electron microscopic data [65] ; none used autoradiographic methods of identifying cells. Yet the subject is so fundamental to basic response to injury, that a paper on membrane regeneration after viral desquamation [100] has become the classic example of the regenerative response in a standard text on pathology. Not only is the knowledge of the regenerative process prerequisite to understanding pathology (response to injury), but the reorientation of the differentiating cells is an outstanding problem in growth and differentiation.

From the very diverse experimental material in these papers we will attempt to develop a proposed sequence of changes. There is general agreement that four or five different responses on the part of the epithelium itself are operative during regeneration: destruction, migration, multiplication of undifferentiated cells, reorientation, and differentiation. The question of dedifferentiation is not yet adequately resolved, but the existing evidence will be reviewed; the role of dedifferentiated migrating cells would be to provide the most immediate cover for a denuded area just as the differentiation of generative cells serves this function. Undifferentiated cells—basal cells—seem to serve as the major source of new cells, and as these multiply and fill in to take the place of the differentiated cells that are lost, it is difficult to "differentiate" them positively from dedifferentiated cells, even by means of sequential studies by light microscopy.

1. Destruction

Death of surface epithelium may be brought about by aging, chemical or physical trauma, or infection. It is always followed either by sloughing of the dead cells into the overlying stream of continuously moving mucus, or by phagocytosis. This disposal of the dead specialized epithelial cells has been studied particularly during the course of viral infections, but there is no reason to suppose that the change in the adhesive properties of these dead cells is limited to infectious processes. Changes in adhesive properties result in loss of contact with similar differentiated cells, or in the increased stickiness for macrophages that precedes phagocytosis. Following physical damage with a brush or cotton, or chemical trauma such as zinc iontophoresis, there is a general sloughing of the epithelium, limited to the area directly involved.

2. Migration from Adjacent Epithelium

Migration of new cells from the undamaged area into the area of denuded basal cells commences within a few hours [108] . This occurs in the presence of inflammation and leukocytic infiltration, neither of which seems to interfere with growth of a new covering epithelium over the inflamed surface area. It has been proposed that this migration is accompanied by dedifferentiation of the existing ciliated and mucous cells, an assumption based on the presence of undifferentiated cells among the surviving ciliated cells at the edge of the damaged area. Wilhelm [108] described early stages of curetted (rat) tracheal epithelium: between 1 and 8 hr after curettage the intact cells at the margin of the 2-mm wound flattened out, lost their cilia, and spread laterally over the denuded area. Only ciliated cells showed this migratory behavior. Quite a number of migrating cells showed cilia at first, but after 24 hr they were rare and were never seen after 36 hr until "the new epithelium was redifferentiated." About 200 μm might be covered in 8 hr. Ciliated cells appeared in the regenerated epithelium by 10-14 days, goblet cells slightly later. Wilhelm interprets the presence of mixed undifferentiated and ciliated cells during the initial lateral migration of ciliated cells as reflecting the urgency to "close the breach" as quickly as possible (personal communication) [110]. Presumably not all cells have had time to shed their cilia at the time of fixation of tissues at this initial phase. Basal cells from the adjacent untraumatized areas are the main source of the new undifferentiated cells which either migrate out as flat covering cells or proliferate to form simple stratified, then completely differentiated mucociliated epithelium.

3. Multiplication of Undifferentiated Cells

Four of the studies of experimental trauma mention the effect of trauma on mitosis. Lane and Gordon [65] studied the effect of slight trauma induced by applying a blunt probe to the rat tracheal epithelium. Excess mitotic figures were seen as early as 2 hr afterward, but were maximum 26-30 hr after injury. Boling indicated excess mitosis in lamb tracheas 6-12 hr after curetting [18]. Wilhelm [108] reported that after tracheal curettage in rats the first wave of mitosis reached a peak 24 hr after curettage and was primarily concentrated in the undamaged half of the trachea. However, since the numbers of epithelial cells in the regenerating layer and the numbers in the undamaged area were not stated, it is difficult quantitatively to determine the relative stimulus to mitosis in the undamaged area. Battista et al. [15] showed maximum mitotic rates (3.3%) 1 day following destruction of tracheal epithelium in the chicken by focal scraping with a brush. Rates returned to normal by 7 days. Despite the difference of techniques in these studies, it is clear that stimulus to mitosis occurs within the first 24 hr. The degree of stimulus is probably related to the degree of trauma.

Lane and Gordon [65] showed that although mitotic indices were highest at 26-30 hr, ciliagenesis and mucigen formation were not seen until the 60th hr, after which there was little cell division. From this they argued that control of cell division is not dependent upon the elaboration or release of a suppressor substance by the mature cells. This argument seems to substitute the word "mature" for "differentiated." It is possible that the initiation of cell differentiation is associated with inhibition of mitosis. Indeed the early appearance of new polarized basal bodies at 60 hr within cells that are beginning to differentiate into ciliated cells [65] suggests that these basal bodies must start to multiply some hours before they are detected.

Boling studied regeneration of nasal epithelium in 40 lambs from whom strips of mucosa were removed surgically. From one side of the fossa a strip was removed from the septum, and from the other side a similar strip was taken from the medial surface of the maxillary concha. The lambs were killed at intervals from 12 hr to 60 days and the mucosal surfaces were studied histologically. Although rates of migration of adjacent epithelial cells into the epithelium varied greatly, flattened cells, presumably immigrating from adjacent intact epithelia, were found on the surface of the wound at 2 days, and a 3-mm area of injury was completely recovered at 3 weeks [18].

4. Reorientation and Differentiation

By the time the new layer of epithelial cells has covered the wound, it has become two to three cells thick at the margin even though most of it is one cell thick in the middle [108]. The burst of mitotic activity is now transferred to this area of new epithelium and increase in the number of cells proceeds to the point where a multilayered cuboidal cell layer covers the surface. In experimental wounds of the nasal membrane in sheep, differentiation of these stratified cuboidal cells into columnar cells begins on the 8th or 9th day with the appearance of droplets of mucigen, and the development of cilia on the free surface of cuboidal epithelium by the 13th or 14th day [18]. The cells assume their polarity with localization of the Golgi apparatus at the upper pole. There is an increase in the number of mitoses. Full mucus cell development sometimes precedes and sometimes follows the appearance of the ciliated cells. In Boling's studies, infection of the mucosa was present in all of 5 animals killed between 18 and 60 days after surgery. The infection was limited to the operative area and led to destruction of the cartilage below the infected area. A variety of substances were used to irrigate the nose during the period of regeneration. Liquid petroleum, water, and physiological solutions of NaCl had no effect, but areas treated with castor oil or cocaine showed a more rapid complete regeneration than untreated control areas.

The sequential events in regeneration of chicken tracheal mucosa are illustrated in Figs. 1-3. The time for complete regeneration of the mucosa in Wil-

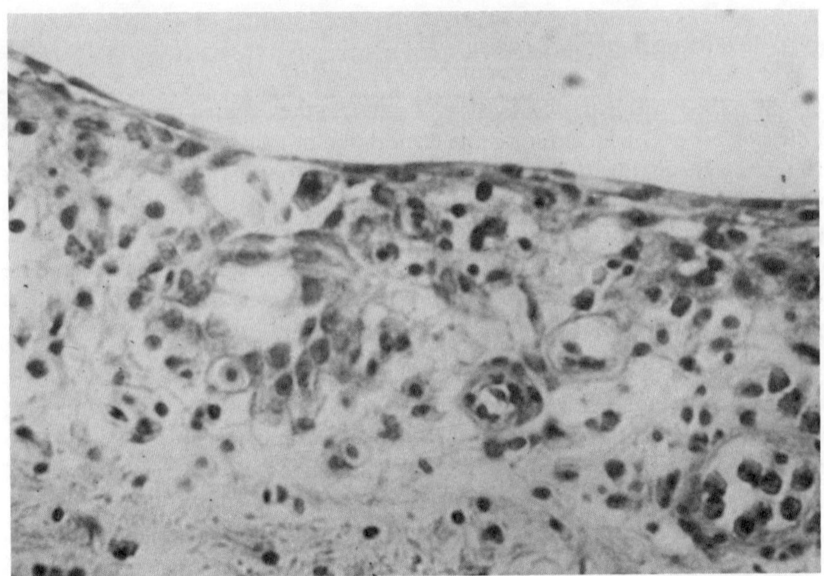

FIGURE 1 Portion of chicken trachea showing the appearance of tracheal mucosa 1 day after mechanical denudation. (From Battista et al. [15]. Reprinted courtesy of the authors and the Editor of *Toxicology and Applied Pharmacology*.)

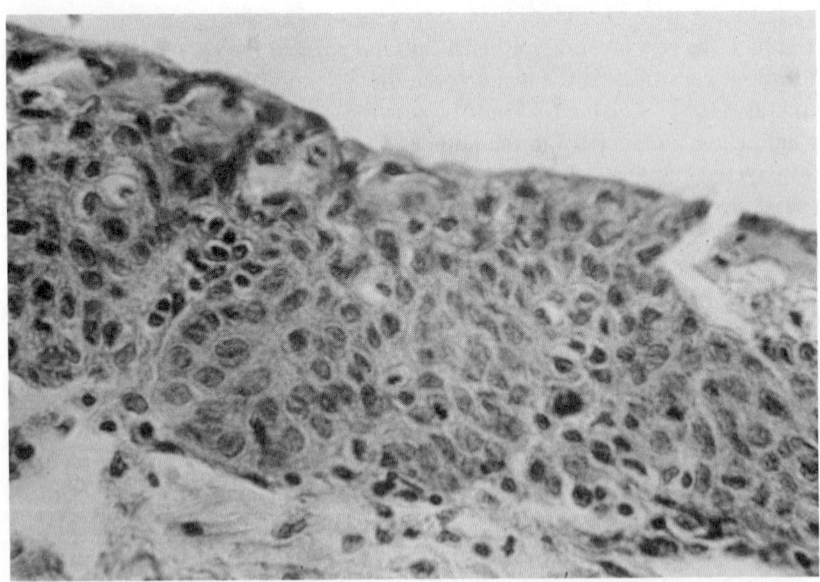

FIGURE 2 Portion of chicken trachea showing the appearance of the mucosa 2 days after mechanical denudation. (From Battista et al. [15]. Reprinted courtesy of the authors and the Editor of *Toxicology and Applied Pharmacology*.)

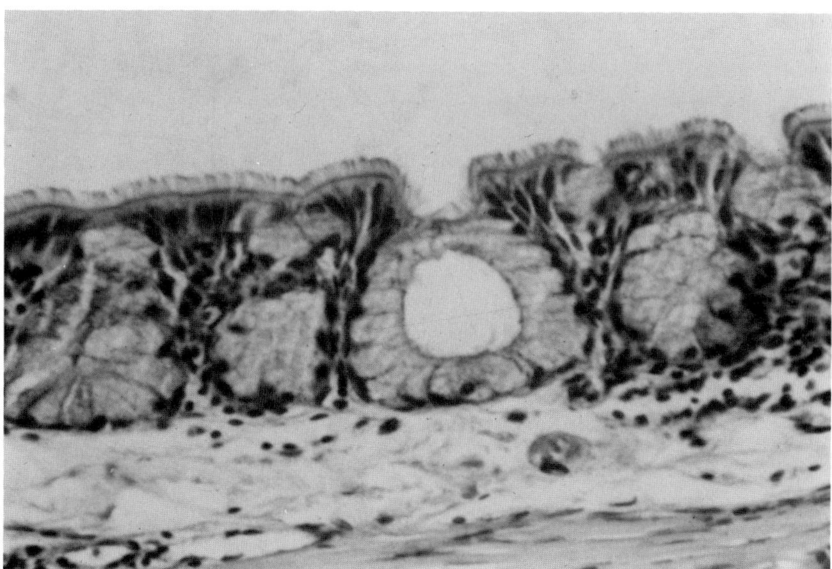

FIGURE 3 Portion of chicken trachea showing the appearance of the mucosa 29 days after mechanical denudation. (From Battista et al. [15]. Reprinted courtesy of the authors and the Editor of *Toxicology and Applied Pharmacology*.)

helm's experiments on curetted rat trachea [108] was 6 weeks; for the chicken trachea less than 4 weeks [15], 5 months in the dog "sinus" [63] and 3 to 5 months in human sinuses [44]. Such an extensive variation was influenced by at least three things other than the difference in species: the extent of the destructive lesion, the presence of secondary bacterial infection, and the difficulty of measuring the endpoint of "complete regeneration," which is likely to vary from observer to observer and from animal to animal. Thus our review does not allow for any estimate of the "average" time taken to reach this endpoint; species, severity, type of mucous membrane, and criteria for complete regeneration are all variable.

The only study in which function of the damaged and regenerating epithelium was measured was that of Battista et al. [15]. In these studies the restoration of mechanically denuded chicken tracheal epithelium was measured by counting ciliary and mucous cells and mucous glands, and by measuring ciliary particle transport in the trachea. Most of the function had returned by 14 days and was complete by 29 days (Fig. 4).

Hilding showed that total experimental desquamation of dog frontal sinuses usually led to obliteration of the sinus by scar tissue [54]. The same

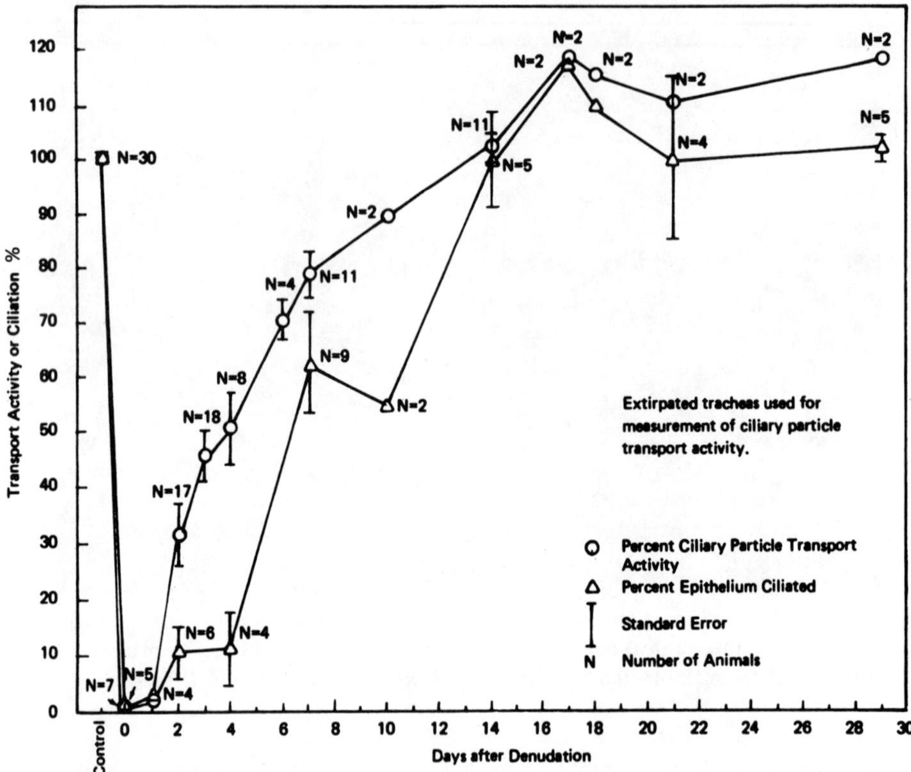

FIGURE 4   In vivo recovery of epithelium and recovery of transport capability in upper part of chick trachea after mechanical denudation. (From Battista et al.. [15]. Reprinted courtesy of the authors and the Editor of *Toxicology and Applied Pharmacology*.)

end result has been observed following desquamation of sinus epithelium by acute virus infection [4].

C.   Regeneration Following Vitamin A Deficiency

1.   *Effects of Vitamin A Deficiency on Respiratory Membranes*

Vitamin A is required for the normal growth and differentiation of specified epithelial systems, including respiratory membranes. The vitamin is now recognized to have a very broad spectrum of biological effects [33]. Wolbach and Howe [111] in 1925 did the first systematic study of the effects of vitamin A deficiency on respiratory membranes and found that normal epithelial cells were gradually replaced by squamous cells which undermined the normal epithelial cells and pushed them off, then the surface squamous cells keratinized. Conversely, in 1953 Fell and Mellanby [35] showed that when an excess of vitamin A was added to cultures of chicken embryonic skin, the normally squamous cells became metaplastic to cuboidal and goblet cells. In 1963 Jackson and Fell [60] in an ultrastructural study of the same phenomenon established that the transformation originated in the basal cells. Thus the basal cells of the epithelium have two pathways of cytodifferentiation: depending on the location in the body and the amount of vitamin A available they can be directed to form squamous cells with the property of keratinization, or to form a mucociliated columnar epithelium.

Wong et al. in 1971 followed the ultrastructural events of metaplasia in tracheal epithelia of rats maintained on vitamin A-deficient diets and also found that metaplastic changes first appeared in basal cells [112]. They concluded also that under the conditions of their experiments there was no dedifferentiation of mature cells. In the nasal epithelia of chicks which were deprived of vitamin A for 21 days after hatching, there was a 50% loss in total numbers of ciliated and goblet cells, and a 50% depression in the rate of mucociliary transport of tracer dye, in comparison with normally fed chicks [5].

2. Regeneration of A-Deficient Membranes After Trauma

Recognition of the specific effects of A deficiency alone raised the question of differential regenerative capability of respiratory membranes in normal and vitamin A-deficient animals. In 1942 Condon [21], a thoracic surgeon, did comparative studies on regeneration of curetted bronchial epithelium in normal and A-deficient rats, and found that regenerative events were much more rapid in the A-deficient than in the normal tracheas. Using colchicine techniques he found that in normal rats the greatest mitotic activity was in basal cells, but that fully differentiated columnar and goblet cells also divided.

Wilhelm [109], who had already studied regeneration of normal rat tracheal epithelium [108], subjected weanling rats to A deficiency. Squamous metaplasia of the mucociliated membrane began after 5-6 weeks. Tracheas of some rats were fully keratinized. There was nearly always infection of the airways. Regeneration began within 1-2 hr after trauma, when simple flat cells from adjacent marginal epithelia and from necks of gland ducts spread over the exposed lamina propria beneath the fibrin clot which covered the wound. Whether curettage was done before or after the rats were made A-deficient, the epithelia became fully keratinized; replacement cells were always squamous cells which subsequently keratinized. The pattern and rate of epithelialization was similar to that found in the normal rats, but unlike normal rats squamous differentiation was not delayed by infection.

3. Regeneration After Viral Infection of A-Deficient Hosts

Viral infection imposes different kinds and degrees of trauma on respiratory membranes, and in vitamin A-deficient animals these effects often work synergistically to produce remarkably rapid changes that differ from those produced by the infection or the deficiency alone [3,6].

Nasal respiratory membranes of chickens which were raised on a diet lacking vitamin A and then intranasally infected with Newcastle disease virus (NDV) showed that initial infection was consistently localized. The shallow membrane of the under-surface of the inferior turbinate was the first area to be infected, and the regenerating membranes were consistently metaplastic to keratinizing squamous epithelium. Throat swabs of these chickens yielded about 100 times more virus on the 5th day after infection than did swabs of NDV-infected normally fed birds [11] but these results have not been consistent [11a].

The combined A-deficiency and infection also had profound effects on the lymphoid cells involved in the immune response, effects not produced by either factor alone. The combination induced rapid loss of lymphocytes from both the bursa of Fabricius and the thymus, and loss of plasma cells from areas usually rich in these immunoglobulin-producing cells. Thus the combination of A deficiency and acute infection causes profound impairment of both the cell-mediated and humoral (T cell and B cell) responses [6,12].

4. Epidemiology of A Deficiency Plus Respiratory Infection

Measles is of course a respiratory disease. The rapid depletion of immunocytes in the chicken model is reminiscent of the often lethal synergistic effects of measles and malnutrition in some parts of Africa [72] where multiple deficiencies and protein deficiencies are common in children from lower socioeconomic families unless vaccination campaigns are instituted [37].

While the chick model showed a rapid synergism between complete A deficiency and directly instilled virus, the field situation is less clear. While vitamin A deficiency is recognized as a major public health problem in many areas of the world today, and reports of the devastating effects of untreated hypovitaminosis A are abundant, there have been few studies of the epidemiology of vitamin A deficiency. One carefully controlled study in an economically poor village has demonstrated distinct seasonal variation in the clinical signs of this deficiency in children 0-4 years old who were followed for a 2-year period [95]. The rate of respiratory infection among these children varied from 44 to 87% in both years. The data have not been completely analyzed, but to date a close correlation has not been found [96]. There is great need for antiphonal field and laboratory studies of the epidemiology of respiratory disease and specific nutritional deficiency.

A final thought on the role of vitamin A: Fell [33] has pointed out that an outstanding question is whether the presently recognized broad spectrum of biological effects of vitamin A can all be explained on a unitary hypothesis. She and her colleagues think that this is not impossible. One hypothesis is that vitamin A may control cell differentiation at the translational level [24].

II. Pathogenesis of Experimental Upper Respiratory Tract Infection

The sequential steps following experimental physical or chemical trauma on the one hand and viral infection of respiratory mucosae on the other are remarkably similar. In each there is destruction, inflammation, migration of undifferentiated cells, multiplication of these cells, and reorientation and differentiation of new cells. In the case of infections there is also another series of events which acts concomitantly (Table 1).

These processes have been studied both in the intact animal and in organ cultures of respiratory mucosae. In this section the sequence of events will be discussed from the standpoint of types of experimental infection, and functions of individual immune systems in the respiratory mucosae.

The difference in the method of virus spread within the mucosae, i.e., between myxoviruses such as influenza or Newcastle disease, and herpes viruses such as laryngotracheitis, leads to major differences in pathogenesis. Those

TABLE 1   Basic Events Common to Destruction and Regeneration After Mechanical and Viral Denudation of Respiratory Epithelium

Mechanical denudation	Viral desquamation
Cell destruction and sloughing	Specificity and continuity of destruction of specific cells
Migration of undamaged cells	Interferon production
Inflammation	Production of local and general antibody
Multiplication of stem cells	Infiltrations of immune cells
Reorientation and differentiation	Hyperplasia

viruses which so change the host cell that it sloughs out of contact with other cells before the agent is able to spread by direct contact into neighboring cells, tend to cause acute self-limited mucosal infections. If, however, death of the cell is accompanied by massive virus release, the agent may rapidly spread out over the mucosa until the immune mechanisms stop it. Thus these infections may vary greatly in virulence and are usually acute events with rapid death or recovery.

On the other hand, if the virus spreads by the direct infection of neighboring cells before these cells change their adhesion to neighboring cells, then the progression of the infection may be somewhat more localized but is preceded by continuous extensive spread from one area to another. In the case of the herpes viruses this method of spread is associated with the ability of the agent to spread from cell to cell in a form which bypasses antibody mechanisms of protection. It is also associated with the presence of a noninfectious form in the cell (DNA) which is susceptible to enzymatic destruction if it leaves the cell.

Experimental infections of the upper respiratory tract (nose and nasopharynx) of animals are rare in contrast with studies of the lower tract (larynx, trachea, and lungs). It is actually more difficult to induce acute experimental infection of the lower respiratory tract than of the nasal tissues if infectious agents are intranasally administered without anesthesia. The original intranasal instillation of human influenza virus into the noses of unanesthetized ferrets was done by Smith, Andrewes, and Laidlaw [98]. This first experimental infection of ferret nasal mucosa with human influenza led to a series of subsequent experimental analyses establishing the selective effect of influenza virus for the ciliated and mucous cells of the epithelium. A classic study of destruction, regeneration, and subsequent resistance of ferret nasal epithelium to influenza virus was done by Francis and Stuart-Harris [38,39]. Organ cultures of epithelium of the ferret showed the same specific susceptibility [13,104]. The mouse, which has been used most frequently in pulmonary influenza research, has rarely been studied in terms of susceptibility of the nasal mucosa to this virus. This seems to be a case of oversight, since the virus may readily be passed from one animal to another by making an extract of the turbinates of infected animals and inoculating unanesthetized animals by placing a drop of the extract on the external nares and allowing the animal to pull it in during normal breathing [59a].

This was the method of intranasal inoculation used in studies on localization of the vaccine strain of NDV in the nasal passages of chicks [19], and in subsequent studies on both laryngotracheitis [4] and a more virulent strain of NDV in chicks [12]. All of these chick studies showed that if a small volume of undiluted virus harvest was inoculated intranasally into an unanesthetized animal, the major infection induced was concentrated in the nasal tissues. The same method was used by Iida [59] in virus transmission experiments: individual "infector" mice were inoculated intranasally with Sendai virus, then were housed with 10 uninfected cage mates; the virus was transmitted to all 10 contacts.

Three different model systems have been investigated in upper respiratory tract pathology; these are:

1. Acute mild infection with myxoviruses (especially NDV) in the chicken nose

2. Acute severe infection with myxoviruses (especially influenza) in the ferret nose

3. Severe desquamating persistent and local infection, exemplified by a herpes virus infection (laryngotracheitis) of the chicken nose

The general principles of the role of interferon production, local antibody and immune cells will be discussed in a separate section. A fourth category of chronic mycoplasmic infections deserves special consideration, but except for the extensive etiologic studies of Nelson [77] little has been done on the mechanism of pathogenesis; a study of pathogenesis in chickens used intratracheal, not nasal, inoculation [22]. The great number of germinal centers which appear in the nasal tissues of chickens infected with mycoplasma suggests that these infections would offer an important area for study (Fig. 5).

A. Newcastle Disease of Chickens

Most experimental studies of this virus have used routes other than intranasal for inoculation, so the results are not applicable to this review. Intranasal inoculation of older chickens with both the mild, or vaccine, strain [19] and the mesogenic strain [9] was followed by local multiplication of the virus in the middle turbinate and by destruction of the mucosa. Destruction was usually localized in the inner lining of the middle turbinate, though scattered areas of destruction were found elsewhere in the nose. This selective destruction is probably related to the route of intranasal flow of mucus, which moves first over the outer surface, then more slowly over the inner surface of the middle turbinate, giving the virus time to penetrate the mucous blanket and infect this shallow epithelium before the blanket moves to the posterior pharynx and is swallowed.

It is thus of considerable interest to determine, in this model of myxovirus infection, how much virus is absorbed to the different tissues immediately after inoculation and before multiplication has taken place. Mucosal tissues removed from chick turbinates at intervals after inoculation of about 10^8 virus particles, and titered before multiplication of the virus has taken place, showed (a) large

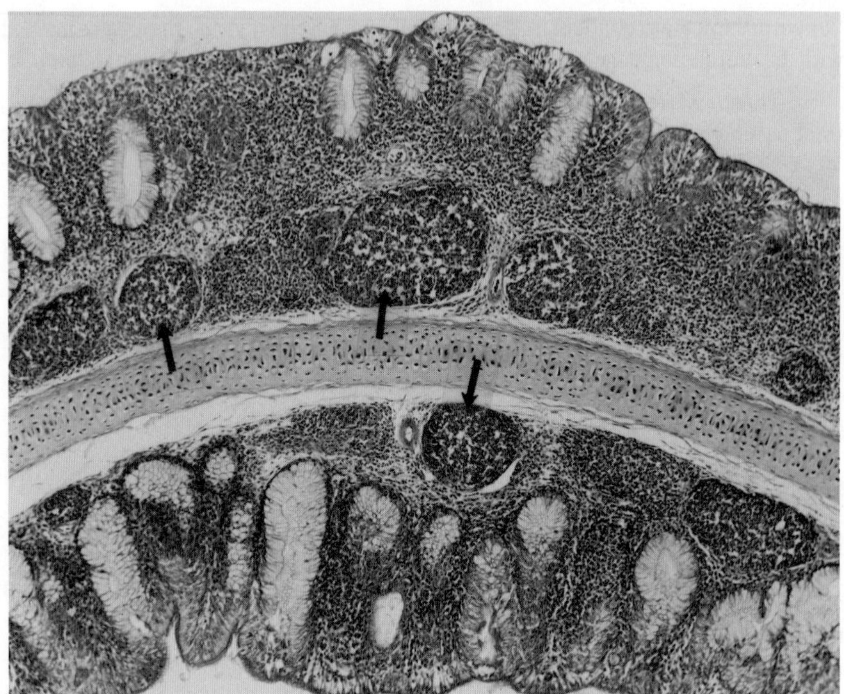

FIGURE 5 Portion of maxillary turbinate of chicken after 7½ weeks infection with mycoplasma, showing active germinal centers (arrows). Hematoxylin and eosin. ×90.

variation in titer from chick to chick, (b) about 1/100 fewer infected cells than after in vitro inoculation of the same tissue, (c) the highest ratio of infected cells 3 hr after infection, and (d) a very low total number of infected cells (10^1-10^3). The latter finding agrees with the fact that intranasal infection of chicks is not routinely successful in producing infection unless about a million or more virus particles are present in the inoculum. The mucus blanket thus seems to be a powerful but somewhat variable protector against infection. Three hours after infection much of the virus may be in the usual "noninfectious" form within the cells of the mucosa, that is: "adherent" to the cells. However, if the chicks are injected with pilocarpine or exposed to severe cold, there is an emptying of the mucous cells and the number of infected cells decreases greatly [10]. This finding, in addition to the presence of virus in the mucus secreted from excised turbinates, suggests that many of the infected mucus gland cells are only temporarily "infected" with virus which is stuck to the mucus granules within the cells. The virus is apparently excreted along with the mucus granules. Many mucous gland cells do get infected.

Following the initial destruction of mucous cells and ciliated cells, there
is rapid replacement of these cells by a cover of very flat thin squamous epithelial
cells. During this early phase there is a continuous process of destruction, sloughing, migration of thin epithelial cells over destroyed inflamed areas, and multiplication of basal replacement cells. All stages may be seen simultaneously in an
affected area, until the regenerative process is fully under way.

There are three sets of data on the effect of infection on the response of
presumably uninfected epithelial cells. The first [78] showed that thymidine
labeling of cells in the mucosa, both on the inner surface and the outer, was reduced below the normal 1 day after infection with the virus. The second study
[88] showed that the labeling index was increased 2 days after infection and
reached a peak at 3 days. A third study contributed rough estimates of the number of mitotic figures (no thymidine uptake) in the regenerating epithelium; most
were found at days 4 and 5 after regeneration [12]. The three studies then may
be interpreted to mean that following initial depression of DNA metabolism,
possibly induced by locally produced interferon, there is a subsequent increase
in DNA synthesis followed promptly by increased mitotic activity.

1. Virus Multiplication

Following the initial establishment of NDV infection, there is a rapid increase in
amount of virus found both in excised turbinate tissues and in throat swabs.
This reaches a titer of 1000 to 100,000 infectious units in throat swab assays by
the first day, and maintains this level for 3 days; from the fourth to sixth day
titers drop rapidly so that positive swabs are rarely obtained beyond the sixth day.

2. Interferon Production

Interferon is produced in chick embryos by all strains of NDV but less by the
lentogenic than by the meso- and velogenic strains [67]. Interferon is produced
during active infection of chicken mucosa with this virus. No data are available
on interferon production in NDV infection in the nose.

3. Antibody Production

Chickens, like most avian species, have no cervical lymph nodes, but the nasal
and paranasal tissues have several types of localized lymphoid cell concentrations
[16,2] : (a) the submucosal lymphocytes, concentrated in the tissues adjacent
to the mucosa and often seemingly invading the area; (b) a system of organized
submucosal lymphoid nodules associated with lymph vessels; (c) collections of

plasma cells around the ducts of the lateral nasal glands; (d) massive concentrations of mature plasma cells in the Harderian gland; and (e) occasional germinal centers in the tonsil area. When NDV invades the mucous membranes some of those lymphoid tissues may be destroyed, but in the normally nourished animal repair is rapid enough to maintain the integrity of the lymphoid tissues.

Heuschele and Easterday [51] found that NDV antigen persisted in scattered histiocytes in the trachea of chicks exposed to aerosols of NDV as long as 120 days after exposure, suggesting that latent infections may persist after introduction of the virus by this method. There are no data on the nasal mucosa. Antibody to NDV was present in mononuclear cells, mucous glands, and the mucus blanket of the trachea in these chickens; since germinal centers were also found in the trachea, antibody is apparently formed and secreted locally within these respiratory tissues.

In other systems it is known that local IgA antibodies form in response to infection of nasal mucous membranes [103], but there are no data on this response in chickens intranasally inoculated with the mesogenic strain of NDV. Data on immunoglobulinemic chickens shed some light on this: Perey and Dent [79] found that bursectomized chickens which produced no humoral antibody were consistently unable to overcome NDV infection, showing 100% mortality after intranasal inoculation of the mesogenic strain. The importance of the T cell system in combating NDV infection has, to date, been less well defined; even though T cell-deficient animals were 100 times more susceptible to NDV in terms of death than control animals [80], the mechanism whereby T cells combat NDV infection remains an open question.

As inflammatory changes develop in the mucosa, there is a clear effect on the cartilage directly underlying the inflamed area [12]. This effect is even more marked in laryngotracheitis, which is a more severe local infection, and its mode of action will be discussed in connection with that infection.

B. Acute Infection with Influenza Viruses in the Ferret

The acute destructive lesions induced by influenza viruses (most studies have been with influenza A) on the respiratory epithelium of the ferret's nose have become standard material for the study of the selective destruction of surface epithelium both in the nose [99] and in organ cultures [13,104]; for the localization of viral antigen first in the nuclei of infected cells [66] then in the cytoplasm; and for studies on the rate and mechanism of recovery of the epithelium [38,57]. Finally, influenza virus has a specific predeliction for the turbinate epithelium, even when the virus is inoculated intravenously [14]. Following these basic studies on ferrets, the careful analysis of lesions of early influenza in man by Hers [49] and by Mulder and Hers [73] established that the same type

TABLE 2 A Comparison of Acute Viral Lesions of the Nasal Mucosa[a]

Influenza in ferrets[b]	Newcastle in chicks[c]
1. Goblet cells empty	1. Disappearance of mucus from acinar cells and goblet cells
2. Necrosis of columnar epithelium; leukocytic infiltration	2. Destruction of epithelium; polymorphonuclear cell infiltration into acini; epithelial blisters and cysts with polymorphonuclear cells inside
3. Beginning regeneration with mitotic figures in thickened epithelium; plasma cells, lymphocytes, fibroblasts	3. Remodeling and repair, associated with continuing destruction; beginning mitosis; heavy lymphoid infiltration
4. Epithelium in full repair; surface cells thin, flat	4. Inflammation reduced; thin surface epithelium, cilia mostly absent
5. Cartilage change; chondroblast destruction	5. Increased mitosis of epithelium; loss of cartilage basophilia
6. Epithelial blisters and polymorphonuclear cells	
7. Developing cilia; goblet cells still poorly developed	
8. Fully ciliated, hyperplastic; intraepithelial cavities lined by cilia	8. Complete regeneration of ciliated cells, mucous cysts lined with ciliated epithelia; plasma cells associated with acini

[a]From Bang et al. [12]. Reprinted courtesy of the Editors of the *American Journal of Pathology*.
[b]From Francis and Stuart-Harris [38].
[c]From Bang et al. [12].

of destruction occurs in human respiratory epithelium. Since most of the studies on humans are concerned with the histopathology of the trachea and bronchi in fatal cases of pneumonia, they will not be discussed in this section. However, the classic description of the sequence of events described by Francis and Stuart-Harris may be compared with the more limited infection of NDV in the chicken. Table 2 summarizes the local events in NDV infection and compares them with influenza infection in ferrets.

C. Laryngotracheitis Infection of Chickens. A Model for Study of Acute Nonfatal Desquamating Rhinitis

When this virus is inoculated intranasally into unanesthetized baby chicks, it remains limited to the nasal fossa, produces acute desquamation of all nasal epithelia, results in functional recovery of the respiratory epithelium, but leaves important residual abnormalities. From the earliest recognizable lesions through 4½ months' convalescence the principle changes are:

1. Initial lesions, or small syncytia of intranuclear "inclusions," first identifiable in the mucociliated cells of the shallowest portion of the epithelium at 21 hours postinoculum (the inner surface of the maxillary turbinate scroll).

2. Acute sloughing (about 3-7 days), marked by (a) spread of lesions from cell to cell via multinucleated "giant cells" which progressively slough and desquamate all surface epithelia from respiratory, olfactory, and sinus mucous membranes, often including epithelial neural elements and blood vessels; (b) appearance of numbers of eosinophilic leukocytes along the basement membrane at the sites of lesions just previous to sloughing; intensive infiltration of the submucosa with small lymphocytes after sloughing begins; (c) histochemical change in the intracellular mucus of the cells which comprise the syncytia (this mucus stains with Alcian blue alone when stained with **AB-PAS**, while the normal mucus is mixed purple).

The cartilages of the maxillary turbinate often become flaccid, and the cartilage cell nuclei and matrix lose both basophilic and Alcian blue staining properties, effects which recede by about the 8th day [4]. It is possible that the characteristic metaplasia and softening of the cartilage which occurs during the acute phase of infection with this virus is mediated by virus-induced activation and release of lysosomal enzymes into the underlying cartilage. This interpretation is supported by in vivo and in vitro studies on cartilage, in which it was shown that the various proteolytic lysosomal enzymes from the cartilage all are capable of breaking down the surrounding cartilage [28]. Similar morphologic changes were apparently observed by Stuart-Harris and Francis in ferret nasal cartilages following acute infection [100].

Schultz [92] has shown that during the first 3 days of infection, there is activation of the lysosomal glycerol acid phosphatase in the mucosal columnar cells, but that this particulate form disappears by the fourth day. Particulate glycerol acid phosphatase and naphthol acid phosphatase are activated within the cartilage cells by day 2-3. β-Glucuronidase, again in particulate form, appears in unfixed infected tissues at the same time, and the appearance and disappearance of these enzymes is closely correlated with the change in the staining characteristics of the cartilage itself.

3. Repair (about 8 to 21 days), marked by rapid initial spread of a sheet of epithelial cells over the infiltrated submucosa, appearance of numbers of plasma cells circulating in the tissues, formation of encapsulated germinal centers, and mucosal adhesions.

4. Convalescence (about 1 to 4½ months when experiments terminated), marked by functional restoration of the mucociliary lining of the nasal fossa.

However, at 4½ months, eight specimens all showed complete metaplasia of the olfactory organ (end nerves, supporting cells, and glands of Bowman) to muco-ciliated epithelium, all showed abnormal formation and alignment of mucous acini, and about 50% had severe persistent sinusitis and/or obliteration of the sinus by granulation tissue [4].

A final note on experimental pathogenesis: Stocks of animals which are to be used for studies on effects of specific factors (such as A deficiency) on the course of respiratory disease must be carefully checked for possible chronic nasal infection. In our own laboratory six of six weanling rats from a particular source were found to have severe chronic nasal infection. Six weanling rats from this source and six from a "clean" stock were simultaneously deprived of vitamin A. When killed after 48 and 61 days of deprivation, 100% of the chronically infected rats had keratinized lacrimal ducts and 50% showed keratinization of respiratory epithelium, while none of the six "clean" rats showed any keratinization. Kelemen has written a series of papers on the hazards of chronic nasal infections of laboratory animals in respiratory research [61].

III. Pathogenesis in Organ Cultures

Organ cultures have been extensively used to explore problems in embryology and tissue relationships. Organ culture techniques were first extensively developed by Honor Fell, who has applied the method to a remarkable variety of specific questions. The methodology is aimed at preserving differentiation in organized tissues and is achieved by transplanting bits of tissues and discarding the migrating cells [34]. Although many of the studies reviewed below deal with cultures of tracheal epithelium, and relatively few with nasal epithelium, the two types of mucous membrane cultures seem nearly identical in their susceptibility to respiratory viruses, and the principles derived from either system apply equally to both. Adaptation of organ culture methods to particular problems (for example, secretory metabolism and virus effects) may have contributed simplified methods of maintaining differentiated respiratory epithelium, but it is likely that some of Fell's carefully disciplined methodology has been lost along the way. However, the following general points have been established in organ cultures.

1. They are of direct value in studying questions of specificity, tissue tropisms and virulence, and in testing possible relationships between susceptibility of specific types of differentiated cells to particular types of viruses. This has been particularly true in the case of respiratory agents.

In the first application of organ cultures to the study of human (and other) viruses, Bang and Niven found that human embryonic nasal and tracheal cultures maintained healthy ciliation and mucus secretion for at least 2 months. These

cultures were not susceptible to chick-adapted human influenza. Cultures of
ferret nasal turbinate epithelium, on the other hand, were rapidly desquamated
by the same stock of influenza [13].

Fell and Mellanby reported on the remarkable conversion of organ cultures
of developing chick embryo epidermis to mucociliated epithelium when grown
in medium containing excess vitamin A [34]. This system was used to study the
differential susceptibility of organ cultures of squamous, and converted mucociliated, embryonic chick skin to influenza and vaccinia. The diverted mucociliated cultures yielded good growth of human influenza virus but not vaccinia
virus, while the normally keratinizing squamous cultures were far less susceptible
to influenza and supported an increase in titer of vaccinia virus [58].

Organ culture studies by Toms and his colleagues [104] used minimal-dose
determinations to show that cultures of ferret nasal epithelium were significantly
easier to infect than cultures of bladder, uterus, and oviduct, and that alimentary
tissues were insusceptible. In another study, organ cultures of ferret and fetal
human tracheas were studied comparatively for effects of different strains of
human influenza: strains of influenza A viruses showed some association between the ability to produce disease in human volunteers and the destruction of
ciliated cells in the human, but not the ferret cultures; there was no association
between effects of three human influenza B viruses on volunteers and on human
organ cultures [48]. A study of the comparative effects of three strains of NDV
on organ cultures of chick trachea did show an association between degrees of
virulence in vivo and in effects on cultures [8]. As Hara et al. suggest [48], it
might be fruitful to apply genetic methods to studies of virulence and specificity
in organ cultures.

2. A simplified method of growing human tissue is useful in growing certain human rhinoviruses; rhinovirus growth seems to be at least partly dependent
on the presence of the differentiated cells [56]. Denny [25] found that a bacterial toxin produced differential ciliastasis when applied to tracheal organ cultures of different animals. He applied supernatant fluids from *Hemophilus influenzae* to cultures of human fetal, chick embryo, and several rodent tracheas
and found that the fluid contained a ciliastatic factor that was active at high dilutions on rat cilia, much less so on the chick and human. The factor caused loss
of cilia, cell damage, and sloughing, was produced early in the growth cycle of the
bacteria, was nondialyzable, heat-stable, and was produced by all strains of *H.
influenzae.* Denny proposes that it may be an endotoxin and might be effective
in producing chronicity and carrier states.

3. Interactions between several viruses and mycoplasmas have been studied
in organ cultures of bovine and porcine trachea by Reed [84]. Prior infection of
the cultures with a rhinovirus, parainfluenza, or influenza viruses enhanced the
growth of *Mycoplasma hyorhinis* and other species of mycoplasma. The effect

was evidently related to the extent of epithelial cell damage caused by the viruses. This paper also reviews studies of clinical mycoplasma and studies of interactions of mycoplasma with bacteria, evaluated in organ cultures and by other methods. Mycoplasmas alone, except for a few of the more pathogenic strains, have been found to cause no damage to cilia in organ cultures, even though the cultures support good growth of the organisms.

4. Absorption of NDV to organ cultures of chick trachea has been determined [9], and has been compared to in vivo absorption [10]. More virus was absorbed to the cultures than to the nose and trachea in vivo. In the cultured tissues there was a remarkable drop (100-fold) in the number of infected cells during the first few hours after absorption was first detectable. This might be ascribed to two events: sloughing of the infected cells and reexcretion of virus from intact mucous gland cells. Sloughing may be greater from cultured than from in vivo epithelia because of greater spreading of cells, but the possibility remains that the gland cells may play a unique role in susceptibility by reexcreting virus in the early stages of infection.

Tracheal organ cultures from 4-day-old hamsters adsorbed about 70% of influenza A (PR8) virus in the first 30 min, and 90% of the total by 2 hr, while explants from 2- and 4-week-old hamsters adsorbed virus much more slowly. Interferon was demonstrated in infected cultures in each of the age groups, in concentrations proportioned to the susceptibility of the given age group [89]. This might have relevance to the differential susceptibility of infant and adult populations to respiratory agents in nature.

Electron microscope studies on the absorption of influenza and Sendai virus to ciliated cells following inoculation of guinea pig tracheal organ cultures showed large amounts of virus present on the cell membrane and even on the surface of the cilia [30]. This study indicates sites where virus could be absorbed, but the model requires such large virus inocula that it yields little data on specificity or pathogenesis.

A. Persistence and Change with Age

Most organ cultures consist primarily of surviving and differentiated tissues. Thus long-term maintenance of such cultures is unusual. However, it has been shown that cultures of pig trachea maintained swine influenza virus for 24 days, and that apparent growth of parainfluenza 3 and Sendai virus proceeded up to 12 to 18 days [83]. Infectious bovine rhinotracheitis virus was maintained in organ cultures of bovine trachea as long as 29 days, and was recovered 1 to 2 months after explantation from nasal and tracheal epithelium of exposed calves [94]. Tracheal organ cultures from 4-day-, 2-week-, and 4-week-old hamsters all showed decreasing susceptibility with age [89], but all supported growth of virus.

B.   Local Immunity as Measured in Organ Cultures

The question of tissue immunity following infection has been examined by a number of authors and found not to exist in the absence of residual lymphocytes. Ferret nasal mucosa maintained in vitro on a chick plasma clot in organ culture was as susceptible to influenza virus if taken from an immune animal as from a susceptible animal [13]. In general, the idea that differentiated cells which undergo the virus infection, such as the ciliated and mucus-secreting cells, become themselves immune to virus infection, has been discarded. However, local tissue immunity, which is mostly, if not entirely, dependent upon lymphocytes and perhaps macrophages, can be studied in explants of tissue taken from immune animals and subjected to challenge with virus [42,43]. The capacity of the lymphocytes to respond will naturally be related to the differential survival of these cells in different media. Thus the importance of the medium.

Immunity in organ culture may be examined from the standpoint of susceptibility to cell destruction (ciliastasis and histology), virus yield, amount of virus needed to induce infection, interferon production, production or persistence of antibody and specificity of the resistance. This has been in part studied for each of several different viruses.

Tracheal ring cultures of chickens were shown to resist the destructive effect of NDV if they were taken from birds which had been immunized by the tracheal route [36]. These cultures were spared from the destructive effect of the virus and yielded a lower titer than did unimmunized controls. Similar cultures showed immunity (lower virus yield) to avian infectious bronchitis when taken from birds which had been immunized to this infection by intranasal inoculation.

Analysis of the interaction of virus and antibody in the chicken itself was made by Heuschele and Easterday [51] through the use of organ cultures taken from animals immunized by the aerosol route against the vaccine strain of Newcastle disease virus. Such birds still had virus in the tracheal explants as long as 85 days after the culture was initiated. The cultures were challenged with a more virulent strain of the virus. Antibody to the virus was measured by plaque neutralization and was found after 2 days of culture but not thereafter. This activity had the electrophoretic characteristics of antibody.

Influenza A virus will infect chick tracheal cultures of immunized birds if 5 to 10 times more virus is inoculated on these cultures than on cultures of non-immune birds. This immunity is specific for the strain of virus inoculated, and virus yield from the successfully inoculated resistant cultures is reduced. Again the immunity is lost within the first few days of culture [90].

In conclusion: Organ cultures of respiratory epithelium have become powerful tools in the analysis of specificity of virus growth in differentiated tissues, and in analysis of host susceptibility and of local tissue resistance—but only during the first days of culture.

IV. Upper Respiratory Tract Infection in Humans, and the "Common Cold"

There is a basic similarity in the pathological sequences involved in limited respiratory tract infection in a variety of different animals, and the available information on the changes in humans suggests that this species has an almost identical sequence of cell death, sloughing, inflammation, regrowth, and new differentiation. It is not surprising then that discussion both by the layman and the medical researcher centers around an entity called the "common cold." Experiments with volunteers, which will not be detailed here, do show that there is a difference in the virulence of different agents and that the clinical syndromes vary somewhat. However, the overlap between different patterns is tremendous [105].

A. Nasal Epithelial Responses

Sloughing of cells altered or killed by viruses invading the upper respiratory tract is readily demonstrated by obtaining from a patient a sample taken directly from a nose blow. This can be blown or dripped directly onto a glass slide and then examined under the microscope. The frequent presence of ciliated cells which maintain their motile cilia so that they swim over the slide shows that pathologically affected cells are lost from the epithelial layer even before the cilia have lost significant activity. Among them are also a variety of mucus cells, and of course leukocytes and lymphocytes in varying numbers.

Ciliary flow rates in the respiratory tract itself may be followed by determining the rate at which tagged particles move through the nasal chamber [82]. This has been done with volunteers inoculated with rhinovirus type 44. The transport rate decelerated in all affected individuals. This occurred at different times after infection and varied from one-half to one-tenth of the original rate [87]. It would be of great interest to determine such rates continuously during the course of infection, measuring the effect of various treatments on the movement of the blanket. This can be done routinely by a two-person team, by determining the time of flow rate of tracer dye from the anterior nares to the pharynx [7].

Despite the fact that it is now impossible to determine the etiology of the different "colds" which were studied histologically by Hilding [53], his evaluation of the pathological changes seen in biopsies of 25 different patients, from the second hour to the second week of their attacks, remains the most complete and careful one available. Sloughing of cells was apparent by the second day and surface regeneration began by the third day and persisted as long as the fourteenth. Early edema was followed by infiltration with polyblasts and polymorphonuclear cells. Necrosis and disintegration of the epithelium, and pathologic cell infiltration continued from the third to sixth day. In some cases the necrosis extended deep into the mucosa, and many, if not all, of the columnar cells disappeared from some areas. As in the experimental infection, the remaining cells proliferated until they were again many layers deep. Repair took place while there was continued increased mucus discharge. Special stains failed to show the penetration of any bacteria, and these were also absent from the cells in scrapings from 70 other patients, even though bacteria were often present in the secretions. This suggests that the major cause of colds is the viral etiology acting alone. Other pathological processes in the upper respiratory tract are reviewed by Rewell [85]. The similarity of these changes to those in organ cultures was pointed out by Tyrrell [105].

B. Interferon

Because of its obvious clinical importance, considerable attention has been paid to the production of and therapeutic usefulness of interferon [102]. Interferon is produced in upper respiratory tract infections of man [55,46], but, as with its presence elsewhere, there is often a direct relationship between the severity of the infection and the amount of interferon produced [75]. For this reason, two other methods of either protecting against infection or altering the course of the disease have been studied. Interferon may be induced by a number of synthetic substances, even on intranasal inoculation [29,62]. Intranasal instillation of the interferon inducer statolon protected mice against lethal doses of influenza; however, this was primarily a study of the pneumonia produced by the virus and may not be relevant to the infection within the nose [62].

In two studies, an amount of induced interferon which was similar to that brought about by natural infection, failed to prevent infection in volunteers [29, 40]. On the other hand, Merigan et al. [71] showed that if large doses of interferon were given a day before inoculation and for 3 days afterward, a significant prevention of symptoms and virus shedding was achieved against adenovirus 4. Thus there remains the possibility that upper respiratory tract infection could be treated and sometimes prevented by this nonspecific method. The need for better knowledge of the physiology of the nasal passages is apparent.

C. Local Secretory Antibody

The importance of, and the rapidly increasing knowledge about, the role of secretory immunoglobulins in respiratory tract infections has produced a very large literature. The major reviews on the subject up to 1974 were noted in Chap. 11.

Intranasal inoculation of animals has been shown to provide better immunization against influenza than parenteral immunization, and there is good evidence that this is also true in man [106]. Richman et al. [86] suggest, however, that in many types of virus infection it may be futile to depend on a broad heterologous local antibody response, and that during epidemics the current epidemic strain of virus should probably be used. Aiuti et al. [1] cultured nasal and conjunctival membranes from different human subjects and then assayed them by immunoelectrophoresis for presence of immunoglobulins: both tissues showed positive synthesis of IgA and only traces of IgM and IgG. Thus there is evidently continuing local synthesis of IgA; but there are many unsolved problems in exploiting local synthesis of antibody against specific viral agents, including those involved in "the common cold" syndrome.

V. Regeneration and Other Events in the Olfactory Membrane

Credence concerning the regenerative capacity of the olfactory membrane has had a three-phase history: a negative phase when it was held that nerve cells once destroyed could not be replaced because postembryonic mitoses of nerve cells did not occur; a phase of conflicting evidence derived from a variety of experimental systems; and the current phase of positive evidence, in several animals, of complete regeneration by newly differentiated cells.

Much of the early interest in the olfactory membrane as a direct route to the brain stemmed from experiments with zinc sulfate iontophoresis as a means of obliterating this pathway of poliovirus entry to the brain. This in turn raised the question whether the effect would be temporary or permanent. Thus a general interest in regeneration of olfactory membranes was aroused.

Since Tagaki [101] has provided a splendid historic review of pertinent experiments, data which seem most relevant to human disease or most challenging of further exploration will be abstracted.

A. Experimental Models

Smith [97a] destroyed rat olfactory epithelium with 1% zinc sulfate solution and observed general sloughing, followed by a period of mixed sloughing and squamous-cell cover, and at 12 days a well-developed cover of typical ciliated cuboidal epithelial cells. He repeated the same experiment in frogs and observed destruction, production of a stratified epithelium of undifferentiated cells by 28 days, and of a complete differentiated olfactory membrane by 70 days [97b]. This had followed destruction of all cells down to the basement membrane. He then punched a disc-shaped piece of membrane out and observed that cells bordering the defect lost their differentiated character, developed mitotic figures, then migrated and closed the gap with differentiated cells by day 31; by day 64 after iontophoresis this new area was fully differentiated. Thus rat membrane showed mucociliated metaplasia, while there was total regeneration of frog membrane following mitosis of undifferentiated cells. These experiments were done 25 years ago.

Graziadei has recently produced a complete accounting of olfactory cell regeneration in frogs, using ultrastructure and autoradiography. Working with Metcalfe [46] he had first established that the normal finite life of the frog olfactory neuron is about 5-8 weeks. Then the succession of events in regeneration after complete (unilateral) surgical resection of the olfactory nerve was followed until complete restoration of the full complement of neurons by 4-6 weeks [45]. In this case the supporting cell population was not directly affected (the normal turnover of these cells is at one-tenth the rate of nerve cells). Replacement of neurons was clearly the work of basal cells, in which mitotic activity began 4 days after ablation, peaked at 2 weeks, and continued until regeneration was complete. Graziadei proposes that the olfactory basal cell is multipotential, since it contributes neurons, supporting cells, and occasionally a metaplastic ciliated-cuboidal cell type. Smith observed "lateral" migration of dedifferentiated cells which redifferentiated to close a gap, and both Smith and Graziadei observed differentiation of new neurons—Graziadei specifically from the basal cells. These events were followed in frog olfactory membranes.

Rabbit olfactory mucosae were destroyed with 1% zinc sulfate by Mulvaney and Heist [74], complete with basal cells. Yet one of four rabbits examined 30 days later showed complete regeneration of all olfactory membrane components. The other three showed complete metaplasia to a ciliated columnar membrane. A large series of rabbits killed at intermediate stages confirmed that no zinc-treated membranes had remained intact; some even showed destruction of the basement membrane. But in all cases the submucosal glands of Bowman, and in most cases part of the ducts, remained intact, and it was from these structures that undifferentiated cells seemed to derive; differentiated neurons first appeared at 4-8 days after iontophoresis. Thus individual variation in regenerative capacity within a species seemed to have been demonstrated.

Monkey olfactory membranes were given the 1% zinc sulfate treatment by Schultz [91], who observed that some replacement cells derived from the gland cells of Bowman, some possibly from the sheath cells of Schwann—the regenerating area having at first the appearance of pseudostratified epithelium at 10 days but assuming generally normal form by 6 to 12 months—a slow recovery from the frog's viewpoint.

Thus the various sources of replacement cells, and their occasional going astray, leave open a series of intriguing questions about growth and differentiation of this membrane. One question is: why is there normal differentiation in some cases, metaplasia in others?

B.   Metaplasia Following Infection

In the experiments described above, metaplasia clearly occurred in some animals following severe mechanical destruction and olfactory nerve oblation. Metaplasia to mucociliated epithelium is also known to occur after severe desquamation resulting from experimental virus infection. Following intranasal infection of chickens with laryngotracheitis virus (a herpes virus which is nonfatal but which causes severe desquamating rhinitis), there is quite often complete desquamation of respiratory and olfactory epithelia, which usually are functionally regenerated by about 3 weeks. However, in all of eight birds which were killed 4½ months after intranasal inoculation with this virus there was complete metaplasia of olfactory membranes to a mucociliated acinar type of respiratory epithelium [4]. In a series of hamsters inoculated with influenza following injections with either benzpyrene or a nitrosamine compound, 4 of 26 showed mucociliated metaplasia of olfactory membranes 56 weeks after intranasal inoculation of virus, and 6 of 26 showed other olfactory membrane changes [93].

There have been a good many descriptions of destruction of olfactory membranes in ferrets [38,81,58], but there were no followup studies to determine whether the tissues regenerated. There are, however, two studies of destruction of olfactory membranes by SO_2, one in mice [41], the other in pigs [68], in which there was severe desquamation of olfactory epithelium and subsequent metaplasia.

C.   Other Aspects of Olfactory Membrane Pathology

The two-way intercourse between central and peripheral neural elements [20,31] has often been suggested as a convenient pathway for neurotropic agents. A combined light and electron microscopic study of entry of amebas into the mouse brain should again stimulate studies on other infective agents. Martinez and his associates [69] reported that following intranasal inoculation of trophozoites of *Naegleria* into mice, the olfactory mucosae first showed an inflammatory response and spaces appeared between the normally tightly juxtaposed sensory, sustentacular, and basal cells; there was some destruction of cells and degeneration of plasma membranes. Amebas directly invaded the sustentacular cells, which lost their villi and became phagocytic. In some cases the cells were destroyed, in others the amebas were destroyed. Intact amebas invaded the intercellular spaces, pierced the basement membrane and lamina propria, reached the submucosal nerve plexuses, invaded the nerve bundles and migrated up the axons and perineural spaces to the subarachnoid space, olfactory bulb, and contiguous central nerve system structures.

Neoplasms (neuroepitheliomas) are selectively produced in hamster olfactory tissues by some carcinogens, especially nitrosamine compounds, whether the carcinogen is introduced intragastrically, intratracheally, or subcutaneously [23]. This selectivity excludes the organ of Jacobson, a fact which adds to the evidence that this is no ordinary "olfactory" membrane.

An ultrastructural study of the nasal tissues of mice after exposure to 10% cigarette smoke twice daily for 6 or 9 days showed distinct differences in effects on two strains of mice, one of which consistently showed selective destruction of olfactory tissue, while the other showed none. The author suggests that susceptibility was genetically determined [41].

Finally, Dimov [26] pointed out that following epidemic influenza 46% of human patients tested had anosmia, 46% hyposmia, 6% parosmia, and 2% hyperosmia, all of which indicated at least temporary pathology; the greatest difficulty was in patients between 50-70 years, which he attributes partly to diminishing acuity with age, and partly to increased susceptibility.

It seems, then, safe to say that regeneration of olfactory epithelium may depend on the method and degree of destruction, the species, the genetic constitution, and possibly the age [76] of the individual involved.

VI. Summary

In summary of this section:

1. Limited current data suggest that normal cell turnover in human nasal mucosae may be 7.5 to 9 days, in mouse nasal mucosae about 28.4 days. Labeling indices in normal human epithelial cells are about 6.1%, in normal chicks 1.0 to 3.0%, in normal mice 1.1%.

2. Regeneration of mucous membranes starts within 1 or 2 hr after trauma, whether mild or severe. It begins with migration of cells adjacent to the wound onto the denuded area and is soon followed by intensive basal cell mitosis.

3. Peak multiplication of cells adjacent to a desquamated area is probably within 24 hr of trauma; peak mitosis of undifferentiated cells in regenerating areas is between 2 and 4 days.

4. Following desquamation by viral infection, complete regeneration (or permanent destruction) may be effected after widely varying time intervals. In the experimental systems reviewed, the area of mucous membrane, severity of destruction, species of host, and criteria for regeneration (whether functional or morphologic) varied.

5. In vitamin A-deficient subjects, regeneration follows the same course, and mitosis of undifferentiated cells is even more rapid. However, in desquamated areas, mucociliated epithelia are not redifferentiated but are replaced entirely by keratinizing squamous cells.

6. Organ cultures are extremely valuable in analysis of resistance and susceptibility of mucous membranes to viral infection, and of specificity of virus multiplication in differentiated cells.

7. "Common cold" rhinoviruses induce local production of interferon and secretory antibody, but therapeutic use of these factors against "the common cold" are just beginning to be exploited.

8. Following destruction of olfactory mucous membranes regeneration may be complete in some species and some individuals, but often take the pathway of permanent metaplasia to mucociliated epithelium.

9. There are outstanding gaps in fundamental knowledge of the normal physiology of nasal tissues and of their responses to mild, severe, acute, and chronic infections.

References

1. F. Aiuti, G. Turbessi, and A. Ugolini, Synthesis in vitro of immunoglobulins produced by different mucous membranes, *Experientia,* **25**:1089-1090 (1969).
2. B. G. Bang and F. B. Bang, Localized lymphoid tissues and plasma cells in paraocular and paranasal organ systems in chickens, *Am. J. Pathol.,* **53**:735-751 (1968).
3. B. G. Bang and F. B. Bang, Replacement of virus-destroyed epithelium by keratinized squamous cells in vitamin A deprived chickens, *Proc. Soc. Exp. Biol. Med.,* **132**:50-54 (1969).
4. B. G. Bang and F. B. Bang, Laryngotracheitis virus in chickens. A model for study of acute nonfatal desquamating rhinitis, *J. Exp. Med.,* **125**:409-428 (1967).

5.   B. G. Bang and F. B. Bang, Experimentally induced changes in nasal mucous secretory systems and their effect on virus infection in chickens. I. Effect on mucosal morphology and function, *J. Exp. Med.*, **130**:105-140 (1969).

6.   B. G. Bang, M. A. Foard, and F. B. Bang, The effect of vitamin A deficiency and Newcastle disease on lymphoid cell systems in chickens, *Proc. Soc. Exp. Biol. Med.*, **143**:1140-1146 (1973).

7.   B. G. Bang, A. L. Mukherjee, and F. B. Bang, Human nasal mucus flow rates, *Johns Hopkins Med. J.*, **121**:38-48 (1967).

8.   F. B. Bang, Physiologic and structural aspects of viral cell pathology, *Lab. Invest.*, **8**:557-573 (1959).

9.   F. B. Bang and M. A. Foard, Interaction of respiratory epithelium of the chick and Newcastle disease virus, *Am. J. Hyg.*, **79**:260-278 (1964).

10. F. B. Bang and M. A. Foard, Experimentally induced changes in nasal mucous secretory systems and their effect on virus infection in chickens. II. Effects on absorption of Newcastle disease virus, *J. Exp. Med.*, **130**: 121-140 (1969).

11. F. B. Bang and M. A. Foard, The effect of acute vitamin A deficiency on the susceptibility of chicks to Newcastle disease and influenza viruses, *Johns Hopkins Med. J.*, **129**:100-109 (1971).

11a. F. B. Bang and M. A. Foard, Unpublished data.

12. F. B. Bang, M. A. Foard, and B. G. Bang, Acute Newcastle viral infection of the upper respiratory tract of the chicken. I. A model for the study of environmental factors on upper respiratory tract infection, *Am. J. Pathol.*, **76**:333-348 (1974).

13. F. B. Bang and J. S. Niven, A study of infection in organized tissue cultures, *Br. J. Exp. Pathol.*, **39**:317-322 (1958).

14. O. Basarab and H. Smith, Quantitative studies on the tissue localization of influenza virus in ferrets after intranasal and intravenous inoculation, *Br. J. Exp. Pathol.*, **50**:612-618 (1969).

15. S. P. Battista, E. P. Denine, and C. J. Kensler, Restoration of tracheal mucosa and ciliary particle transport activity after mechanical denudation in the chicken, *Toxicol. Appl. Pharmacol.*, **22**:59-69 (1972).

16. P. M. Biggs, The association of lymphoid tissue with the lymph vessels in the domestic chicken (*Gallus domesticus*), *Acta Anat.*, **19**:36-47 (1957).

17. W. K. Blenkinsopp, Proliferation of respiratory tract epithelium in the rat, *Exp. Cell Res.*, **46**:144-154 (1967).

18. L. R. Boling, Regeneration of nasal mucosa, *Arch. Otolaryngol.*, **22**:689-724 (1935).

19. T. Burnstein and F. B. Bang, Infection of the upper respiratory tract of the chick with a mild (vaccine) strain of Newcastle disease virus. I. Initiation and spread of the infection. II. Studies on the pathogenesis of the infection, *Bull. Johns Hopkins Hosp.*, **102**:127-157 (1958).

20. W. E. Le Gros Clarke, Anatomical investigation into routes by which infections pass from the nasal cavity into the brain, Reports on Public Health and Medical Subjects, No. 54, London, 1929.

21. W. B. Condon, Regeneration of tracheal and bronchial epithelium, *J. Thorac. Surg.*, **11**:333-346 (1942).

22. R. E. Corstvet and W. W. Sadler, A comparative study of single and multiple respiratory infections in the chicken: single infections (with *Mycoplasma gallisepticum* and Newcastle disease virus), *Am. J. Vet. Res.*, **27**:1721-1733 (1966).

23. C. J. Dawe, Neoplasms induced by polyoma virus in the upper respiratory tracts of mice. In *Cancer of the Nasopharynx*. Edited by C. S. Muir and K. Shanmugaratam. Copenhagen, Munksgaard, 1967.

24. L. DeLuca and G. Wolf, Vitamin A and protein synthesis in mucous membranes, *Am. J. Clin. Nutrit.*, **22**:1050-1062 (1969).

25. F. W. Denny, Effect of a toxin procuded by *Haemophilia influenzae* on ciliated respiratory epithelia, *J. Infect. Dis.*, **129**:93-100 (1974).

26. D. Dimov, Les troubles olfactifs d'origine grippale, *J. Fr. Otorhinolaryngol.*, **22**:53-56 (1973).

27. J. T. Dingle, H. B. Fell, and D. S. Goodman, The effect of retinol and of retinol binding protein on embryonic skeletal tissue in organ culture, *J. Cell. Sci.*, **2**:393-402 (1972).

28. J. T. Dingle, J. A. Lucy, and H. B. Fell, Studies on the mode of action of excess of vitamin A. I. Effect of excess of vitamin A on the metabolism and composition of embryonic chick limb cartilage grown in organ culture, *Biochem. J.*, **79**:497-500 (1961).

29. R. G. Douglas and R. F. Betts, Effect of induced interferon in experimental rhinovirus infections in volunteers, *Infect. Immun.*, **9**:506-510 (1974).

30. R. R. Dourmashkin and D. A. J. Tyrrell, Attachment of two myxoviruses to ciliated epithelial cells, *J. Gen. Virol.*, **9**:77-88 (1970).

31. W. M. Faber, The nasal mucosa and the subarachnoid space, *Am. J. Anat.*, **62**:121-148 (1937).

32. J. I. Fabrikant and J. Cherry, The kinetics of cellular proliferation in normal and malignant tissues. X. Cell proliferation in the nose and adjoining cavities, *Ann. Otol. Rhinol. Laryngol.*, **79**:572-575 (1970).

33. H. B. Fell, Introduction to session III, International Symposium on Metabolic Function of Vitamin A, *Am. J. Clin. Nutrit.*, **22**:975-977 (1969).

34. H. B. Fell and E. Mellanby, The effect of hypervitaminosis A on embryonic limb-bones cultivated in vitro, *J. Physiol. (Lond.)*, **116**:320-349 (1952).

35. H. B. Fell and E. Mellanby, Metaplasia produced in cultures of chick ectoderm by high vitamin A, *J. Physiol. (Lond.)*, **119**:470-488 (1953).

36. M. S. Finkelstein, M. McWilliams, and C. G. Huizenga, Use of chicken tracheal organ cultures for studying resistance of respiratory epithelium to viral infection, *J. Immunol.*, **108**:183-194 (1972).

37. W. H. Foege, Famine, infections and epidemics. In *Famine*. Edited by G. Blix, Y. Hofvander, and B. Vahlqvist. Stockholm, Almqvist & Wiksell, 1971.

38. T. Francis, Jr. and C. H. Stuart-Harris, Studies on the nasal histology of epidemic influenza virus infection in the ferret. I. The development and repair of the nasal lesion, *J. Exp. Med.*, **68**:789-802 (1938).

39.   T. Francis and C. H. Stuart-Harris, Studies on the nasal histology of epidemic influenza virus infection in the ferret. III. Histological and serological observations on ferrets receiving repeated inoculations of epidemic influenza virus, *J. Exp. Med.,* **68**:813-831 (1938).

40.   B. G. Gatmaitan, E. Stanley, and G. G. Jackson, The limited effect of nasal interferon induced by rhinovirus and a topical chemical inducer on the course of infection, *J. Infect. Dis.,* **127**:401-407 (1973).

41.   W. E. Giddens and G. A. Fairchild, Effects of sulfur dioxide on the nasal mucosa of mice, *Arch. Environ. Health,* **25**:166-173 (1972).

42.   A. Globerson and R. Auerbach, Primary antibody response in organ cultures, *J. Exp. Med.,* **124**:1001-1016 (1966).

43.   L. Gomez and L. G. Raggi, Local immunity to avian infectious bronchitis in tracheal organ culture, *Avian Dis.,* **18**:346-368 (1974).

44.   C. B. Gorham and J. A. Bacher, Regeneration of the human maxillary antral lining, *Arch. Otolaryngol.,* **11**:763-771 (1930).

45.   P. P. C. Graziadei, Cell dynamics in the olfactory mucosa, *Tissue Cell,* **5**: 113-131 (1973).

46.   P. P. C. Graziadei and J. F. Metcalf, Autoradiographic and ultrastructural observations on the frog's olfactory mucosa, *Z. Zellforsch.,* **116**:305-318 (1971).

47.   I. Gresser and H. B. Dull, A virus inhibitor in pharyngeal washings from patients with influenza, *Proc. Soc. Exp. Biol. Med.,* **115**:192-196 (1964).

48.   K. Hara, A. S. Beare, and D. A. J. Tyrrell, Growth and pathogenicity of influenza viruses in organ cultures of ciliated epithelium, *Arch. Ges. Virusforsch.,* **44**:227-236 (1974).

49.   J. H. Ph. Hers, *The Histopathology of the Respiratory Tract in Human Influenza,* Leiden, Stenfert Kroese N.V., 1954.

50.   W. P. Heuschele and B. C. Easterday, Local immunity and persistence of virus in the tracheas of chickens following infection with Newcastle disease virus. I. Organ culture studies, *J. Infect. Dis.,* **121**:486-496 (1970).

51.   W. P. Heuschele and B. C. Easterday, Local imminity and persistence of virus in the tracheas of chickens following infection with Newcastle disease virus. II. Immunofluorescent and histopathologic studies, *J. Infect. Dis.,* **121**:497-504 (1970).

52.   A. C. Hilding, The common cold, *Arch. Otolaryngol.,* **12**:133-150 (1930).

53.   A. C. Hilding, Regeneration of respiratory epithelium after minimal surface trauma, *Ann. Otol. Rhinol. Laryngol.,* **74**:903-914 (1965).

54.   A. C. Hilding, Experimental sinus surgery. IV. Effects of operative windows on normal sinuses, *Ann. Otol. Rhinol. Laryngol.,* **50**:379-393 (1941).

55.   D. A. Hill, S. Baron, J. C. Perkins, M. Worthington, J. E. VanKirk, J. Mills, A. Z. Kapikan, and R. M. Chanock, Evaluation of an interferon inducer in viral respiratory disease, *JAMA,* **219**:1179-1184 (1972).

56.   B. Hoorn and D. A. J. Tyrrell, Effects of some viruses on ciliated cells, *Am. Rev. Resp. Dis.,* **93**:156-161 (1966).

57.   G. Hotz and F. B. Bang, Electron microscope studies of ferret respiratory cells infected with influenza, *Bull. Johns Hopkins Hosp.,* **101**:175-208 (1957).

58. J. S. Huang and F. B. Bang, The susceptibility of chick embryo skin organ cultures to influenza virus following excess vitamin A, *J. Exp. Med.*, **120**: 129-147 (1964).

59. T. Iida, Experimental study on the transmission of Sendai virus in specific pathogen-free mice, *J. Gen. Virol.*, **14**:69-75 (1972).

59a. T. Iida and F. B. Bang, Infection of the upper respiratory tract of mice with influenza virus, *Am. J. Hyg.*, **77**:169-176 (1963).

60. S. F. Jackson and H. B. Fell, Epidermal fine structure in embryonic chicken skin during atypical differentiation induced by vitamin A in culture, *Dev. Biol.*, **7**:396-419 (1963).

61. G. Kelemen, The nasal and paranasal cavities of the rabbit in experimental work, *AMA Arch. Otolaryngol.*, **61**:497-512 (1955).

62. W. J. Kleinschmidt and F. Streightoff, Inhibition of influenza virus and interferon response intranasally with statolon, *Infect. Immun.*, **4**:738-741 (1971).

63. C. D. Knowlton and G. W. McGregor, How and when the mucous membrane of the maxillary sinus regenerates, *Arch. Otolaryngol.*, **8**:647-656 (1928).

64. E. Koburg, Autoradiografische Untersuchungen zur Zellneubildungsrate an den Epithelium des oberen Respirations- und Verdauungstraktes, *Arch. Ohren. Nasen. Kehlkopfheilk.*, **180**:616-621 (1962).

65. B. P. Lane and R. Gordon, Regeneration of rat tracheal epithelium after mechanical injury. I. The relationship between mitotic activity and cellular differentiation, *Proc. Soc. Exp. Biol. Med.*, **145**:1139-1144 (1974).

66. C. Liu, Studies on influenza infection in ferrets by means of fluorescein-labelled antibody. I. The pathogenesis and diagnosis of the disease, *J. Exp. Med.*, **101**:665-676 (1955).

67. B. Lomniczi, Studies on interferon production and interferon sensitivity of different strains of Newcastle disease virus, *J. Gen. Virol.*, **21**:305-313 (1973).

68. S. W. Martin and R. A. Willoughby, Effect of sulfur dioxide on the respiratory tract of swine, *J. Am. Vet. Med. Assoc.*, **159**:1518-1522 (1971).

69. A. J. Martinez, R. J. Duma, E. C. Nelson, and F. L. Moretta, Experimental Naegleria meningoencephalitis in mice, *Lab. Invest.*, **29**:121-133 (1973).

70. D. H. Matulionis, Ultrastructure of olfactory epithelia in mice after smoke exposure, *Ann. Otol. Rhinol. Laryngol.*, **83**:192-201 (1974).

71. T. C. Merigan, S. E. Reed, T. S. Hall, and D. A. J. Tyrrell, Inhibition of respiratory virus infection by locally applied interferon, *Lancet*, **1**:563-567 (1973).

72. D. Morley, Measles in Nigeria, *Am. J. Dis. Child.*, **103**:230-233 (1962).

73. J. Mulder and J. F. Ph. Hers, *Influenza*, Groningen, Wolters, Noordhoff, 1972.

74. B. D. Mulvaney and H. Heist, Regeneration of rabbit olfactory epithelium, *Am. J. Anat.*, **131**:241-252 (1971).

75. B. R. Murphy, S. Baron, E. G. Chalhub, C. P. Uhlendorf, and R. Chanock, Temperature-sensitive mutants of influenza virus. IV. Induction of inter-

feron in the nasopharynx by wild-type and a temperature-sensitive recombinant virus, *J. Infect. Dis.*, **128**:488-493 (1973).

76. R. Naessen, An enquiry on the morphological characteristics and possible changes with age in the olfactory region of man, *Acta Otolaryngol. (Stockh.)*, **71**:49-62 (1971).

77. J. B. Nelson, Infectious catarrh of the albino rat, *J. Exp. Med.*, **72**:645-662 (1940).

78. W. K. Ota, B. G. Bang, and F. B. Bang, Tritiated thymidine labeling of normal and NDV-infected chick nasal epithelial cells, *Proc. Soc. Exp. Biol. Med.*, **137**:1278-1282 (1971).

79. D. Y. E. Perey and P. B. Dent, Host resistance mechanisms to Newcastle disease virus in immunodeficient chickens, *Proc. Soc. Exp. Biol. Med.*, (1974).

80. D. Y. E. Perey, G. B. Cleland, and P. B. Dent, Newcastle disease in normal and immunodeficient chickens, *J. Am. Vet. Res. [Suppl.]*, (1974).

81. T. L. Perrin and J. W. Oliphant, Pathologic histology of experimental virus influenza in ferrets, *Publ. Health Rep.*, **55**:1077-1086 (1940).

82. M. F. Quinlan, S. D. Salman, D. L. Swift, H. N. Wagner, and D. F. Proctor, Measurement of mucociliary function in man, *Am. Rev. Respir. Dis.*, **99**:13-23 (1969).

83. S. E. Reed, Persistent respiratory virus infection in tracheal organ cultures, *Br. J. Exp. Pathol.*, **50**:378-388 (1969).

84. S. E. Reed, Interaction between mycoplasmas and respiratory viruses studied in tracheal organ cultures, In *Pathogenic Mycoplasmas*. A Ciba Foundation Symposium. New York, Elsevier, 1972.

85. R. E. Rewell, *Pathology of the Upper Respiratory Tract*, Edinburgh and London, Livingstone, 1963.

86. D. D. Richman, B. R. Murphy, E. L. Tierney, and R. M. Chanock, Specificity of the local secretory antibody to influenza A virus infection, *J. Immunol.*, **113**:1654-1656 (1974).

87. Y. Sakakura, Y. Togo, H. N. Wagner, R. B. Hornick, A. R. Schwartz, and D. F. Proctor, Mucociliary function during experimentally induced rhinovirus infection in man, *Ann. Otol.*, **82**:203-211 (1973).

88. Y. Sakakura, Personal communication, 1974.

89. L. J. Schiff, Studies on the mechanisms of influenza virus infection in hamster trachea organ culture, *Arch. Ges. Virusforsch.*, **44**:195-204 (1974).

90. R. C. Schmidt and H. F. Maassab, Local immunity to influenza virus in chicken tracheal organ cultures, *J. Infect. Dis.*, **129**:637-643 (1974).

91. E. W. Schultz, Repair of the olfactory mucosa, *Am. J. Pathol.*, **37**:1-19 (1960).

92. W. W. Schultz, Virus induced lysosomal enzyme dissolution of chick nasal cartilage, Thesis: Johns Hopkins Univ. School of Hygiene and Public Health, 1972; W. W. Schultz and F. B. Bang, *Am. J. Pathol.* (1977).

93. K. V. Shah and B. G. Bang, (unpublished data).

94. E. L. Shroyer and B. C. Easterday, Growth of infectious bovine rhino-tracheitis virus in organ cultures, *J. Am. Vet. Assoc.,* **29**:1355-1362 (1968).

95. D. Sinha, Seasonal variation in signs of vitamin-A deficiency in rural West Bengal children, *Lancet,* **2**:228-231 (1973).

96. D. Sinha, Personal communication, 1974.

97a. C. G. Smith, Changes in the olfactory mucosa and the olfactory nerves following intranasal treatment with one per cent zinc sulfate, *Can. Med. J.,* **39**:138-140 (1938).

97b. C. G. Smith, Regeneration of sensory olfactory epithelium and nerves in adult frogs, *Anat. Rec.,* **109**:661-671 (1951).

98. W. Smith, C. A. Andrewes, and P. A. Laidlaw, A virus obtained from in-fluenza patients, *Lancet,* **2**:66 (1933).

99. C. H. Stuart-Harris, *Influenza and Other Virus Infections of the Respira-tory Tract,* 2nd ed. London, Edward Arnold, 1965.

100. C. H. Stuart-Harris and T. Francis, Studies on the nasal histology of epi-demic influenza infection in the ferret. II. The resistance of regenerating respiratory epithelium to re-infection and to physico-chemical injury, *J. Exp. Med.,* **68**:803-812 (1938).

101. S. F. Tagaki, Degeneration and regeneration of the olfactory epithelium. In *Handbook of Sensory Physiology,* Vol. IV. Edited by L. M. Beidler. Berlin, Springer-Verlag, 1971, pp. 75-94.

102. J. D. Todd, F. J. Volenec, and I. M. Paton, Interferon in nasal secretions and sera of calves after intranasal administration of avirulent infectious bovine rhinotracheitis virus: Association of interferon in nasal secretions with early resistance to challenge with virulent virus, *Infect. Immun.,* **5**: 699-706 (1972).

103. T. B. Tomasi, Jr. and J. Bienenstock, Secretory immunoglobulins, *Adv. Immunol.,* **9**:1-96 (1968).

104. G. L. Toms, I. Rosztoczy, and H. Smith, The localization of influenza virus: Minimal infectious dose determinations and single cycle kinetic studies on organ cultures of respiratory and other ferret tissues, *Br. J. Exp. Pathol.,* **55**:116-129 (1974).

105. D. A. J. Tyrrell, *Common Colds and Related Diseases,* London, Edward Arnold, 1965, Chap. 5.

106. R. H. Waldman and W. J. Coggins, Influenza immunization: Field trial on a university campus, *J. Infect. Dis.,* **126**:242-248 (1972).

107. A. B. Wells, The kinetics of cell proliferation in the tracheobronchial epithelia of rats with and without chronic respiratory disease, *Cell Tissue Kinet.,* **3**:185-206 (1970).

108. D. L. Wilhelm, Regeneration of tracheal epithelium, *J. Pathol. Bacteriol.,* **65**:543-550 (1953).

109. D. L. Wilhelm, Regeneration of the tracheal epithelium in the vitamin A deficient rat, *J. Pathol. Bacteriol.,* **67**:361-365 (1954).

110. D. L. Wilhelm, Personal communication, 1974.

111. S. B. Wolbach and P. R. Howe, Tissue changes following deprivation of fat-soluble vitamin A, *J. Exp. Med.*, **42**:753-777 (1925).
112. Y. C. Wong and R. C. Buck, An electron microscope study of metaplasia of the rat tracheal epithelium in vitamin A deficiency, *Lab. Invest.*, **24**:55-66 (1971).

DATE DUE

NOV 1 4 1979			
FEB 0 1 1983 Feb, 1, 1983			
GAYLORD			PRINTED IN U.S A